COMPARATIVE GENOMICS

Basic and Applied Research

Edited by James R. Brown

CRC Press
Taylor & Francis Group
Boca Raton London New York

CRC Press is an imprint of the
Taylor & Francis Group, an **informa** business

CRC Press
Taylor & Francis Group
6000 Broken Sound Parkway NW, Suite 300
Boca Raton, FL 33487-2742

Printed in the United States of America on acid-free paper
10 9 8 7 6 5 4 3 2 1

International Standard Book Number-13: 978-0-8493-9216-0 (Hardcover)

Library of Congress Cataloging-in-Publication Data

Comparative genomics : basic and applied research / editor, James R. Brown.
p. ; cm.
Includes bibliographical references and index.
ISBN-13: 978-0-8493-9216-0 (hardcover : alk. paper)
ISBN-10: 0-8493-9216-0 (hardcover : alk. paper)
1. Genomics. 2. Physiology, Comparative. I. Brown, J. R. (James Raymond), 1956- II. Title.
[DNLM: 1. Genomics. 2. Physiology, Comparative. QU 58.5 C7375 2008]

QH447.C6517 2008
572.8'6--dc22 2007024832

Visit the Taylor & Francis Web site at
http://www.taylorandfrancis.com

and the CRC Press Web site at
http://www.crcpress.com

Contents

Part II Applied Research in Comparative Genomics

Preface

Since beginning my career in pharmaceutical research and development over 10 years ago, I have seen a remarkable acceleration in the merger of basic and applied genomic research. The pharmaceutical industry, and indeed much of the academic biomedical research community, initially viewed comparative genomics as a limited venture confined to the "holy trinity species" of medical research: mouse, rat, and human. Of course, an exception has always been infectious diseases, for which comparative genomics plays a vital role in understanding viral, bacterial, and parasitic pathogens — although the importance of looking at nonpathogenic, evolutionary immediate species was often a tough sell. However, that view is changing. Through rigorous comparative analysis, the genomes of cold-blooded vertebrate, avian, and other mammalian species are providing new understandings of the human genome. Moreover, genomic sequences are becoming available for several species that are important for drug research, such as dogs and primates, as well as more specialized applications such as bovine models for osteoarthritis and zebrafish as a model for a variety of developmental and neurological conditions.

The chapters in this book are roughly equally distributed between two sections: "Basic Research in Comparative Genomics" and "Applied Research in Comparative Genomics." My goal for organization is not to create further stratifications or subdisciplines in the field but rather to point out the commonalities and synergies across the broad field of comparative genomics. Database administrators and software engineers would ask me to select or prioritize the public genomic sequences for integration into our internal bioinformatics environment. Much to their chagrin, my stock response for selection was, "All of them!" Fortunately, that message was soon accepted and embraced. Because of the growing repertoire of species genomes, comparative genomic analysis, in particular molecular evolutionary approaches, is increasingly important in drug discovery. I hope those readers in the applied sciences see the important opportunities for mining species genomes beyond those of immediate practical utility to their field and are enlightened about technological advances in DNA sequencing and phylogenetic methods as well as understanding the impact of comparative genomics on shaping conceptual thought on the evolution of species and populations. Conversely, those with a perspective focused on more basic evolutionary issues might gain an appreciation of the utility of comparative genomics in biomedical and agricultural research.

Rather than a comprehensive volume covering every aspect of comparative genomics, this book embodies the diverse interests of prominent researchers in the field. The first section, "Basic Research in Comparative Genomics," begins with three chapters covering different challenges in the field and the methodologies used to address them. Appropriately, Michael Metzker leads with a review of the revolutionary advances in DNA sequencing technology that promise to tremendously accelerate the generation of new genomic data. Next, expanding our insight into evolution relationships among species is one of the key benefits of comparative genomics, yet the organization and analysis of large phylogenetic data sets will require bold new approaches such as those described by Bernard Moret. The virology community has been dealing with comparative genomic data analysis longer than any other group, so the description by Chris Upton and Elliot Lefkowitz on the organization and methods applied to viruses is an example of a mature and sophisticated bioinformatics genomics resource.

The remaining four chapters in the first section cover the impact of comparative genomics on our basic understanding of the evolution and genomics of several key groups of organisms. William Martin, Tal Dagan, and Katrin Henze discuss theories derived from comparative genomics on one of the most important and controversially areas of "deep" evolution study — the evolution of the eukaryotic cell and the mitigating role of organelle biogenesis. As the most diverse and largest metazoan group, the genomics of invertebrates is now poised to provide insights into their evolution as well as the origin of vertebrates, which is discussed by Takeshi Kawashima, Eiichi Shoguchi, Yutaka Satou, and Nori Satoh. The DNA sequencing projects for many additional vertebrate species are either in progress or in the planning stage, and James Thomas provides an overview of resources and fundamental principles that are the basis for contemporary studies in comparative vertebrate genomics. Completing the basic research section is a chapter by Michael Barnes on human populations that has two messages: the application of comparative genetic analysis at the intraspecific level and insights into genetic polymorphisms linked to diseases, which is a natural segue into the second section of this book.

In the section "Applied Research in Comparative Genomics," I open with a general overview, with some examples, on the utility of comparative genomics in pharmaceutical research. The next three chapters concern the role of comparative genomics in the treatment of infectious agents. Diarmaid Hughes discusses the relevance of bacterial pathogen genomes in the renewed and urgent efforts toward novel antimicrobial drugs. Malaria and other eukaryotic parasites are the most deadly killers in the developing world, but genomic sequence data hold the promise of finding new therapies as described by Emilio Merino, Steven Sullivan, and Jane Carlton. Philippe Lemey, Koen Deforche, and Anne-Mieke Vandamme discuss the application of comparative genomics of human immunodeficiency virus (HIV) in support of acquired immunodeficiency syndrome (AIDS) research, with particular emphasis on the critical concern of drug resistance.

The next four chapters concern other human diseases and drug safety issues. Cancer cells are highly polymorphic, and understanding the patterns of mutations and chromosomal aberrations among tumor types is another application of comparative genomics as described by Timon Buys, Wan Lam, and colleagues. Another chapter by Alice Kuo, Wan Lam, and colleagues covers the emerging field of epigenomics, with an emphasis on the role of DNA methylation in cancer and the opportunities for epigenomic-based drug therapies. Understanding the universe of human drug targets and their role in disease is of critical importance to the pharmaceutical industry, and Steven Foord discusses in depth the genomics of G protein-coupled receptors with respect to neurological diseases. Evaluation of the safety of drugs and chemicals involves different model organisms, and the role of increasingly sophisticated comparative analysis of multispecies transcriptomic data for safety assessment and toxicology studies is described by Joshua Kwekel, Lyle Burgoon, and Timothy Zacharewski.

Of course, comparative genomics has wider applications beyond biomedical and pharmaceutical research. The final two chapters examine the field of genomics in agricultural research. Michael Francki and Rudi Appels review the increasing number of plant genomics projects and their role in advancing the improvement of important crop species. Leif Andersson provides an overview of advances in domestic animal genomics that are bolstering the thousands of years of selective animal breeding for desirable traits.

Space and time did not permit comprehensive coverage of all areas of comparative genomics in this volume. In addition to environmental metagenomics, the impact of comparative genomics on bioremediation and bioprocessing is missing. Researchers for other human diseases are using genomic data from multiple species to advance their work as well. These topics are fertile grounds for some future review.

The various contributions in this book should give the sense that there is already a healthy cross-disciplinary interaction among researchers working on applied and fundamental aspects of comparative genomics. Every advance in science is built on the foundations laid earlier. If this book serves to further enlighten only a few about the excitement of comparative genomics as well as the crucial interaction and interdependency of applied and basic research, then it will have overwhelmingly achieved its objectives.

James R. Brown

Editor

Dr. James Brown is currently an associate director in molecular discovery research informatics with the global pharmaceutical company GlaxoSmithKline (GSK) and is based in Collegeville, Pennsylvania. He is responsible for coordinating bioinformatics analyses in support of diverse therapeutic areas, including antibiotics, antivirals, tropical diseases, musculoskeletal diseases, and cancer. In his work in the pharmaceutical industry, Dr. Brown has placed special emphasis on novel applications of evolutionary biology and phylogenetic analyses in drug discovery. Prior to joining GSK in 1996, he was a Medical Research Council of Canada postdoctoral fellow studying archaebacteria and the universal tree of life in the laboratory of Dr. W. Ford Doolittle at Dalhousie University, Halifax, Canada. His master of science and doctor of philosophy degrees, with thesis research on oyster aquaculture and sturgeon molecular population genetics, respectively, were granted from Simon Fraser University, Vancouver, Canada. He was granted a bachelor of science in marine biology from McGill University, Montreal, Canada, and has been involved in fieldwork throughout the Great Lakes and Canadian Arctic. Dr. Brown is an author of over 70 peer-reviewed publications and book chapters.

Contributors

Leif Andersson
Department of Medical Biochemistry and Microbiology
Uppsala University
Department of Animal Breeding and Genetics
Swedish University of Agricultural Sciences
Uppsala, Sweden

Rudi Appels
Department of Agriculture and Food Western Australia
South Perth, Australia
Murdoch University and Molecular Plant Breeding
Cooperative Research Centre
Murdoch, Western Australia, Australia

Michael R. Barnes
Molecular Discovery Research Informatics
GlaxoSmithKline Pharmaceuticals
Harlow, Essex, United Kingdom

Carolyn J. Brown
University of British Columbia
Vancouver, British Columbia, Canada

James R. Brown
Molecular Discovery Research Informatics
GlaxoSmithKline
Collegeville, Pennsylvania

Lyle D. Burgoon
Michigan State University
Department of Biochemistry and Molecular Biology
East Lansing, Michigan

Timon P. H. Buys
British Columbia Cancer Research Centre
Vancouver, British Columbia
Canada

Jane M. Carlton
Department of Medical Parasitology
New York University School of Medicine
New York, New York

Raj Chari
British Columbia Cancer Research Centre
Vancouver, British Columbia
Canada

Bradley P. Coe
British Columbia Cancer Research Centre
Vancouver, British Columbia
Canada

Tal Dagan
Institute of Botany
University of Düsseldorf
Düsseldorf, Germany

Jonathan J. Davies
British Columbia Cancer Research Centre
University of British Columbia
Vancouver, British Columbia
Canada

Koen Deforche
Rega Institute
Katholieke Universiteit Leuven
Leuven, Belgium

Steven M. Foord
Molecular Discovery Informatics
GlaxoSmithKline
Medicines Research Centre
Stevenage, Hertfordshire
United Kingdom

Michael Francki
Department of Agriculture
and Food Western Australia
South Perth, Australia
Value Added Wheat Cooperative
Research Centre
North Ryde, New South Wales
Australia

Cathie Garnis
British Columbia Cancer
Research Centre
Vancouver, British Columbia
Canada

Katrin Henze
Institute of Botany
University of Düsseldorf
Düsseldorf, Germany

Diarmaid Hughes
Department of Cell and Molecular
Biology
Uppsala University
Uppsala, Sweden

Takeshi Kawashima
Center for Integrative Genomics
Department of Cell and Molecular
Biology
University of California at Berkeley
Berkeley, California

Jennifer Y. Kennett
British Columbia Cancer Research
Centre
Vancouver, British Columbia
Canada

Alice N. C. Kuo
British Columbia Cancer Research
Centre
University of British Columbia
Vancouver, British Columbia
Canada

Joshua C. Kwekel
Michigan State University
Department of Biochemistry
and Molecular Biology
East Lansing, Michigan

Wan L. Lam
British Columbia Cancer
Research Centre
University of British Columbia
Vancouver, British Columbia
Canada

Eric H. L. Lee
British Columbia Cancer
Research Centre
University of British Columbia
Vancouver, British Columbia
Canada

Elliot J. Lefkowitz
Department of Microbiology
University of Alabama at Birmingham
Birmingham, Alabama

Philippe Lemey
Department of Zoology
University of Oxford
Oxford, United Kingdom
Rega Institute
Katholieke Universiteit Leuven
Leuven, Belgium

William W. Lockwood
British Columbia Cancer
Research Centre
Vancouver, British Columbia
Canada

Calum MacAulay
British Columbia Cancer
Research Centre
University of British Columbia
Vancouver, British Columbia
Canada

William Martin
Institute of Botany
University of Düsseldorf
Düsseldorf, Germany

Emilio F. Merino
Department of Medical
Parasitology
New York University School
of Medicine
New York, New York

Michael L. Metzker
Human Genome Sequencing Center
and Department of Molecular
and Human Genetics
Baylor College of Medicine
Houston, Texas
LaserGen, Inc.
Houston, Texas

Bernard M. E. Moret
Laboratory for Computational
Biology and Bioinformatics
The School of Computer and
Communication Sciences
Ecole Polytechnique Fédérale
de Lausanne
Lausanne, Switzerland

Nori Satoh
Department of Zoology
Graduate School of Science
Kyoto University
Kyoto, Japan

Yutaka Satou
Department of Zoology
Graduate School of Science
Kyoto University
Kyoto, Japan

Ashleen Shadeo
British Columbia Cancer
Research Centre
Vancouver, British Columbia, Canada

Eiichi Shoguchi
Department of Zoology
Graduate School of Science
Kyoto University
Kyoto, Japan

Steven A. Sullivan
Department of Medical Parasitology
New York University School
of Medicine
New York, New York

James W. Thomas
Department of Human Genetics
Emory University
Atlanta, Georgia

Ivy F. L. Tsui
British Columbia Cancer
Research Centre
Vancouver, British Columbia, Canada

Chris Upton
Department of Biochemistry
and Microbiology
University of Victoria
Victoria, British Columbia, Canada

Anne-Mieke Vandamme
Rega Institute
Katholieke Universiteit Leuven
Leuven, Belgium

Emily Vucic
British Columbia Cancer
Research Centre
University of British Columbia
Vancouver, British Columbia, Canada

Ian M. Wilson
British Columbia Cancer
Research Centre
University of British Columbia
Vancouver, British Columbia, Canada

Tim R. Zacharewski
Michigan State University
Department of Biochemistry and
Molecular Biology
East Lansing, Michigan

1 Introduction *The Broad Horizons of Comparative Genomics*

James R. Brown

CONTENTS

ABSTRACT

Since the publication in 1977 of the first complete genome sequence, that of a simple bacteriophage, the field of comparative genomics has been of growing importance to evolutionary, biomedical and agricultural studies. With the advent of new sequencing technologies, advances in functional genomics, and more powerful informatics, the field is now poised for an unprecedented era of growth. Here, we provide a brief retrospective of the area and discuss emerging trends in comparative gonomics research.

1.1 INTRODUCTION

All science is comparative. Throughout the ages, the very definition of any advancement in knowledge is the significance of contrasts between the familiar and the novel. Thc foundational scientific tenet, the null hypothesis, involves comparison of the null or known existing state to new results arising as a consequence of specific experimental manipulations. Although early naturalists categorized newly discovered specimens by fastidious comparisons to well-characterized species, they did not coin the term *comparative taxonomy*. Diverse groups of scientists, such as ecologists, astronomers, physicists, and physicians, all utilize the power of comparative analysis in their work. Yet, a growing group of molecular biologists, molecular evolutionary biologists, and bioinformatics scientists who work with large-scale genome-wide data sets have defined their particular area of expertise as *comparative* genomics. What warrants this special emphasis on the comparative?

The genome is an attractive entity for study since it represents both an end and a new beginning. The DNA content of any individual is finite. Once DNA sequencing of an entire genome is completed down to the last nucleotide (which is seldom the case), one could claim that all the basic elements in the genetic "program" determining the fate of that individual, of any species, have been revealed. The project is finished and makes a tidy tale for the subsequent genome publication, end of story. However, we are still far from understanding all the subtleties of genome function. Having the DNA sequence of an individual opens new vistas on their evolution, biochemistry, behavior, and development. The irony of the genome is that while it alone defines the uniqueness of an individual, the ubiquity of DNA also connects all inhabitants of the earth, past, present, and future, into a single fabric. Only through comparative genomic analysis can we begin to discern those genetic elements that define individuality from those that provide genetic commonality among various life-forms. Comparative genomics can be applied at many levels, from a single pair of individuals to larger collections spanning populations, species, or phyla. Comparative genomics is also used to discern differences between healthy and diseased individuals as well as groups that are either sensitive or resistant to drugs or pathogens. The fundamental importance of these scientific questions perhaps lends justification for defining comparative genomics as a major discipline in its own right.

The major landmarks in genomics can be best viewed in terms of the first decoded genomes from key organisms. The first complete "organism" sequence was the 5,368 nucleotide genome of the bacteriophage phiX174 published by Sanger and coworkers in 1977.[1] In 1995, the bacterium *Haemophilus influenzae* was the first cellular organism to have its entire genomic DNA sequence determined.[2] Metazoan genomics was ushered in by the completion of the genomes of the nematode *Caenorhabditis elegans*[3] and fruit fly *Drosophila melanogaster*.[4] Plant genomics are marked by the completion of the thale crest *Arabidopsis thaliana* genome,[5] while both mycologists and molecular biologists heralded the completion of the first fungal genome, *Saccharomyces cerevisiae*.[6] Perhaps viewed through overtly anthropomorphic lenses, the pinnacle of genomics was the joint publication of the human genome by both public[7] and private[8] ventures in 2001. However, genomics, like all science, is built on the shoulders of earlier discoveries that spanned many fields. Many advances in molecular biology and informatics, such as recombinant DNA techniques, DNA sequencing,[9] the polymerase chain reaction (PCR),[10] the institution of public sequence databases in the 1980s, and the invention of the BLAST (basic local alignment search tool) algorithm[11,12] laid the necessary foundations to attain the present status of this discipline.

The ultimate purpose of any genomics study is to further understand the relationships between genotypes and phenotypes. Of course, just reading the DNA sequence of a species provides little insight into the execution of that genetic plan. Unfolding the interpretation, implementation, and activation of that "blueprint" is the realm of functional genomics, which uses the DNA sequence as the starting point in the design of genome-wide interrogation experiments. We now have exquisite tools for probing the internal workings of a cell at the molecular level, such as DNA microarrays, RNA interference (RNAi) methods, and proteomics technologies. Layered onto the genomic data are information on specific protein–protein interactions for revealing cell signaling cascades and protein–nucleotide interactions for mapping regulatory transcriptional networks. Advances in structural biology have led to a rapid increase in the number of proteins with available three-dimensional (3D) structures. Other specialized information is overlaid on genomic data, such as small molecule or drug interaction maps derived from data on binding to specific gene targets and modulation of certain biochemical pathways. The management and mining of these extensive and vibrant data sources is the challenging remit of bioinformatics.

Despite these new technologies and data types, we are only beginning to understand the complexity and intricacies about the implementation of the DNA blueprint in even the simplest organisms. However, already there are several examples of significant shifts in our thinking about genetics, the organization of biological systems, and evolution that can be directly attributed to the rapidly growing field of comparative genomics.

1.2 THE NATURE OF GENETIC DIVERSITY

A casual review of the literature reveals that the extent of genetic change in terms of genomic variation is not directly correlated with the magnitude of phenotypic change. Selection pressures conferring specific point mutations in a single gene, FOXP2, in humans might account for our species' unique acquisition of language among primates and all other species.[13] Yet, the genome size differences between phenotypically similar strains of the humble bacterium *Escherichia coli* can vary by as much as 1 million base pairs or 25% of its total DNA.[14] The lack of correlation between organism complexity and genome size has been long known as the *C-value paradox*.[15] While comparative genomics has not resolved all the mechanisms behind the C-value paradox, it has illuminated a multitude of mechanisms driving genome evolution.

Gene acquisition, duplication, divergence, and loss are the primary agents of genome evolutionary change and hence are determinants of phenotype and speciation. Comparisons of genomes from various species of yeast show that duplications of genes and larger chromosomal regions tempered with concurrent massive gene loss have occurred multiple times during the evolution of these fungi.[16] Vertebrates and mammals have also seen multiple rounds of gene duplication, which might have been massive, involving two to three whole-genome events in early vertebrate evolution.[17] While most vertebrate genomes have genes that are either novel or have homologs in other species, several gene families that are otherwise universally conserved in animals have been lost in mammals and other chordates.[18]

Over a decade of prokaryote genome sequencing has revealed that, in addition to gene duplication and loss, the acquisition of genes from distantly related species has also widely occurred.[14,19] Before genomics, lateral or horizontal gene transfer (HGT) was identified as a means by which one bacterial species acquired genes conferring resistance to antibiotics from another species, mediated by vectors such as phage and extrachromosomal plasmids. Early comparative genomics and phylogenetic analysis revealed further examples of HGT both within and between species of the major groups of life, eukaryotes, eubacteria (called Bacteria), and archaebacteria (termed Archaea).[20] In the late 1990s, on the eve of genomics, it was suggested that eukaryotes, Bacteria, and Archaea share perhaps at least 100 genes.[21] However, as more genome sequences became available, the estimates of universal conserved genes rapidly dropped, and the number of potential HGT events dramatically increased.[22]

HGT is now recognized as a major force in not only the evolution of prokaryotes but also the emergence of the eukaryotic cell. Considerable evidence exists for ancient HGT involving the transfer of genes from putative bacterial endosymbiont ancestors of organelles, namely, mitochondria and chloroplasts, to the eukaryotic host nuclear genome. Some groups of single-cell eukaryotic protists, such as Apicomplexa, which includes the human malarial parasite *Plasmodium falciparum*, evolved from multiple endosymbiosis and engulfment events (for review, see Brown[23]). The extensive occurrence of potential HGT events has challenged the concept of species classification for prokaryotes as well as the prospects for reconstructing a universal tree of life.[24,25] Comparative genomics has shown HGT to be, at the very least, a potentially significant mechanism of genome modification with an impact on nearly all species at some point in their evolutionary history.

1.3 NOT-SO-JUNK DNA

Genes encoding proteins and RNAs, such as ribosomal and transfer RNAs, were traditionally thought to be the key functional elements of the genome. While regulatory elements in noncoding DNA such as promoters and enhancers were recognized as crucial, other noncoding regions of DNA were thought to be "space fillers" or traps for selfish, parasitic DNA segments such as transposons. However, this so-called junk DNA has been shown to control critical cellular functions largely through the application of comparative genomic analyses. High-density tiling DNA arrays have revealed that most of the human genome is actively transcribed, even non-protein-coding regions.[26, 27] Studies have unveiled the critical roles that RNAi mediated by small noncoding RNAs (ncRNAs) play in the regulation of eukaryotic genes. A particular important ncRNA class is microRNA (miRNA), single-stranded, 19- to 23-nucleotide long RNAs that repress translation by binding to specific messenger RNA target sites. The miRNA were first discovered in *C. elegans* but subsequently were found to be widespread throughout metazoans.[28] The miRNAs differ from short interfering RNAs (siRNAs) in that they are derived from single-stranded rather than double-stranded RNA precursors. Yet, like siRNAs, miRNAs can under some circumstances also effect messenger RNA degradation and generally share a common route to biogenesis. Computational predictions of miRNA genes and their target sites suggest that most metazoan and plant genomes encode at least several hundred, if not thousands, of miRNA genes, and that a large proportion of protein-coding genes have putative miRNA regulatory binding sites (reviewed in Brown and Sanseau[29]). Many crucial cellular processes are regulated by miRNAs, including tissue morphogenesis[30] and metabolic pathways.[31] The miRNAs are also implicated in various disease pathologies, including cancer[32] and host–virus interactions.[33]

Other ncRNAs have been discovered, particularly a novel class of small RNAs isolated from mouse testis libraries; these ncRNAs are called *PIWI-interacting RNAs* or piRNAs based on their processing proteins.[34,35] The piRNAs are encoded by specific genomic regions, also conserved in rat and human, and appear to play a role in the suppression of transposon activation.[36,37] These exciting discoveries, facilitated by comparative genomics, have unveiled an important mechanism of cellular regulation by indigenous antisense RNAs.

1.4 EMERGING TRENDS IN COMPARATIVE GENOMICS

With genomes from a variety of species sequenced at a breathtaking rate along with innovations in genomic investigation technologies, it is difficult to project the future for comparative genomics. However, some recent trends in genomics will likely accelerate and become more prominent over the next few years.

In March 2007, the National Cancer and Blood Institute (NCBI) reported 471 genomes of prokaryotes, 435 of which were Bacteria (eubacteria) and 36 were Archaea (archaebacteria). A total of 345 eukaryotic genome projects were cited at various stages of completion (26 genomes), assembly (128 genomes), or in progress (191 genomes). Among eukaryotes, 50 genome projects alone involved mammalian species, 2 of which were recorded as complete, with the remainder equally split between assembly and in-progress phases. Of course, the viruses have the largest representation in the sheer number of genomes, with 2,731 reference sequences available for 1,782 viral genomes and 36 reference sequences for smaller viroids.

The selection of species for genomic determination has undergone an interesting evolution. The criteria for choosing some of the initial subjects, such as *H. influenzae*, was mainly based on the small size and tractability of their genomes for complete DNA sequence determination. Additional consideration was given to model organisms that had a long history of genetic investigation, such as the nematode, fruit fly, mouse, and rat. Biomedical relevance drove the human genome project and, to a large extent, determined the priority of microbial pathogens for bacterial genome sequencing.

However, since about 2001, with the advent of more cost-efficient DNA sequencing technologies and increasingly sophisticated informatics, key species associated with pivotal evolutionary events rose in priority for genome sequencing projects. An example is the origin of cellular organisms and the prokaryote–eukaryote transition, for which insights are being gained from genomic sequences of species of Archaea, Bacteria, and, in particular, eukaryotic protists lacking rudimentary mitochondria or having analogous organelles.[38] Another example of pivotal evolutionary events being addressed by genomics is the origin of vertebrates, with DNA sequences from species such as urochordates (tunicates), fish, amphibians, and mammals providing insights into vertebrate evolution and developmental biology.[39] Over the next few years, additional evolutionary questions at all levels of life will be framed in the terms of genomic investigation.

Another trend in genomics is the increasing depth of sequences available within a single species. Again, the virus community pioneered this area with the sequencing of multiple isolates such as 2,003 different avian and human influenza virus strains. A review of the NCBI Web site revealed that several key bacterial pathogens have also been resequenced across multiple isolates, such as *E. coli* (22 strains, including the "lab-rat" strain K12), *Staphylococcus aureus* (12 strains), and *Streptococcus pneumoniae* (14 strains). Understanding intraspecies variability in bacteria and viruses is particularly important given their propensity for recombination and HGT.

The advent of faster and more cost-effective DNA sequencing technologies as well as opportunities for personalized medicine is driving similar tactics in human genomics. A comparison of 13,023 genes across 11 breast and 11 colorectal cancers to identify tumorigenic changes offered a glimpse at the future for human population and disease genomics.[40] Beyond single-nucleotide polymorphisms, comparative genomics have revealed extensive structural changes between the genomes of normal human individuals, with one study revealing 297 sites of size variation, mostly encompassing from 8 to 40 kilobases (kb) but others spanning deletions of several hundred kilobases and inversions in the megabase realm.[41] A survey of copy number variants in the human genome revealed that these regions included many genes of functional importance associated with olfaction, immunity, and protein secretion.[42] Thus, the human genome itself might be a more dynamic entity than first imagined.[43]

A third trajectory of genomics, which woefully is not covered in this book, is environmental metagenomics. The vast majority of microbial organisms cannot be cultured in the laboratory; hence, traditional environmental surveys of microbial diversity that relied on culture isolation techniques grossly underestimated species diversity. Genomic techniques that can amplify large DNA genomic regions in situ without culturing the organisms are now used to investigate microbial communities sampled from their natural environments. Although still in the early days, a wide scope of environments has been sampled, including open ocean microbial plankton,[44] the Sargasso Sea,[45] and acidic mine drainages.[46] Closer to home have been studies of the human distal gut microbiome[47] and the guts of lean versus obese mice, the latter of which were shown to have distinct microbial genomic signatures.[48] These reports illustrate the growing awareness of the critical roles of internal microbial communities likely play in maintaining our own health.

There is little doubt that comparative genomics will find increasing applications in biomedical research. The genomes from other species are essential for further understanding the human genome. In particular, cold-blooded vertebrates and invertebrate sequences are often helpful in sorting paralogous and orthologous relationships within large multigene families of drug targets such as kinases and G protein-coupled receptors (GPCRs). As a minor example, we performed an evolutionary analysis of Aurora kinases, a potential anticancer target, which provided the context for the transference of knowledge from model systems to humans as well as pointed out a potential opportunity for targeting the adenosine triphosphate (ATP)-binding pockets of multiple kinases with a single inhibitor.[49] Discovery of drug targets against the malarial parasite *P. falciparum* benefits from the recognition of the unique evolutionary history of its genome, which involved the acquisition of bacterial, fungal, as well as plant gene homologs via multiple serial endosymbiosis events.[50] There are other potential applications of comparative genomics to biomedical research; for example, the triad of chimpanzee, macaque, and human genomes will be important for the identification of noncoding regulatory regions as well as defining human-specific disease-associated variants.[51]

1.5 CONCLUSION

There is little doubt that genomics will be the foundation of the biological sciences for decades to come. The future horizons of genomics from the DNA sequencing perspective alone are vast since only a tiny fraction of species have had their genomes sequenced. But, beyond the issues of data acquisition and analytical methodologies, the genomics community must be aware of their growing bioethical and social responsibilities. Positive involvement in public discussions emphasizing the value to society of properly conducted genomic research for biomedical, agricultural, conservational, and educational purposes should also be on the agenda of comparative genomics researchers.

ACKNOWLEDGMENTS

This work was supported by Informatics, Molecular Discovery Research, GlaxoSmithKline. I wish to thank Amber Donley, Marsha Hecht, and Judith Speigel of Taylor and Francis for their excellent editorial and production assistance.

REFERENCES

1. Sanger, F. et al. Nucleotide sequence of bacteriophage phi X174 DNA. *Nature* **265**, 687–695 (1977).
2. Fleischmann, R.D. et al. Whole-genome random sequencing and assembly of *Haemophilus influenzae* Rd. *Science* **269**, 496–512 (1995).
3. The *C. elegans* Sequencing Consortium. Genome sequence of the nematode *C. elegans*: a platform for investigating biology. *Science* **282**, 2012–2018 (1998).
4. Adams, M.D. et al. The genome sequence of *Drosophila melanogaster. Science* **287**, 2185–2195 (2000).
5. The Arabidopsis Initiative. Analysis of the genome sequence of the flowering plant *Arabidopsis thaliana. Nature* **408**, 796–815 (2000).
6. Goffeau, A. et al. Life with 6,000 genes. *Science* **274**, 546, 563–546, 567 (1996).
7. Lander, E.S. et al. Initial sequencing and analysis of the human genome. *Nature* **409**, 860–921 (2001).
8. Venter, J.C. et al. The sequence of the human genome. *Science* **291**, 1304–1351 (2001).
9. Sanger, F., Nicklen, S. & Coulson, A.R. DNA sequencing with chain-terminating inhibitors. *Proc. Natl. Acad. Sci. U. S. A.* **74**, 5463–5467 (1977).
10. Mullis, K. et al. Specific enzymatic amplification of DNA in vitro: the polymerase chain reaction. *Cold Spring Harb. Symp. Quant. Biol.* **51 Pt. 1**, 263–273 (1986).
11. Altschul, S.F. et al. Gapped BLAST and PSI-BLAST: a new generation of protein database search programs. *Nucleic Acids Res.* **25**, 3389–3402 (1997).
12. Altschul, S.F., Gish, W., Miller, W., Myers, E.W. & Lipman, D.J. Basic local alignment search tool. *J. Mol. Biol.* **215**, 403–410 (1990).
13. Enard, W. et al. Molecular evolution of FOXP2, a gene involved in speech and language. *Nature* **418**, 869–872 (2002).
14. Binnewies, T.T. et al. Ten years of bacterial genome sequencing: comparative-genomics-based discoveries. *Funct. Integr. Genomics* **6**, 165–185 (2006).
15. Gregory, T.R. Coincidence, coevolution, or causation? DNA content, cell size, and the C-value enigma. *Biol. Rev. Camb. Philos. Soc.* **76**, 65–101 (2001).
16. Goffeau, A. Evolutionary genomics: seeing double. *Nature* **430**, 25–26 (2004).
17. Blomme, T. et al. The gain and loss of genes during 600 million years of vertebrate evolution. *Genome Biol.* **7**, R43 (2006).
18. Danchin, E.G., Gouret, P. & Pontarotti, P. Eleven ancestral gene families lost in mammals and vertebrates while otherwise universally conserved in animals. *BMC Evol. Biol.* **6**, 5 (2006).
19. Abby, S. & Daubin, V. Comparative genomics and the evolution of prokaryotes. *Trends Microbiol.* **15**, 135–141 (2007).

20. Smith, M.W., Feng, D.F. & Doolittle, R.F. Evolution by acquisition: the case for horizontal gene transfers. *Trends Biochem. Sci.* **17**, 489–493 (1992).
21. Brown, J.R. & Doolittle, W.F. Archaea and the prokaryote-to-eukaryote transition. *Microbiol. Mol. Biol. Rev.* **61**, 456–502 (1997).
22. Koonin, E.V. Comparative genomics, minimal gene-sets and the last universal common ancestor. *Nat. Rev. Microbiol.* **1**, 127–136 (2003).
23. Brown, J.R. Ancient horizontal gene transfer. *Nat. Rev. Genet.* **4**, 121–132 (2003).
24. Doolittle, W.F. & Papke, R.T. Genomics and the bacterial species problem. *Genome Biol.* **7**, 116 (2006).
25. Doolittle, W.F. & Bapteste, E. Pattern pluralism and the tree of life hypothesis. *Proc. Natl. Acad. Sci. U. S. A.* **104**, 2043–2049 (2007).
26. Carninci, P. et al. The transcriptional landscape of the mammalian genome. *Science* **309**, 1559–1563 (2005).
27. Willingham, A.T. & Gingeras, T.R. TUF love for "junk" DNA. *Cell* **125**, 1215–1220 (2006).
28. He, L. & Hannon, G.J. MicroRNAs: small RNAs with a big role in gene regulation. *Nat. Rev. Genet.* **5**, 522–531 (2004).
29. Brown, J.R. & Sanseau, P. A computational view of microRNAs and their targets. *Drug Discov. Today* **10**, 595–601 (2005).
30. Cobb, J. & Duboule, D. Tracing microRNA patterns in mice. *Nat. Genet.* **36**, 1033–1034 (2004).
31. Mersey, B.D., Jin, P. & Danner, D.J. Human microRNA (miR29b) expression controls the amount of branched chain alpha-ketoacid dehydrogenase complex in a cell. *Hum. Mol. Genet.* **14**, 3371–3377 (2005).
32. Calin, G.A. & Croce, C.M. MicroRNA–cancer connection: the beginning of a new tale. *Cancer Res.* **66**, 7390–7394 (2006).
33. Sullivan, C.S. & Ganem, D. MicroRNAs and viral infection. *Mol. Cell* **20**, 3–7 (2005).
34. Aravin, A. et al. A novel class of small RNAs bind to MILI protein in mouse testes. *Nature* **442**, 203–207 (2006).
35. Girard, A., Sachidanandam, R., Hannon, G.J. & Carmell, M.A. A germline-specific class of small RNAs binds mammalian Piwi proteins. *Nature* **442**, 199–202 (2006).
36. Carmell, M.A. et al. MIWI2 is essential for spermatogenesis and repression of transposons in the mouse male germline. *Dev. Cell* **12**, 503–514 (2007).
37. Aravin, A.A., Sachidanandam, R., Girard, A., Fejes-Toth, K. & Hannon, G.J. Developmentally regulated piRNA clusters implicate MILI in transposon control. *Science* **316**, 744–747 (2007).
38. Simpson, A.G. & Roger, A.J. Eukaryotic evolution: getting to the root of the problem. *Curr. Biol.* **12**, R691–R693 (2002).
39. Dehal, P. & Boore, J.L. Two rounds of whole genome duplication in the ancestral vertebrate. *PLoS. Biol.* **3**, e314 (2005).
40. Sjoblom, T. et al. The consensus coding sequences of human breast and colorectal cancers. *Science* **314**, 268–274 (2006).
41. Tuzun, E. et al. Fine-scale structural variation of the human genome. *Nat. Genet.* **37**, 727–732 (2005).
42. Nguyen, D.Q., Webber, C. & Ponting, C.P. Bias of selection on human copy-number variants. *PLoS. Genet.* **2**, e20 (2006).
43. Lee, C. Vive la difference! *Nat. Genet.* **37**, 660–661 (2005).
44. DeLong, E.F. et al. Community genomics among stratified microbial assemblages in the ocean's interior. *Science* **311**, 496–503 (2006).
45. Venter, J.C. et al. Environmental genome shotgun sequencing of the Sargasso Sea. *Science* **304**, 66–74 (2004).

46. Tringe, S.G. et al. Comparative metagenomics of microbial communities. *Science* **308**, 554–557 (2005).
47. Gill, S.R. et al. Metagenomic analysis of the human distal gut microbiome. *Science* **312**, 1355–1359 (2006).
48. Turnbaugh, P.J. et al. An obesity-associated gut microbiome with increased capacity for energy harvest. *Nature* **444**, 1027–1031 (2006).
49. Brown, J.R., Koretke, K.K., Birkeland, M.L., Sanseau, P. & Patrick, D.R. Evolutionary relationships of Aurora kinases: implications for model organism studies and the development of anti-cancer drugs. *BMC Evol. Biol.* **4**, 39 (2004).
50. Gardner, M.J. et al. Genome sequence of the human malaria parasite *Plasmodium falciparum*. *Nature* **419**, 498–511 (2002).
51. Harris, R.A., Rogers, J. & Milosavljevic, A. Human-specific changes of genome structure detected by genomic triangulation. *Science* **316**, 235–237 (2007).

Part I

Basic Research in Comparative Genomics

2 Advances in Next-Generation DNA Sequencing Technologies

Michael L. Metzker

CONTENTS

ABSTRACT

The Human Genome Project has facilitated the sequencing of many species, with demand for revolutionary technologies that deliver fast, inexpensive, and accurate information on the rise. Several next-generation sequencing devices have been introduced to the marketplace following sizable awards by the National Human Genome Research Institute and joint ventures, mergers, and acquisitions of large corporations. An unprecedented contest, the Archon X PRIZE for Genomics, further spotlights interest in next-generation technologies. In this review, DNA polymerase-dependent strategies of single-nucleotide addition (SNA) and cyclic reversible termination (CRT), along with the DNA ligase-dependent strategy of sequencing by ligation (SBL), are discussed to highlight recent advances and potential challenges in genome sequencing.

2.1 INTRODUCTION

Next-generation sequencing technologies stand to change the way we think about scientific approaches in basic, applied, and clinical research. Numerous reviews have highlighted different strategies, with the goal of delivering accurate, inexpensive, and complete information of whole genomes.[1–7] The broadest application for these next-generation technologies is medical resequencing of human genomes, which could unravel genetic causes of common diseases and cancer, assist doctors in prescribing personalized medicine, and provide predictive indicators of disease prior to onset, opening the door for preventive therapies. The impetus for research and development of emerging technologies is largely credited to the National Human Genome Research Institute (NHGRI). Since 2004, the NHGRI has awarded $83 million to academic and corporate investigators for development of next-generation sequencing technologies[8]; these awards have facilitated much of the progress to date.

The vitality of this emerging field can also be gauged by recent joint ventures, mergers, and acquisitions. Recently, the corporate landscape has changed dramatically, with giants in the genomics reagent and instrumentation market joining forces with or acquiring smaller technology developers. In 2005, the company 454 Life Sciences, based on a pyrosequencing platform,[9] entered into a joint venture with Roche Applied Sciences, a division of Roche Diagnostics, to distribute its instrument and reagents worldwide.[10] In July 2006, Applied Biosystems acquired Agencourt Personal Genomics, along with its sequencing-by-ligation (SBL) platform,[11] for US $120 million.[12] More recently, Illumina Inc. announced a US $650 million merger with Solexa Inc.[13] to further advance their reversible terminator platform,[5, 14] also under development by Helicos Biosciences Corporation,[15] Intelligent Bio-Systems Inc.,[16] and LaserGen Inc.[2,17] Presumably, more deals are in the pipeline, with an estimated US $1 billion market expected to grow even larger by 2015.

Marking the 50th anniversary of the discovery of the structure of DNA,[18] the International Human Genome Sequencing Consortium reported completion of the human genome sequence in 2004, with approximately 99% coverage and an error rate of about 1 in 100,000 bases.[19] This milestone was accomplished using Sanger sequencing at a cost of more than US $300 million and 10 years of effort. The Archon X PRIZE for genomics, the second contest conceived by the X PRIZE Foundation, is offering a $10 million purse to the first team to sequence 100 human genomes in 10 days or less.[20] The winner must sequence at least 98%, with an error rate of 1 in 100,000 bases, at a cost of US $10,000 or less per genome. The identity of the 100 subjects will be kept anonymous; however, a second group, called the Genome 100, includes celebrities such as Google Inc. cofounder Larry Page; Microsoft Corporation cofounder Paul G. Allen; the Milken Institute founder Michael Milken; physicist Stephen Hawkings; and CNN's talk show host Larry King.[21] Participation in such a group is evidence of our desire to understand the genetic fabric that makes us who we are.

Sanger sequencing remains the most widely used technology platform in research today, although it is too expensive, labor intensive, and time consuming to accomplish large-scale medical resequencing of numerous human genomes.[2] For many years the sole technology source to turn to, it is probably unrealistic that a single technology can meet the needs of all sequencing applications today. Whereas a comparative study of highly related genomes would require an inexpensive, ultrathroughput, short-read technology, a blended sequencing approach may be better suited for production of a de novo, high-quality, finished assembly of a given genome. Several next-generation sequencing technologies will likely occupy the genomics marketplace, offering researchers the flexibility to choose the platform that best fits their application.

This review focuses on near-term technologies that promise to bring sequencing devices to the market within the next five years. Many of these approaches are commonly referred to as *sequencing by synthesis* (SBS), which does not clearly delineate the different mechanics of sequencing DNA.[2,7] Here, the DNA polymerase-dependent strategies are classified as single-nucleotide addition (SNA) and cyclic reversible termination (CRT) to describe pyrosequencing and reversible terminator platforms, respectively. An approach by which DNA polymerase is replaced by DNA ligase is referred to as SBL. Chemistry platforms for SNA, SBL, and CRT are all described along with their supporting instruments. It is important to note that other approaches representing long-term endeavors are also under development but are not covered in this chapter. Those include real-time and nanopore sequencing, both of which promise tens of thousands of bases in single reads from individual DNA molecules. Real-time technology efforts are under development at Pacific Biosciences,[22] VisiGen Biotechnologies, and Li-Cor Biosciences. Advances in nanopore sequencing have been highlighted in several recent reviews.[6,23,24]

2.2 SINGLE-NUCLEOTIDE ADDITION: PYROSEQUENCING

The most successful non-Sanger method developed to date is pyrosequencing, first described by Hyman in 1988.[25] *Pyrosequencing* is a nonelectrophoretic, nonfluorescent method that measures the release of inorganic pyrophosphate (PPi), which is proportionally converted into visible light by a series of enzymatic reactions.[9,26] Unlike other sequencing approaches that use modified nucleotides to terminate DNA synthesis, the pyrosequencing assay manipulates DNA polymerase by single addition of a 2′-deoxyribonucleotide (dNTP) in limiting amounts. DNA polymerase extends the primer upon incorporation of the complementary dNTP and then pauses. DNA synthesis is reinitiated following the addition of the next complementary dNTP in the dispensing cycle. The light generated by the enzymatic cascade is recorded as a series of peaks called a *pyrogram* (454 Life Sciences calls them *flowgrams*). The order and intensity of the light peaks reveal the underlying DNA sequence. One primary limitation of the pyrosequencing method is that homopolymer repeats greater than five nucleotides cannot be quantitatively measured.[2]

The company 454 Life Sciences has integrated their PicoTiterPlate (PTP) platform[27] with the pyrosequencing method.[28] Coupled with their approach is a solution-based emulsion PCR strategy to clonally amplify single DNA molecules onto beads. Genomic DNA is fragmented, ligated to common adaptors, separated into single strands (Figure 2.1A), and captured onto beads to perform the emulsion PCR step[29] (Figure 2.1B). The PTP is manufactured by anisotropic etching of a fiber-optic face plate to create well sizes of approximately 40 μm, into which only one DNA-amplified bead will fit (Figure 2.1C). This fiber-optic slide contains about 1.6 million wells, although the company recommends filling about half of them to minimize well-to-well cross talk (i.e., interfering light signals from an adjacent well). Following loading of the DNA-amplified beads into individual PTP wells, additional beads, coupled with PPi converting enzymes, are added (Figure 2.1D). The fiber-optic slide is mounted in a flow chamber, enabling the delivery of sequencing reagents to the bead-packed wells. The back side of the fiber-optic slide is directly attached to a

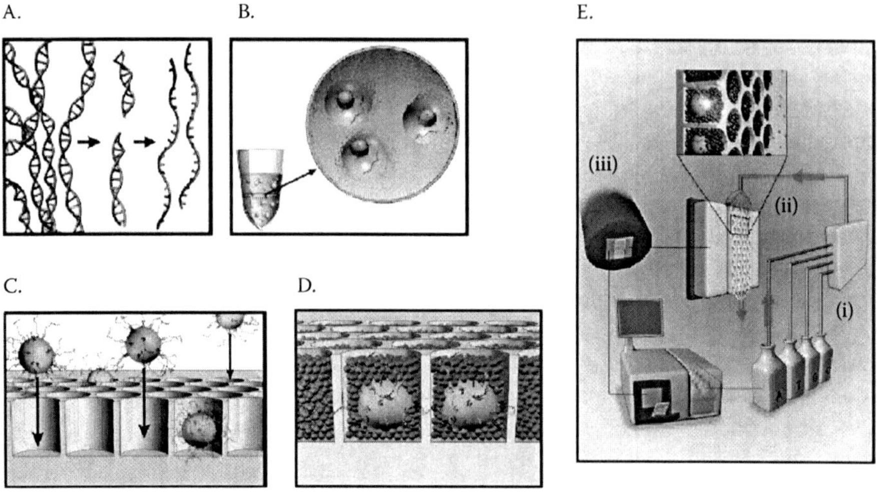

FIGURE 2.1 (See color figure in the insert following page 48.) 454 Life Sciences sequencing. (A) DNA preparation: Isolated genomic DNA is fragmented, ligated to adaptors, and separated into single strands. (B) Emulsion PCR: Single-stranded DNAs are bound to beads under conditions that favor one DNA molecule per bead. An oil-PCR reaction mixture is added to encapsulate bead–DNA complexes into single oil droplets, onto which PCR amplification is performed to create beads containing several million copies of the same template sequence. (C) Deposition of the PCR-amplified beads into individual wells in the PTP is followed by the addition of smaller beads immobilized with ATP surfurylase and luciferase (D), which convert inorganic pyrophosphate into a light signal. (E) Schematic of the GS20 instrument, which consists of the following subsystems: (i) fluidic assembly for delivery of dATP, dCTP, dGTP, and dTTP reagents; (ii) PTP; and (iii) CCD camera. Figure reprinted from Margulies et al., *Nature* 437, 376–380, 2005, by permission from Macmillan Publishers Ltd., copyright (2005).

high-resolution charged coupled device (CCD) camera, permitting detection of the light generated from each PTP well undergoing the pyrosequencing reaction (Figure 2.1E). With a pass rate of ~50% and a read length of 100 bases, one run will produce about 30–40 million bases of sequence data in 4–5 hours.

The Genome Sequencer 20 (GS20) instrument was launched by 454 Life Sciences in 2005. More than 40 articles have since been published on the GS20 platform, describing sequencing of bacterial genomes,[28,30–34] surveying microbial environments (i.e., metagenomics),[35–40] profiling expressed sequence tags (ESTs),[41–44] and whole-genome surveys of ancient DNA.[45–47] Many of these studies highlight the advantages and disadvantages of the GS20, depending on the intended goals of the research effort. For example, Hofreuter et al. reported the sequencing and characterization of the highly pathogenic *Campylobacter jejuni* strain 81-176.[34] Two 454 Life Sciences runs were performed, generating 60,905,794 high-quality bases from 558,331 successful reads (i.e., the average read length was 109 bases). A de novo assembly produced a genome with 34x coverage (i.e., on average, each nucleotide in the assembly was called by 34 different reads) in 43 contigs (contiguous sequence represented by two or more reads in the alignment). The majority of the gaps were closed by traditional PCR and Sanger sequencing methods. In a simulated study to evaluate de novo assemblies using short reads, Chaisson et al. analyzed the highly related *C. jejuni* strain NCTC11168 using error-free, 70-base read lengths with coverage of 30x the genome.[48] This simulated assembly produced fewer contigs (21 vs. 43), with the higher number presumably attributed to errors in the 454 Life Sciences sequence data set. Goldberg et al. evaluated a blended approach, with Sanger and 454 Life Sciences read data, using six marine microbial genomes, which provided a representative spectrum of assembly characteristics.[32] The authors found that a hybrid approach produced more accurate de novo assemblies than either approach alone and concluded that Sanger data should reign primary, with 454 Life Sciences data complementing the process.

Genome survey experiments, on the other hand, may be well suited for ultrathroughput, short-read sequencing technologies. Ancient DNA isolated from an exceptionally well-preserved woolly mammoth bone specimen produced 302,692 reads from a single 454 Life Sciences run. Comparative genome studies revealed that 137,527 of those reads aligned with the African elephant genome, a distant relative, identifying the reads as that of mammoth DNA. Alignment of the two genome sequences revealed an identity of approximately 98.5%, consistent with the evolutionary divergence of the two mammals that occurred approximately 5–6 million years ago.[46] Not all fossil samples, however, are as well preserved. Green et al. reported sequence analysis of the Neanderthal genome, providing valuable insights into this distinct hominid group.[47] Two 454 Life Sciences runs yielded only about 1 million bases of Neanderthal sequence. A majority of the sequences (79%) derived from the fossil extract did not reveal any significant matches to database sequences, supporting the finding that most of the DNA recovered from ancient samples is *exogenous* (i.e., colonized by microbes after death of the organism and/or introduced by investigator handling and laboratory procedure). Next-generation technologies can easily compensate for overwhelming contaminated sequences by the sheer volume of sequencing throughput.

Goldberg et al. noted in their study that short read lengths, a lack of paired-end templates, and lower read accuracy were deficiencies of the 454 Life Sciences platform in de novo assemblies of bacterial genomes.[32] Several advances, however, may overcome these shortcomings. For instance, 454 Life Sciences launched their second instrument, the GS FLX. Early specifications reported improved read-through to 250 bases, yielding about 100 million bases in 8–9 hours. Moreover, Ng et al. developed a method to create paired-end template libraries to facilitate de novo assemblies of genomes.[49] New releases of 454 Life Sciences' base-calling algorithms continue to improve the quality of assembled contig data as well. As we observed with developing Sanger technology, advances are expected to continue with longer read lengths, higher throughput, and improved accuracy.

2.3 SEQUENCING BY LIGATION

Sequencing by ligation (SBL) shares many common features with the SNA and CRT platforms. All require a priming oligonucleotide to initiate the sequencing chemistry and are performed in a cyclic manner. Template preparation of SBL can be performed using emulsion PCR[29] as with SNA, and the sequencing assay can be multiplexed in four colors as with CRT. Unlike the SNA and CRT platforms, however, DNA polymerase is replaced by DNA ligase,[50] and the four nucleotides are substituted with a library of degenerate oligonucleotides. Specificity of the SBL method is determined by hybridization of a second, complementary oligonucleotide (derived from the degenerate library) adjacent to the priming oligonucleotide site, such that the DNA ligase catalyzes formation of the phosphodiester bond between the two nucleic acids.

Shendure et al. applied this method in high-throughput DNA sequencing using a degenerate library of nonamers, with the middle base associated with a particular fluorescent dye (Figure 2.2A).[11] A genomic library from a modified strain of *Escherichia coli* MG1655 was prepared by circularizing randomly sheared genomic DNA, which was gel purified to yield approximately 1-kb fragments, with a universal linker containing *Mme*I sequence sites (Figure 2.2B). *Mme*I, a type II restriction enzyme, cleaves DNA 18 bases from its recognition site, generating a linear template construct with genomic paired ends. Following ligation of adaptors to the ends of the construct, emulsion PCR is performed to clonally amplify individual DNA constructs onto beads.[29] Millions of beads are then immobilized in a polyacrylamide gel onto a standard microscope slide. Following the ligation step of the complementary, fluorescently labeled nonamer, the slide is imaged using epifluorescence microscopy at four different emission wavelengths (Figure 2.2C). The anchor primer, dye-labeled nonamer complex is then stripped from the template-bound beads, and a different anchor primer (i.e., A2, A3, or A4) is hybridized to begin the SBL cycle again. This strategy creates discontinuous sequence data. For each SBL cycle, fluorescence intensities for each bead are extracted from the image and normalized to a 4D unit vector. Base calls are assigned from the maximum intensities to this vector, resulting in spatial clustering (Figure 2.2D). A custom-designed software algorithm maps the discontinuous reads back to the reference *E. coli* genome. Two instrument runs produced about 48 million high-quality bases, which mapped to approximately 70% of the *E. coli* MG1655 genome.[11]

Applied Bioysystems is now developing a modified version of the SBL platform, called Support Oligonucleotide Ligation Detection (SOLiD). Instrument development is under way and projected to launch in October 2007. A key improvement in the SBL chemistry is the development of a cleavable, fluorescently labeled nonamer. Upon four-color imaging, the bond between the fifth and sixth bases of the nonamer is cleaved, and the dye-labeled portion of the nonamer is washed away. This reaction yields a 3′-PO_4 group at the end of the ligated nonamer, which serves as the substrate for the next SBL cycle of ligation, imaging, and cleavage. Five SBL cycles are performed in toto, creating a discontinuous sequence, with every fifth base being called. The anchor primer, dye-labeled nonamer complex is stripped from the template-bound beads, an $n - 1$ anchor primer (Figure 2.2E) is hybridized, and the query position is reset one base to the right of that shown in Figure 2.2A. Subsequent rounds of SBL with $n - 2$, $n - 3$, and $n - 4$ anchor primers, with the query position reset accordingly, allow for phasing of the five discontinuous reads into a single continuous read of 25 bases. Early specifications reported production of approximately 1 billion high-quality bases in about two days.

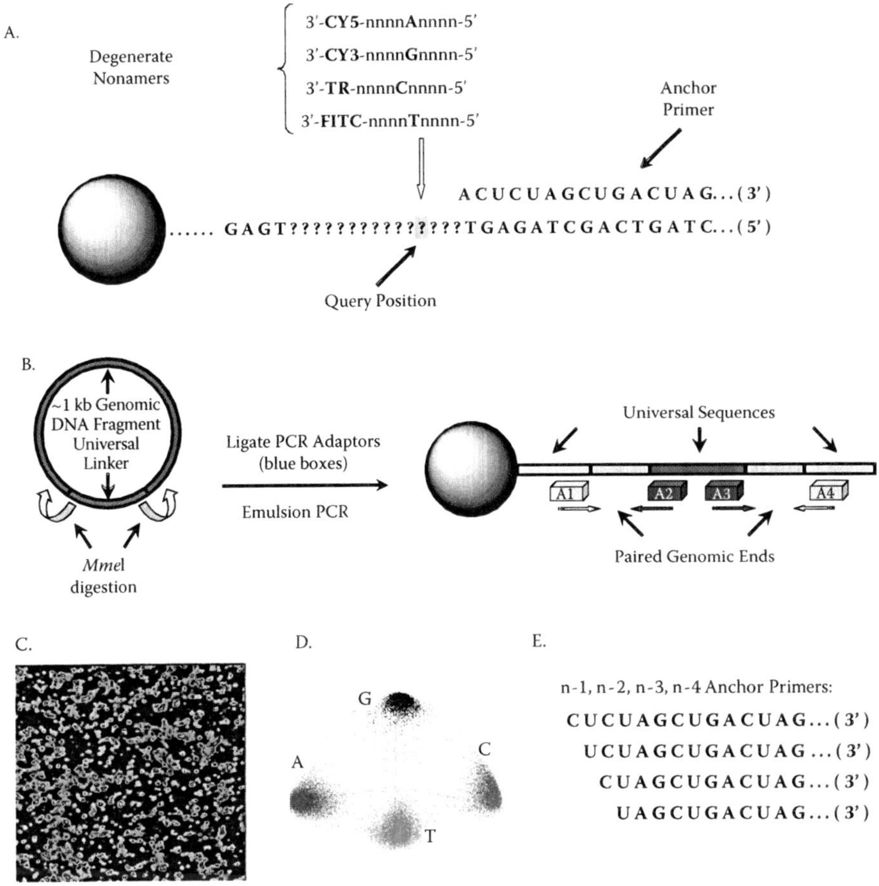

FIGURE 2.2 (See color figure in the insert following page 48.) Sequencing by ligation. (A) Basic chemistry step, which involves hybridization of an anchor primer to a bead-bound template (created by emulsion PCR; see Figure 2.1B legend), followed by ligation of the complement, dye-labeled nonamer from the degenerate library. The "n" represents all four nucleobases (i.e., A, C, G, and T), which yield a library of 262,144 unique nonamers (i.e., 4^9 sequences). (B) Creation of the paired-end library by emulsion PCR. Boxes, denoting A1 through A4, are anchor priming sites. (C) A four-color image obtained using epifluorescence microscopy. (D) The four-color data are displayed in a tetrahedral plot in which each spot in image C represents a single bead shown in Figure 2.2A. The four-color cluster corresponds to the four base calls. Following imaging, the anchor primer, dye-labeled nonamer complex is stripped; another anchor primer is hybridized; and the SBL cycle is repeated. (E) SOLiD sequencing. Instead of stripping the primer–nonamer complex, the dye-labeled nonamer is cleaved just 3′ to the query base, releasing the fluorescent dye and generating a 3′-PO_4 group. This group serves as the substrate for subsequent SBL cycles, resulting in every fifth base being called. Following four additional SBL cycles, anchor primer–nonamer complexes are stripped from the bead-bound template. A new $n - 1$ anchor primer is hybridized to reset the query position one base to the right. SBL is repeated until all anchor primers have been cycled. Contiguous DNA sequence information is then phased together using discontinuous reads from the different anchor primer data. (Figures 2.2A through 2.2D were reprinted from Shendure et al., *Science* 309, 1728–1732, 2005; modified with permission from AAAS.)

2.4 CYCLIC REVERSIBLE TERMINATORS

The CRT cycle is comprised of three steps: incorporation, imaging, and deprotection.[2] Reversible terminators are modified nucleotides that terminate DNA synthesis after incorporation of one modified nucleotide by DNA polymerase. These modified nucleotides contain a blocking group at the 3′-end of the ribose group, resulting in termination of DNA synthesis.[14,16,51–53] Subtle modifications to this position, such as reducing the group from the hydroxyl group (OH) to a hydrogen atom (H), (i.e., a 2′,3′-dideoxynucleotide), adversely effect the kinetic properties of DNA polymerases.[54–56] As such, a large body of literature has been devoted to mutagenesis experiments that reengineer DNA polymerases to improve the kinetic properties for 2′,3′-dideoxynucleotide substrates.[54–60] The case for reversible terminators is more challenging because the 3′-blocking groups are larger than the OH group, causing further bias against incorporation with DNA polymerase. Fluorescent dyes are therefore attached to the nucleobase structures to limit the size of the 3′-blocking groups.

Several blocking groups for reversible terminators, including the 3′-*O*-anthranyloyl,[52] 3′-*O*-allyl,[14,16,51,53,61] and 3′-*O*-(2-nitrobenzyl),[51] have been described in published articles and patents. As reported at the 2007 Advances in Genome Biology and Technology (AGBT) meeting,[62] however, efforts by the LaserGen team to replicate the published synthesis and characterization of the latter 3′-*O* blocking group was unsuccessful.[17] Ju and colleagues have published several fluorescently labeled 3′-*O*-allyl-dNTP structures, with different dyes attached to the four nucleobases.[16,53] These reversible terminators require dual deprotection steps to cleave the fluorophore from the nucleobase and restore the 3′-OH group. Following deprotection, a 3-aminopropynyl (AP3) linker remains attached to the nucleobase, creating a molecular scar, which accumulates with subsequent CRT cycles. In the field of molecular evolution, numerous groups have examined the effects of base-modified nucleotides in PCR.[63] Depending on the DNA polymerase, molecular scars, represented by singly substituted 5-(AP3)-dUTP[64,65] or 5-(AP3)-dCTP[66] with their corresponding natural nucleotides, have been shown to lower yield of full-length PCR products. The degree of PCR product yield is inversely proportional to target length,[65] with combinations of modified nucleotides further decreasing yields.[67] This evidence suggests that accumulation of these scars on the growing primer strand may limit read length for CRT sequencing.

Figure 2.3A shows a 13-base, four-color CRT sequence read using the fluorescently labeled 3′-O-allyl-dNTPs.[16] These 3′-*O*-allyl analogs are incorporated with a mutant 9°N(exo-) DNA polymerase,[68] which contains the A485L and Y409V amino acid variants. These substitutions are analogous to those described for Vent(exo-) DNA polymerase,[58] with the Y409V residue acting as a "steric" gate for incorporation of ribonucleotides (NTPs).[58,69–71] This gate discriminates against the 2′-hydroxyl group of NTPs, and substitution of the smaller valine residue permits DNA polymerase to incorporate NTPs and, apparently, fluorescently labeled 3′-*O*-allyl dNTPs. While the Illumina reversible terminator chemistry has not been published in detail, patents[14,72] reveal interesting similarity of structures with that published by Ju and colleagues.[61] Sharing considerable overlap in chemical functionality of 3′-blocking groups and nucleobase linkers, both groups also reported use of the mutant A485L/Y409V 9°N(exo-) DNA polymerase.[16,53,73]

A.

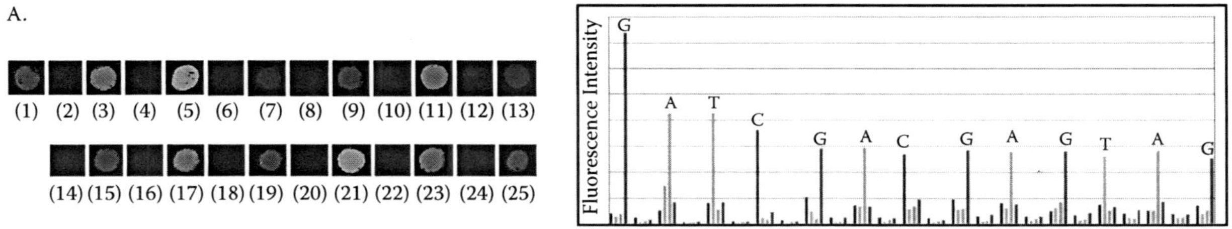

FIGURE 2.3 (See color figure in the insert following page 48.) Cyclic reversible termination: (A) 13-base CRT sequencing using the 3′-*O*-allyl terminators developed by Ju and colleagues,[16] illustrating fluorescence scanned data and four-color intensity histogram plot. The template was immobilized to a solid support using the self-priming method (not shown). (B) Five panels illustrate Illumina's single-molecule array (SMA) technology.[5] In panel 1, isolated genomic DNA is fragmented and ligated with adaptors, which are then made single-stranded and attached to the solid support. Bridge amplification (panel 2) is performed to create double-stranded templates (panel 3), which are denatured (panel 4) and bridge amplified several more times to create template clusters (panel 5). (C) Nine-base CRT sequencing highlighting two different template sequences. The series of images was obtained from a 40-million cluster SMA (not shown). (Panel A was reprinted from Ju et al., *Proc. Natl. Acad. Sci. U. S. A.* 103, 19635–19640, 2006, by permission of the National Academy of Sciences, U. S. A., copyright 2006. Figures 2.3B and 2.3C were obtained by permission from Illumina Inc.)

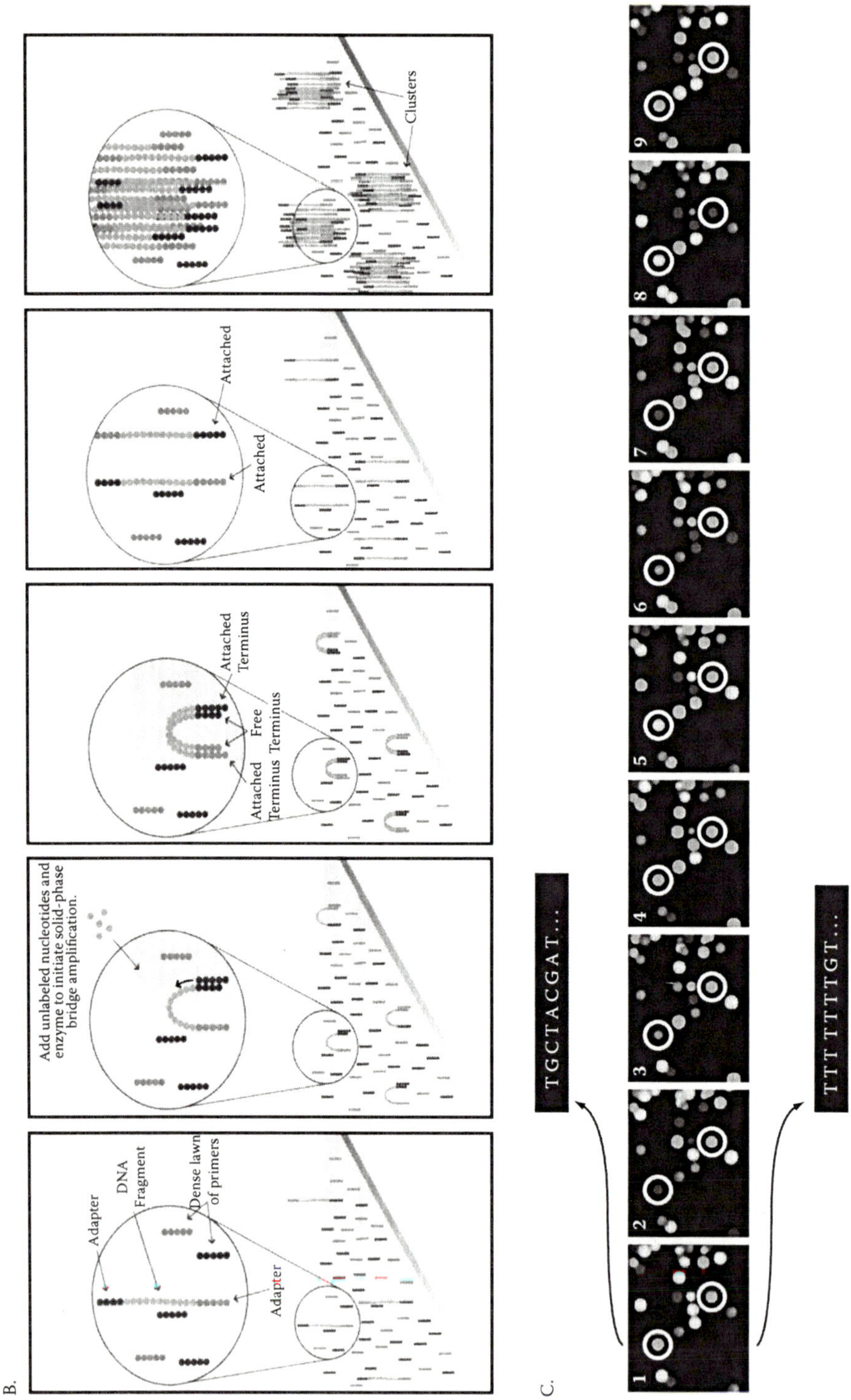

FIGURE 2.3 (Continued).

Illumina Inc. released the Genome Analyzer instrument in 2006 utilizing a strategy of template preparation called single-molecule arrays (SMAs)[5] that generates random arrays of millions of single-template clusters from fragmented genomic DNA (Figure 2.3B). The SMAs are formatted on an eight-channel flow cell (not shown), allowing eight independent experiments simultaneously. Up to 40 million template clusters can be generated per flow cell, and with a read length of 25 bases, the Genome Analyzer can produce approximately 1 billion high-quality bases in about two days.

At the 2007 AGBT meeting,[62] LaserGen reported a novel paradigm in reversible terminator chemistry: unblocked 3′-OH nucleotides that can terminate DNA synthesis without leaving molecular scars.[17] Advantages of this chemistry platform over 3′-blocked terminators (Figure 2.4) are as follows:

1. An unblocked 3′-OH group provides more favorable enzyme incorporation properties, unlike a 3′-blocked nucleotide, which requires high-throughput screening of mutant polymerase libraries to identify the desired biological properties.

A.

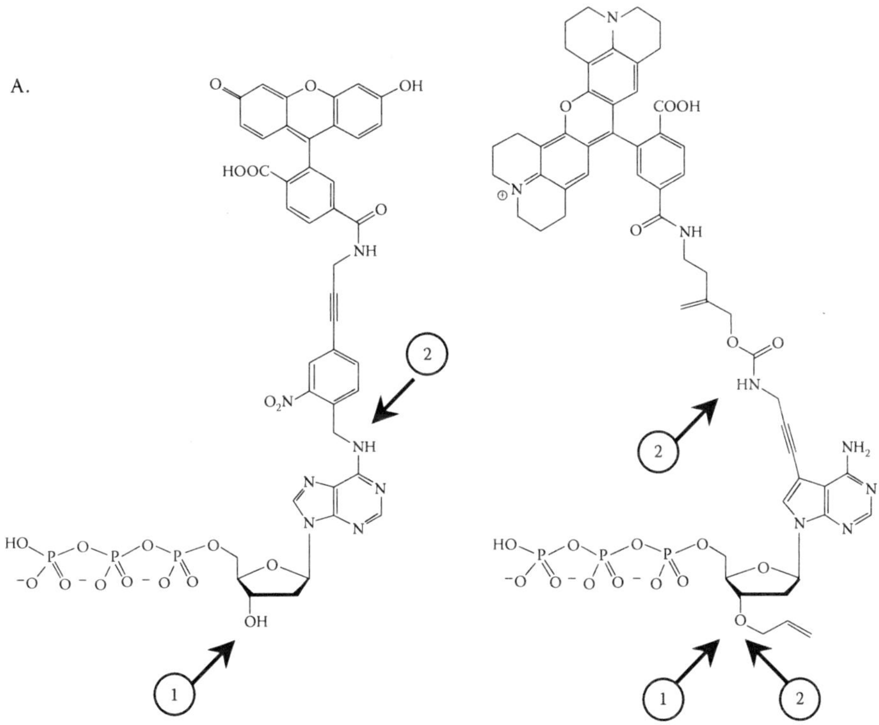

FIGURE 2.4 Comparison of dye-labeled 2′-deoxy adenosine terminators. (A) Chemical structures highlighting the 3′-unblocked nucleotide with a single attachment site for the terminating and dye groups compared with that of Ju et al.[16] (B) Three-dimensional model of three bases from the stepwise extension and deprotection using both terminator types shown in Figure 2.4A. The template strand is not shown to simplify the illustration of resulting natural nucleotides for the LaserGen terminators (*) compared with the accumulation of "molecular scars" (arrows) found with the 3′-*O*-allyl terminators.

B.

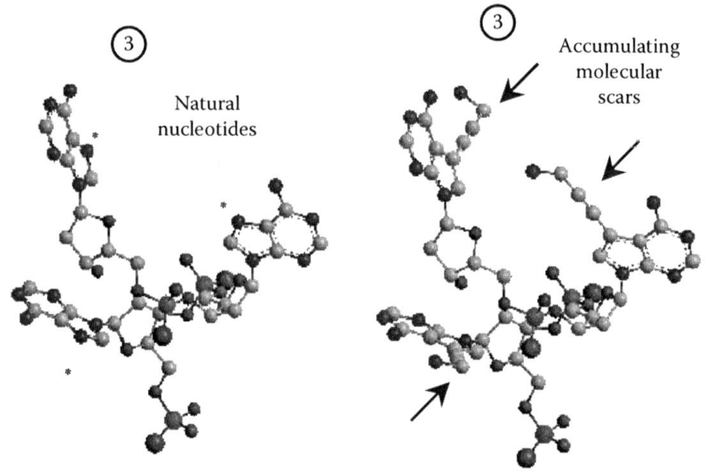

FIGURE 2.4 (Continued).

2. A single attachment step in removing the terminating and fluorescent dye groups provides more efficient deprotection, unlike doubly substituted nucleotides, of which the deprotection efficiency is a product of the individual sites.
3. The modified nucleotide is transformed back to its natural state, unlike that of other terminators, which leave an accumulating molecular scar with each sequencing cycle.

The challenge inherent to this technology is creating the appropriate modifications to the 2-nitrobenzyl group that cause termination of DNA synthesis after a single base addition while maintaining specificity of accurate DNA sequence data. This is important because an unblocked 3′-OH group is the natural substrate for DNA synthesis. Manuscripts are in preparation to describe this work in greater detail, and instrument development of LaserGen's CRT chemistry, coupled with its proprietary Pulsed-Multiline Excitation technology,[73] is under way.

At the 2007 AGBT meeting,[62] Helicos Biosciences and Intelligent Bio-Systems also presented progress on their instrument development efforts, with launches projected in the next one to two years. With several CRT technologies coming to market in the near future, competition will flourish, providing the researcher with multiple technology platforms for specific applications.

2.5 CLOSING REMARKS

Since 2005, tremendous progress has been made in next-generation technology development. One billion bases of sequence information can be produced by a single instrument run in just a few days, which is remarkable feat, indeed, although insufficient to meet the mark of complete genome sequencing that is accessible and affordable to all. Efforts to meet the NHGRI goal of the US $1,000 genome will involve multiple approaches that will spawn as-yet-unimagined applications. The many flavors of next-generation technologies will allow researchers to choose from a virtual menu, further expanding potential applications. More corporate giants will certainly appear with continuing advances from technology developers, further increasing the fluidity of the genomics marketplace.

ACKNOWLEDGMENT

I am extremely grateful to NHGRI for their support from grants R01 HG003573, R41 HG003072, R41 HG003265, and R21 HG002443.

REFERENCES

1. Shendure, J., Mitra, R. D., Varma, C. & Church, G. M. Advanced sequencing technologies: methods and goals. *Nat. Rev. Genet.* 5, 335–344 (2004).
2. Metzker, M. L. Emerging technologies in DNA sequencing. *Genome Res.* 15, 1767–1776 (2005).
3. Chan, E. Y. Advances in sequencing technology. *Mutat. Res.* 573, 13–40 (2005).
4. Bai, X., Edwards, J. & Ju, J. Molecular engineering approaches for DNA sequencing and analysis. *Expert Rev. Mol. Diagn.* 5, 797–808 (2005).
5. Bennett, S. T., Barnes, C., Cox, A., Davies, L. & Brown, C. Toward the $1,000 human genome. *Pharmacogenomics* 6, 373–382 (2005).
6. Bayley, H. Sequencing single molecules of DNA. *Curr. Opin. Chem. Biol.* 10, 628–637 (2006).
7. Fan, J.-B., Chee, M. S. & Gunderson, K. L. Highly parallel genomic assays. *Nat. Rev. Genet.* 7, 632–644 (2006).
8. National Human Genome Research Institute. NHGRI aims to make DNA sequencing faster, more cost effective (2006). http://www.nih.gov/news/pr/oct2006/nhgri-04b.htm.
9. Ronaghi, M., Uhlén, M. & Nyrén, P. A sequencing method based on real-time pyrophosphate. *Science* 281, 363, 365 (1998).
10. 454 Life Sciences. 454 Life Sciences and Roche announce commercial launch of. http://www.454.com/news-events/press-releases.asp?display=detailandid=36 (2005).
11. Shendure, J. et al. Accurate multiplex polony sequencing of an evolved bacterial genome. *Science* 309, 1728–1732 (2005).
12. Applied Biosystems. Applied Biosystems completes acquisition of Agencourt Personal Genomics, developer of genetic analysis technologies. http://press.appliedbiosystems.com/corpcomm/applerapress.nsf/ABIDisplayPress/65863C0773312370882571A700826263?OpenDocumentandtype=abi (2006).
13. Illumina Inc. Illumina signs definitive agreement to acquire Solexa. http://investor.illumina.com/phoenix.zhtml?c=121127andp=irol-newsArticleandID=929959andhighlight= (2006).
14. Barnes, C., Balasubramanian, S., Liu, X., Swerdlow, H. & Milton, J. Labelled nucleotides. U.S. patent 7,057,026 B2, 2006.
15. Braslavsky, I., Hebert, B., Kartalov, E. & Quake, S. R. Sequence information can be obtained from single DNA molecules. *Proc. Natl. Acad. Sci. U. S. A.* 100, 3960–3964 (2003).
16. Ju, J. et al. Four-color DNA sequencing by synthesis using cleavable fluorescent nucleotide reversible terminators. *Proc. Natl. Acad. Sci. U. S. A.* 103, 19635–19640 (2006).

17. Wu, W. et al. Termination of DNA synthesis by N^6-alkylated, not 3′-*O*-alkylated, photocleavable 2′-deoxyadenosine triphosphates. *Nucleic Acids Res.* (in press).
18. Watson, J. D. & Crick, F. H. Molecular structure of nucleic acids; a structure for dexoyribose nucleic acid. *Nature* 171, 737–738 (1953).
19. International Human Genome Sequencing Consortium. Finishing the euchromatic sequence of the human genome. *Nature* 431, 931–945 (2004).
20. X PRIZE Foundation. X PRIZE Foundation announces largest medical prize in history. http://genomics.xprize.org/newsevents/press_releases_2006-10-04_Archon_X_PRIZE_for_Genomics.html (2006).
21. Regalado, A. Celebrity Genome Project? $10 million may speed decoding. *Wall Street Journal,* October 4, 2006.
22. Levene, M. J. et al. Zero-mode waveguides for single-molecule analysis at high concentrations. *Science* 299, 682–686 (2003).
23. Rhee, M. & Burns, M. A. Nanopore sequencing technology: research trends and applications. *Trends Biotechnol.* 24, 580–586 (2006).
24. Yan, H. & Xu, B. Towards rapid DNA sequencing: detecting single-stranded DNA with a solid-state nanopore. *Small* 2, 310–312 (2006).
25. Hyman, E. D. A new method of sequencing DNA. *Anal. Biochem.* 174, 423–436 (1988).
26. Ronaghi, M., Karamohamed, S., Pettersson, B., Uhlén, M. & Nyrén, P. Real-time DNA sequencing using detection of pyrophosphate release. *Anal. Biochem.* 242, 84–89 (1996).
27. Leamon, J. H. et al. A massively parallel PicoTiterPlate™ based platform for discrete picoliter-scale polymerase chain reactions. *Electrophoresis* 24, 3769–3777 (2003).
28. Margulies, M. et al. Genome sequencing in microfabricated high-density picolitre reactors. *Nature* 437, 376–380 (2005).
29. Dressman, D., Yan, H., Traverso, G., Kinzler, K. W. & Vogelstein, B. Transforming single DNA molecules into fluorescent magnetic particles for detection and enumeration of genetic variations. *Proc. Natl. Acad. Sci. U. S. A.* 100, 8817–8822 (2003).
30. Andries, K. et al. A diarylquinoline drug active on the ATP synthase of *Mycobacterium tuberculosis. Science* 307, 223–227 (2005).
31. Velicer, G. J. et al. Comprehensive mutation identification in an evolved bacterial cooperator and its cheating ancestor. *Proc. Natl. Acad. Sci. U. S. A.* 103, 8107–8112 (2006).
32. Goldberg, S. M. D. et al. A Sanger/pyrosequencing hybrid approach for the generation of high-quality draft assemblies of marine microbial genomes. *Proc. Natl. Acad. Sci. U. S. A.* 103, 11240–11245 (2006).
33. Oh, J. D. et al. The complete genome sequence of a chronic atrophic gastritis *Helicobacter pylori* strain: evolution during disease progression. *Proc. Natl. Acad. Sci. U. S. A.* 103, 9999–10004 (2006).
34. Hofreuter, D. et al. Unique features of a highly pathogenic *Campylobacter jejuni* strain. *Infect. Immun.* 74, 4694–4707 (2006).
35. Leininger, S. et al. Archaea predominate among ammonia-oxidizing prokaryotes in soils. *Nature* 442, 806–809 (2006).
36. Turnbaugh, P. J. et al. An obesity-associated gut microbiome with increased capacity for energy harvest. *Nature* 444, 1027–131 (2006).
37. Angly, F. E. et al. The marine viromes of four oceanic regions. *PLoS Biol.* 4, 2121–2131 (2006).
38. Krause, L. et al. Finding novel genes in bacterial communities isolated from the environment. *Bioinformatics* 22, e281–e289 (2006).
39. Sogin, M. L. et al. Microbial diversity in the deep sea and the underexplored "rare biosphere." *Proc. Natl Acad. Sci. U. S. A.* 103, 12115–12120 (2006).
40. Edwards, R. et al. Using pyrosequencing to shed light on deep mine microbial ecology. *BMC Genomics* 7, 57 (2006).

41. Cheung, F. et al. Sequencing *Medicago truncatula* expressed sequenced tags using 454 Life Sciences technology. *BMC Genomics* 7, 272 (2006).
42. Bainbridge, M. et al. Analysis of the prostate cancer cell line LNCaP transcriptome using a sequencing-by-synthesis approach. *BMC Genomics* 7, 246 (2006).
43. Emrich, S. J., Barbazuk, W. B., Li, L. & Schnable, P. S. Gene discovery and annotation using LCM-454 transcriptome sequencing. *Genome Res.* 17, 69–73 (2007).
44. Gowda, M. et al. Robust analysis of 5′-transcript ends (5′-RATE): a novel technique for transcriptome analysis and genome annotation. *Nucleic Acids Res.* 34, e126 (2006).
45. Stiller, M. et al. Inaugural article: patterns of nucleotide misincorporations during enzymatic amplification and direct large-scale sequencing of ancient DNA. *Proc. Natl Acad. Sci. U. S. A.* 103, 13578–13584 (2006).
46. Poinar, H. N. et al. Metagenomics to paleogenomics: large-scale sequencing of mammoth DNA. *Science* 311, 392–394 (2006).
47. Green, R. E. et al. Analysis of 1 million base pairs of Neanderthal DNA. *Nature* 444, 330–336 (2006).
48. Chaisson, M., Pevzner, P. & Tang, H. Fragment assembly with short reads. *Bioinformatics* 20, 2067–2074 (2004).
49. Ng, P. et al. Multiplex sequencing of paired-end ditags (MS-PET): a strategy for the ultra-high-throughput analysis of transcriptomes and genomes. *Nucleic Acids Res.* 34, e84 (2006).
50. Tomkinson, A. E., Vijayakumar, S., Pascal, J. M. & Ellenberger, T. DNA ligases: structure, reaction mechanism, and function. *Chem. Rev.* 106, 687–699 (2006).
51. Metzker, M. L. et al. Termination of DNA synthesis by novel 3′-modified deoxyribonucleoside triphosphates. *Nucleic Acids Res.* 22, 4259–4267 (1994).
52. Canard, B. & Sarfati, R. DNA polymerase fluorescent substrates with reversible 3′-tags. *Gene* 148, 1–6 (1994).
53. Ruparel, H. et al. Design and synthesis of a 3′-*O*-allyl photocleavable fluorescent nucleotide as a reversible terminator for DNA sequencing by synthesis. *Proc. Natl. Acad. Sci. U. S. A.* 102, 5932–5937 (2005).
54. Tabor, S. & Richardson, C. C. A single residue in DNA polymerases of the *Escherichia coli* DNA polymerase I family is critical for distinguishing between deoxy- and dideoxyribonucleotides. *Proc. Natl. Acad. Sci. U. S. A.* 92, 6339–6343 (1995).
55. Astatke, M., Grindley, N. D. F. & Joyce, C. M. How *E. coli* DNA polymerase I (Klenow fragment) distinguishes between deoxy- and dideoxynucleotides. *J. Mol. Biol.* 278, 147–165 (1998).
56. Brandis, J. W. Dye structure affects *Taq* DNA polymerase terminator selectivity. *Nucleic Acids Res.* 27, 1912–1918 (1999).
57. Joyce, C. M. Choosing the right sugar: How polymerases select a nucleotide substrate. *Proc. Natl. Acad. Sci. U. S. A.* 94, 1619–1622 (1997).
58. Gardner, A. F. & Jack, W. E. Determinants of nucleotide sugar recognition in an archaeon DNA polymerase. *Nucleic Acids Res.* 27, 2545–2553 (1999).
59. Hamilton, S. C., Farchaus, J. W. & Davis, M. C. DNA polymerases as engines for biotechnology. *Biotechniques* 31, 370–383 (2001).
60. Arezi, B., Hansen, C. J., & Hogrefe, H. H. Efficient and high fidelity incorporation of dye-terminators by a novel archaeal DNA polymerase mutant. *J. Mol. Biol.* 322, 719–729 (2002).
61. Ju, J., Li, Z., Edwards, J. R. & Itagaki, Y. Massive parallel method for decoding DNA and RNA. U.S. patent 6,664,079 B2, 2003.
62. Advances in Genome Biology and Technology meeting. http://www.agbt.org (2007).
63. Bittker, J. A., Phillips, K. J. & Liu, D. R. Recent advances in the in vitro evolution of nucleic acids. *Curr. Opin. Chem. Biol.* 6, 367–374 (2002).

64. Battersby, T. R. et al. Quantitative analysis of receptors for adenosine nucleotides obtained via in vitro selection from a library incorporating a cationic nucleotide analog. *J. Am. Chem. Soc.* 121, 9781–9789 (1999).
65. Lee, S. E. et al. Enhancing the catalytic repertoire of nucleic acids: a systematic study of linker length and rigidity. *Nucleic Acids Res.* 29, 1565–1573 (2001).
66. Roychowdhury, A., Illangkoon, H., Hendrickson, C. L. & Benner, S. A. 2′-Deoxycytidines carrying amino and thiol functionality: synthesis and incorporation by Vent (exo-) polymerase. *Org. Lett.* 6, 489–492 (2004).
67. Gourlain, T. et al. Enhancing the catalytic repertoire of nucleic acids. II. Simultaneous incorporation of amino and imidazolyl functionalities by two modified triphosphates during PCR. *Nucleic Acids Res.* 29, 1898–1905 (2001).
68. Southworth, M. W., Kong, H., Kucera, R. B., Jannasch, H. W. & Perler, F. B. Cloning of thermostable DNA polymerases from hyperthermophilic marine Archaea with emphasis on *Thermococcus* sp. 9 degrees N-7 and mutations affecting 3′-5′ exonuclease activity. *Proc. Natl. Acad. Sci. U. S. A.* 93, 5281–5285 (1996).
69. Gao, G., Orlova, M., Georgiadis, M. M., Hendrickson, W. A. & Goff, S. P. Conferring RNA polymerase activity to a DNA polymerase: a single residue in reverse transcriptase controls substrate selection. *Proc. Natl. Acad. Sci. U. S. A.* 94, 407–411 (1997).
70. Astatke, M., Ng, K., Grindley, N. D. F. & Joyce, C. M. A single side chain prevents *Escherichia coli* DNA polymerase I (Klenow fragment) from incorporating ribonucleotides. *Proc. Natl. Acad. Sci. U. S. A.* 85, 3402–3407 (1998).
71. Fa, M., Radeghieri, A., Henry, A. A. & Romesberg, F. E. Expanding the substrate repertoire of a DNA polymerase by directed evolution. *J. Am. Chem. Soc.* 126, 1748–1754 (2004).
72. Milton, J., Ruediger, S. & Liu, X. Labelled nucleotides. WO 2004/108493 A1, 2004.
73. Lewis, E. K. et al. Color-blind fluorescence detection for four-color DNA sequencing. *Proc. Natl. Acad. Sci. U. S. A.* 102, 5346–5351 (2005).

3 Large-Scale Phylogenetic Reconstruction

Bernard M. E. Moret

CONTENTS

ABSTRACT

Phylogenies, the (reconstructed evolutionary histories of groups of organisms or other biological units, have becomc ubiquitous in biological and biomedical research. As high-throughput methods find their way into every area of the life sciences, large-scale analyses are rapidly becoming a necessity; phylogenetic analysis is no exception. Indeed, renewed attention to the reconstruction of the Tree of Life, a phylogeny of all species on this planet, has served to stress the need for more accurate, robust, and efficient computational approaches to phylogenetic reconstruction. This chapter reviews the basics of phylogenetic reconstruction, highlights the scaling issues we are facing today, discusses the most promising solutions currently under development, and invites reflection on questions of modeling and assessment in computational molecular biology.

3.1 PHYLOGENETIC RECONSTRUCTION: WHAT AND WHY?

A casual search of PubMed revealed nearly 20,000 citations to phylogenetic reconstruction packages, with steeply increasing counts over the last several years. Thus, the biomedical, biological, and pharmaceutical communities are making ever-increasing use of phylogenetic reconstruction; indeed, if journals in various areas of the life sciences are examined, we see phylogenies describing the relationships between predators and prey, the main families of chemical receptors, the geographical distribution of an infectious disease over time, categories of conserved protein folds, the sensitivity of patients to a specific drug, and many other uses over a bewilderingly varied range of data, subjects, and mechanisms. What are these phylogenies, and why have they assumed such importance in recent years?

3.1.1 PHYLOGENIES

A *phylogeny* is the evolutionary history of a group of related entities. In the most obvious case, we can think of the evolutionary history of a collection of related organismal species; thus, for instance, Figure 3.1 shows a phylogeny (after Montague and Hutchinson[1]) of the main herpesviruses that attack humans. This particular example takes the form of an unrooted tree, and indeed, most published phylogenies take the form of a tree, rooted or not. (There are exceptions to this form, but they remain rare to date, in part due to the lack of reliable methodologies for inferring more complex relationships.)

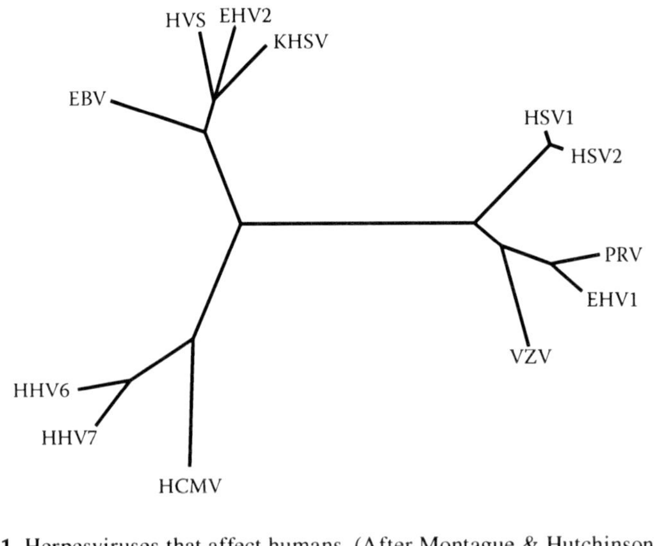

FIGURE 3.1 Herpesviruses that affect humans. (After Montague & Hutchinson, Gene content and phylogeny of herpesviruses, *Proceedings of the National Academy of Sciences of the United States of America*, 97:5334–5339, 2000.)

Evolution is an all-encompassing concept, so we encounter phylogenies describing coevolution of parasites and hosts, evolution of drug-resistance mechanisms within a few strains of the same bacterial species, evolution of a particular protein domain across many proteins with similar functionality, evolution across space as well as time of an infectious disease, and so on. It is the very pervasiveness of evolution throughout life that makes phylogenies so important — in 1973, Dobzhansky famously wrote a paper, "Nothing in Biology Makes Sense Except in the Light of Evolution,"[2] in which he wrote, in conclusion,

> Seen in the light of evolution, biology is, perhaps, intellectually the most satisfying and inspiring science. Without that light it becomes a pile of sundry facts, some of them interesting or curious but making no meaningful picture as a whole.

Phylogenies have thus become one of the main tools of modern biology in making sense of data — especially in the case of the enormous amounts of data generated by various high-throughput molecular methods.

Herein, though, lies a paradox: We can observe the contemporary results of evolution and, in relatively rare cases, collect some data on earlier manifestations (such as human records of diseases, paleological data from fossils, or more indirectly, dating methods, evidence of migrations, etc.), but how can we use a phylogeny to help us understand the data when the phylogeny is missing and, in any case, appears to imply greater understanding of the data than may be needed to answer the question at hand? The resolution of this paradox is that, of course, we do not use the true evolutionary history of the group under study but an estimate of that history obtained through reconstruction based on modern data. In other words, phylogenetic reconstruction, not phylogenies per se, is what is powering modern biological research.

3.1.2 Phylogenetic Reconstruction

Ever since Darwin published his seminal work, scientists have proposed phylogenies for various groups of organisms. Even before the widespread adoption of computers, scientists proposed methods for reconstructing phylogenies. Since then, dozens of software packages have been built and thousands of papers published, each proposing a slightly different way of reconstructing phylogenies. All such methods, however, are based on a few common principles: All begin with the extraction of so-called characters from the raw data, all proceed to operate on the characters only (and not the raw data), and all are based on some local or global optimization (or approximation thereof) according to one's preferred (and usually highly simplified) model of evolution for the chosen characters. For instance, much phylogenetic reconstruction in systematic biology until the 1980s was based on *morphological characters*, that is, discrete encodings of specific morphological traits of organisms — one may think of a child counting the number of leg pairs on an arthropod or of a paleontologist measuring fossil bones. The chosen characters must reflect the evolutionary relationships that one is attempting to reconstruct, so that many characters must typically be used in judicious combinations. Over the last few decades, the data of choice have been molecular sequences, more commonly protein-coding sequences; in such cases, the characters could be the nucleotide positions within the sequences, with each character assuming one of four possible states. More recently, interest in higher-level molecular characters has led to a focus on the ordering of genes within the whole genome, in which case the entire ordering forms a single character, which can then assume an enormous number of possible states.

Armed with a collection of characters, one can proceed to the stage of reconstruction, which includes two problems: modeling and algorithm design. Modeling comes into play because the changes in each character are dictated by evolutionary pressures; algorithm design is then required to provide a computational method for inverting the model — for reconstructing an evolutionary scenario from its outcomes. Models are naturally uncertain ground, so one may attempt to proceed in the most model-independent manner possible to design the simplest possible models or to parameterize models to fit the model to the data. All of these approaches have been used and are briefly described in this chapter.

3.1.3 Data Used in Phylogenetic Reconstruction

I have already alluded not only to the bewildering variety of data used in phylogenetic reconstruction, but also to the fact that molecular data have become favored over the last few decades. Molecular data, in the form of nucleotide sequences, amino acid sequences, protein sequences, structural information, whole-genome gene composition and ordering, and yet other forms, have a number of advantages: (1) They are extracted directly from the genome, which is the unit of propagation for genetic material and thus the vehicle of evolution; (2) they are typically discrete and thus offer the possibility of extracting exact data, not the noisy approximations typical of continuous data; (3) they are generated today in high-throughput settings in enormous quantities, enabling one to use not only combinatorial but also statistical methods to study them; and (4) they are much simpler to model than higher levels of data, such as morphological characters.

Yet, there are striking differences between various kinds of molecular data. For instance, nucleotide sequences based on a chosen gene provide 500–2,000 nucleotide characters, each capable of assuming one of four states, while gene orderings of, say chloroplast organelles with 120 genes, provide a single character (the oriented ordering) with up to $2^{120} \cdot 120!$ possible states. The first kind of characters is easy and inexpensive to gather in large numbers, but its very small number of possible states means that it is quite possible that, in the course of evolution, the character has passed through the same state more than once, making it very difficult to discern what happened to it from just the modern data, whereas, in contrast, it is basically impossible for the second type of character to assume the same state more than once. On the other hand, modeling the evolution of a single nucleotide is obviously far easier than modeling the evolution of the gene content and ordering of an entire genome.

Another example is provided by derived molecular characters used in a study by Yang et al.[3] in which the authors used the absence or presence of protein domain architectures (in effect, fold superfamilies) as characters to reconstruct a phylogeny for 174 complete genomes. Binary characters such as these can only take one of two states and are thus particularly prone to reverting to an earlier state, and modeling their appearance or disappearance is not well understood; yet the study, using an i.i.d. (identically and independently distributed) model, showed rather good accuracy across a broad range of organisms.

The choice of data is thus a complex issue: We want data that are relatively easy to collect in abundance, inexpensive to refine, characteristic of evolution on appropriate scales, internally consistent, and easy to model. Needless to say, these objectives are usually in conflict. The fact that nucleotide sequences have become the data of choice over the last 10 years is due mostly to the first two factors: high availability and low cost.

3.1.4 Scaling Issues

Biological and biomedical research have historically been constrained by low throughput. Since it took much time and effort to collect just a few data, investigations tended to be on a small scale — most published phylogenies in the 20th century have fewer than 50 leaves. High-throughput methods have turned research in the life sciences upside down: The main choke point today is often the analysis as data are pouring out of sequencers, mass spectrometers, microarrays, and the like. Trees published in the last five years often have over 100 leaves, and some, published in online appendices, have several hundred to a thousand leaves. There is no reason to believe that this tendency will abate: New high-throughput data production methods are announced regularly in other areas (metabolomics is a recent addition, for instance), and existing ones are refined to reduce the cost, the time, and the error rate. For instance, whereas it took the community 20 years to sequence the complete genomes of a couple dozen bacteria, there are now predictions that several thousand more will be fully sequenced within a few years. The day is thus not that far away when phylogenetic methods will be applied to data sets of thousands, perhaps even tens of thousands, of leaves. Current methods, however, are not ready for this challenge.

Broadly speaking, there are three major problems facing a designer of methods for phylogenetic reconstruction[4]: (1) How accurate is the method? (2) how fast is the method? and (3) how reliable is the method? Accuracy is of course the primary goal of any method; systematists in particular have been known to run a reconstruction method for a year on one data set to obtain the best-possible answer.[5] Accuracy, however, is hard to assess: On data sets obtained from nature, we do not know the "correct" answer (assuming one exists) and so have difficulty assessing the quality of a reconstruction; while it is easy to compare a reconstruction with the true answer in the case of simulated data sets, the value of the result is only as good as the simulations themselves, which brings up another serious problem. Accuracy has also been construed as limited to the data set at hand, an attitude that brings with it a host of problems since the most accurate and efficient "algorithm" for reconstruction of a fixed data set is simply the one that prints the best recorded answer; indeed, this particular aspect is a major reason for the third facet, reliability. Speed is pretty much a function of accuracy: Anyone can print a bad phylogeny quickly but producing a good one is time consuming as most optimization criteria are nondeterministic polynomial-time hard (NP-hard). *Reliability*, the ability of a reconstruction method to return accurate answers on entirely new data sets rather than just those on which it has been tested (and often developed), remains largely unexplored; while systematists are accustomed to getting so-called bootstrap scores for their tree edges or estimates of distributions of trees from their Markov chain Monte Carlo (MCMC) methods, the predictive value of the reconstruction methods and the significance on any given sample data set of these quality measures remain mostly unknown.

Surprises have been encountered time after time as the scale of reconstruction increased; thus, current methods, even if reliably accurate within their current ranges (something we do not know), are not likely to remain so as we move to larger scales.

3.1.5 Reconstructing the Tree of Life

Many biologists have been calling for some time for a community effort to attempt the reconstruction of the tree of life, the phylogeny of all organisms on this planet. Such an endeavor naturally has no end since evolution is an ongoing process and is not particularly well defined since thousands of organisms become extinct every year, if not every day. The scale is truly daunting: While we have methods that can reconstruct phylogenies for up to a thousand leaves (and scale poorly beyond that), there are well over a million described species of organisms, and estimates of the existing number vary from ten million to several hundred millions. Finally, it is not clear that we need a single giant phylogeny; many of the branches of this phylogeny are well identified and broadly accepted and so could be investigated mostly independently of all others. Yet, the tree of life should hold a special place in the heart of every human: It describes the wonderful diversity of life on this planet, helps us understand where we humans come from and what is our place within the larger scheme of life, and most importantly, gives us a basis to understand where we are all heading. The project to reconstruct this phylogeny also motivates the community to revisit many aspects of phylogenetic analysis, particularly those that have to do with scaling and reliability. After all, there is only one tree of life for this planet, so there will not soon be a chance to compare our reconstruction with one done for another tree of life elsewhere.

In the United States, the National Science Foundation initiated the Assembling the Tree of Life program that has funded, to date, well over 30 groups collecting, filtering, and analyzing data on all branches of the tree. Through another program, it has also enabled the Cyberinfrastructure for Phylogenetic Research (CIPRES) project (www.phylo.org), with the aim to develop the informatics infrastructure (software framework, databases, analysis modules, workflow, and hardware platform) necessary to attack the computational problems that the community will face in attempting a reconstruction of the tree of life. Many other research groups throughout the world are working on the tree of life in some form. The resulting surge of interest in large-scale phylogenetic reconstruction from combinatorialists, statisticians, algorithm designers, high-performance computing specialists, and of course, biologists and biomedical researchers has begun to yield spectacular results.

3.2 RECONSTRUCTION METHODS

In this section we review the main computational approaches to phylogenetic reconstruction, with particular attention to their scaling properties. We begin by a discussion of phylogenetic distances since every method for reconstruction makes use of distance or similarity measures, and some methods are based exclusively on such measures.

3.2.1 Phylogenetic Distances

A fundamental property of a tree is that, given any two of its nodes, there exists a unique path connecting the two. Thus, we can define the true evolutionary distance between two nodes in the tree (whether current data or ancestral data) as the length of the unique path connecting the two nodes. How length is measured, however, is a matter of choice. Along each edge in the path, we might want to measure elapsed time, number of evolutionary events (as chosen from a defined collection of possible events), or perhaps best of all, "amount of evolution," which we can formalize through a model of changes that takes into account the frequency and perhaps even the functional significance of each change. For nucleotide data, for instance, we can study the 4×4 nucleotide substitution matrix and assign different values (probabilities or costs) to each entry according to biochemical principles or experimental data. Getting an accurate value for the amount of evolution between any two leaves of the input set, what is usually called the *true evolutionary distance*, would give us invaluable information from which to rebuild the phylogeny; several methods are guaranteed to return the true tree if given the true evolutionary distances between leaves.

Naturally, however, we can only hope to estimate these values according to a chosen model. The basis for computation is instead the *edit distance* between two leaves, that is, the least-cost series of changes that transforms the data at one leaf into the data at the other. Under a given cost model, this is a well-defined measure that is subject to computation; for instance, in the case of two nucleotide sequences and under a model that uses gaps (indels) and nucleotide substitutions, we can use a dynamic programming approach (as in the well-known Smith-Waterman algorithm[6]) to compute this edit distance. (Note that this distance computation involves an alignment of the sequences; the latter is an indispensable prelude to the former.) In other settings, the edit distance may be much more difficult to compute; for instance, it took nearly 20 years to obtain a linear-time algorithm to compute the inversion-based edit distance between two gene orders[7] using what remains one of the most sophisticated theoretical results in computational molecular biology.[8,9]

However, nature is not efficient in the sense of always deriving new forms through the least-cost series of changes; as any simulation quickly reveals, most new forms are derived through more expensive paths. Given a particular model of evolution, it is sometimes possible to invert this model to produce an estimate of the true evolutionary distance from the edit distance. A common example is the so-called Jukes-Cantor correction (see, e.g., Swofford et al.[10]) for edit distances between two nucleotide sequences, derived on purely model-theoretic grounds; another is the so-called empirically derived estimator (EDE) correction[11] derived empirically to obtain a more accurate estimate of the actual number of inversions used to reorder a genome. These *corrected distances* give us a statistical estimate, under the chosen model, of the true evolutionary distance but at the cost of ever-increasing variance in the estimate: Since the edit distance cannot exceed a fixed (usually linear) function of the input size but the true evolutionary distance is unbounded as the edit distance approaches its maximum, the estimate must diverge. Figure 3.2 illustrates the situation for the EDE correction.

These three types of distances, the true evolutionary distance, the edit distance, and the corrected distance, are used throughout the rest of this chapter. In fact, we begin with reconstruction methods that work solely on the pairwise (edit or corrected) distance matrix. The prototype for such methods is the *neighbor-joining* (NJ) method.[12] Using the matrix of pairwise distances, it identifies a nearest-neighbor pair; joins the two subtrees (initially, the two leaves) into a subtree; replaces the two matrix rows and columns for these two subtrees by a single new row and column for the new, larger subtree (a process that entails computing new pairwise distances between the new subtree and the remaining, unaffected $n - 2$ subtrees); and repeats the process until only three subtrees are left, at which point it joins them into a star. Ties, if any, are broken arbitrarily. The algorithm is easy to implement and runs in cubic time even in a naïve implementation. It always produces binary trees, that is, trees where the degree of every internal node is 3, and does not root the final tree, even though the subtrees produced along the way are, in effect, rooted. It is known that NJ will return the true tree if given a matrix of true pairwise evolutionary distances; it is also known that, for nucleotide sequence data under the simplest of models, NJ will converge to the true tree if the sequences from which the distance matrix is computed are of length exponential in the number of taxa.[13] However, experience (see, e.g., Nakhleh et al.[14]) has shown NJ to be particularly sensitive to the value of the *evolutionary diameter* of the data set, that is, the ratio of the largest pairwise distance to the smallest one — a ratio that is bound to increase quickly in most cases as the size of the data set increases and one that is extremely large in the case of the tree of life. Thus, while its speed scales reasonably well, its accuracy does not. Much the same can be said of other distance-based methods. Since much of the problem accrues from the requirement that the method produce a tree, however ill-equipped it is to reconstruct certain edges, a recent article explored the possibility of returning a forest rather than a tree, with significant reported improvements.[15]

3.2.2 Criterion-Based Methods

Criterion-based methods are all based on a measurable and optimizable surrogate for the "truth" — our unmeasurable goal. Of the many methods in this general category, two are of particular note: maximum parsimony (MP) methods and methods that attempt to estimate (conditional) likelihood of trees under some model.

3.2.2.1 Maximum Parsimony

Given a fixed tree and the character sequences associated with its leaves, we can seek to associate character sequences with internal nodes of the tree to minimize, summed over all edges of the tree, the number of changes in each character position. This problem, sometimes known as the *little parsimony problem*, is easily solvable in linear time through a tree traversal, propagating possibilities up from the leaves

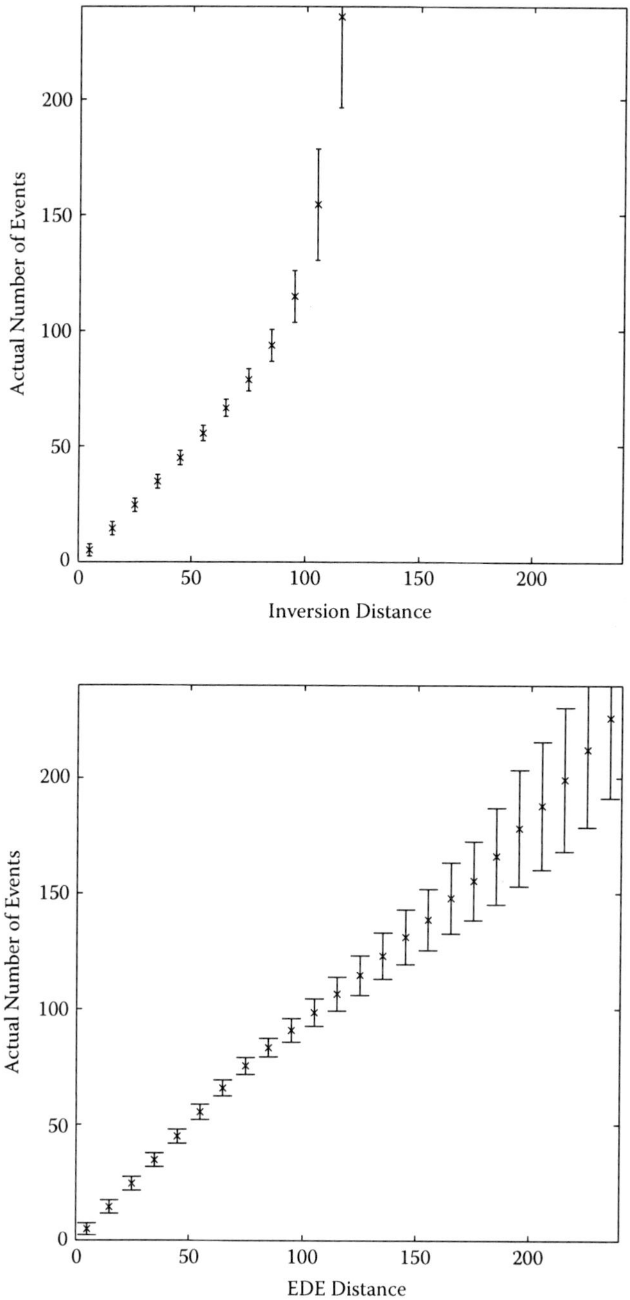

FIGURE 3.2 Edit and corrected distances: On the left, true evolutionary distance versus inversion edit distance; on the right, true evolutionary distance versus corrected (EDE) inversion distance.

and then reflecting constraints down from the root; this algorithm was first given in 1977 by Fitch.[16] Since, however, we do not know the tree, the full MP problem is to identify the tree, along with its internal character sequences, that minimizes the sum of changes. In sharp contrast to the little parsimony problem, MP is NP-hard,[17] and the best exact algorithms strain to get beyond 20 to 30 taxa; heuristic approaches abound, with the best software in current distribution Goloboff's TNT,[18] which can routinely handle within reasonable time 500 to 1,000 taxa.

An interesting recent finding[19] about the parsimony criterion is its relationship to "correctness," that is, how it correlates to the true topology. While MP scores remain fairly high, improvements (i.e., decreases in scores) correlate strongly with improvements in the accuracy of the tree topology, but once MP scores come close to optimal, this correlation is lost, and additional improvements in MP scores have nearly random effects on the tree topology. The other interesting fact that came out of this study is where this transition takes place: On the test sets used in the study, of sizes varying from a few hundred to a few thousand taxa, "close to optimal" for the MP score was within 0.01% of the best score found, yet at that level the tree topology was only about 95% accurate. This finding serves as a sharp reminder of the benefits and perils of using surrogate criteria: They do indeed guide the computation toward better solutions, but the details of optimization for the surrogate criterion and for the desired tree are likely to be quite different.

3.2.2.2 Maximum Likelihood and Bayesian Estimators

Maximum likelihood (ML) methods are based on a specific model choice and attempt to identify the tree that is most likely, under the chosen model, to have given rise to the observed data. In the process, they estimate all model parameters, which usually include the types and numbers of evolutionary events on each tree edge. In principle at least, any model could be used, with any number of parameters, so that ML methods should be able to deal with any data set, however difficult to analyze; in practice, of course, overparameterization leads to overfitting, complex models are computationally too expensive, and the choice of model itself becomes a very complex, as well as crucial, issue. Even for a fixed tree topology, estimating all parameters to obtain a likelihood score, what might be called the small likelihood problem, is an NP-hard problem[20] (in sharp contrast to its MP version). In consequence, until recently, ML methods were limited to very small data sets; over the last few years, however, two new methods have emerged that rival the best MP methods in terms of scalability and accuracy: GARLI (genetic algorithm on rapid likehood interference)[21] and RAxML (randomized A(x)ccelerated maximum likelihood)[22] (the latter scales gracefully to 1,000 taxa).

Advocates of Bayesian methods make no claim to return the best tree but instead attempt to characterize (in a limited way) the distribution of trees (or characteristics thereof) in a neighborhood of high interest. Again, a model must be selected, as well as a prior on the distribution, and again these choices are crucial to the behavior of the algorithm and the quality of the solution. (The pitfalls are perhaps worse than advocates of the method had originally suspected,[23] although recent implementations take suitable precautions.) MCMC methods used to implement Bayesian estimation are unavoidably slow as they must accumulate sufficient numbers of visits to specific states to derive reliable answers; the best software available for Bayesian phylogenetic estimation, MrBayes,[24] scales reasonably well to several hundred taxa.

3.2.3 Metamethods

Since none of the methods described above is suitable to data sets with tens of thousands of taxa, to say nothing of a data set on the scale of the tree of life, computer scientists have sought to apply algorithm design to overcome the various limitations of distance- and criterion-based methods. The earliest attempt was in fact due to biologists, who sought to reconstruct a tree based on reconstruction of trees for each of the $\binom{n}{4}$ possible subsets, called *quartets*, of the data set. The rationale was that building good trees for subsets of four taxa should be easy, and that, assuming enough of these trees were built, they should contain among themselves everything needed to reconstruct the true tree. The problem was what to do with quartets that produced contradictory trees. Tree-puzzling,[25] this first effort, simply added noncontradictory quartets in a random order until a tree was built; later efforts from computer scientists added the ability to filter out "bad" quartets and eventually established the theoretical feasibility of building true trees from quartet data.[26] None of these methods did well in practice, however. Yet, the basic idea of divide and conquer is a very powerful one in this case: Running existing methods on smaller data sets avoids running time or accuracy issues, while controlling the decomposition makes it easier to reassemble the subtrees into a single tree.

A different take on assembling a big tree is the approach collectively known as *supertree* methods.[27] Here, one assumes that many trees will have been produced independently on various data sets, and that assembling them all into one large tree will yield the desired big tree. This approach can be viewed as an "uncontrolled" divide and conquer in which we have no control over the decomposition (each group chooses their own data set) and usually no access to the original data and so we must reassemble the trees themselves as best as we can. While the approach makes sense for assembling the entire tree of life, it does not help us build larger component subtrees and says nothing about scaling. Detailed experiments conducted by Warnow's and my groups[28] indicate that, as might be expected, the accuracy of such an approach is inferior to that of a well-designed prior decomposition.

Just such a solution has been developed over the last several years by Warnow's and my groups: the family of disk-covering methods (DCMs).[26,29–32] Methods in this family control the size, evolutionary diameter, and other attributes of the subsets into which they break the original data set to match the subsets to the characteristics of the analysis methods. Because the subsets are much larger than quartets, the subtrees used in assembling the answer are less numerous and more informative (in the sense that they indicate combinations of edges, not a single edge at a time); because larger subtrees can share a significant number of nodes, assembling them into a larger tree can be done more reliably; and because the decomposition matches the subsets to the characteristics of the underlying methods used on the subsets, challenging data sets can be tackled with the best possible tools. The DCM methods have been used to extend gene-order reconstruction from 16 taxa to simulated data sets of over a thousand taxa[32] and have been applied for MP reconstruction to nucleotide sequence data for over 20,000 taxa.[31] New DCM methods are being derived to improve on existing applications and to tackle computational tasks heretofore considered intractable, such as simultaneous sequence alignment and phylogeny reconstruction (the so-called Sankoff problem[33]) or ML reconstructions on a very large scale.

3.3 DISK-COVERING METHODS

The principle of a DCM is divide and conquer: Divide the data set into smaller subsets, solve the subsets, and assemble these subsolutions into a solution to the original data set. This approach has proved one of the most successful in algorithmic design, leading to very fast algorithms. In a sense, of course, such an approach does not solve the application problem; what it does, in a manner typical of good algorithmic design, is reduce the solution of the entire problem to a collection of simpler tasks. We still need one or more *base methods*, that is, methods to tackle the simpler tasks and provide the needed subsolutions. Fast algorithms for sorting data, for building geometric structures in modeling, for various tasks in geographic information systems, and many other applications all use this approach with great success.

Use of divide and conquer in phylogenetic reconstruction, however, requires much care. The subsets must obey a collection of potentially conflicting constraints. First, they must overlap if there is to be any hope of assembling the subtrees into a single tree — in fact, a substantial overlap is desirable. However, the subsets should also be well separated from each other so that reconstruction on one subset is as independent as possible from reconstruction on another. Next, depending on the reconstruction method to be used on a subset, that subset should have a limited size (for methods such as ML and MP) or a low evolutionary diameter (for a distance-based method). We also need to design a method for reassembling the subtrees that can exploit the structure put into place at the decomposition stage.

In the article that introduced the first DCM,[29] Warnow and her group proposed basing the decomposition on a triangulated threshold graph; each taxon becomes a node of the graph, and two taxa are connected by an edge in the graph whenever their pairwise distance does not exceed a prescribed threshold. In view of the need for overlap, we want the resulting graph to be connected, which puts a lower bound on the value of the threshold; while the threshold cannot be determined in advance, there are at most $\binom{n}{2}$ thresholds and so conceivably every choice could be tested. The resulting graph is then triangulated (with some greedy heuristic) because many crucial graph structures, such as cliques and separators, can be found in polynomial time on triangulated graphs, but are NP-hard otherwise. The maximal cliques of this graph are then identified; they form the subsets to be solved separately. For any nontrivial problem, there will be more than one clique, and any clique will overlap with at least one other because the graph is connected. Because every taxon in a clique is connected only to taxa at distances not exceeding the prescribed threshold, the evolutionary diameter of the subset is typically much lower than that of the original data set. Finally, unless the threshold is very high, the data set will be decomposed into several cliques, thereby reducing the size of each problem to be solved. The matching assembly algorithm, which takes a tree for each subset and assembles these trees into a tree for the

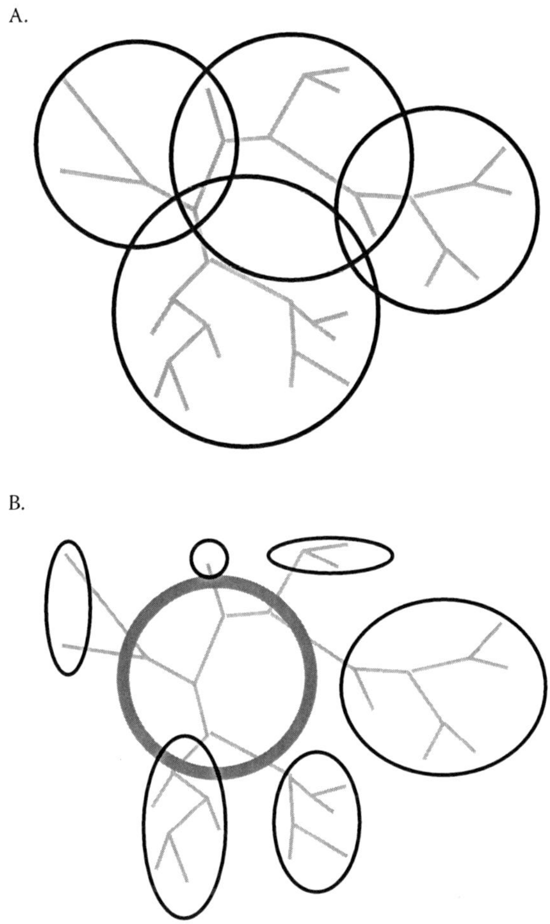

FIGURE 3.3 A schematic view of DCM1 (A) and DCM2 (B, with the graph separator outlined more heavily).

original data set, is a *strict consensus merger*, that is, a method that retains from each given subtree only edges with which every subtree that overlaps with the given subtree also agrees. The process is symbolized in Figure 3.3A. This particular approach can be shown to converge to the true tree when the base method does.

Because the dominant feature of this particular DCM, which we denote DCM1, is the clique and thus tight constraints on pairwise distances, DCM1 works well with a base method such as NJ. On the other hand, DCM1, while it ensures overlap between some pairs of subsets, does not ensure that all subsets will overlap in pairwise fashion or provide any guarantee on the amount of overlap. Warnow and her group[30] thus designed DCM2 to focus on overlap properties. The first steps are the same but instead of finding maximal cliques, the next step finds a *maximal separator*, that is, a subgraph that, when removed, disconnects the triangulated graph into two or more pieces. The subsets are then each composed of one of the disconnected pieces plus the separator, thereby ensuring that all subsets have a pairwise intersection exactly equal to the graph separator, which is typically quite large. The controlled overlap comes at a price, though: The number of induced subsets is often small, and each subset tends to be large, usually half or more of the original set. The resulting approach works best with relatively fast base methods since the reduction in the size of the problem is not very significant. The process is symbolized in Figure 3.3B.

Another interesting aspect of DCMs is their ability to improve convergence. Most phylogenetic reconstruction methods that can be proved to converge to the true tree when given sufficient data appear to require an amount of data that is exponential in the number of taxa — for instance, the length of the DNA sequences needs to double for each additional taxon to preserve the quality of reconstruction. In contrast, a fast-converging method would only require some constant increase in the length of the DNA sequences. Warnow's and my groups[26,34] showed that a slightly different version of DCM (called DCM*) could turn any slow-converging method into a fast-converging one, and that fast approximations for DCM* did well in practice. Given that nature cannot provide arbitrarily long sequences, this result is crucial for scaling to truly large (10^5 or more) data sets.

I also used DCM in a computationally much more demanding setting: reconstruction from gene-order data. In this setting, the base method, GRAPPA,[35]could handle at most 15 taxa; even DCM1, with its tight subsets, could often not find a threshold that ensured graph connectivity and yet decomposed the data set into small enough cliques for the purpose. Tang and Moret decided to apply the approach recursively, another standard methodology from algorithm design: Whenever the clique remained too large, it would be subjected to the same DCM1 process again, but with a reduced range of thresholds. This approach worked remarkably well, enabling the analysis of as many as 1,000 taxa on simulated data.[32]

The conflicting advantages and problems of DCM1 and DCM2 made it clear that better DCMs could be designed, and that the decomposition stage was crucial to the success of the method. Yet, this decomposition stage is determined entirely by the distance matrix and a threshold, and as discussed, experience shows that basing everything on just the distance matrix (an entirely static structure) ignores too much useful information. A third version, DCM3, was then designed to enable iterative improvements in the decomposition; in this approach, the decomposition, while still using a threshold graph, is guided by a tree, which is simply the best reconstruction to date and thus, with every change, may enable a yet better decomposition. Combining this approach with the recursive one just mentioned yielded a recursive and iterative DCM, Rec-I-DCM3, which combined very well with MP base methods, TNT in particular. In experiments using very large real data sets (up to roughly 20,000 taxa), Rec-I-DCM3-TNT easily outperformed any other MP method in terms of both speed and accuracy.[31]

The DCM3 method uses, in effect, only one edge of the best tree so far in guiding the new decomposition — the median edge, which can be viewed as the most trusted partitioning edge because it is farthest from the leaves. As the tree is refined, surely more edges become trustworthy and could also be used in a new decomposition; moreover, using more edges would enable a finer decomposition and save on levels of recursion and potential error propagation. Various groups are at work on devising new DCMs that combine the ideas sketched in this section. Needless to say, progress on these DCMs should not discourage work on the base methods; a DCM is just a way to scale up, and as the recursive approach makes clear, the better performing the base method, the easier the task of scaling it up is.

3.4 AN EXPERIMENTAL METHODOLOGY

Any discussion of large-scale computational efforts needs to take into account testing and assessment. Testing and assessment are even more important than usual in a context like that of the tree of life, for which we have only one instance of the problem and must somehow contrive to convince ourselves of the accuracy of our methods when they are applied to this single instance, yet do so on the basis of tests conducted on far simpler and smaller data sets.

3.4.1 Why Do We Need Experimentation?

An algorithm designer is accustomed to providing an analysis of any proposed algorithm; if that algorithm is an approximation algorithm rather than an exact one, then the algorithm designer also provides performance guarantees for the approximation. Thus, to a large degree, both the running time and the quality of solutions returned by the algorithm are characterized so that, historically, little importance has been placed on actual experimentation in many areas of algorithm design. However, most algorithms for phylogenetic reconstruction are heuristics, with no performance guarantees beyond, at best, a proof that in the limit, with enough data, and under strong independence conditions, the algorithm will return the true tree with high probability — obviously not a very significant guarantee for any given finite instance. In the area of heuristics for NP-hard optimization problems, experimentation has been the main tool for the assessment of new algorithms (see, e.g., D. S. Johnson's work with the TSP[36–38] or with simulated annealing[39,40]). Moreover, algorithmic studies normally assume that the criterion to be optimized is actually the one of interest, whereas as we have seen, parsimony and likelihood criteria are just standing in for topological accuracy and adherence to the truth. Because of the surrogate nature of our criteria, an experimental evaluation would be necessary even for an algorithm known to return the optimal solution in low polynomial time — not so much to evaluate the algorithm as to evaluate the surrogate criterion.

3.4.2 Real and Simulated Data

If we are to conduct experimentation for assessment, then which test suites should we run? In classical optimization problems such as the TSP, there exist libraries of test cases, special challenge problems, and most important, test instance generators. Most, if not all, of these instances are artificial, constructed to test specific aspects of algorithms or to ensure that difficult parts of the problem space are explored. In phylogeny, however, most publications in the area have been authored by biologists and focused on a few real data sets (sometimes even just one) — and frequently the study of these data sets was the motivation for and entire validation of the algorithmic development. Simulation has been advocated as a study tool by leading biologists,[41] but many biology researchers remain suspicious of simulations, citing insufficient realism in the models as well as differences in the computational behavior of algorithms on simulations and on real data sets.

In his seminal article, Hillis[41] mentioned simulations first among four assessment tools; the others are known phylogenies, statistical analyses, and congruence analyses. Known phylogenies and congruence studies (agreement among multiple studies, preferably using different data, for the same set of taxa) can make direct use of real data but are sharply limited in terms of size and availability. Their main use, as Hillis suggested, is in testing predictions from simulation studies. Statistical analyses are best at distinguishing valid conclusions from random noise; in other uses, they require models and so tend to suffer from many of the same problems as some of the methods (ML, Bayesian inference) that they may be used to evaluate. To these four, one might add the use of "comparable computational behavior" between simulated and real data sets (especially when one does not have much information about good answers for the real data).

In any case, the conclusions are clear: Simulations are much more useful than real data for assessing the behavior and accuracy of algorithms because simulations are based on an underlying "true tree" to which reconstructions can be compared, because they can be steered to test various aspects of the algorithms, because they can create data sets of carefully graded sizes and complexity to test scalability, and because they can create large populations of instances to ensure repeatability and statistical significance. Real data sets do not come in such handily graded sizes, rarely have accepted answers for all tree branches, and exist in only relatively small numbers.

On the other hand, real data sets embody the essence of the problem we really care about and often display unexpected complexities that our best models cannot re-create; simulated datasets are only as good as the model and parameter values that created them, which given the relatively simplistic level of current model, may not be a compliment. For instance, experience has shown that typical simulated evolution of sequence data, even under the most complex model for nucleotide substitution, tends to generate overly easy data sets when compared to real data; in contrast, even the simplest model of gene-order evolution through uniformly distributed inversions tends to generate overly difficult data sets when compared to real data. Moreover, the focus on the more easily quantifiable aspects of molecular evolution, such as the model of nucleotide substitution, has obscured what are proving to be far more challenging and influential parts, such as the model of speciation, which has all too often been assumed to be a simple memoryless birth-death process (whereas some branches are well known to be speciose and others bereft of quantifiable evolution for hundreds of thousands of years).

We thus need to work on improving the realism and complexity of current simulations while taking advantage of existing real data sets and of the best possible simulation approaches to assess new algorithms.

3.4.3 Increasing Realism and Size for Simulations

To improve our assessments of algorithms for reconstruction, we thus need to improve the quality of our simulations; we need to do so even more crucially for large data sets since data sets on the scale of the tree of life will not follow any single model or any single set of parameters, no matter how complex, but will involve very complex mixtures of models at all levels — from speciation down to nucleotide substitutions. There have been early attempts at formulating better models of speciation[42,43] and of the resulting tree shapes.[44] A better understanding of where the phylogenetic information lies hidden within the input data would be of tremendous help in designing better simulators — much of what we simulate today is most likely noise, not signal.[45] Likelihood models are capable, at least in principle, of accounting for dependencies of arbitrary nature among characters — and moving beyond the current i.i.d. view of character evolution is surely a prerequisite for more realistic models. RNA secondary structure is relatively well understood and can form the basis for early efforts at characterizing distant interdependencies among sites in nucleotide sequence evolution; the forthcoming Crimson database (led by J. Kim from the CIPRES project) for the assessment of phylogenetic reconstruction algorithms uses such a strategy, among many others.

Increasing size certainly means mixing models, rates, and all other parameters. It then becomes questionable to generate individual data sets; indeed, taking inspiration from the single tree of life and its many reflections in our limited and error-prone samplings of it, the best approach may well be to generate a single enormous data set according to constantly varying mixes of models and parameters and to provide sampling tools to extract subsets according to models, to rates, to clades, to other stratification criteria, or purely at random. Again, this is the strategy used by the Crimson simulation database.

3.4.4 The Predictive Value of Experimentation

Finally, as we embark on a course of computational experiments, it may be a good idea to reflect on the predictive value of the eventual results. After all, it is well known that the landscape of any NP-hard optimization problem must include regions of nearly unpredictable irregularity. What if the solutions identified happen to lie within such a region? Would it not render the results nearly meaningless — after all, they would certainly have little, if any, predictive value? And, even if the solutions happen to lie within a reasonably smooth region, what if it is the "wrong" one — what if, somewhere far removed in the solution space, there exists another smooth region with better solutions? Both possibilities are very real when dealing with an NP-hard problem; the question is how serious an occurrence of either would be for us.

Fortunately for us, the surrogate nature of our criteria this time comes to our rescue. We have evidence that seeking the absolute best solution to the MP problem (and the same applies to the ML problem) does not ensure that we will get the true tree; in fact, given the definition of parsimony, it is intuitively obvious that the true tree is not very likely to be the most parsimonious one. We thus must rely on the assumption that the true tree lies in the neighborhood of the most parsimonious (or likely) one; otherwise, our surrogates are useless. Hence, rather than worry about the shape of our optimization space for MP or ML, we should worry about the correlation between these criteria and the topological accuracy of the reconstruction. This is actually a question that we can explore experimentally, at least in simulations, both forward (by building the best MP or ML trees we can and comparing them with the true tree) and backward (by scoring trees in the neighborhood of the true tree and observing variations in parsimony or likelihood scores). The one study of this type to date[19] had reassuring news, at least for MP: MP scores did correlate well with topological accuracy, and when the correlation was lost in the neighborhood of the most parsimonious trees, all trees examined were quite close to the true tree. Clearly, however, more of that type of work is sorely needed, especially with more refined simulations and, if possible, with real data sets.

3.5 CONCLUSION

The enormous growth in the use of phylogenies in biomedical and biological research and the increased interest in a reconstruction of the tree of life have focused attention on scalability issues in phylogenetic reconstruction. In this review, we outlined the problems and sketched some possible avenues of solution. The state of the art in this area is changing faster now than it has in the past 30 years; that much remains to be done is not in doubt, but that exciting progress is being made, with the promise of resolving many of the problems discussed here, is equally clear.

REFERENCES

1. M.G. Montague & C.A. Hutchinson III. Gene content and phylogeny of herpesviruses. *Proceedings of the National Academy of Sciences of the United States of America*, 97:5334–5339, 2000.
2. T. Dobzhansky. Nothing in biology makes sense except in the light of evolution. *The American Biology Teacher*, 35:125–129, 1973.
3. S. Yang, R. Doolittle & P. Bourne. Phylogeny determined by protein content. *Proceedings of the National Academy of Sciences of the United States of America*, 102(2):373–378, 2005.
4. B.M.E. Moret. Computational challenges from the tree of life. In *Proc. 7th SIAM Workshop on Algorithm Engineering and Experiments (ALENEX'05)*, pp. 3–16, SIAM Press, Philadelphia, 2005.
5. K. Rice, M. Donoghue & R. Olmstead. Analyzing large datasets: rbcL500 revisited. *Systematic Biology*, 46:554–563, 1997.
6. T.F. Smith & M.S. Waterman. Identification of common molecular subsequences. *Journal of Molecular Biology*, 147:195–197, 1981.
7. D.A. Bader, B.M.E. Moret & M. Yan. A fast linear-time algorithm for inversion distance with an experimental comparison. *Journal of Computational Biology*, 8(5):483–491, 2001.
8. S. Hannenhalli & P.A. Pevzner. Transforming cabbage into turnip (polynomial algorithm for sorting signed permutations by reversals). In *Proceedings of the 27th Annual ACM Symposium on the Theory of Computing (STOC'95)*, pp. 178–189, ACM Press, New York, 1995.

9. S. Hannenhalli & P.A. Pevzner. Transforming mice into men (polynomial algorithm for genomic distance problems). In *Proceedings of the 36th Annual IEEE Symposium on the Foundations of Computer Science (FOCS'95)*, pp. 581–592, IEEE Press, Piscataway, NJ, 1995.
10. D.L. Swofford, G.J. Olsen, P.J. Waddell & D.M. Hillis. Phylogenetic inference. In D.M. Hillis, B.K. Mable, & C. Moritz, Eds., *Molecular Systematics*, pp. 407–514, Sinauer Associates, Sunderland, MA, 1996.
11. B.M.E. Moret, J. Tang, L.-S. Wang & T. Warnow. Steps toward accurate reconstructions of phylogenies from gene-order data. *Journal of Computer Systems Science*, 65(3):508–525, 2002.
12. N. Saitou & M. Nei. The neighbor-joining method: a new method for reconstructing phylogenetic trees. *Molecular Biology and Evolution*, 4:406–425, 1987.
13. K. Atteson. The performance of the neighbor-joining methods of phylogenetic reconstruction. *Algorithmica*, 25(2/3):251–278, 1999.
14. L. Nakhleh, B.M.E. Moret, U. Roshan, K. St. John & T. Warnow. The accuracy of fast phylogenetic methods for large datasets. In *Proceedings of the 7th Pacific Symposium on Biocomputing (PSB'02)*, pp. 211–222, World Scientific, 2002.
15. C. Daskalakis, C. Hill, A. Jaffe, R. Mihaescu, E. Mossel & S. Rao. Maximal accurate forests from distance matrices. In *Proceedings of the 10th International Conference on Research in Computational Molecular Biology (RECOMB'06)*, Vol. 3909 of *Lecture Notes in Computer Science*, pp. 281–295, Springer-Verlag, New York, 2006.
16. W.M. Fitch. On the problem of discovering the most parsimonious tree. *American Naturalist*, 111:223–257, 1977.
17. W.H.E. Day & D. Sankoff. Computational complexity of inferring phylogenies by compatibility. *Systematic Zoology*, 35(2):224–229, 1986.
18. P. Goloboff. Analyzing large datasets in reasonable times: solutions for composite optima. *Cladistics*, 15:415–428, 1999.
19. T.L. Williams, D.A. Bader, M. Yan & B.M.E. Moret. High-performance phylogeny reconstruction under maximum parsimony. In A.Y. Zomaya, Ed., *Parallel Computing for Bioinformatics and Computational Biology*, pp. 369–394, Wiley, New York, 2006.
20. S. Roch. A short proof that phylogenetic tree reconstruction by maximum likelihood is hard. *ACM/IEEE Transactions on Computational Biology and Bioinformatics*, 3(1), 2006.
21. D. Zwickl. GARLI. Available at www.zo.utexas.edu/faculty/antisense/Garli.html.
22. A. Stamatakis, T. Ludwig & H. Meier. RAxML-III: a fast program for maximum likelihood-based inference of large phylogenetic trees. *Bioinformatics*, 21(4): 456–463, 2005.
23. E. Mossel & E. Vigoda. Limitations of Markov chain Monte Carlo algorithms for Bayesian inference of phylogeny [short report]. *Science*, 309(5744):2207–2209, 2005.
24. J.P. Huelsenbeck & F. Ronquist. MrBayes: Bayesian inference of phylogeny. *Bioinformatics*, 17:754b, 2001. Available at morphbank.ebc.uu.se/mrbayes/.
25. K. Strimmer & A. von Haeseler. Quartet puzzling: a quartet maximum likelihood method for reconstructing tree topologies. *Molecular Biology and Evolution*, 13:964–969, 1996.
26. T. Warnow, B.M.E. Moret & K. St. John. Absolute convergence: true trees from short sequences. In *Proc. 12th Annual ACM/SIAM Symposium on Discrete Algorithms (SODA'01)*, pp. 186–195, SIAM Press, 2001.
27. O.R.P. Bininda-Edmonds, Ed. *Phylogenetic Supertrees: Combining Information to Reveal the Tree of Life*, Kluwer Academic, Dordrecht, 2004.
28. U. Roshan, B.M.E. Moret, T. Warnow & T.L. Williams. Performance of supertree methods on various dataset decompositions. In O.R.P. Bininda-Edmonds, Ed., *Phylogenetic Supertrees: Combining Information to Reveal the Tree of Life*, pp. 301–328, Kluwer Academic, Dordrecht, 2004.

29. D. Huson, S. Nettles & T. Warnow. Disk-covering, a fast converging method for phylogenetic tree reconstruction. *Journal of Compututational Biology*, 6(3):369–386, 1999.
30. D. Huson, L. Vawter & T. Warnow. Solving large scale phylogenetic problems using DCM-2. In *Proceedings of the 7th International Conference on Intelligent Systems for Molecular Biology (ISMB'99)*, pp. 118–129, AAAI Press, Menlo Park, CA, 1999.
31. U. Roshan, B.M.E. Moret, T.L. Williams & T. Warnow. Rec-I-DCM3: a fast algorithmic technique for reconstructing large phylogenetic trees. In *Proceedings of the Third IEEE Computational Systems Bioinformatics Conference CSB'04*, pp. 98–109, IEEE Press, Piscataway, NJ, 2004.
32. J. Tang & B.M.E. Moret. Scaling up accurate phylogenetic reconstruction from gene-order data. In *Proc. 11th Int'l Conference on Intelligent Systems for Molecular Biology (ISMB'03)*, Vol. 19 of *Bioinformatics*, pp. i305–i312, Oxford University Press, New York, 2003.
33. D. Sankoff. Minimal mutation trees of sequences. *SIAM Journal of Applied Mathematics*, 28(1):35–42, 1975.
34. B.M.E. Moret, U. Roshan & T. Warnow. Sequence length requirements for phylogenetic methods. In *Proceedings of the 2nd International Workshop on Algorithms in Bioinformatics (WABI'02)*, Vol. 2452 of *Lecture Notes in Computer Science*, pp. 343–356, Springer-Verlag, New York, 2002.
35. B.M.E. Moret, S.K. Wyman, D.A. Bader, T. Warnow & M. Yan. A new implementation and detailed study of breakpoint analysis. In *Proceedings of the 6th Pacific Symposium on Biocomputing (PSB'01)*, pp. 583–594, World Scientific, 2001.
36. D.S. Johnson, G. Gutin, L.A. McGeoch, A. Yeo, W. Zhang & A. Zverovitch. Experimental analysis of heuristics for the ATSP. In G. Gutin & A.B. Punnen, Eds., *The Traveling Salesman Problem and Its Variations*, Vol. 12 of *Combinatorial Optimization*, pp. 445–487, Springer-Verlag, New York, 2002.
37. D.S. Johnson & L.A. McGeoch. The traveling salesman problem: a case study. In E. Aarts & J.K. Lenstra, Eds., *Local Search in Combinatorial Optimization*, pp. 215–310, Wiley, New York, 1997.
38. D.S. Johnson & L.A. McGeoch. Experimental analysis of heuristics for the STSP. In G. Gutin & A.B. Punnen, Eds., *The Traveling Salesman Problem and Its Variations*, Vol. 12 of *Combinatorial Optimization*, pp. 369–443, Springer-Verlag, New York, 2002.
39. C.R. Aragon, D.S. Johnson, L.A. McGeoch & C. Shevon. Optimization by simulated annealing: an experimental evaluation; part II, graph coloring and number partitioning. *Operations Research*, 39(3):378–406, 1991.
40. D.S. Johnson, C.R. Aragon, L.A. McGeoch & C.J. Shevon. Optimization by simulated annealing: an experimental evaluation; part I, graph partitioning. *Operations Research*, 37(6):865–892, 1989.
41. D. M. Hillis. Approaches for assessing phylogenetic accuracy. *Systematic Biology*, 44:3–16, 1995.
42. S.B. Heard. Patterns in phylogenetic tree balance with variable and evolving speciation rates. *Evolution*, 50:2141–2148, 1996.
43. A.O. Mooers & S.B. Heard. Inferring evolutionary process from phylogenetic tree shape. *Quarterly Review of Biology*, 72:31–54, 1997.
44. D.J. Aldous. Stochastic models and descriptive statistics for phylogenetic trees, from Yule to today. *Statistical Science*, 16:23–34, 2001.
45. S. Angelov, B. Harb, S. Kannan, S. Khanna & J. Kim. Efficient enumeration of phylogenetically informative substrings. In *Proceedings of the 10th International Conference on Research in Computational Molecular Biology (RECOMB'06)*, Vol. 3909 of *Lecture Notes in Computer Science*, pp. 248–264, Springer-Verlag, New York, 2006.

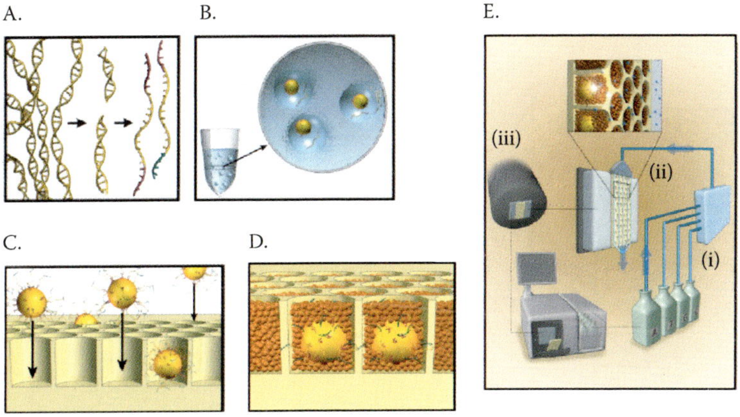

COLOR FIGURE 2.1 (See caption on page 16.)

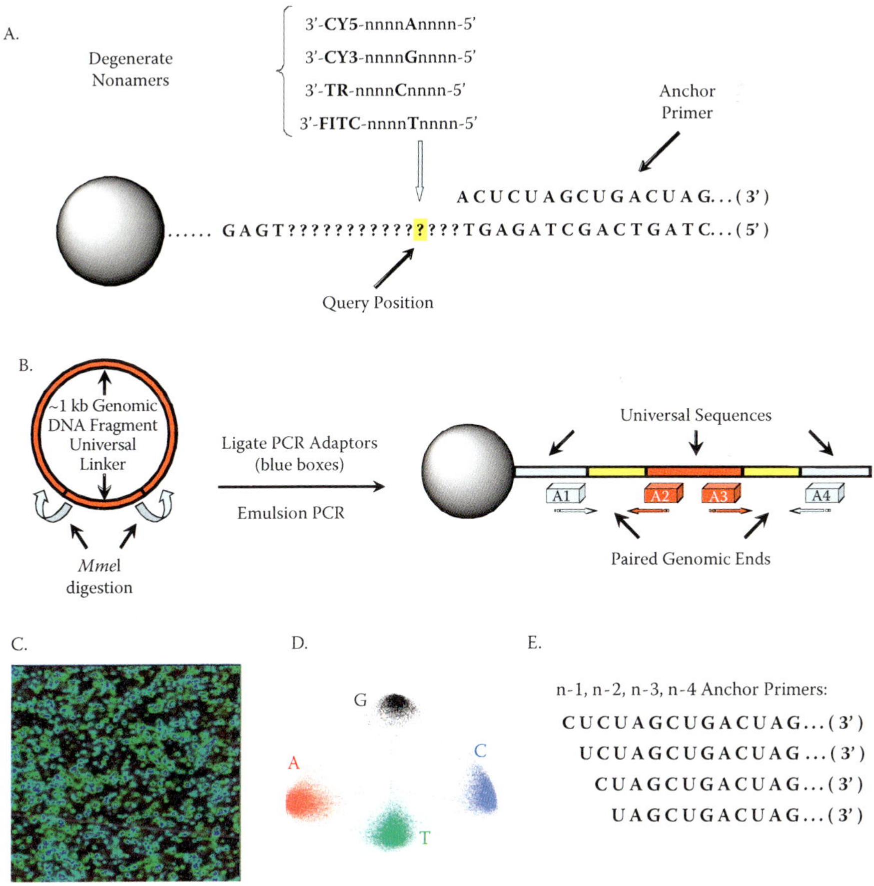

COLOR FIGURE 2.2 (See caption on page 19.)

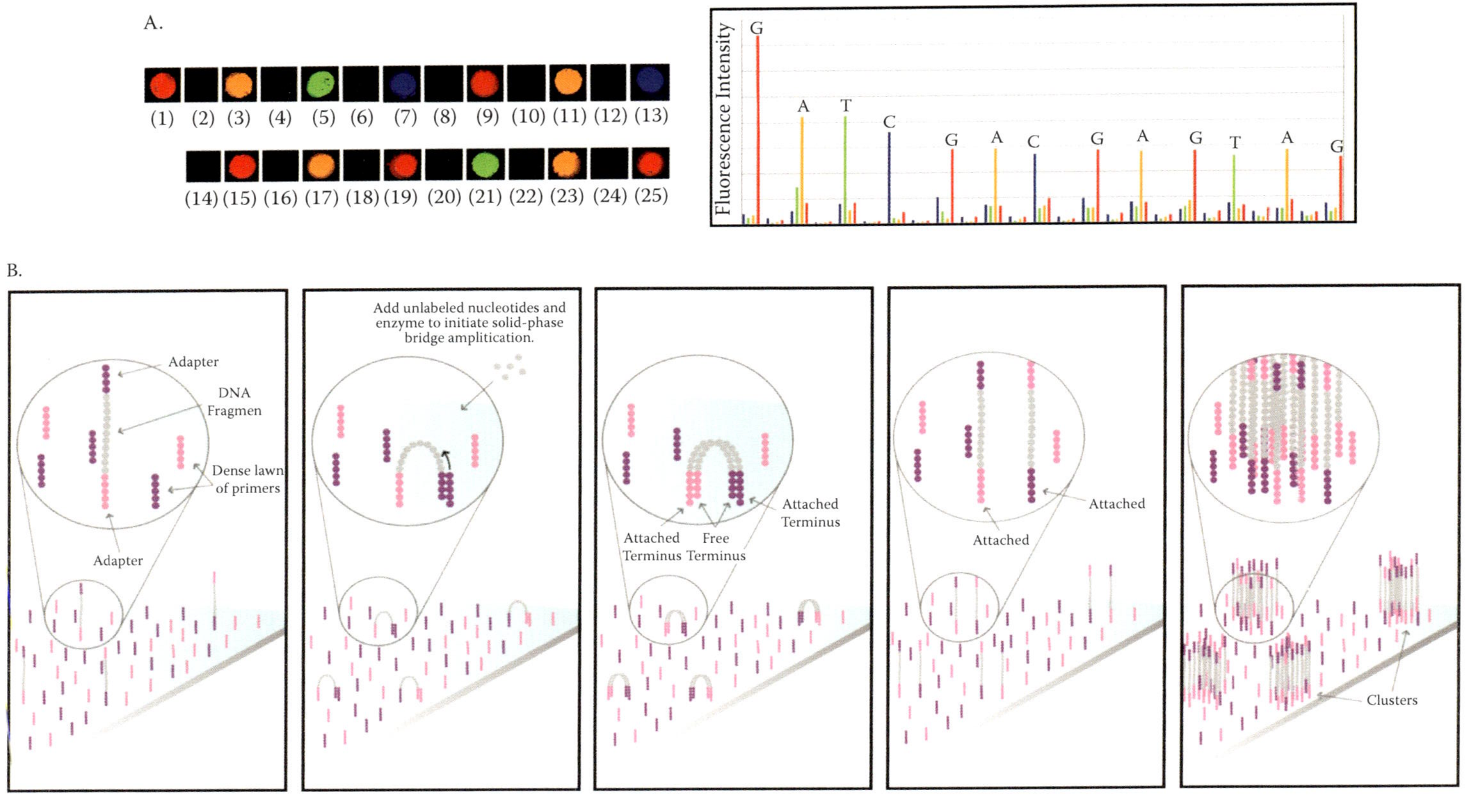
A.
(1) (2) (3) (4) (5) (6) (7) (8) (9) (10) (11) (12) (13)
(14) (15) (16) (17) (18) (19) (20) (21) (22) (23) (24) (25)
Fluorescence Intensity
G A T C G A C G A G T A G
B.
Adapter
DNA Fragmen
Dense lawn of primers
Adapter
Add unlabeled nucleotides and enzyme to initiate solid-phase bridge amplitication.
Attached Terminus
Free Terminus
Attached Terminus
Attached
Attached
Clusters

C.

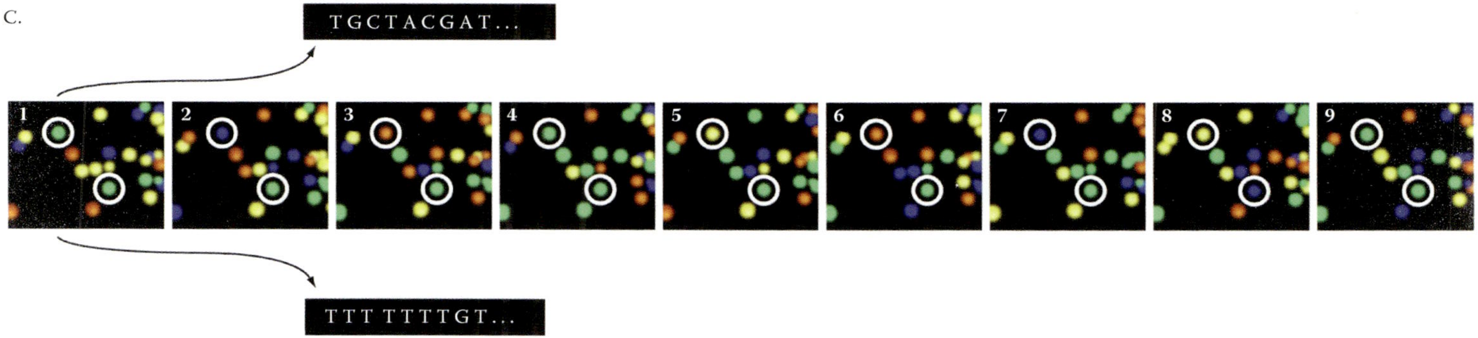

COLOR FIGURE 2.3 Cyclic reversible termination: (A) 13-base CRT sequencing using the 3′-*O*-allyl terminators developed by Ju and colleagues,[16] illustrating fluorescence scanned data and four-color intensity histogram plot. The template was immobilized to a solid support using the self-priming method (not shown). (B) Five panels illustrate Illumina's single-molecule array (SMA) technology.[5] In panel 1, isolated genomic DNA is fragmented and ligated with adaptors, which are then made single-stranded and attached to the solid support. Bridge amplification (panel 2) is performed to create double-stranded templates (panel 3), which are denatured (panel 4) and bridge amplified several more times to create template clusters (panel 5). (C) Nine-base CRT sequencing highlighting two different template sequences. The series of images was obtained from a 40-million cluster SMA (not shown). (Panel A was reprinted from Ju et al., *Proc. Natl. Acad. Sci. U. S. A.* 103, 19635–19640, 2006, by permission of the National Academy of Sciences, U. S. A., copyright 2006. Figures 2.3B and 2.3C were obtained by permission from Illumina Inc.)

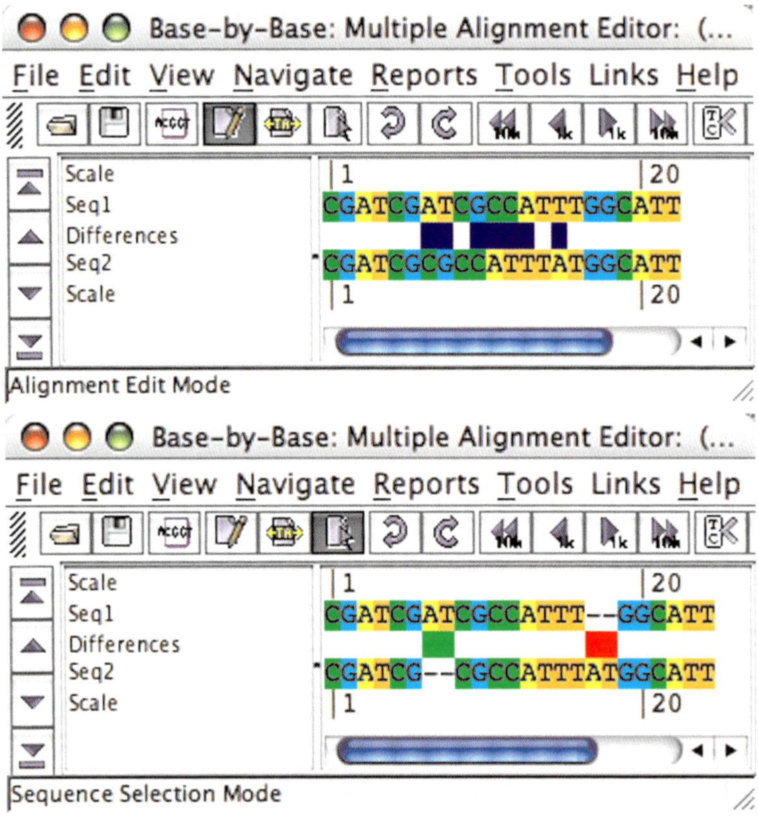

COLOR FIGURE 4.7 Detection of errors in an MSA using Base-By-Base. Top panel: an alignment of two DNA sequences containing seven mismatches, which are indicated by blue boxes in the *differences* row. Bottom panel: insertion of two gaps (indicated by green and red boxes in the *differences* row) results in sequence realignment, eliminating all mismatches.

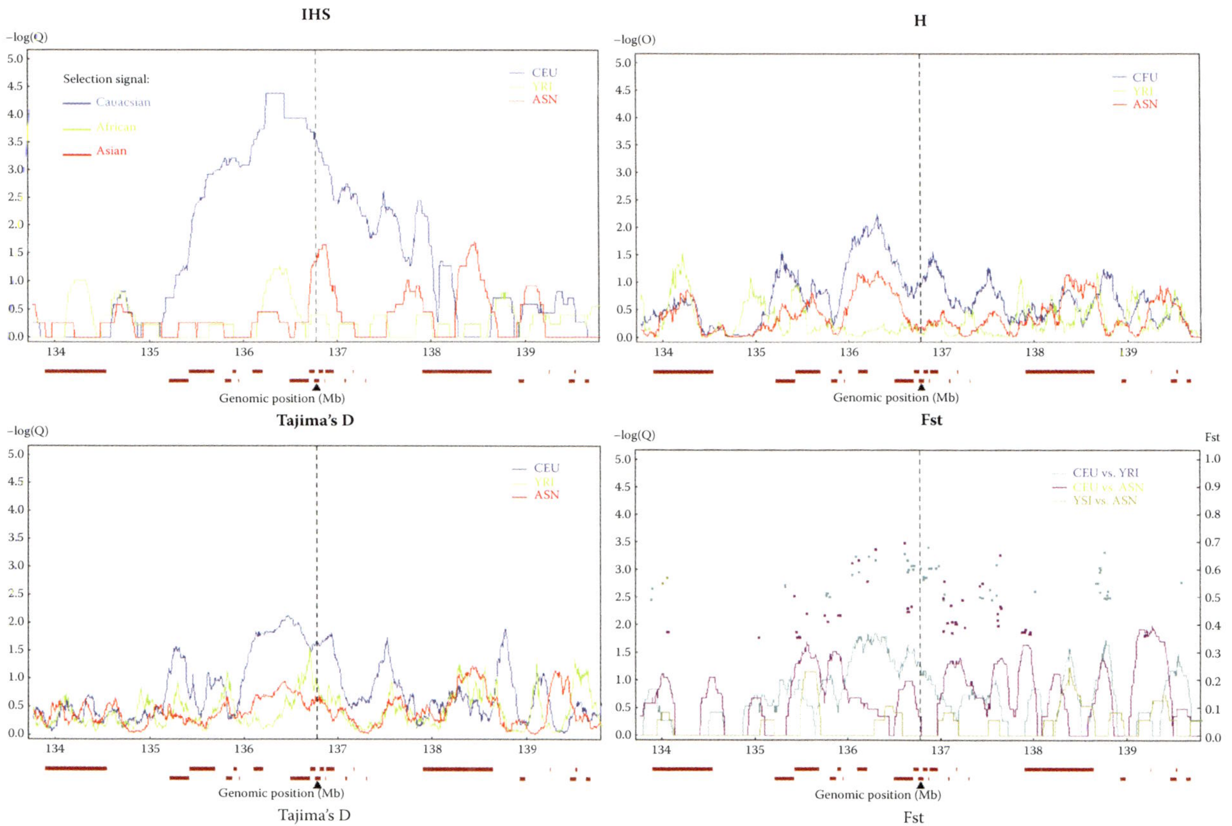

COLOR FIGURE 8.4 Haplotter output across the LCT locus. Results of four different molecular selection analysis methods (iHS, H, Tajima's D, Fst) are presented across the LCT locus.

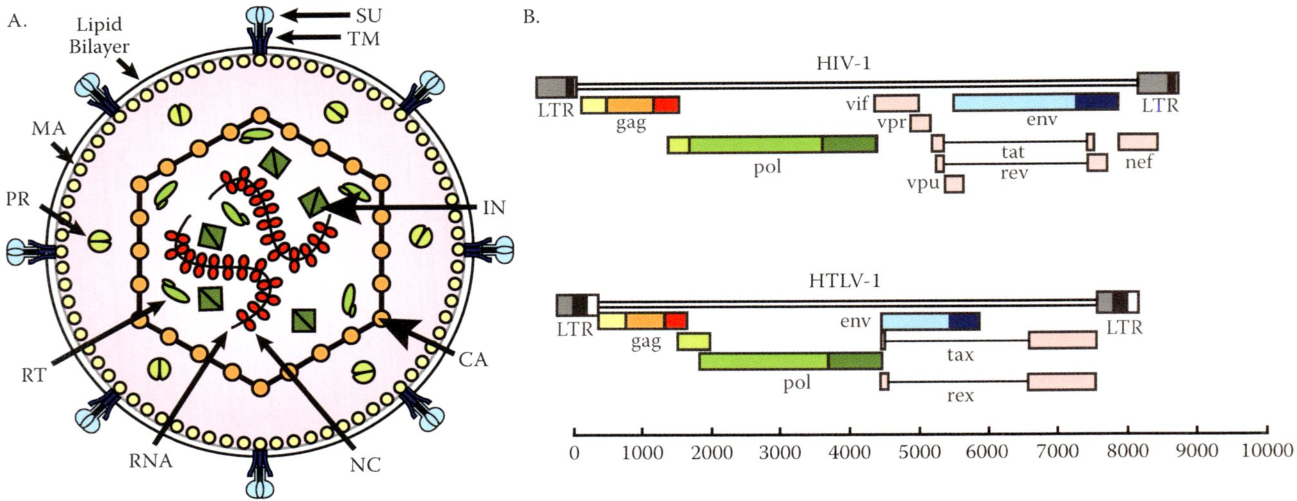

COLOR FIGURE 12.1 (A) Schematic cross section through a retroviral particle. CA, capsid; IN, integrase; MA, matrix; NC, nucleocapsid; PR, protease; RT, reverse transcriptase; SU, surface unit; TM, transmembrane. (B) Schematic organization of the HIV genome. As a comparison, the genome of another complex retrovirus, HTLV-1, is depicted. The color codes in the genomes correspond to the encoded proteins in the particle. (Adapted from Voght, P. K., in *Retroviruses,* Eds. Coffin, J. M., Hughes, S. H., & Varmus, H. E., Cold Spring Harbor Laboratory Press, New York, 1997.)

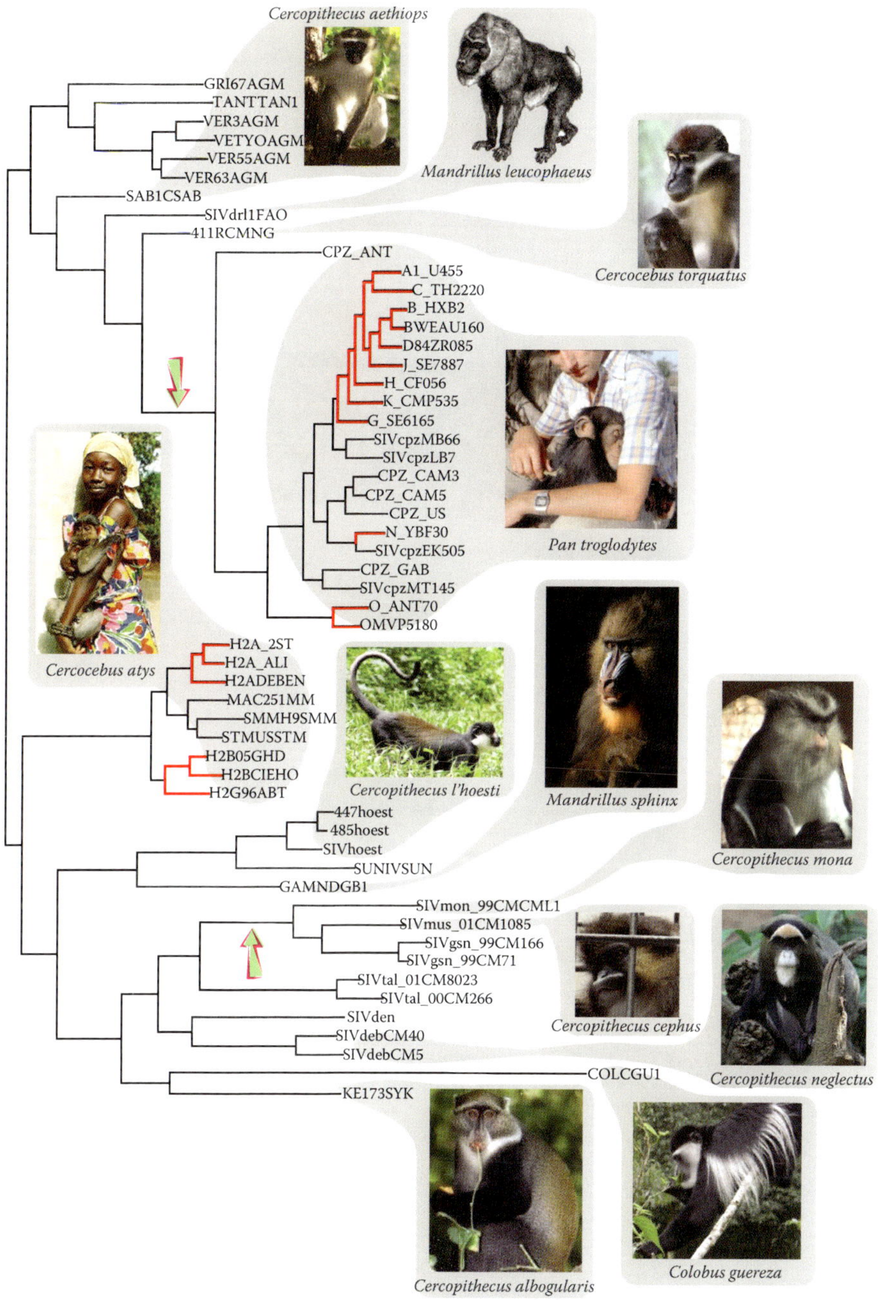

COLOR FIGURE 12.3 (See caption on page 226.)

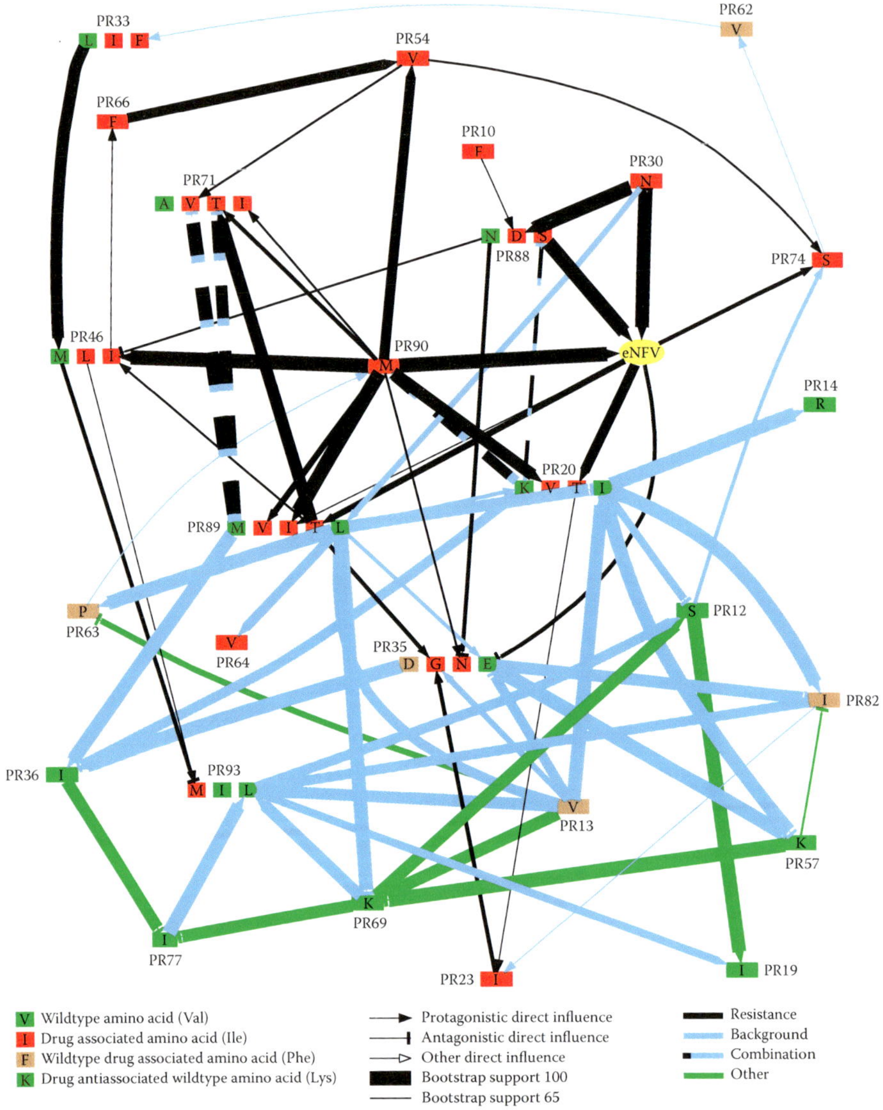

COLOR FIGURE 12.6 Bayesian network model for drug resistance against nelfinavir visualizes relationships between exposure to treatment (eNFV), drug resistance mutations (red), and background polymorphisms (green).[117]

4 Comparative Genomics of Viruses Using Bioinformatics Tools

Chris Upton and Elliot J. Lefkowitz

CONTENTS

ABSTRACT

The comparative genomics of viruses is a broad topic, in part because of the great variation in viral genome structures and associated replication strategies. This chapter, however, tries to focus on the value of comparative methods in the hope that it will be generally applicable. Since the volume of genomics data is ever increasing, we also emphasize the importance of bioinformatics tools in managing and analyzing genomic data in an efficient manner. For examples, we have drawn on our background with the Viral Bioinformatics Resource Center (VBRC; www.vbrc.org).

4.1 INTRODUCTION

The virosphere encompasses an extremely diverse group of organisms. Genomic variations among viral species include differences in large-scale genome structure, genome size, nucleotide composition, and coding strategy. Examples of the various types of genomic structure include the following: influenza virus (negative-sense single-stranded RNA [ssRNA]), poliovirus (positive-sense ssRNA), rotavirus (double-stranded RNA [dsRNA]), HIV (positive-sense ssRNA, requiring a dsDNA intermediate), variola virus (dsDNA), and parvovirus (ssDNA). Viral genomes may be nonsegmented or segmented (e.g., Bunyaviridae) and may contain genes on either a single strand or both strands (some RNA virus genomes may even be ambisense with open reading frames (ORFs) encoded on both strands). Genome size is another widely differing characteristic; most RNA viruses range in size from approximately 3 to 20 kb (coronaviruses are unusual, with genomes of ~30 kb). DNA viruses show even more variation, ranging from small (e.g., parvovirus, < 10 kb) through medium (adenovirus, ~35 kb) and large (poxviruses, 150–350 kb) to the recently discovered "supersize" mimivirus, approximately 1,200 kb dsDNA virus. Expression strategies also differ; viruses may or may not utilize RNA processing or editing to produce functional messenger RNA (mRNA) transcripts. These differences translate into a wide variety of viral strategies for the basic processes of genome replication, transcription, and protein translation/maturation; the study of such differences is known as *comparative virology*.

Accordingly, the analytical procedures used to explore such genomic differences will often vary depending on the genome size and coding strategy; analyses routinely used in the characterization of one virus family may be meaningless when applied to a different family. For example, gene content is an important parameter in the comparative study of larger viruses (such as poxviruses, herpesviruses, baculoviruses, and coronaviruses) but is not useful when applied to a smaller virus such as poliovirus. Large viruses often contain nonessential "virulence" genes that can be lost in various strains to create attenuated phenotypes without affecting *in vitro* viral replication. In contrast, smaller viruses such as poliovirus retain the same gene content in all strains, with single-nucleotide mutations (causing minor amino acid or gene regulation differences) acting as attenuation markers instead.[1] Yet another complicating factor in viral genomic analyses is the diversity of hosts that are infected by viruses.

This chapter attempts to address these complications, approaching the study of comparative genomics of viruses from a techniques-and-tools standpoint. Real-life examples are provided, using data from a variety of virus families to illustrate the analysis techniques under discussion. If possible, these examples use tools that are freely accessible via the Internet, and although a variety of bioinformatics resources are discussed, we have drawn heavily on our own experiences in developing the VBRC (Table 4.1). Given our research background, this chapter abounds with examples and references specific to poxviruses; however, readers should feel free to substitute their own favorite group of large DNA viruses as appropriate (e.g., herpesviruses, baculoviruses, iridoviruses, phycodnaviruses, asfaviruses, or even phage).

As discussed in this chapter, one of the most prevalent problems that molecular virologists encounter in bioinformatics lies in the initial choice of software tools. Sometimes, it can seem that many near-identical applications exist to perform a single task; other times, no tools are available to do exactly what one wants. To address the first issue, similar applications will often not give identical results for a single task; subtle differences between applications (e.g., computer platform, browser type, execution speed, input and output formats, etc.), which are not immediately apparent,

TABLE 4.1
List of URLs for Bioinformatics Resources

Resource	Internet URL
All the Virology on the WWW	http://www.virology.net
BioDirectory	http://www.biodirectory.com
BioEdit	http://www.mbio.ncsu.edu/BioEdit/bioedit.html
BioHealthBase	http://www.biohealthbase.org/GSearch
Bionet	http://www.bio.net
COGs	http://www.ncbi.nlm.nih.gov/COG
Descriptions of Plant Viruses	http://www.dpvweb.net
ExPASy	http://www.expasy.org/tools
HCV	http://hcv.lanl.gov
HIV	http://hiv-web.lanl.gov
ICTV	http://www.ncbi.nlm.nih.gov/ICTVdb
IMV	http://virology.wisc.edu/virusworld
LAJ	http://www.bx.psu.edu/miller_lab
LANL	http://www.lanl.gov/science/pathogens
Mauve	http://gel.ahabs.wisc.edu/mauve
NCBI	http://www.ncbi.nlm.nih.gov
NCBI, genotyping	http://www.ncbi.nlm.nih.gov/projects/genotyping
NCBI, taxonomy	http://www.ncbi.nlm.nih.gov/Taxonomy
NCBI, viruses	http://www.ncbi.nlm.nih.gov/genomes/VIRUSES/viruses.html
Open Source software	http://www.opensource.org
PATRIC	https://patric.vbi.vt.edu
PubMed	http://www.pubmed.org
RefSeq	http://www.ncbi.nlm.nih.gov/RefSeq
R'MES	http://genome.jouy.inra.fr/ssb/rmes
Synteny tool	http://www.vbrc.org/synteny.asp
Universal Virus Database	http://www.ncbi.nlm.nih.gov/ICTVdb
VB-Ca	http://www.virology.ca
VBRC	http://www.vbrc.org
VIDA	http://www.biochem.ucl.ac.uk/bsm/virus_database/VIDA3/VIDA.html
Viper	http://viperdb.scripps.edu
VOCs database	http://www.virology.ca
VOG	http://www.ncbi.nlm.nih.gov/genomes/VIRUSES/vog.html
Wikiomics	http://www.wikiomics.org
Wikipedia	http://www.wikipedia.org/wiki/Database

can nonetheless affect the results obtained. The second problem is often caused by the fact that a desired analysis tool can be embedded within a larger application and thus hidden from a novice's first exploration of the software. However, in our experience, the bioinformatics community is generally helpful, and software authors are usually happy to help others use their software and to respond to bug reports (or, sometimes, to explain to the researcher that these apparent bugs are actually useful features). Therefore, the novice should not hesitate to either contact the software authors or ask questions via public forums such as Bionet, Wikiomics, or BioDirectory (Table 4.1).

4.2 VIRUS-SPECIFIC BIOINFORMATICS RESOURCES

The Internet contains a wide variety of bioinformatics tools, databases, and general information sites intended to assist with comparative analysis of viral genomes (see Table 4.1 for URLs [uniform resource locators] of Web sites discussed in this section). This review discusses only a few of the more comprehensive sites available in Fall 2006. These bioinformatics resources often differ in the information they contain and the types of analyses they support. A resource may (1) simply supply raw data (e.g., a database that accepts a query via a Web interface and returns a list of DNA sequences); (2) carry out analyses on selected data and present the results (in text or graphic form) to the user via a Web interface; or (3) provide information connected, by a vast array of links, to related items in different databases (e.g., PubMed). In the above models, the user interacts with the resource via a Web browser, and much of the analysis occurs through *canned routines* on the resource's server. However, in our VBRC and Virus Bioinformatics-Canada (VB-Ca) Web resources, we have tried a fourth model that uses a *client–server* approach. Our servers provide a variety of databases, and although the initial interaction with the user is via a Web page, client software is seamlessly downloaded to the user's local computer and is then used to perform a variety of analyses on the data. Each of the above systems has its own merits and evolves in response to the type of data it supports and the needs of its users.

Most researchers today are familiar with the PubMed literature database, the Entrez genome/protein sequence databases, and the suite of similarity search (basic local alignment tool; BLAST[2]) software, all located on the National Center for Biotechnology Information (NCBI) Web page. The page also contains specialized resources for the HIV, severe acute respiratory syndrome (SARS), and influenza viruses, as well as a section devoted to viral genomes. Another good resource for all things virological is All the Virology on the WWW; this site is a compendium of links to other virology-related sites of all types and is especially useful for obtaining basic information and educational material. Authoritative information on virus classification can be found at the Universal Virus Database of the International Committee on Taxonomy of Viruses (ICTV). Three of the eight Bioinformatics Resource Centers (BRCs) funded by the National Institutes of Health (NIH) have mandates to support databases on specific virus families. VBRC supports Pox-, Flavi-, Toga-, Arena-, Bunya-, Filo-, and Paramyxoviridae; PATRIC supports Calici-, Corona-, and Rhabdoviridae as well as hepatitis A and E viruses; and BioHealthBase supports influenza virus. Although these BRCs focus primarily on potential agents of biowarfare/bioterrorism or those viruses that represent emerging or reemerging disease threats, other viruses commonly used as biological models in the same families are also included. The Virus Database at University College London (VIDA) supports Herpes-, Pox-, Papilloma-, Corona- and Arteriviridae databases and provides information on ortholog families and functionally related proteins. The Descriptions of Plant Viruses site contains virus classifications and genomic data. The Los Alamos National Laboratory (LANL) provides databases on a variety of human pathogens, including HIV (genome sequences, resistance, immunology, vaccine trials); hepatitis C virus (HCV) (genome sequences and immunology); influenza; and oral pathogens/sexually transmitted diseases (STDs), including papillomaviruses and herpesviruses. VB-Ca supports Adeno-, Asfar-, Baculo-, Herpes-, Irido- and Coronaviridae in addition to the families covered by the VBRC site. For information on virion structure, the reader is directed to the Virus Particle Explorer (Viper) and the Institute for Molecular Virology (IMV), which provide descriptions of icosahedral virus capsid structures along with tools for structural and computational analysis. Finally, users should not hesitate to use the Google search engine, which can work surprisingly well.

As noted elsewhere in this review, researchers should not take information from these databases blindly; just as the quality of all of the available genome sequences is not equal, sequence annotations and analysis tools vary as well. It is most certainly a case for *caveat emptor,* and the cost of a program or software package is often not directly proportional to its quality. This is not meant to imply that most available databases are flooded with bad data, but rather that all resources tend to be, to some degree, incomplete (after all, the researchers creating these databases will naturally focus on their particular areas of interest). Large genome and protein databases, such as those at the NCBI, often contain (out of necessity) many computer-generated annotations, which tend to be less accurate and specific than those provided by expert human researchers (found in a curated database such as RefSeq). Analysis tools located on different Web sites may have different default parameters; also, multiple tools that at first appear to provide a common function (e.g., multiple sequence alignment, MSA) in practice often fail to provide identical results (they may be designed for different sequence types or lengths or may use different algorithms or have different parameter settings).

As well, even with the wide variety of existing software, it is not always possible to find bioinformatics tools that are capable of performing a desired task (as previously discussed). The input format may be incompatible with your software, the output can often be difficult to interpret meaningfully, or a Web server-based tool may only be able to process your 1,000 protein sequences one at a time. How can a researcher deal with these types of stumbling blocks? The simple answer is: Collaborate. The developers of these tools want them to be both *useful* and *used* and are therefore usually eager for user feedback — including requests for more comprehensive documentation or enhancements of their software.

Although it is difficult for the virologist to manage some of these problems, one area in which individuals can play a big role is annotation. The NCBI would certainly welcome assistance in annotating RefSeq entries, or a knowledgeable user might offer to assist on a curation/annotation project organized by members of a particular research community. Examples of such projects include the *Pseudomonas* Genome Project, The Institute for Genomic Research (TIGR) Rice Genome Annotation, and the *Saccharomyces* Genome Database. The NIH BRCs are involved in the curation of virus pathogens and would also welcome community input to support their annotation processes.

4.3 SO WHAT'S WITH THE *COMPARATIVE* STUFF?

Classical, "wet-lab" biochemistry is often time consuming, expensive, and challenging; however, *comparative genomics* is not easy either, a fact that perhaps needs more recognition. In the 21st century, both approaches are necessary; each has its own strengths and weaknesses. Bioinformatics is often thought of as merely a preliminary data-crunching/-mining process to generate hypotheses that must ultimately be tested at the bench. However, *comparative genomics/analyses,* when used together with appropriate statistical analyses, can generate solid, highly useful inferences about molecular structure and function. It is true that these remain only inferences that must still be subjected to rigorous laboratory confirmation, but the predictive power of these models for generating useful hypotheses is considerable. We are reminded of Douglas Adams's quotation: "If it looks like a duck, and quacks like a duck, we have at least to consider the possibility that we have a small aquatic bird of the family *Anatidae* on our hands." Thus, *in silico* analysis may be extremely useful; save time, effort, and money in solving biochemical problems; and substantially narrow the range of hypotheses for further testing — but *in silico* analysis is certainly not infallible.

A good example is that of the poxvirus uracil DNA glycosylase (UNG)[3]; standard BLASTP[2] and FASTA[4] programs failed to detect the very weak similarities between the poxvirus UNG proteins (at that time proteins of unknown function) and several UNGs previously identified in other organisms. Although subsequent protein database searches with the Needleman-Wunsch global alignment algorithm[5] did detect some of these weak similarities, it was only the presence of multiple UNGs from several very diverse organisms (*Escherichia coli, Bacillus subtilis,* herpesvirus, human) that suggested the results were significant. Figure 4.1A shows part of the alignment between the vaccinia virus and *E. coli* UNGs, and Figure 4.1B shows a percent identity matrix for a selection of UNG proteins. In this case, it was the comparison of multiple database search results and the generation of a MSA that ultimately demonstrated that a small number of significant amino acids were highly conserved across this diverse group (Figure 4.1C). The vaccinia virus protein in question was subsequently expressed and shown to possess UNG activity.[3] In analyzing these results (both computational and experimental) along with other known facts, an initial surprise came from the fact that there was previous genetic evidence showing the gene encoding the vaccinia virus UNG had a temperature-sensitive allele, suggesting that the protein may be essential for virus replication, while in all other organisms tested (eukaryotes and prokaryotes), UNG activity was not essential. A second surprise came when researchers were able to generate site-specific mutations that inactivated the UNG enzymatic function of the vaccinia protein without disrupting virus replication, providing evidence for two different functional roles encoded by this gene. Thus, the apparent requirement for UNG enzymatic activity in vaccinia virus was actually a requirement for its previously unknown role in replication as a

FIGURE 4.1 (Opposite) Analysis of poxvirus uracil DNA glycosylases (UNGs). Panel A, alignment of a region of vaccinia virus D4R protein (query) with the *E. coli* UNG (subject); panel B, percent identity matrix for a series of diverse UNGs; panel C, multiple sequence alignment (MSA) and consensus for a series of UNGs showing one of the most conserved regions of the enzyme.

A

```
Uracil DNA glycosylase [Escherichia coli F11]
Identities = 44/170 (25%)
Query  57   LRNKRVCVCGIDPY--PKDGTGVPFE------SPNFTKKSIKEIASSISRLTGVIDYKGY  108
            L + +V + G DPY  P    G+ F        +P        KE+ ++I    T      GY
Sbjct  53   LGDVKKVVILGQDPYHGPGQAHGLAFSVRPGIATPPSLLNMYKELENTIPGFTR--PNHGY  110

Query  109  NLNII-DGVIPWNYYLSCKLGETKSHA-----IYWDKISKLLLQHITKHVSVLYCLGKTD  162
              +    GV+  N  L+ + G+  SHA      + DK+  L+ QH     V +L+
Sbjct  111  LESWARQGVLLLNTVLTVRAGQAHSHASLGWETFTDKVISLINQHRKGVVFLLWGSHAQK  170

Query  163  FSNIRAKLESPVTTIVGYHPAARDRQFEKDRSFEIINVLLELDNKVPINW  212
               I K    V       P +  R F     F + N  LE   + PI+W
Sbjct  171  KGAIIDKQRHHVLKAPHPSPLSAHRGFFGCNHFVLANQWLEQRGETPIDW  220
```

B

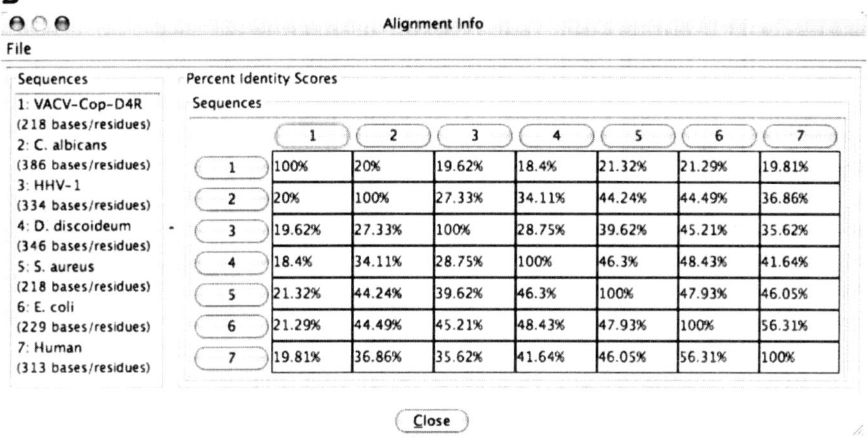

	1	2	3	4	5	6	7
1	100%	20%	19.62%	18.4%	21.32%	21.29%	19.81%
2	20%	100%	27.33%	34.11%	44.24%	44.49%	36.86%
3	19.62%	27.33%	100%	28.75%	39.62%	45.21%	35.62%
4	18.4%	34.11%	28.75%	100%	46.3%	48.43%	41.64%
5	21.32%	44.24%	39.62%	46.3%	100%	47.93%	46.05%
6	21.29%	44.49%	45.21%	48.43%	47.93%	100%	56.31%
7	19.81%	36.86%	35.62%	41.64%	46.05%	56.31%	100%

C

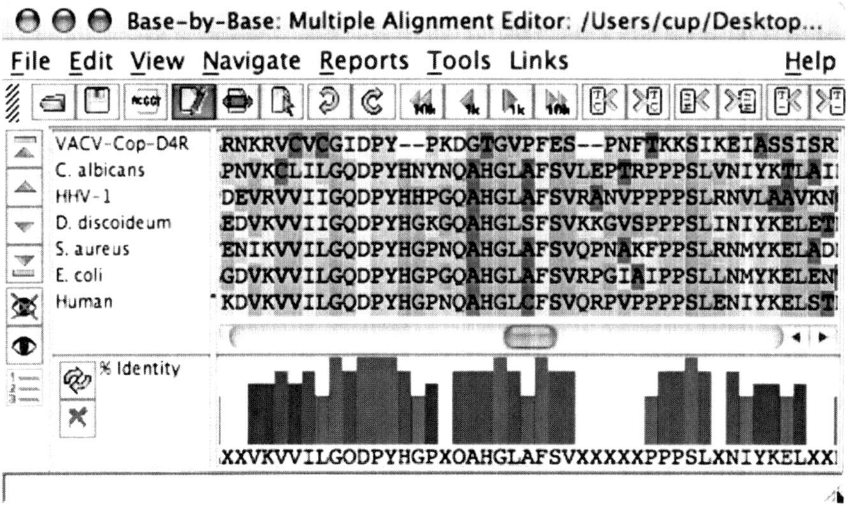

processivity factor.[6,7] Thus, comparative genomics evidence was instrumental in discerning one of these functions, while classical biochemistry/genetics was required to fill in the other part of the story. Over the last 15 to 20 years, improvements to search algorithms, including the development of position specific iterative-BLAST (PSI-BLAST),[2] along with the accumulation of huge amounts of additional genomics data have resulted in it not only being much easier to recognize the connection between the poxvirus UNG proteins and the UNGs from other organisms, but also to more generally predict functional conservation by identifying significant sequence similarities in this gray zone of very weak (<25%) similarity.

A more recent trend is to use structural similarity to support weak sequence similarity matches. For example, using profile-hidden Markov models, the program HHsearch[8] searches the database of protein sequences with known structures (PDB[9]) so that subsequent homology modeling can be used to look for corroborating data. As an aside, it is noteworthy that the success of any similarity search procedure is dependent on genome annotation (as a source for protein sequences), and that in turn, annotation often relies heavily on comparative analyses for coding sequences (CDS) prediction. A good example of this principle is the prediction of small exons in eukaryotic DNA, which can be spotted within large genes by comparison of isogenic regions of mouse and human DNA. Although less applicable to smaller RNA viruses, comparative analysis is valuable in annotating the large DNA viruses.[10]

4.4 SO YOU WANT TO COMPARE THESE GENOMES? TRY A DOTPLOT

One of the simplest and easiest ways to compare two large DNA sequences is to generate a dotplot.[11] A *dotplot* is essentially a matrix comparison of two sequences that is created by moving a relatively short sequence window along the two sequences. When a match is observed between the sequence windows, a dot is recorded in the matrix at the appropriate position. This type of output is a visual representation of the overall similarity between two genomes and provides information that cannot be derived from a phylogenetic tree or percent identity statistics. Figure 4.2 shows two dotplots, which clearly highlight the locations of highest similarity between two very different human coronaviruses (SARS and OC43; Figure 4.2A) and, conversely, the small regions of dissimilarity between two closely related coronaviruses (human SARS and bat SARS; Figure 4.2B). These plots were generated with JDotter[12]; this is essentially a Java interface to Dotter,[11] but it can also link to our VOCs (Virus Ortholog Clusters) database (Table 4.1) and display precalculated dotplots and gene annotations. One of the advantages of the Dotter algorithm is that window size and score cutoff criteria can be quickly changed without recalculating the entire plot. Another use of dotplots is to identify repeated sequences and regions of unusual DNA composition in large (genome-size) sequences by plotting the same DNA sequence on both axes; specialized software can also assist with this task.[13] Figure 4.2C shows a self-plot of the crocodile poxvirus (CRV) genome[14]; the scoring threshold has been lowered (relative to Figure 4.2), and there is a significant background visible (in addition to the black diagonal line that represents the 100% identical self vs. self-alignment). Although the bulk of this genome has a GC content greater than 65%, some smaller

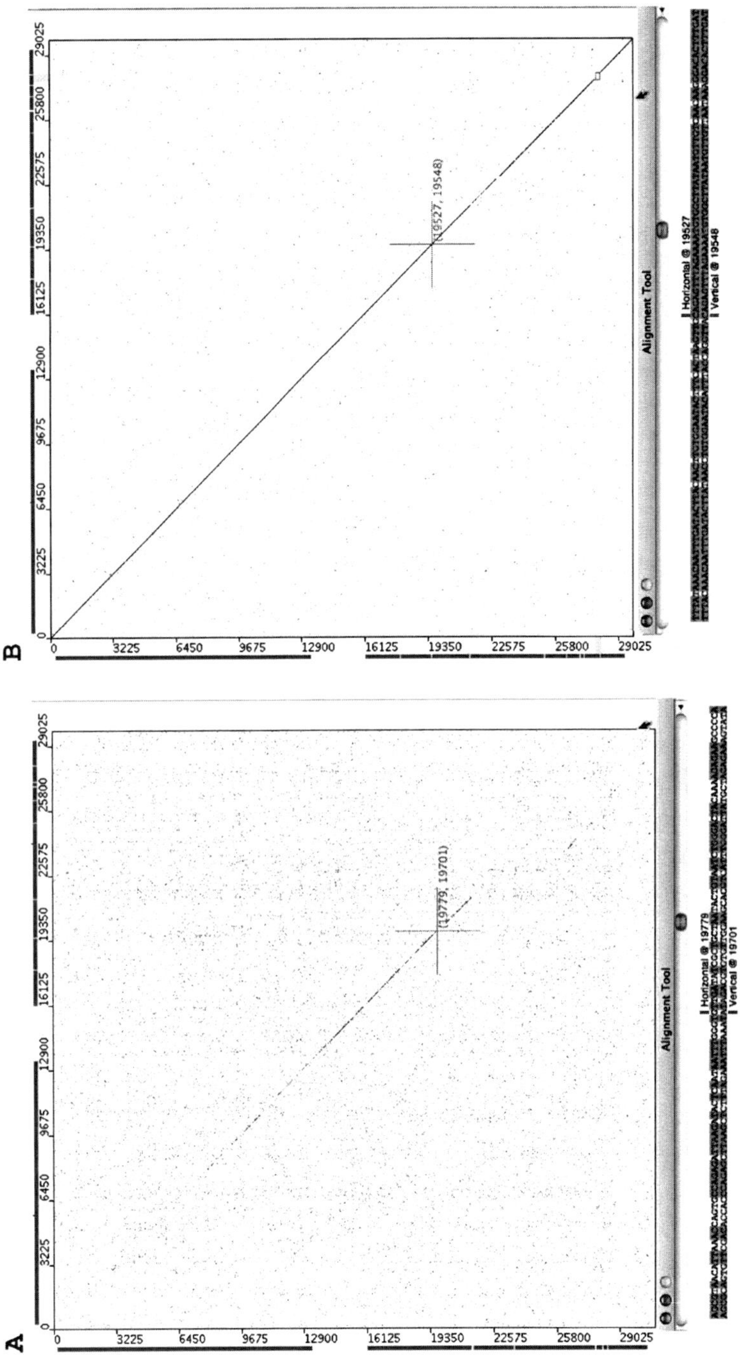

FIGURE 4.2 Dotplots created with JDotter. Panel A, human SARS strain Tor2 (horizontal) and human coronavirus OC43 (vertical); panel B, human SARS strain Tor2 (horizontal) and bat SARS strain HKU3-1 (vertical). The crosshairs are positioned on the diagonal of identity, as shown in the DNA sequence alignment in the bottom window. The small box within the plot in panel B indicates a gap in the sequence alignment. Panel C, self-plot of the crocodile poxvirus genome. The crosshairs show a region of repeated DNA sequence; the bars outside the plot represent annotated genes. Panel D, a zoomed-in view of the boxed region in panel C; the crosshairs mark a faint diagonal line that represents the weak similarity between genes 33 and 35.

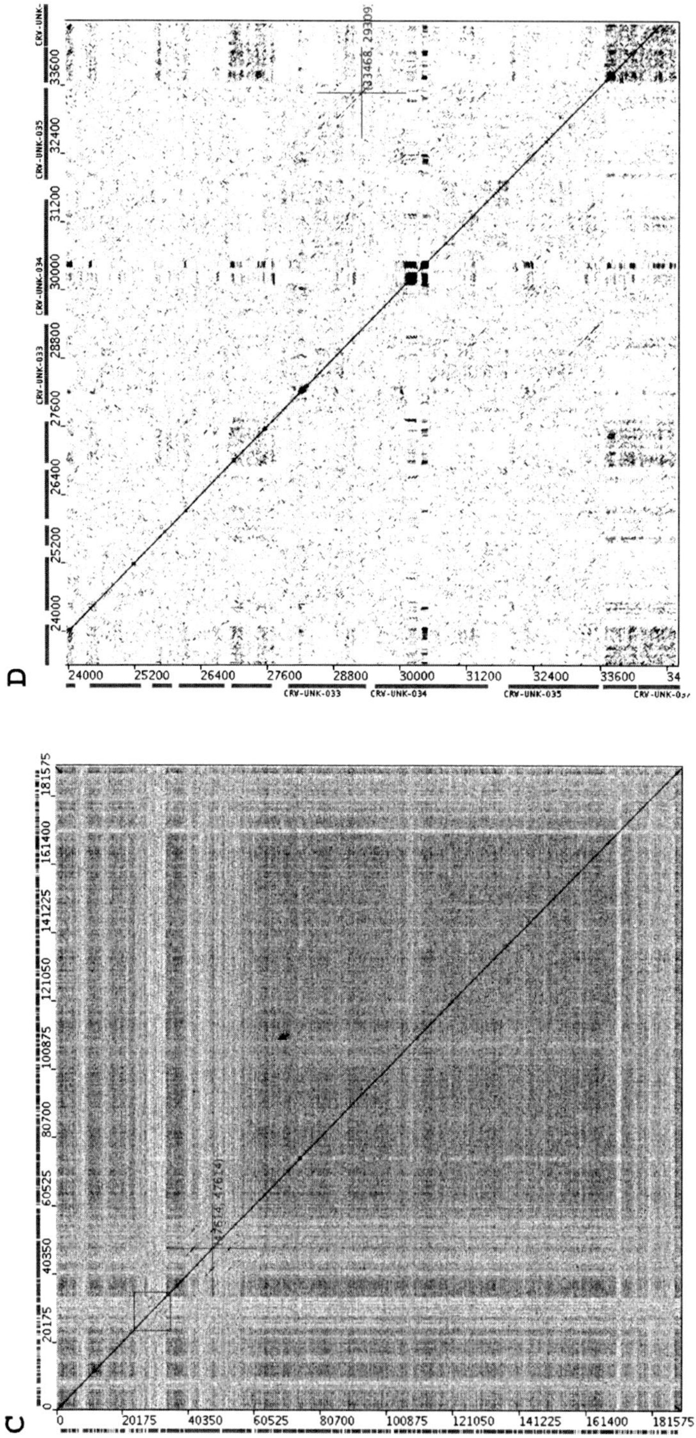

FIGURE 4.2 (Continued).

regions are significantly more AT rich; these show up as light gray background stripes in the plot (fewer random nucleotide matches, due to the greater diversity in base content, lead to a lighter background). Figure 4.2D shows a zoomed-in view of one of these stripes (the boxed region in Figure 4.2C.) As predicted, most of the genes in this region have a lower-than-average GC content; genes 33, 34, and 35 contain 45%, 48%, and 47% GC, respectively. Interestingly, these three genes are related to each other but not to any other known poxvirus gene, and they likely represent an acquisition event unique to crocodilepox of host DNA, followed by gene duplication similar to that observed in molluscum contagiosum.[15]

Dotplot-like figures for whole genomes can also be generated based on whole-gene similarity rather than individual or short stretches of nucleotides. Figure 4.3 shows a comparison between the predicted gene sets of two poxvirus genomes (variola and fowlpox viruses). This gene synteny analysis tool is available online from the VBRC Web site (Table 4.1). The program uses a precomputed set of BLASTP[2] comparison results between the gene sets of all species in the VBRC poxvirus database; these results are displayed in the form of a gene similarity dotplot that reveals the proteins shared between the two genomes. This particular comparison shows that a significant number of fowlpox virus genes have been inverted relative to the variola

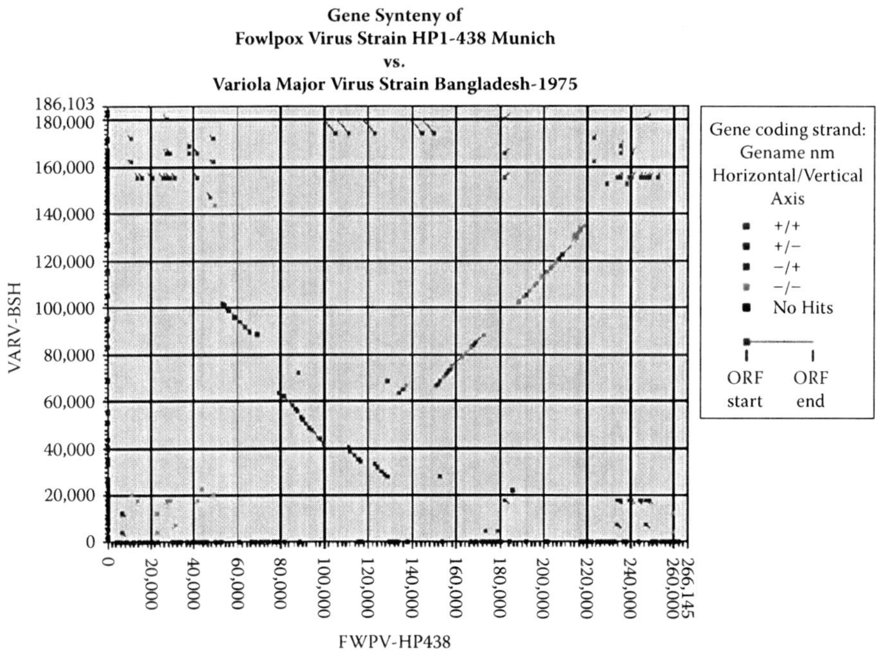

FIGURE 4.3 A gene synteny plot of fowlpox virus (horizontal) versus variola virus (vertical). All predicted proteins coded for by each virus were compared to each other using BLASTP2. Each pair of proteins with some degree of similarity (a BLAST expect [E] value < 10^{-5}) is shown in the figure, plotted according to location of the genes within the two genomes. The color (not shown in this image) of a given point reflects the coding strands of the two genes (described in the figure legend). Black points located along either of the two axes represent proteins unique to that genome.

virus genome (a feature that all avipoxviruses share). This inversion can be seen in two ways: (1) the diagonal line with a negative slope (indicates inversion of gene direction); and (2) the green/red (not visible in this grayscale figure) coloring of this diagonal line (also indicates that the genes in question are on opposite strands).

Another variation on the dotplot theme is generated by Local Alignment Java (LAJ) (Table 4.1). This tool creates a plot of two large DNA sequences from a series of local alignments detected by BLASTZ (BLAST modified for long, gapped alignments)[16]; an example, using two distantly related coronaviruses, is shown in Figure 4.4. The user can zoom in to the plot and examine the local alignments, which are shown at the bottom of the window. A useful feature of LAJ is its display of the individual alignment scores in a percent identity plot (PIP) just above the local alignment window; this highlights small regions that have unusually high identity.

A key principle of the dotplot or LAJ plot is that a single, definitive alignment is not generated; rather, a gestalt view of the data is provided to the researcher, and relationships between spatially distant DNA sequences can be easily observed. This feature allows researchers to determine easily if regions of DNA have been duplicated, transposed, or inverted (e.g., Figure 4.2C, crosshairs). Also, in regions

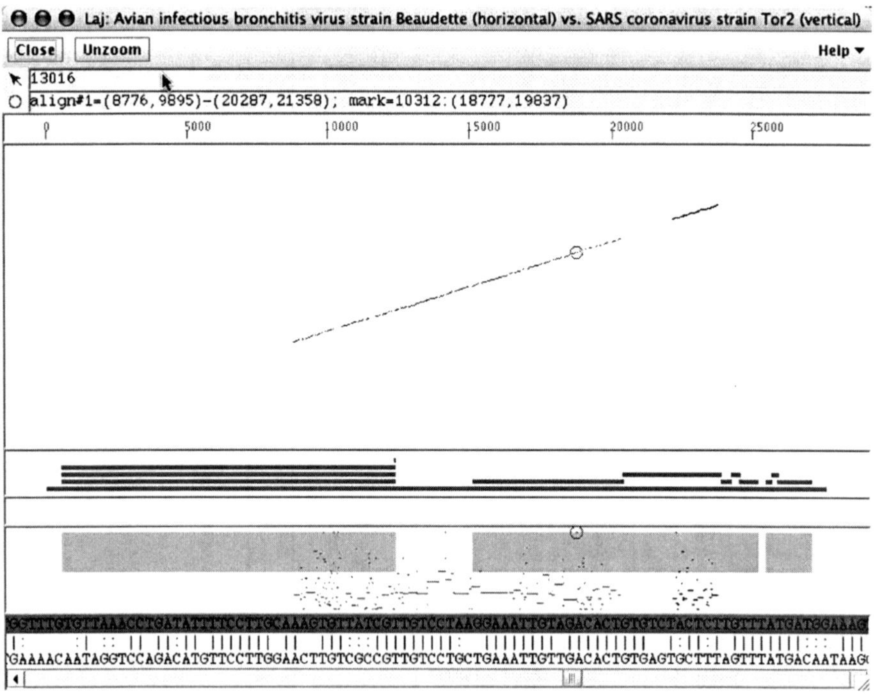

FIGURE 4.4 An LAJ plot of an avian coronavirus against the distantly related SARS virus. Window sections, from bottom to top: display of local sequence alignment; percent identity plot (higher-scoring alignments are shown by dots/lines placed higher in the plot); gene annotations (if included with the sequences); main dotplot window; information on the selected local alignment (location shown by a circle in other windows). Colors (not shown in this figure) are used to indicate active local alignments.

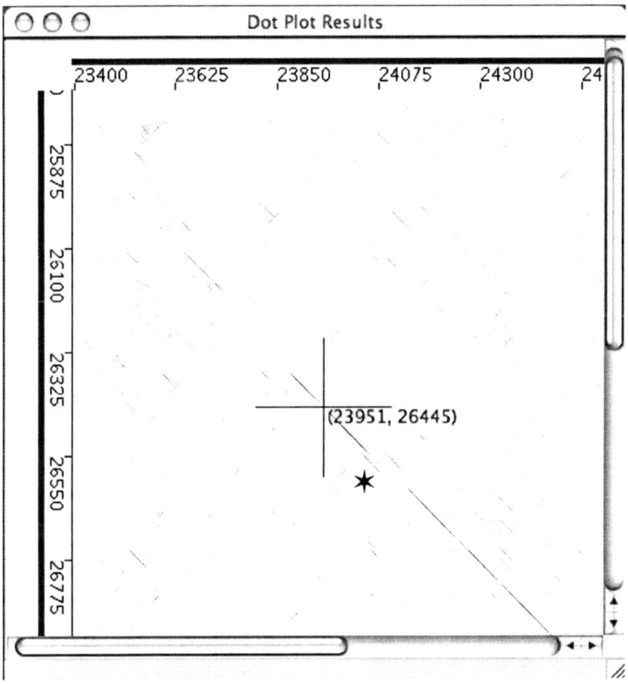

FIGURE 4.5 Potential alternate alignments in a region of weak similarity; zoomed region from plot in Figure 4.2B. Parallel diagonal lines (near star) represent different alignments with very similar scores.

of weak similarity, alternative alignments for the same subsequence can be readily visualized (Figure 4.5, starred region). This feature is very useful when examining distantly related proteins in which several alignments (with very similar or identical scores) appear equally likely to be correct. A Java tool for displaying sets of near-optimal alignments is available[17] and is provided as an option in the VOCs database tools (Table 4.1).

4.5 ANOTHER BIRD'S-EYE VIEW: WHAT DOES THE VIRUS ENCODE?

When dealing with a newly sequenced, entirely unannotated, virus — especially one about which we have little or no experimental data — some of the first questions that should be asked are, What genes does the virus encode? What proteins are expressed? At first glance, this appears to be a simple problem, easily solvable by using a closely related, annotated virus (assuming that one is available) to annotate the new species. For many small viruses, especially those with RNA genomes, this can be a successful strategy; these viruses show little variability in their coding strategies and share most of their ORFs at the family or genus level (as most of these genes are essential for replication). However, the huge amount of variation in the virosphere[18] means that it can be very difficult to make accurate gene predictions for other viral species. This problem is most severe for the larger DNA viruses but can also exist for some RNA viruses (such as the minor mRNA splice variants of HIV[19] and the nonessential ORFs of SARS and other coronaviruses that may be deleted without seriously compromising virus replication *in vitro* culture[17, 20, 21]). The large DNA viruses (such as herpesvirus, baculovirus, and poxvirus) contain a significant number of genes (virulence factors) that are not required for replication in tissue culture; instead, they encode proteins that enhance virus infection, replication, pathogenesis, and transmission in its natural host. The processes of mutation, virus evolution, and host-range restriction have led to the loss of some of these nonessential genes in certain species and their retention in others. Alternatively, duplication of these ORFs can form sets of paralogous genes, allowing for gene divergence and acquisition of new functions. Thus, when a group of closely related large DNA viruses (such as the cowpox, camelpox, and variola poxviruses) are compared, one generally finds considerable differences in gene content.

These possibilities lead to another disputed question, that of *pseudogene annotation*. Some researchers annotate every ORF (including those on the opposite DNA strand to a well-characterized gene), while others avoid annotating ORFs that appear to be gene fragments and thus not transcribed or translated into functional proteins. The many questions that arise from this debate include the following: When should a gene with a 3′ truncation be labeled a pseudogene? Should a gene that has lost its initiating methionine codon be labeled a pseudogene? Should a gene that has lost a functional promoter (as determined by computational analysis) be labeled a pseudogene? These problems can lead to somewhat misleading results; for example, the number of genes assigned to the various vaccinia virus strains range from 163 to 284 genes. Although there are indeed significant differences between these strains (including sizable deletions between the two viruses at the extreme ends of this numerical range), the number of functional genes missing from the smaller virus is in actual fact probably as small as 30, with the rest of the apparent difference caused by varying annotation procedures.

When gene annotation based on a related strain is, for one reason or another, not a viable option, gene prediction becomes the next logical step. Because mechanisms of gene expression (transcription and translation) vary extensively between virus families, gene prediction must be tailored appropriately. At its crudest, it is little more than ORF detection; however, it is also possible to examine the presence/absence of promoters or functional amino acid motifs, codon use, base composition, amino acid composition and isoelectric point of predicted proteins, and similarity/ ortholog search results.[22] Accurate gene prediction is important for many reasons, including comparative analysis of basic genotypic-phenotypic relationships present in the virus genomic sequences. To facilitate the comparison of viruses at the gene content level, we have developed a database system (VOCs) that not only stores genome, gene, and protein sequence information, but also categorizes genes into *families* of conserved orthologs. This is similar to other existing databases of orthologous proteins, such as the Clusters of Orthologous Groups (COGs) (Table 4.1) and Viral Orthologous Groups (VOG) (Table 4.1) databases at NCBI. In VOCs, the assignment of genes to families is a two-step process; first, automated BLASTP[2] searches are used to find clearly related genes; a human annotator confirms these results and searches for more distant relatives. Currently, we maintain 13 databases at VB-Ca, each associated with a taxonomic virus family. This system allows the user to formulate complex queries and retrieve specific gene information, such as the following: (1) Which gene families are present in all variola viruses (47 genomes) but absent from all monkeypox viruses (9 genomes)? Result: 7 gene families. (2) Which gene families are present in every sequenced poxvirus (105 genomes)? Result: 49 gene families (Figure 4.6A).

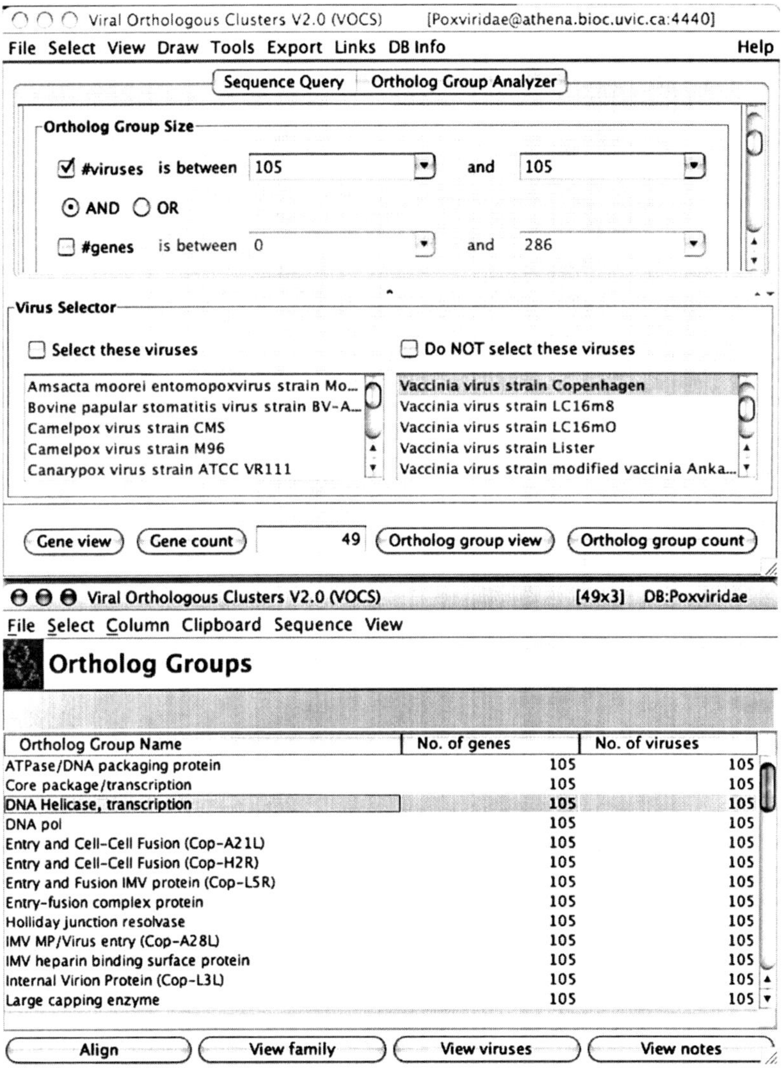

FIGURE 4.6 Analyses using the VOCs database. Top panel: query to determine which ortholog groups are present in all poxviruses. Bottom panel: partial list of results from query in the top panel.

These searches take only a few seconds to set up and run from an intuitive graphical user interface; results are returned to the user in table form (Figure 4.6B). Comparing genomes at this level of detail will often provide useful clues regarding which genes may be responsible for virulence or attenuation; however, more detailed analyses may then be required, including genome comparison at the nucleotide level (see Section 4.6).

An important point to remember when using these publicly available bioinformatics resources is that results are always dependent on the accuracy of the raw data as well as the subsequent annotations of these sequences. Some annotation systems involve solely automated, computational processes, while others, like VOCs, include assessment by a human annotator. Of course, neither should be relied on blindly. In the VOCs system, the user has access to the same tool set as the annotator, so that these tools can be used to try to substantiate what might be an unexpected result. For example, VOCs contains BLASTN/BLASTP/TBLASTN[2] tools, allowing searches of a given gene/protein sequence against the entire VOCs sequence database. These tools allow the user to determine whether a gene has been genuinely lost by deletion or mutation or only "lost" due to an error in annotation.

4.6 SEQUENCE ALIGNMENTS, THE HEART OF COMPARATIVE GENOMICS

In the context of comparative genomics, sequence alignments usually encompass large regions of DNA containing multiple genes; for viruses, such alignments may include complete genomes. Alignment construction may be complex, depending on the lengths, similarities, and number of sequences. The generation of large MSAs can be greatly limited by computational constraints. The choice of alignment algorithm must be carefully considered by the researcher, bearing in mind that although significant advances continue to be made in both computer hardware and software capabilities, the "garbage in, garbage out" principle still applies. For example, alignment tools will readily generate a whole-genome "alignment" of variola and fowlpox virus genomes. However, due to extensive rearrangements of the fowlpox virus genome, large parts of the conserved genome cores are not collinear; thus, much of the alignment will be meaningless (see Figure 4.3). Similarly, an alignment of the complete cowpox and variola virus genomes will have large unreliable regions due to their widely differing terminal inverted repeats (TIRs). Therefore, the generation of a dotplot is a useful first step if one is unfamiliar with the relationships between the genomes to be aligned. If rearrangements are suspected, the user should try the Mauve (Table 4.1) software package.[23] This program identifies locally collinear blocks present in multiple genomic-size sequences; the output, presented graphically, assists the user in interpreting complex rearrangement patterns. Alternatively, some tools generate alignments in two steps,[24] using a series of high-quality anchor alignments extracted from an initial global alignment as a framework for subsequent local alignments (using a different algorithm) of the regions between the anchors.

A number of existing alignment programs are suitable for generating whole-genome MSAs for viral (and other) genomes; many of these are listed on the ExPASy Web site (Table 4.1). An important tip is that often it is much quicker *not* to recalculate a very large MSA when adding only a few new genomes; for example, the average molecular virologist may find it easier, because of the high similarity, to update an MSA of 500 HIV genomes by manually adding a few new genomes rather than by rerunning such a large alignment. This requires an MSA editor such as Base-By-Base (BBB[25]) or BioEdit (Table 4.1). Some alignment programs will also accept a preexisting MSA and a single sequence as input and then align the single sequence to the MSA. MUSCLE is an example of one such program.[26]

It is also important to note that the alignment parameters, such as the penalties imposed on the alignment score for opening and extending gaps, may lead to alignment errors if multiple gaps are required in close proximity. Thus, it is important to carefully check MSAs for minor alignment errors such as the one shown in Figure 4.7; however, this is not a trivial undertaking when a sequence alignment is several hundred kilobases in length. Solving this problem was one of the driving forces behind the development of the BBB[25] editor. It has several features that are used in the checking/correction process. First, it can edit very large MSAs; second, it is able to highlight differences between aligned sequences in a way that is very easy for the user to identify (Figure 4.7); third, it is possible to navigate through long alignments

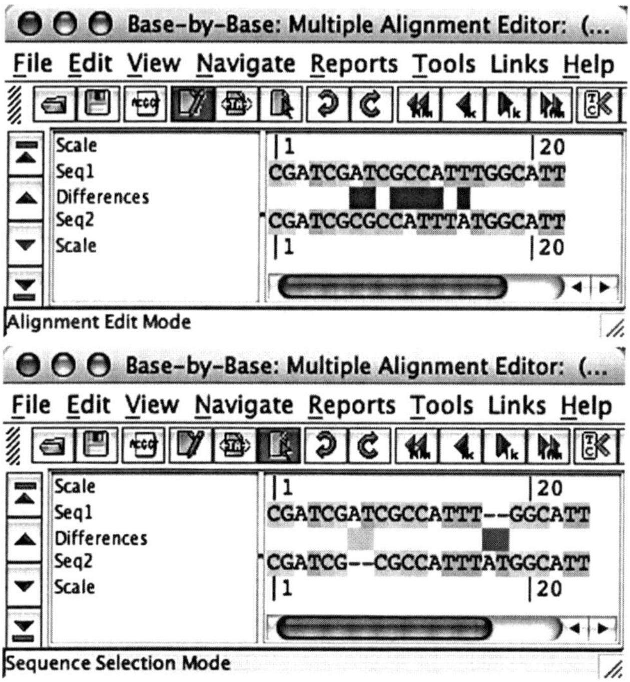

FIGURE 4.7 (See color figure in the insert following page 48.) Detection of errors in an MSA using Base-By-Base. Top panel: an alignment of two DNA sequences containing seven mismatches, which are indicated by blue boxes in the *differences* row. Bottom panel: insertion of two gaps (indicated by green and red boxes in the *differences* row) results in sequence realignment, eliminating all mismatches.

rapidly by easily jumping to the next mismatches or gap; and fourth, local regions of an MSA can be realigned independently from the rest of the MSA. Other MSA editors such as Jalview[27] and CINEMA,[28] which were designed primarily for protein MSAs, lack the above features, and BioEdit is restricted to the Windows operating system. Because most phylogenetic analysis programs ignore alignment columns that contain gaps, the correction of regions such as that shown in Figure 4.7 in which seven mismatches are replaced by two gaps could have a significant effect.

4.7 PHYLOGENY AND MORE

The Universal Virus Database (Table 4.1) is authorized by the ICTV and provides a list of approved virus names linked to descriptions. The ICTV produces a *consensus taxonomy* from the family to the species level based on sequence analysis and classical taxonomic characteristics.[18,29,30] Taxonomic information is also available from NCBI, which lists 269 viral genera and 3,701 viral species at the time this chapter was compiled (NCBI, taxonomy; Table 4.1). Note that the NCBI taxonomy is not always congruent with that of the current Eighth Report of the ICTV. The ICTV report should be considered the official reference, and efforts are under way to align the NCBI taxonomy with that of the ICTV.

Assignment of a new virus isolate to a particular family, genus, and species is the logical next step following an initial comparative analysis — a necessary prerequisite to fully understand the biology of the isolate and its role in the virosphere. In addition to this obvious role for comparative genomics in the identification and classification of new viruses, this type of analysis is also becoming essential to the field of viral diagnostics; new laboratory techniques give comparative genomics a central role in the process of rapid virus detection and characterization. Current *virus chips* attempt to identify viruses from all known families in a single pass[30,31] using microarray technologies, and for certain pathogens, they may also be able to distinguish between species that differ significantly in virulence. Manipulation of DNA oligonucleotide probe specificity offers this ability to screen for novel viruses or to focus on known isolates.[32–34] For example, diagnosis of orthopoxvirus infections by traditional techniques (lesion pathology, symptoms, and microscopic techniques) is problematic; a variety of species may give similar results, and other parameters (e.g., inoculum size, vaccination, and the health status of the host) can affect the diagnosis.[35–37] If a poxvirus outbreak were to occur, it would be extremely important to be able to quickly and accurately distinguish not only among smallpox, monkeypox, and less-dangerous poxviruses, but also among virus strains with varying capacities for virulence.[38,39] Virus chips can also assist in the discovery of viruses that may be difficult or impossible to culture; for example, xenotropic murine leukemia virus[34] was discovered using a DNA microarray designed to detect all known viral families.[40]

Other uses of phylogenetic-like comparison of virus sequences include the genotyping of viruses for epidemiological analysis of virus outbreaks[41] and drug resistance spread,[42] or the subtyping of viruses to help in the determination of treatment regimes.[43,44] Tools for genotyping a variety of viruses are available at NCBI, and other applications may be found at some virus-specific databases, such as those for HIV and HCV (Table 4.1).

4.8 THE IMPORTANCE OF DATA ORGANIZATION

Wikipedia defines a *database* as "an organized collection of data" (Table 4.1) and provides an excellent description of various database models. Most people are familiar with databases in one form or another; for example, the indexing of file names and file contents can help us find a particular e-mail message on our desktop computer or a reference in PubMed (Table 4.1). However, for a database to be most useful, it should not only provide rapid and easy access to the raw data it stores but also assist the user in further data manipulation. This can be accomplished to some extent by linking the data items to relevant sources of information (e.g., PubMed provides links to related articles). Another way to add value to a database is to provide utilities that can return raw data in different formats (e.g., NCBI Entrez, which allows the user to retrieve viral genomic data in a variety of formats). Since the collection of all the sequences required for a large multiple alignment can be tedious, some databases preprocess these queries (HIV Sequence Database and HCV Database; Table 4.1). However, these solutions tend to lack flexibility and are limited in scope. With the design of the VOCs database, we have tried to provide (1) quick-and-easy access to data, retrievable in various formats; (2) flexible, user-driven querying of the data; (3) retrieval of the data directly into analysis tools. Thus, it is straightforward to perform analyses such as the following:

1. Retrieve all genes from the vaccinia virus genome; sort by %(A + T) (time required < 30 s).
2. Collect DNA polymerase protein sequences from all poxvirus genomes; select one from each genus, align and return in an MSA editor for minor manual adjustments; generate a percent identity table for all pairwise alignments (time required < 90 s).
3. Find all poxvirus proteins that have a {KHL}DEL endoplasmic reticulum retention signal at the carboxy terminus; collect orthologs of all these proteins; align and compare to determine if there is variability in this motif sequence (time required < 60 s).
4. Retrieve "apoptosis inhibitor" protein sequences from all orthopoxvirus genomes; select five proteins of interest; generate a 5 × 5 dotplot to view the repeat sequences in these proteins (time required < 60 s).

Although these tasks are straightforward, the hands-on time required to process them manually would be prohibitive. The ability to easily access and analyze genomic data — using VOCs or a similar system — thus allows researchers to work with the data in new and more complex ways.

Since DNA sequence databases are growing at an exponential rate, it is often essential for bioinformatics researchers to repeat similarity searches at frequent intervals. However, such searches are often performed with large query sets (many sequences or even whole genomes). This, together with the ever-increasing size of result sets, makes such searches a tedious task. ReHAB (Recent Hits Acquired from BLAST) is a tool for tracking new protein hits in repeated PSI-BLAST searches.[45] It is designed to simplify the analysis of large numbers of database matches and is therefore especially suited to comparative genomics. Results are presented in a user-friendly graphical interface with simple-to-navigate tables, and new hits are indicated by highlighted text. Since ReHAB maintains its own database of sequence hits, it allows simple selection of sequences from the BLAST hits for piping directly into a multiple alignment tool and finally viewing in the MSA editor BBB. ReHAB databases are maintained for a variety of virus families at VBRC. A similar tool for managing multiple InterProScan[46] searches is also available. This tool, Java GUI for InterPro Scan (JIPS),[47] also allows the user to compare the results of InterProScan searches using orthologs.

4.9 OTHER COMPARATIVE ANALYSES

It is sometimes difficult to distinguish the borders separating the various-omics sciences. Therefore, it would be remiss not to mention some of the other areas that could be construed as touching comparative genomics of viruses. One such field is the analysis of regulatory sequences, which encompasses the study of promoter sequences, enhancer elements, splice junctions, and translational frame-shifting sequences. Comparisons of similarly functioning regulatory sequences in a single virus (e.g., late promoters within a baculovirus) or of a single sequence found in many related viruses (e.g., all poxvirus DNA polymerase promoters) can generate a consensus sequence revealing short key motifs within such elements. A common theme among such analyses is that the essential motifs are relatively small and are usually embedded in a non-conserved sequence (e.g., each baculovirus late promoter is associated with a different gene and will therefore be surrounded by different sequences). Some alignment programs, including BBB, can generate simple graphics highlighting conserved residues within a sequence, but LOGO[48,49] is capable of more precise representations.

When examining genomic sequences, a researcher may notice unusual patterns of bases. Such patterns can be tested for statistical significance using the R'MES program[50] (Table 4.1); this software can also detect "words" (short nucleotide strings) that have unexpected frequencies within a sequence. However, these results must be interpreted with caution; although the unexpected frequency of a given pattern may be suggestive of an associated function, the inverse is not automatically true — not all functional sequences occur at unusual frequencies. To refer to an earlier example, the baculovirus late promoter core sequence is underrepresented in the genome. One might presume that this is a deliberate mechanism to prevent spurious late transcription. However, it may also be simply a statistical consequence of the low number of late genes in the baculovirus genome. Another comparative technique frequently used in viral genomics is to scan aligned genomes for recombination events.[51–53] One method (they are numerous) is to use a sliding window (set to a specific number of nucleotides) that moves along the entire length of the alignment in incremental steps, calculating a distance/similarity score at each location. The result is a numerical comparison between the query sequence and the other sequences over the entire alignment. The distance/similarity data are plotted in graphical form; recombination breakpoints can then be located at the crossover points between two lines on the graph.[54,58]

4.10 SUMMARY

Comparative genomics of viruses is a relatively new area of virology and one that will continue to evolve rapidly as new technologies continuously generate more and more genomic data. As well, new data types and advancements in available technologies will allow novel comparative studies to be performed on viruses. Some of these data will undoubtedly come from the -omics fields (e.g., high-throughput generation of transcriptome, proteome, or protein structure data); however, improvements in computer technology will almost certainly result in a large contribution from the computer modeling field as well. This will include epidemiological modeling of disease transmission, molecular modeling of protein–protein and protein–DNA interactions and model-based approaches to antiviral drug design.

To exploit this wealth of new information to its fullest potential, there will be an ever-growing need both for continual improvement of bioinformatics tools to organize and archive such data in a useful format and for trained wet-lab investigators capable of using such tools. Today's virologists must take on the responsibility of learning about the ever-changing variety of computer-based databases and tools, just as they work to keep their knowledge of laboratory techniques current. It is unavoidable that some initial confusion will result as both computer programmers and virologists will struggle with unfamiliar concepts and terminology in this interdisciplinary field. However, researchers should never hesitate, when in doubt, to seek out their local bioinformatician, statistician, or computer scientist and to strike up a new collaboration.

ACKNOWLEDGMENTS

We would like to acknowledge the many programmers who have contributed to the VBRC over the years, especially Angelika Ehlers at the University of Victoria and Curtis Hendrickson and Jim Moon at the University of Alabama, Birmingham; Vasily Tcherepanov, Catherine Galloway, and Cristalle Watson for reviewing the manuscript; and other authors of Open Source software (Table 4.1). This work was supported by a NIH/National Institute of Allergy and Infections Diseases contract HHSN266200400036C to E. J. L. and C. U. and by Natural Sciences and Engineering Research Council of Canada Strategic and Discovery grants to C. U.

REFERENCES

1. Westrop, G. D., Wareham, K. A., Evans, D. M., Dunn, G. et al. Genetic basis of attenuation of the Sabin type 3 oral poliovirus vaccine. *J Virol* **63,** 1338–1344 (1989).
2. Altschul, S. F., Madden, T. L., Schaffer, A. A., Zhang, J. et al. Gapped BLAST and PSI-BLAST: a new generation of protein database search programs. *Nucleic Acids Res* **25,** 3389–3402 (1997).
3. Upton, C., Stuart, D. T. & McFadden, G. Identification of a poxvirus gene encoding a uracil DNA glycosylase. *Proc Natl Acad Sci U S A* **90,** 4518–4522 (1993).
4. Pearson, W. R. Flexible sequence similarity searching with the FASTA3 program package. *Methods Mol Biol* **132,** 185–219 (2000).

5. Needleman, S. B. & Wunsch, C. D. A general method applicable to the search for similarities in the amino acid sequence of two proteins. *J Mol Biol* **48,** 443–453 (1970).
6. Stanitsa, E. S., Arps, L. & Traktman, P. Vaccinia virus uracil DNA glycosylase interacts with the A20 protein to form a heterodimeric processivity factor for the viral DNA polymerase. *J Biol Chem* **281,** 3439–3451 (2006).
7. De Silva, F. S. & Moss, B. Vaccinia virus uracil DNA glycosylase has an essential role in DNA synthesis that is independent of its glycosylase activity: catalytic site mutations reduce virulence but not virus replication in cultured cells. *J Virol* **77,** 159–166 (2003).
8. Soding, J. Protein homology detection by HMM-HMM comparison. *Bioinformatics* **21,** 951–960 (2005).
9. Berman, H. M., Westbrook, J., Feng, Z., Gilliland, G. et al. The Protein Data Bank. *Nucleic Acids Res* **28,** 235–242 (2000).
10. Brunetti, C. R., Amano, H., Ueda, Y., Qin, J. et al. Complete genomic sequence and comparative analysis of the tumorigenic poxvirus Yaba monkey tumor virus. *J Virol* **77,** 13335–13347 (2003).
11. Sonnhammer, E. L. & Durbin, R. A dot-matrix program with dynamic threshold control suited for genomic DNA and protein sequence analysis. *Gene* **167,** GC1–G10 (1995).
12. Brodie, R., Roper, R. L. & Upton, C. JDotter: a Java interface to multiple dotplots generated by dotter. *Bioinformatics* **20,** 279–281 (2004).
13. Taneda, A. Adplot: detection and visualization of repetitive patterns in complete genomes. *Bioinformatics* **20,** 701–708 (2004).
14. Afonso, C. L., Tulman, E. R., Delhon, G., Lu, Z. et al. Genome of crocodilepox virus. *J Virol* **80,** 4978–4991 (2006).
15. Senkevich, T. G., Koonin, E. V., Bugert, J. J., Darai, G. & Moss, B. The genome of molluscum contagiosum virus: analysis and comparison with other poxviruses. *Virology* **233,** 19–42 (1997).
16. Schwartz, S., Kent, W. J., Smit, A., Zhang, Z. et al. Human-mouse alignments with BLASTZ. *Genome Res* **13,** 103–107 (2003).
17. Smoot, M. E., Guerlain, S. A. & Pearson, W. R. Visualization of near-optimal sequence alignments. *Bioinformatics* **20,** 953–958 (2004).
18. Fauquet, C. M., Ball, L. A., Desselberger, U., Maniloff, J., & Mayo, M. A. *Virus Taxonomy: Classification and Nomenclature of Viruses; Eighth Report of the International Committee on Taxonomy of Viruses* (Academic Press, New York, 2005).
19. Neumann, M., Harrison, J., Saltarelli, M., Hadziyannis, E. et al. Splicing variability in HIV type 1 revealed by quantitative RNA polymerase chain reaction. *AIDS Res Hum Retroviruses* **10,** 1531–1542 (1994).
20. Brian, D. A. & Baric, R. S. Coronavirus genome structure and replication. *Curr Top Microbiol Immunol* **287,** 1–30 (2005).
21. Inberg, A. & Linial, M. Evolutional insights on uncharacterized SARS coronavirus genes. *FEBS Lett* **577,** 159–164 (2004).
22. Upton, C. Screening predicted coding regions in poxvirus genomes. *Virus Genes* **20,** 159–164 (2000).
23. Darling, A. C., Mau, B., Blattner, F. R. & Perna, N. T. Mauve: multiple alignment of conserved genomic sequence with rearrangements. *Genome Res* **14,** 1394–1403 (2004).
24. Wang, C. & Lefkowitz, E. J. Genomic multiple sequence alignments: refinement using a genetic algorithm. *BMC Bioinformatics* **6,** 200 (2005).
25. Brodie, R., Smith, A. J., Roper, R. L., Tcherepanov, V. & Upton, C. Base-By-Base: single nucleotide-level analysis of whole viral genome alignments. *BMC Bioinformatics* **5,** 96 (2004).

26. Edgar, R. C. MUSCLE: multiple sequence alignment with high accuracy and high throughput. *Nucleic Acids Res* **32,** 1792–1797 (2004).
27. Clamp, M., Cuff, J., Searle, S. M. & Barton, G. J. The Jalview Java alignment editor. *Bioinformatics* **20,** 426–427 (2004).
28. Parry-Smith, D. J., Payne, A. W., Michie, A. D. & Attwood, T. K. CINEMA — a novel colour INteractive editor for multiple alignments. *Gene* **221,** GC57–GC63 (1998).
29. Buchen-Osmond, C. The Universal Virus Database ICTVDB. *Comput Sci Eng* **5,** 16–25 (2003).
30. Bryant, P. A., Venter, D., Robins-Browne, R. & Curtis, N. Chips with everything: DNA microarrays in infectious diseases. *Lancet Infect Dis* **4,** 100–111 (2004).
31. Wang, D., Coscoy, L., Zylberberg, M., Avila, P. C. et al. Microarray-based detection and genotyping of viral pathogens. *Proc Natl Acad Sci U S A* **99,** 15687–15692 (2002).
32. Chou, C. C., Lee, T. T., Chen, C. H., Hsiao, H. Y. et al. Design of microarray probes for virus identification and detection of emerging viruses at the genus level. *BMC Bioinformatics* **7,** 232 (2006).
33. Urisman, A., Fischer, K. F., Chiu, C. Y., Kistler, A. L. et al. E-Predict: a computational strategy for species identification based on observed DNA microarray hybridization patterns. *Genome Biol* **6,** R78 (2005).
34. Urisman, A., Molinaro, R. J., Fischer, N., Plummer, S. J. et al. Identification of a novel gammaretrovirus in prostate tumors of patients homozygous for R462Q RNASEL variant. *PLoS Pathog* **2,** e25 (2006).
35. Di Giulio, D. B. & Eckburg, P. B. Human monkeypox: an emerging zoonosis. *Lancet Infect Dis* **4,** 15–25 (2004).
36. Gooze, L. L. & Hughes, E. C. Smallpox. *Semin Respir Infect* **18,** 196–205 (2003).
37. Lewis-Jones, S. Zoonotic poxvirus infections in humans. *Curr Opin Infect Dis* **17,** 81–89 (2004).
38. Chen, N., Li, G., Liszewski, M. K., Atkinson, J. P. et al. Virulence differences between monkeypox virus isolates from West Africa and the Congo basin. *Virology* **340,** 46–63 (2005).
39. Dumbell, K. R. & Huq, F. The virology of variola minor. Correlation of laboratory tests with the geographic distribution and human virulence of variola isolates. *Am J Epidemiol* **123,** 403–415 (1986).
40. Wang, D. Urisman, A., Liu, Y. T., Springer, M. et al. Viral discovery and sequence recovery using DNA microarrays. *PLoS Biol* **1,** E2 (2003).
41. Aitken, C. K., McCaw, R. F., Bowden, D. S., Tracy, S. L. et al. Molecular epidemiology of hepatitis C virus in a social network of injection drug users. *J Infect Dis* **190,** 1586–1595 (2004).
42. Eyer-Silva, W. A. & Morgado, M. G. A genotyping study of human immunodeficiency virus type-1 drug resistance in a small Brazilian municipality. *Mem Inst Oswaldo Cruz* **100,** 869–873 (2005).
43. Ong, H. T., Duraisamy, G., Kee Peng, N., Wen Siang, T. & Seow, H. F. Genotyping of hepatitis B virus in Malaysia based on the nucleotide sequence of preS and S genes. *Microbes Infect* **7,** 494–500 (2005).
44. Sansom, C. Genotyping HIV isolates paves the way to effective treatment regimes. *Mol Med Today* **5,** 417 (1999).
45. Whitney, J., Esteban, D. J. & Upton, C. Recent Hits Acquired by BLAST (ReHAB): a tool to identify new hits in sequence similarity searches. *BMC Bioinformatics* **6,** 23 (2005).
46. Quevillon, E., Silventoinen, V., Pillai, S., Harte, N. et al. InterProScan: protein domains identifier. *Nucleic Acids Res* **33,** W116–W120 (2005).

47. Syed, A. & Upton, C. Java GUI for InterProScan (JIPS): a tool to help process multiple InterProScans and perform ortholog analysis. *BMC Bioinformatics* **7,** 462 (2006).
48. Crooks, G. E., Hon, G., Chandonia, J. M. & Brenner, S. E. WebLogo: a sequence logo generator. *Genome Res* **14,** 1188–1190 (2004).
49. Schneider, T. D. & Stephens, R. M. Sequence logos: a new way to display consensus sequences. *Nucleic Acids Res* **18,** 6097–6100 (1990).
50. Schbath, S. An efficient statistic to detect over- and under-represented words in DNA sequences. *J Comput Biol* **4,** 189–192 (1997).
51. Zhang, X. W., Yap, Y. L. & Danchin, A. Testing the hypothesis of a recombinant origin of the SARS-associated coronavirus. *Arch Virol* **150,** 1–20 (2005).
52. Etherington, G. J., Dicks, J. & Roberts, I. N. Recombination Analysis Tool (RAT): a program for the high-throughput detection of recombination. *Bioinformatics* **21,** 278–281 (2005).
53. Chen, J., Powell, D. & Hu, W. S. High frequency of genetic recombination is a common feature of primate lentivirus replication. *J Virol* **80,** 9651–9658 (2006).
54. Lole, K. S., Bollinger, R. C., Paranjape, R. S., Gadkari, D. et al. Full-length human immunodeficiency virus type 1 genomes from subtype C-infected seroconverters in India, with evidence of intersubtype recombination. *J Virol* **73,** 152–160 (1999).
55. Siepel, A. C., Halpern, A. L., Macken, C. & Korber, B. T. A computer program designed to screen rapidly for HIV type 1 intersubtype recombinant sequences. *AIDS Res Hum Retroviruses* **11,** 1413–1416 (1995).

5 Archaebacteria and the Prokaryote-to-Eukaryote Transition (and the Role of Mitochondria Therein)

William Martin, Tal Dagan, and Katrin Henze

CONTENTS

ABSTRACT

The process through which prokaryotes are related to eukaryotes is still the subject of much debate. No genome-wide analyses have been published that would resolve the issue to everyone's satisfaction. Methods of genome analysis that can recover non-Darwinian processes of genome evolution, such as lateral gene transfer and endosymbiosis, are needed to obtain an overview of the history of microbial life, but such methods are only just now in development. The ubiquity of mitochondria among all eukaryotes studied so far suggests that endosymbiosis might have had more to do with the prokaryote-to-eukaryote transition than is currently assumed.

5.1 INTRODUCTION

This book is mostly for people who are not primarily concerned with early evolution. A nonspecialist might come into this chapter thinking that, with all the information available from genomes, the origin of eukaryotes, the role of organelles therein, and the overall shape of the tree of life ought to be well-resolved issues about which one just needs to write a few simple words, like fresh icing on an old cake. This is not so. Several fundamentally different views of the prokaryote-to-eukaryote transition are current, and they are hotly debated. Most of the debate is among specialists and hence is not always in the breaking news. Notably, all current views about prokaryote–eukaryote relationships arose in their more or less modern formulations before there were any genome sequences available. Put another way, genomics has not generated any fundamentally new ideas about eukaryote origins, the more widely recognized importance of lateral gene transfer (LGT) in genome evolution notwithstanding.

The title of this chapter paraphrases Brown and Doolittle's 1997 work.[1] Because biologists in this field are still debating the same issues as they were debating 10 years ago, this chapter is not terribly different in terms of bottom-line content from theirs. The reader might ask: Haven't genomes made a big difference in the way that most biologists view the prokaryote-to-eukaryote transition? The answer is: "No, not really," which is interesting in its own right.

Genomes contain information from which we can distill some sequence comparison results. Those results can then be contrasted to the expectations and predictions that follow from various alternative views about early evolution. This chapter presents what we think are the main current views about the prokaryote-to-eukaryote transition, and an attempt is made to contrast those views to some available observations from genomes.

We hope you notice that the current views on the prokaryote-to-eukaryote transition differ radically. This is because the views stem from very different schools of thought that weigh the available evidence differently. The results from genome comparisons have not been so clear-cut as to convince any proponents that this or that view about early evolution should be abandoned or to convince opponents that this or that view is right after all. A peculiarity unique to the field of early evolution is that people *tend* to believe what they have always believed about early evolution, regardless of what any forms of scientific evidence say. That is important. It will help in understanding how diametrically opposed and mutually incompatible theories can coexist in the modern literature in the face of exactly the same data. Each camp weighs parts of the data heavily (the part that supports their own views) while disregarding or otherwise explaining away the rest.

The following sections briefly summarize some current views about the relatedness of prokaryotes and eukaryotes, with an attempt to explain whence we suppose the views stem and — importantly in our view — what evolutionary significance each view attaches to organelles (mitochondria and chloroplasts) regarding the process of eukaryote origins.

5.2 THE rRNA TREE

For nonspecialists, the classical ribosomal RNA tree of life as forged since the late 1970s by Carl Woese and coworkers[2–7] conveys the most widespread and the most visible picture of the prokaryote-to-eukaryote transition (Figure 5.1). The tree is based in sequence comparisons of ribosomal RNA (rRNA), but other components of the information storage-and-retrieval machinery (informational genes[8]) show a very similar picture.[9]

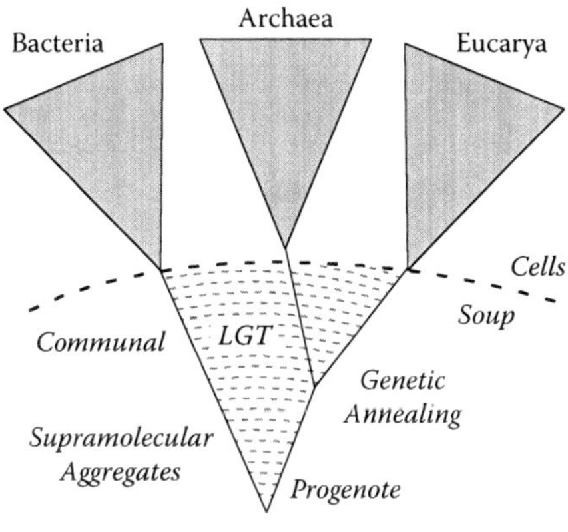

FIGURE 5.1 The rRNA tree as rooted with ancient paralogues.

In its current interpretations, the rRNA tree suggests that the prokaryote-to-eukaryote transition occurred before the evolution of cellular lineages.[2,5] The universal ancestor of all life (the progenote)[2] is seen as a communal collection of information-storing and -processing entities that are not yet organized as cells.[5] Lateral gene transfer is seen as the main mode of genetic novelty at the early stages of evolution, and the process of vertical inheritance arises only with the process of genetic annealing from within this mixture, at which point the emerging cellular lineages of prokaryotes and eukaryotes became refractory to LGT and thus traversed a kind of Darwinian threshold from the organizational state of supramolecular aggregates to the organizational state of cells. Traversing that threshold is seen as equivalent to the primary emergence, from the broth in which life itself arose, of the three kinds of cells that we recognize today: archaebacteria, eubacteria, and eukaryotes.[6] These groups were suggested to be renamed as Archaea, Bacteria, and Eucarya,[3] respectively, but for reasons explained elsewhere[10–12] the older names are preferable because they have precedence and designate the same groups.

The rRNA tree assumed its current shape in 1990, when independent studies of anciently diverged protein-coding genes suggested the root of the universal tree to lie on the eubacterial branch.[1,13,14] It goes without saying that the rRNA tree view of the prokaryote-to-eukaryote transition admits that chloroplasts and mitochondria did arise via endosymbiosis, but it sees no role for mitochondria or any other kind of symbiosis in the emergence of the eukaryotic lineage, and their genetic contribution to eukaryotes is seen as detectable but negligible in terms of evolutionary or mechanistic significance.[15] As Woese[6] has put it: "Because endosymbiosis has given rise to the chloroplast and mitochondrion, what else could it have done in the more remote past? Biologists have long toyed with an endosymbiotic (or cellular fusion) origin for the eukaryotic nucleus, and even for the entire eukaryotic cell" (p. 8742). Proponents of the rRNA tree contend that eukaryotes, eubacteria, and archaebacteria are of equal rank at the highest taxonomic level,[16] and that the term *prokaryote* is misleading and hence should be banned from the scientific literature.[7] Accordingly, those proponents would contend that there never was a prokaryote-to-eukaryote transition to begin with because the three primary lineages are seen as emerging from the primordial soup as the more or less ready-made cellular lineages that we see today.

The rRNA tree is taken by some to indicate that eukaryotes are in fact sisters of archaebacteria at the level of the whole genome,[7] a view that comes from simply extrapolating from the rooted version of the rRNA tree[3] to the rest of the genome, but without actually looking at the data. Others see this close relationship of the gene expression machinery in eukaryotes and archaebacteria as reflecting an archaebacterial ancestry for only a component of the eukaryote genome.[17] The rRNA tree stems from a long tradition of work on ribosomes, the genetic code, and translation. These characters are more heavily weighted in this tree than, for example, genes or characters involved in metabolism or biosynthesis.

5.3 THE INTRONS EARLY TREE

At about the same time that archaebacteria were discovered, introns in eukaryotic genes were discovered.[18] It was not long until Walter Gilbert suggested that exon shuffling might be an important tool for gene evolution,[19] and W. Ford Doolittle suggested that the ancestral state of genes might be "split" and that some introns in eukaryotic genes might be holdovers from the primordial assembly of protein-coding regions.[20] In that case, the organizational state of eukaryotic genes (having introns) would represent the organizational state of the very first genomes,[21] and the intronless prokaryotic state would hence be a derived condition (Figure 5.2), a view that was called *introns early*.[22] The assumed process of intron loss in prokaryotes was initially called *streamlining* but later was labeled *thermoreduction*.[23]

Although Doolittle invented the introns-early view and later abandoned it,[24,25] it has found other proponents, who draw on different lines of evidence in support of their view, that they do not, however, call introns early, but introns first instead.[26] They agree that the eubacterial root of the rRNA tree suggested by the ancient duplicated genes is questionable, and that a eukaryote root of the rRNA tree is more likely.[27–32] Some proponents interpret various other aspects of RNA processing in

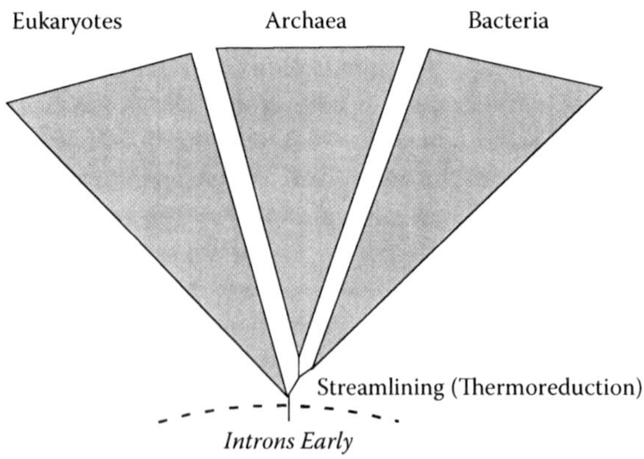

FIGURE 5.2 The introns-early tree.

eukaryotes such as rRNA modification through small nucleolar RNAs or snoRNAs, in addition to introns, as direct holdovers from the RNA world and hence as evidence for eukaryote antiquity.[26,31,33,34]

As in the case of the rRNA tree, there is no prokaryote-to-eukaryote transition in the introns-early tree because prokaryotic genome organization is seen as a very early derivative of eukaryotic gene organization. Accordingly, the relationship of eukaryotes and prokaryotes is depicted largely as a more or less unresolved trichotomy,[15] and the contribution of organelles or symbiosis to eukaryote evolution is viewed as existing, but negligible. Characters involved in RNA processing are more heavily weighted in this tree than, for example, genes or characters involved in information storage, metabolism, or biosynthesis.

5.4 THE NEOMURAN TREE

The neomuran tree (Figure 5.3) stems from Tom Cavalier-Smith.[10,35–38] No theory, current or otherwise, is more explicit on the prokaryote-to-eukaryote transition in terms of mechanistic details (see Cavalier-Smith[38]). In the main, it suggests that the common ancestor of all cells was a free-living eubacterium (in the most recent formulation, a *Chlorobium*-like anoxygenic photosynthesizer), and that eubacteria were the only organisms on Earth up until about 900 million years ago, at which time a member of the eubacteria, in recent formulations an actinobacterium, lost its murein-containing cell wall and was faced with the task of reinventing a new cell wall (hence the Latin name: *neo*, new; *murus*, wall). This led to the origin of a group of rapidly evolving organisms called the neomura.

The loss of the cell wall precipitated an unprecedented process of descent with modification. During a short period of time (perhaps 50 million years), the characters that are shared by archaebacteria and eukaryotes arose (for a list of those characters, see Cavalier-Smith[36]). This lineage (the neomura) then underwent diversification into two lineages with another long list of evolutionary changes in each. One lineage invented isoprene ether lipid synthesis and gave rise to archaebacteria. One lineage

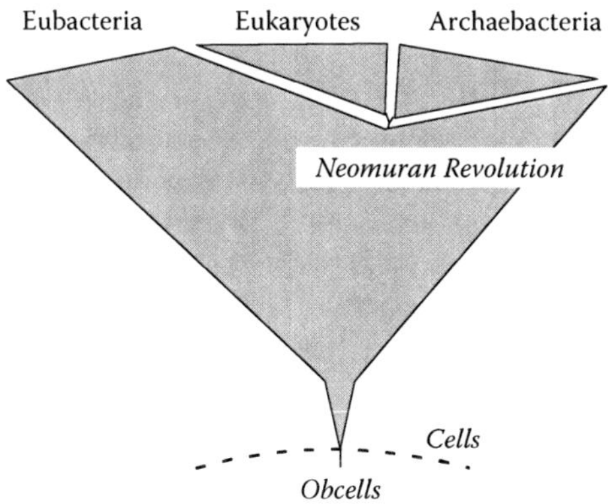

FIGURE 5.3 The neomuran tree.

became phagotrophic and gave rise to the eukaryotes. In the eukaryote lineage, the endoplasmic reticulum (ER) arose from the plasma membrane of the phagocytosing neomuran prokaryote; the nucleus then arose from the ER.

In older formulations, some eukaryote lineages branched off before the mitochondrion was acquired; these lineages were once called the archezoa.[39] In newer formulations, the mitochondrion comes into the eukaryote lineage before any archezoa can arise. The genetic mechanism of the prokaryote-to-eukaryote transition is mutation. No evolutionary intermediates between actinobacteria, neomura, archaebacteria, and mitochondrion-containing eukaryotes persist among modern biota. The nucleus arose before the mitochondrion,[40] simultaneously with the mitochondrion,[10] or after the mitochondrion,[38] depending on the version of the theory. Such variation on the theme may seem disturbing to some, but by contrast, neither the origin of the nucleus nor the origin of the mitochondrion are really an issue for the rRNA tree or for the introns-early tree, so the neomuran tree has the advantage of at least addressing those issues. A constant in all versions of the neomuran theory is that the origin of phagocytosis is seen as an adamantine prerequisite for the origin of mitochondria[38]: "All theories for the host being a prokaryote are unsound" (p. 982). The neomuran tree focuses on characters and downweights sequence similarity as a measure of overall relatedness of lineages.

5.5 THE SYMBIOTIC TREE WITH A EUKARYOTE HOST

In 1967, Lynn Margulis (as Lynn Sagan[41]) repopularized the early 1900s theories of Mereschkowsky,[42] Portier (as expained by Sapp),[43] and Wallin[44] for the endosymbiotic origin of chloroplasts and mitochondria. Those old and contentious ideas had been repeatedly condemned as nonsense[45,46] but not completely forgotten by the botanists[47] into the 1960s. So, at about the same time that archaebacteria and introns in eukaryotic genes were discovered, biologists were still fiercely debating the issue of whether mitochondria and chloroplasts were once free-living prokaryotes. In the mid-1970s, there were those who weighed in with data in favor of endosymbiotic theory[48,49] and those who weighed in with hefty arguments against it.[50,51]

One could say that when Woese challenged the field with his tripartite tree,[52] he was challenging the symbiotic view of eukaryote origins as championed by Margulis,[53] which had by that time gained enough momentum to be labeled as the "conventional tree of life."[52] No one ever challenged the significance and uniqueness of archaebacteria, but there was much debate about their place in endosymbiotic theory in terms of their relationship to the host that acquired mitochondria (for a discussion, see Brown,[1] Woese,[2] Doolittle,[21] and Gray[54]). At the same time, Margulis's version of endosymbiotic theory was hardly the conventional tree of life that it was made out to be because it contained from its inception, and still contains,[55] an additional partner at eukaryote origins in which no specialists other than Margulis have ever taken any stock at all: the spirochaete origin of eukaryotic flagella. From the standpoint of modern data, the spirochaete origin of eukaryotic flagella can be seen as both unsupported and unnecessary.[56]

As it became clear that archaebacteria and eukaryotes do have quite a bit in common, a modified version of Margulis's symbiotic theory that lacked the spirochaete and

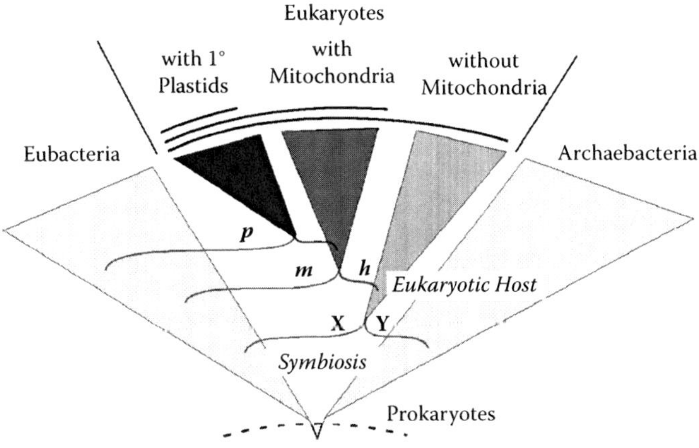

FIGURE 5.4 The symbiotic tree with a eukaryote host.

had a host that was related to archaebacteria came into play.[57] Quite a few gene comparisons later, it also became clear that eukaryotes are not just grown-up archaebacteria because they contain too many eubacterial genes for comfort.[1,8,58,59] Moreover, the eubacterial genes started cropping up in the archezoa, the eukaryotes that were supposed never to have had mitochondria.[60–62] That left a few possibilities for the symbiotic tree to evolve. Either (1) there was an additional symbiosis that preceded the mitochondrion but was not a spirochaete[63]; or (2) the mitochondrion had a more diverse collection of genes than was previously assumed, donated more genes to eukaryotes than was previously assumed, and was present in the common ancestor of all eukaryotes[64]; or that (3) LGT is the general solution to that and a whole slate of other problems that had been gnawing on the tree of life for quite a while anyway,[65] as recently reviewed elsewhere.[66]

The eukaryote host version of the symbiotic tree as one could construe it at the moment is shown in Figure 5.4. The term *eukaryote host* is used here to designate a collection of views concerning the kinds of symbioses that led to eukaryotes and that are fundamentally different in terms of the kinds of partners and the polarity of symbiosis involved. These views are unified, however, by one important aspect: They all posit that there was a symbiosis of bona fide prokaryotes that led to a nucleated but mitochondrion-lacking cell that was the founder of the eukaryotic lineage and that gave rise to the host that acquired the mitochondrion (hence eukaryote host). The partners X and Y that are presumed by the different versions of the eukaryote host tree can designate (1) an unspecified eubacterial partner and an archeabacterium in an indescript symbiosis,[67] (2) an archaebacterial origin of the nucleus as a symbiont in a eubacterial host,[63,68–71] or (3) a spirochaete origin of flagella (and the nucleus) in an archaebacterial host.[55,72]

5.6 THE SYMBIOTIC TREE WITH A PROKARYOTE HOST

The rRNA tree, the neomuran tree, the introns-early tree, and the various eukaryote host versions of the symbiotic tree all assume that the host that acquired the mitochondrion was a eukaryote. If that assumption is true, then the exciting prediction follows that there should still be some eukaryotes out there that never came into contact with mitochondria.[39] In the 1990s, that idea sent molecular biologists scrambling to study contemporary eukaryotes that were thought to lack mitochondria. That work unearthed findings of the most unexpected kind. First, all of the suspected primitive and mitochondrion-lacking lineages were not demonstrably primitive because the trees that had suggested them to be early branching were replete with phylogenetic artifacts.[73] But, there was more: The lineages in question did not even lack mitochondria. The mitochondria are there after all, but they do not use oxygen,[74,75] they are small and hence easily overlooked,[76] and some do not even produce adenosine triphosphate (ATP).[77] These "new" members of the mitochondrial family among eukaryotic anaerobes (and some parasitic aerobes[78]) are called hydrogenosomes and mitosomes (reviewed in van der Giezen[79]).

Such findings pointed to the antiquity of mitochondria[60,61] and opened the possibility that the host that acquired the mitochondrion might have just been an archaebacterium outright.[80,81] Several prokaryote host hypotheses have been published in Martin,[64] Searcy,[80] and Vellai[82] (these are reviewed in Martin,[11] Embley and Martin,[66] and Martin et al.[83]), some of which account for the common ancestry of mitochondria and hydrogenosomes[64] and some of which account for the origin of the nucleus.[84]

In the prokaryote host tree (Figure 5.5), the many differences that distinguish eukaryotes from prokaryotes are interpreted as having arisen after (rather than before) the acquisition of mitochondria. The main difference between the eukaryote host and the prokaryote host versions of the symbiotic tree concerns the predictions regarding the number of symbiotic partners involved at eukaryote origins (>2 vs. 2, respectively) and the existence or nonexistence, respectively, of primitively amitochondriate eukaryotes. The prokaryote host tree suggests that the main source of genes among eukaryotes comes from two prokaryotes: the host (an archaebacterium) at the origin of mitochondria and the mitochondrial endosymbiont, with an additional cyanobacterial component at the origin of plastids in the plant lineage.[85] Endosymbiotic gene transfer, the process through which endosymbionts donate genes to

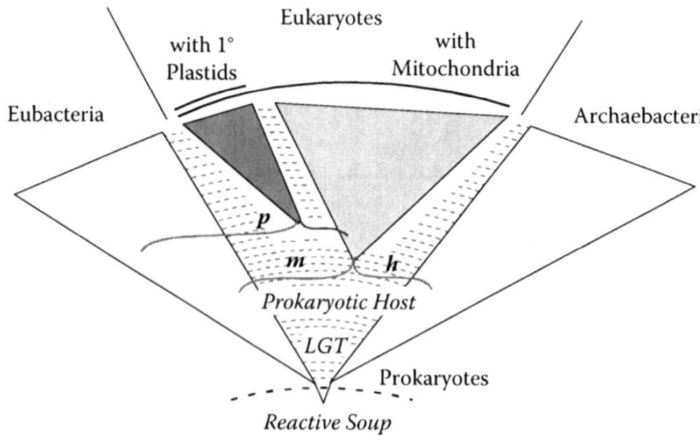

FIGURE 5.5 The symbiotic tree with a prokaryote host.

their host,[86,87] plays a central and quantitatively important role in this view. The LGT between prokaryotes is also essential to the symbiotic tree because it is an important mechanism of natural variation among prokaryotes that helped to shape the genomes of the symbiotic partners involved in eukaryote origins.

The process of secondary symbiosis, in which a eukaryote acquires a photosynthetic eukaryote as a symbiont that subsequently undergoes reduction to become a plastid surrounded by three or four membranes, has not been considered in any of the models outlined here. Such symbioses have occurred at least three times during eukaryote evolution, twice involving green algal endosymbionts, and at least once involving a red algal endosymbiont.[88,89] Secondary symbioses show that symbiosis is a real and tangible biological mechanism that generates novel taxa at higher levels, but secondary symbiosis does not address the issue of how eukaryotes arose.

5.7 WHAT DO THE DATA SAY?

It turns out that one can bring individual aspects of the available genome data into agreement with any of the models outlined. For that reason, each camp is able to maintain the argument that its model is preferable to the others, as one could argue citing many recent articles that support each of the alternatives in favor of the others. Clearly, individual genes tell different stories about the prokaryote-to-eukaryote transition, which was known before the age of genomes,[1] but it is not clear why that is so, which was also the case before the age of genomes. The role of LGT has come to play a more prominent role in thinking about the prokaryote-to-eukaryote transition, but depending on what slant one takes on the issue, that role could be seen as (1) many eukaryote genes come from organelles[64,86,87,90]; (2) LGT has affected so many (or all) genes that there is no single tree of life that is reflected as a nontransferable "core"[65,91]; or (3) LGT mysteriously generates by itself some kind of interpretable phylogenetic signal.[92] Before the genome era, LGT also played a role in thinking about early evolution, but only on a gene-for-gene basis.[93–95] Now, the issue is to try to look at the prokaryote-to-eukaryote transition on a genome-for-genome basis in a manner that would discriminate between some of the competing alternatives on the issue, and that has proven to be harder to do than most of us would have expected.[66,86,96]

One thing seems certain at this point: Because of all the conflicting data in genomes, a single bifurcating tree is not going to do.[17,65,91] This insight has sent those mathematically inclined scrambling to develop methods of evolutionary inference that produce graphs that are more complicated than simple trees. This seems a reasonable thing to do because the evolutionary process connecting prokaryotes and eukaryotes is clearly more complicated than any single bifurcating tree. These new methods include procedures that deliver rings[17] and networks.[97] Supertree methods[98] would also seem to have some applicability to the analysis of genome data, but only recently have bioinformaticians explored supertree analyses in a way that would address the prokaryote-to-eukaryote transition.[100] Simple comparisons of genome-wide sequence similarity indicate that eukaryotes possess far more eubacterially related genes than they possess archaebacterial related genes,[91,99] which is not what most of us would have expected 10 years ago.

5.8 CONCLUSION

The prokaryote-to-eukaryote transition is a controversial topic, and consensus is not likely to be reached any time soon. Genome sequences have challenged the field of molecular evolution to find new approaches to data analysis that could shed light on the issue. The circumstance that mitochondria have turned out to be ubiquitous among eukaryotes precludes the need to assume that there ever were any primitively amitochondriate eukaryotes,[66,79] a circumstance that proponents of the prokaryote host tree could offer in support of their view were they so inclined.

REFERENCES

1. Brown, J. R. & Doolittle, W. F. Archaea and the prokaryote-to-eukaryote transition. *Microbiol. Mol. Biol. Rev.* **61**, 456–502 (1997).
2. Woese, C. R. & Fox, G. E. The concept of cellular evolution. *J. Mol. Evol.* **10**, 1–6 (1977).
3. Woese, C. R., Kandler, O. & Wheelis, M. L. Towards a natural system of organisms: proposal for the domains Archaea, Bacteria and Eucarya. *Proc. Natl. Acad. Sci. U. S. A.* **87**, 4576–4579 (1990).
4. Woese, C. R. Bacterial evolution. *Microbiol. Rev.* **51**, 221–271 (1987).
5. Woese, C. R. The universal ancestor. *Proc. Natl. Acad. Sci. U. S. A.* **95**, 6854–6859 (1998).
6. Woese, C. R. On the evolution of cells. *Proc. Natl. Acad. Sci. U. S. A.* **99**, 8742–8747 (2002).
7. Pace, N. R. Time for a change. *Nature* **441**, 289 (2006).
8. Rivera, M. C., Jain, R., Moore, J. E. & Lake, J. A. Genomic evidence for two functionally distinct gene classes. *Proc. Natl. Acad. Sci. U. S. A.* **95**, 6239–6244 (1998).
9. Ciccarelli, F. D. et al. Toward automatic reconstruction of a highly resolved tree of life. *Science* **311**, 1283–1287 (2006).
10. Cavalier-Smith, T. The phagotrophic origin of eukaryotes and phylogenetic classification of Protozoa. *Int. J. Syst. Evol. Microbiol.* **52**, 297–354 (2002).
11. Martin, W. Archaebacteria (Archaea) and the origin of the eukaryotic nucleus. *Curr. Opin. Microbiol.* **8**, 630–637 (2005).
12. Martin, W. & Embley, T. M. Early evolution comes full circle. *Nature* **431**, 134–136 (2004).
13. Iwabe, N., Kuma, K.-I., Hasegawa, M., Osawa, S. & Miyata, T. Evolutionary relationship of archaebacteria, eubacteria and eukaryotes inferred from phylogenetic trees of duplicated genes. *Proc. Natl. Acad. Sci. U. S. A.* **86**, 9355–9359 (1989).
14. Gogarten, J. P. et al. Evolution of the vacuolar H^+-ATPase: implications for the origin of eukaryotes. *Proc. Natl. Acad. Sci. U. S. A.* **86**, 6661–6665 (1989).
15. Kurland, C. G., Collins, L. J. & Penny, D. Genomics and the irreducible nature of eukaryote cells. *Science*, **312**, 1011–1014 (2006).
16. Woese, C. R. Default taxonomy: Ernst Mayr's view of the microbial world. *Proc. Natl. Acad. Sci. U. S. A.* **95**, 11043–11046 (1998).
17. Rivera, M. C. & Lake, J. A. The ring of life provides evidence for a genome fusion origin of eukaryotes. *Nature* **431**, 152–155 (2004).
18. Breathnach, R., Mandel, J. L. & Chambon, P. Ovalbumin gene is split in chicken DNA. *Nature* **270**, 314–319 (1977).
19. Gilbert, W. Why genes in pieces? *Nature* **271**, 501 (1978).
20. Doolittle, W. F. Genes in pieces: were they ever together? *Nature* **272**, 581–582 (1978).

21. Doolittle, W. F. Revolutionary concepts in evolutionary biology. *Trends Biochem. Sci.* **5**, 146–149 (1980).
22. Doolittle, W. F. The origin and function of intervening sequences in DNA: a review. *Am. Nat.* **130**, 915–928 (1987).
23. Forterre, P. Thermoreduction, a hypothesis for the origin of prokaryotes. *C. R. Acad. Sci. III* **318**, 415–422 (1995).
24. Roger, A. J. & Doolittle, W. F. Why introns-in-pieces? *Nature* **364**, 289–290 (1993).
25. Stoltzfus, A., Spencer, D. F., Zuker, M., Logsdon, J. M. & Doolittle, W. F. Testing the exon theory of genes: the evidence from protein structure. *Science* **265**, 202–207 (1994).
26. Poole, A. M., Jeffares, D. C. & Penny, D. The path from the RNA world. *J. Mol. Evol.* **46**, 1–17 (1998).
27. Forterre, P. et al. The nature of the last universal ancestor and the root of the tree of life, still open questions. *Biosystems* **28**, 15–32 (1992).
28. Forterre, P. & Philippe, H. Where is the root of the universal tree of life? *Bioessays* **21**, 871–879 (1999).
29. Philippe, H. & Forterre, P. The rooting of the universal tree of life is not reliable. *J. Mol. Evol.* **49**, 509–523 (1999).
30. Lopez, P., Forterre, P. & Philippe, H. The root of the tree of life in the light of the covarion model. *J. Mol. Evol.* **49**, 496–508 (1999).
31. Jeffares, D. C., Poole, A. M. & Penny, D. Relics from the RNA world. *J. Mol. Evol.* **46**, 18–36 (1998).
32. Brinkmann, H. & Philippe, H. Archaea sister group of Bacteria? Indications from tree reconstruction artifacts in ancient phylogenies. *Mol. Biol. Evol.* **16**, 817–825 (1999).
33. Penny, D. An interpretative review of the origin of life research. *Biol. Philos.* **20**, 633–671 (2005).
34. Poole, A., Penny, D. & Sjoberg, B. M. Confounded cytosine! Tinkering and the evolution of DNA. *Nat. Rev. Mol. Cell Biol.* **2**, 147–151 (2001).
35. Cavalier-Smith, T. The origin of eukaryote and archaebacterial cells. *Ann. N. Y. Acad. Sci.* **503**, 17–54 (1987).
36. Cavalier-Smith, T. The neomuran origin of archaebacteria, the negibacterial root of the universal tree and bacterial megaclassification. *Int. J. Syst. Evol. Microbiol.* **52**, 7–76 (2002).
37. Cavalier-Smith, T. Only six kingdoms of life. *Proc. R. Soc. Lond. B.*, **271**, 1251–1262 (2004).
38. Cavalier-Smith, T. Cell evolution and Earth history: stasis and revolution. *Philos. Trans. R. Soc. Lond. B Biol. Sci.* **361**, 969–1006 (2006).
39. Cavalier-Smith, T. Eukaryotes with no mitochondria. *Nature* **326**, 332–333 (1987).
40. Cavalier-Smith, T. Only six kingdoms of life. *Proc. Roy Soc. Lond. B* **271**, 1251–1262 (2004).
41. Sagan, L. On the origin of mitosing cells. *J. Theoret. Biol.* **14,** 225–274 (1967).
42. Mereschkowsky, C. Über Natur und Ursprung der Chromatophoren im Pflanzenreiche. *Biol. Centralbl.* **25**, 593–604 (1905). English translation in Martin, W. & Kowallik, K. V. *Eur. J. Phycol.* **34**, 287–295 (1999).
43. Sapp, J. *Evolution by Association: A History of Symbiosis.* Oxford University Press, New York (1994).
44. Wallin, I. E. *Symbionticism and the origin of species.* Bailliere, Tindall & Cox, London (1927).
45. Wilson, E. B. *The Cell in Development and Heredity.* 3rd rev. ed. Macmillan, New York (1928). Reprinted by Garland, New York (1987).
46. Buchner, P. *Endosymbiose der Tiere mit pflanzlichen Mikroorganismen.* Birkhäuser, Basel (1953).

47. Ris, H. & Plaut, W. Ultrastructure of DNA-containing areas in the chloroplasts of Chlamydomonas. *J. Cell Biol.* **12**, 383–391 (1962).
48. Bonen, L. & Doolittle, W. F. Prokaryotic nature of red algal chloroplasts. *Proc. Natl. Acad. Sci. U. S. A.* **72**, 2310–2314 (1975).
49. John, P. & Whatley, F. R. *Paracoccus denitrificans* and the evolutionary origin of the mitochondrion. *Nature* **254**, 495–498 (1975).
50. Bogorad, L. Evolution of organelles and eukaryotic genomes. *Science* **188**, 891–898 (1975).
51. Cavalier-Smith, T. The origin of nuclei and of eukaryotic cells. *Nature* **256**, 463–468 (1975).
52. Woese, C. R. Archaebacteria. *Sci. Am.* **244**, 98–122 (1981).
53. Margulis, L. Symbiosis and evolution. *Sci. Am.* **225**, 48–57 (1971).
54. Gray, M. W. & Doolittle, W. F. Has the endosymbiont hypothesis been proven? *Microbiol. Rev.* **46**, 1–42 (1982).
55. Margulis, L., Chapman, M., Guerrero, R. & Hall, J. The last eukaryotic common ancestor (LECA): acquisition of cytoskeletal motility from aerotolerant spirochetes in the Proterozoic eon. *Proc. Natl. Acad. Sci. U. S. A.* **103**, 13080–13085 (2006).
56. Jekely, G. & Arendt, D. Evolution of intraflagellar transport from coated vesicles and autogenous origin of the eukaryotic cilium. *Bioessays* **28**, 191–198 (2006).
57. van Valen, L. M. & Maiorana, V. C. The archaebacteria and eukaryotic origins. *Nature* **287**, 248–250 (1980).
58. Doolittle, W. F. Fun with genealogy. *Proc. Natl. Acad. Sci. U. S. A.* **94**, 12751–12753 (1997).
59. Martin, W. & Schnarrenberger, C. The evolution of the Calvin cycle from prokaryotic to eukaryotic chromosomes: a case study of functional redundancy in ancient pathways through endosymbiosis. *Curr. Genet.* **32**, 1–18 (1997).
60. Clark, C. G. & Roger, A. J. Direct evidence for secondary loss of mitochondria in *Entamoeba histolytica. Proc. Natl. Acad. Sci. U. S. A.* **92**, 6518–6521 (1995).
61. Henze, K., Badr, A., Wettern, M., Cerff, R. & Martin, W. A nuclear gene of eubacterial origin in *Euglena gracilis* reflects cryptic endosymbioses during protist evolution. *Proc. Natl. Acad. Sci. U. S. A. 92*, 9122–9126 (1995).
62. Martin, W. Is something wrong with the tree of life? *BioEssays* **18**, 523–527 (1996).
63. Lake, J. A. & Rivera, M. C. Was the nucleus the first endosymbiont? *Proc. Natl. Acad. Sci. U. S. A.* **91**, 2880–2881 (1994).
64. Martin, W. F. & Müller, M. The hydrogen hypothesis of the first eukaryote. *Nature* **392**, 37–41 (1998).
65. Doolittle, W. F. Phylogenetic classification and the universal tree. *Science* **284**, 2124–2128 (1999).
66. Embley, T. M. & Martin, W. Eukaryotic evolution, changes and challenges. *Nature* **440**, 623–630 (2006).
67. Zillig, W. et al. Did eukaryotes originate by a fusion event? *Endocyt. C. Res.* **6**, 1–25 (1989).
68. Gupta, R. S. & Golding, G. B. The origin of the eukaryotic cell. *Trends. Biochem. Sci.* **21**, 166–171 (1996).
69. Horiike, T., Hamada, K., Kanaya, S. & Shinozawa, T. Origin of eukaryotic cell nuclei by symbiosis of Archaea in Bacteria is revealed by homology hit analysis. *Nature Cell Biol.* **3**, 210–214 (2001).
70. Horiike, T., Hamada, K., Miyata, D. & Shinozawa, T. The origin of eukaryotes is suggested as the symbiosis of *Pyrococcus* into γ-proteobacteria by phylogenetic tree based on gene content. *J. Mol. Evol.* **59**, 606–619 (2004).
71. Lopez-Garcia, P. & Moreira, D. Selective forces for the origin of the eukaryotic nucleus. *Bioessays* **28**, 525–533 (2006).

72. Margulis, L., Dolan, M. F. & Guerrero, R. The chimeric eukaryote: origin of the nucleus from the karyomastigont in amitochondriate protists. *Proc. Natl. Acad. Sci. U. S. A.* **97,** 6954–6959 (2000).
73. Embley, T. M. & Hirt, R. P. Early branching eukaryotes? *Curr. Opin. Genet. Dev.* **8**, 655–661 (1998).
74. Müller, M. The hydrogenosome. *J. Gen. Microbiol.* **139**, 2879–2889 (1993).
75. Müller, M. Energy metabolism. Part I: Anaerobic protozoa. In: *Molecular Medical Parasitology* (Ed. Marr, J.), pp. 125–139. Academic Press, London (2003).
76. Tovar, J., Fischer, A. & Clark, C. G. The mitosome, a novel organelle related to mitochondria in the amitochondrial parasite *Entamoeba histolytica*. *Mol. Microbiol.* **32**, 1013–1021 (1999).
77. Tovar, J. et al. Mitochondrial remnant organelles of *Giardia* function in iron-sulphur protein maturation. *Nature* **426**, 172–176 (2003).
78. Williams, B. A., Hirt, R. P., Lucocq, J. M. & Embley, T. M. A mitochondrial remnant in the microsporidian *Trachipleistophora hominis*. *Nature* **418**, 865–869 (2002).
79. van der Giezen, M., Tovar, J. & Clark, C. G. Mitochondrion-derived organelles in protists and fungi. *Int. Rev. Cytol.* **244**, 175–225 (2005).
80. Searcy, D. G. Origins of mitochondria and chloroplasts from sulphur-based symbioses. In: *The Origin and Evolution of the Cell* (Eds. Hartman, H. & Matsuno, K.), pp. 47–78. World Scientific, Singapore (1992).
81. Doolittle, W. F. Some aspects of the biology of cells and their possible evolutionary significance. In: *Evolution of Microbial Life* (ed. Roberts, D., Sharp, P., Alserson, G. & Collins, M.), pp. 1–21. 54th Symp. Soc. Gen. Microbiol. Cambridge University Press, Cambridge, UK (1996).
82. Vellai, T., Takács, K. & Vida, G. A new aspect on the origin and evolution of eukaryotes. *J. Mol. Evol.* **46**, 499–507 (1998).
83. Martin, W., Hoffmeister, M., Rotte, C. & Henze, K. An overview of endosymbiotic models for the origins of eukaryotes, their ATP-producing organelles (mitochondria and hydrogenosomes), and their heterotrophic lifestyle. *Biol. Chem.* **382**, 1521–1539 (2001).
84. Martin, W. & Koonin, E. V. Introns and the origin of nucleus-cytosol compartmentalization. *Nature* **440,** 41–45 (2006).
85. Martin, W. et al. Evolutionary analysis of *Arabidopsis*, cyanobacterial, and chloroplast genomes reveals plastid phylogeny and thousands of cyanobacterial genes in the nucleus. *Proc. Natl. Acad. Sci. U. S. A.* **99**, 12246–12251 (2002).
86. Brown, J. R. Ancient horizontal gene transfer. *Nat. Rev. Genet.* **4**, 121–132 (2003).
87. Timmis, J. N., Ayliffe, M. A., Huang, C. Y. & Martin, W. Endosymbiotic gene transfer: organelle genomes forge eukaryotic chromosomes. *Nat. Rev. Genet.* **5**, 123–135 (2004).
88. Stoebe, B. & Maier, U.-G. One, two, three: nature's toolbox for building plastids. *Protoplasma* **219**, 123–130 (2002).
89. Rogers, M. B., Gilson, P. R., Su, V., McFadden, G. I. & Keeling, P. J. The complete chloroplast genome of the chlorarachniophyte *Bigelowiella natans*: evidence for independent origins of chlorarachniophyte and euglenid secondary endosymbionts. *Mol. Biol. Evol.* **24,** 54–62 (2006).
90. Henze, K. & Martin, W. How do mitochondrial genes get into the nucleus? *Trends Genet.* **17**, 383–387 (2001).
91. Dagan, T. & Martin, W. The tree of 1%. *Genome Biol.* **7,** 118 (2006).
92. Huang, J. & Gogarten, J. P. Ancient horizontal gene transfer can benefit phylogenetic reconstruction. *Trends Genet.* **22**, 361–366 (2006).
93. Martin, W. & Cerff, R. Prokaryotic features of a nucleus encoded enzyme: cDNA sequences for chloroplast and cytosolyic glyceraldehyde-3-phosphate dehydrogenases from mustard (*Sinapis alba*). *Eur. J. Biochem.* **159**, 323–331 (1986).

94. Doolittle, R. F., Feng, D, F., Anderson, K. L. & Alberro, M. R. A naturally occurring horizontal gene transfer from a eukaryote to a prokaryote. *J. Mol. Evol.* **31**, 383–388 (1990).
95. Smith, M. W., Feng, D.-F. & Doolittle, R. F. Evolution by acquisition: the case for horizontal gene transfers. *Trends Biochem. Sci.* **17**, 489–493 (1992).
96. Doolittle, W. F. et al. How big is the iceberg of which organellar genes in nuclear genomes are but the tip? *Phil. Trans. R. Soc. Lond. B Biol. Sci.* **358**, 39–58 (2003).
97. Huson, D. H. & Bryant, D. Application of phylogenetic networks in evolutionary studies. *Mol. Biol. Evol.* **23**, 254–267 (2006).
98. Wilkinson, M. et al. The shape of supertrees to come: tree shape related properties of fourteen supertree methods. *Syst. Biol.* **54**, 419–431 (2005).
99. Esser, C. et al. A genome phylogeny for mitochondria among α-proteobacteria and a predominantly eubacterial ancestry of yeast nuclear genes. *Mol. Biol. Evol.* **21**, 1643–1660 (2004).
100. Pisani, D., Cotton, J. A., & McInerney, J. O. Supertrees disentangle the chimeric origin of eukaryotic genomes. *Mol. Biol. Evol.* **24,** 1752–1760 (2007).

6 Comparative Genomics of Invertebrates

Takeshi Kawashima, Eiichi Shoguchi,
Yutaka Satou, and Nori Satoh

CONTENTS

ABSTRACT

An organism's genome contains all of its genetic information, and thus sequenced genomes provide the basis for the entire field of biological sciences. At the end of 2006, genomes of six groups of invertebrates had been decoded, including two species of nematode worms, two species of insect flies, an insect mosquito, an insect silkworm, a social honeybee, an echinoderm sea urchin, and an urochordate ascidian. We review here comparative and characteristic features of the genome of each animal and discuss the significant role of genome information in exploring various problems in animal biology.

6.1 INTRODUCTION

Taxonomists have identified and described approximately 1,320,000 species of multicellular animals or metazoans to date. Comparative studies of morphology of larvae and adults and mode of embryogenesis as well as molecular phylogenetic analyses reveal that metazoans are categorized into approximately 34 major groups or phyla.[1] As shown in Figure 6.1, multicellular animals are first subgrouped into

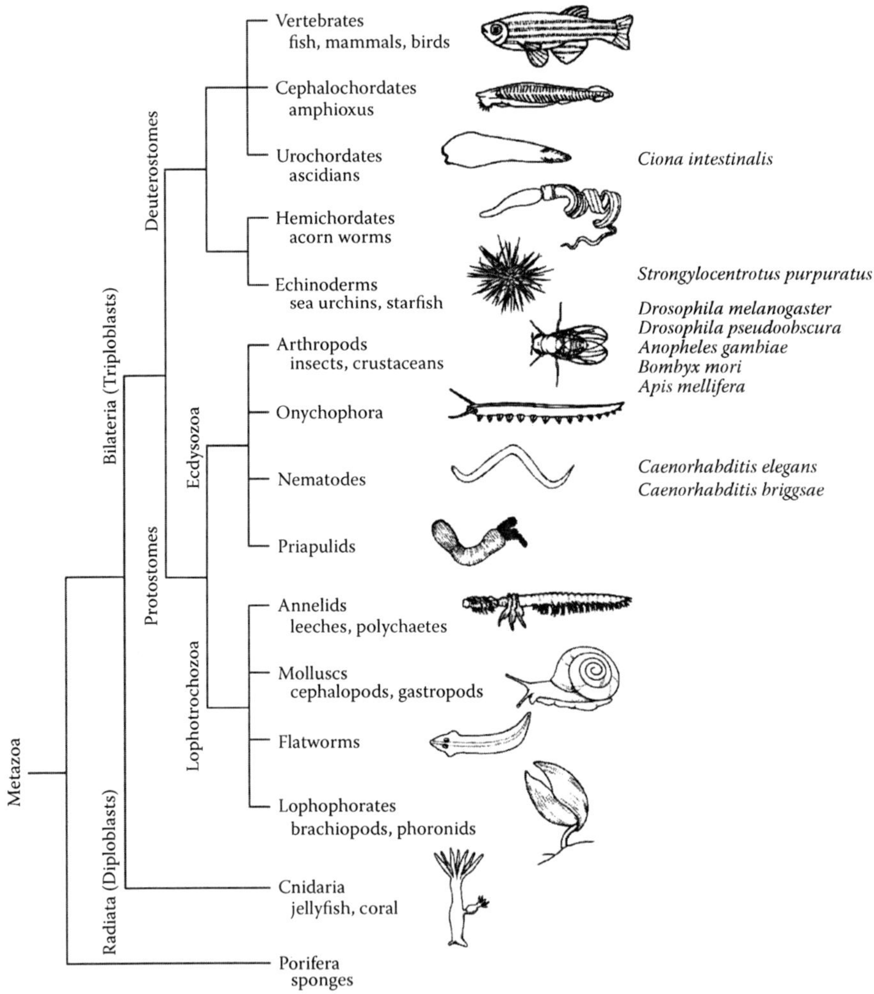

FIGURE 6.1 A schematic drawing to show the phylogenetic relationships among Metazoan phyla, mainly resolved by molecular phylogenetic studies. In bilaterians, three primary clades exist: the deuterostomes, including echinoderms, hemichordates, and chordates (urochordates, cephalochordates, and vertebrates); the ecdysozoans, including arthropods, priapulids, and nematodes; and the lophotrochozoans, including annelids, mollusks, and lophophorates. On the other hand, radiates are the Cnidaria, including jellyfish and anemones, and the Porifera. Animal species for which the genome has been sequenced are shown at the right. (Modified from Carroll, S. B., et al., *From DNA to Diversity. Molecular Genetics and the Evolution of Animal Design*, Blackwell Science, MA, 2001.)

two major clades: diploblasts (also called radiates, including cnidarians and ctenophores; porifera [sponges] with less tissue-organization body is sometimes included in this clade) and triploblasts (also called bilaterians, including most of the other animals). The bilaterian body consists of three germ layers: outer ectoderm, inner endoderm, and intermediate mesoderm. The diploblast body lacks the mesoderm. Bilaterians are further subdivided into protostomes and deuterostomes, depending on whether the blastopore gives rise to mouth (protostomes) or anus (deuterostomes) (Figure 6.1).

Previously, about 27 phyla of protostomes were categorized based on the mode of the formation of body cavity. However, recent molecular phylogenetic studies have demonstrated that protostomes might be comprised of ecdysozoans and lophotrochozoans, the former including plathelminthes, nematodes, and arthropods, and the latter including annelids, mollusks, and lophophorates[2–4] (Figure 6.1). On the other hand, deuterostomes comprise echinoderms, hemichordates, and chordates. Multicellular animals are sometimes subgrouped in general into those with backbone (vertebrates) and those without backbone (invertebrates). Because the primordial organ of vertebrates, the notochord, is also possessed by the urochordates (or tunicates) and cephalochordates, an animal phylogeny supports a view that vertebrates are not a discrete group that constitutes a phylum, but they are a subgroup of the phylum Chordata, together with urochordates and cephalochordates; these three groups also share a dorsal hollow neural tube (or nerve cord), gill slits, endostyle, and other features.[5] Therefore, the term *invertebrates* does not represent a monophyletic group, and urochordates and cephalochordates are included in this review. Fossil records suggest that all the invertebrate groups evolved from a common ancestor prior to or during the Cambrian explosion in the period of 650 to about 520 million years ago.

The genomes of invertebrates are different from those of the vertebrates in the redundancy of genes encoded there. It has been thought that, in the course of vertebrate evolution after the split of vertebrates/tunicates, two series of genome-wide duplication events (whole-genome duplications or genome-wide gene duplications) occurred.[6,7] Invertebrate genomes therefore contain fewer genes than those of vertebrates with less redundancy, but they are very complex with profound genetic information.

In late 1998, the genome of a nematode, *Caenorhabditis elegans*, was decoded as the first from a multicellular organism,[8] followed in 2000 by decoding of the genome of a fruit fly, *Drosophila melanogaster*.[9] At the end of 2006, genomes of six groups of invertebrates had been decoded, including nematode worms *Caenorhabditis elegans* and *Caenorhabditis briggsae*; insect flies *Drosophila melanogaster* and *Drosophila pseudoobscura*; an insect mosquito, *Anopheles gambiae*; an insect silkworm, *Bombyx mori*; a social honeybee, *Apis mellifera*; an echinoderm sea urchin, *Strongylocentrotus purpuratus*; and an urochordate ascidian, *Ciona intestinalis* (Figure 6.1). National Center for Biotechnology Information (NCBI) genome information data show that, in addition to the above-mentioned animals, the genome projects of more than 20 animal species are now in progress, and nearly 40 are now targeted for future studies (Table 6.1). Each of the invertebrates with a sequenced genome has a distinct reason behind its genome project. Here, we review comparative and characteristic features of the genome of each animal and then discuss the significant role of genome information in exploring various problems in animal biology.

TABLE 6.1
Sequenced Genomes of Invertebrates

Species	Group	Genome Size (Mb)	No. of Genes	Haploid Chromosomes	Status	NCBI Project ID	Online Repositories
Caenorhabditis elegans	Roundworms	100	19,735	6	Complete	9548	http://www.wormbase.org
Drosophila melanogaster	Insects	180	14,461	4	Complete	9554	http://www.flybase.org/
Caenorhabditis briggsae	Roundworms	104	19,500	6	Draft assembly	9547	http://www.wormbase.org
Drosophila pseudoobscura	Insects	120	14,400	4	Draft assembly	12559	http://species.flybase.net/
Anopheles gambiae	Insects	220	14,000	3	Draft assembly	9553	http://www.malaria.mr4.org
Apis mellifera	Insects	200	10,000	16	Draft assembly	9555	http://www.hgsc.bcm.tmc.edu/projects/honeybee/
Bombyx mori	Insects	530	18,500	28	Draft assembly	10637	http://papilio.ab.a.u-tokyo.ac.jp/lep-genome/index.html
Strongylocentrotus purpuratus	Echinoderms	800	23,000		Draft assembly	10728	http://www.hgsc.bcm.tmc.edu/projects/seaurchin/
Ciona intestinalis	Tunicates	160	15,852	14	Draft assembly	9556	http://genome.jgi-psf.org/ciona4/ciona4.home.html http://ghost.zool.kyoto-u.ac.jp/indexr1.html
Caenorhabditis remanei	Roundworms				Draft assembly	12669	
Drosophila ananassae	Insects	150		4	Draft assembly	12632	
Drosophila erecta	Insects	150		4	Draft assembly	12660	
Drosophila grimshawi	Insects	150		4	Draft assembly	12675	
Drosophila mojavensis	Insects	150		4	Draft assembly	12680	
Drosophila simulans	Insects	150		4	Draft assembly	12463	

Drosophila virilis	Insects	150	4	Draft assembly	12687
Drosophila willistoni	Insects	150	4	Draft assembly	12663
Drosophila yakuba	Insects	180	4	Draft assembly	12265
Aedes aegypti	Insects	800	3	Draft assembly	9551
Aplysia californica	Insects	1,800	17	Draft assembly	13634
Tribolium castaneum	Insects	200	10	Draft assembly	12539
Ciona savignyi	Tunicates	180	14	Draft assembly	9585
Acyrthosiphon pisum	Insects	300	4	In progress	13646
Bicyclus anynana	Insects	490		In progress	13881
Biomphalaria glabrata	Crustaceans	930	18	In progress	12878
Brugia malayi	Roundworms	110	6	In progress	9549
Culex pipiens	Insects	540	3	In progress	12963
Daphnia pulex	Crustaceans			In progress	12755
Drosophila americana	Insects	150	4	In progress	12762
Drosophila hydei	Insects	150	4	In progress	12780
Drosophila miranda	Insects	150	4	In progress	12758
Nasonia vitripennis	Insects	330	5	In progress	13647
Oikopleura dioica	Tunicates	70		In progress	12900
Pediculus humanus	Insects			In progress	16222
Rhodnius prolixus	Insects	670	11	In progress	13645
Saccoglossus kowalevskii	Hemichordates			In progress	12886
Schistosoma mansoni	Worms	270	8	In progress	12599
Spisula solidissima	Mollusks	1,200		In progress	12959

Only representative species are shown from those of the genome project in progress.

6.2 CHARACTERISTICS OF GENOMES OF INVERTEBRATES

6.2.1 GENOME OF *CAENORHABDITIS ELEGANS*

The genome project of a nematode, *Caenorhabditis elegans*, was undertaken in the early 1980s by construction of a clone-based physical map. The map of overlapping cosmids and later yeast artificial chromosomes (YAC), along with large-scale expressed sequence tags (ESTs), accomplished the decoding of its genome in late 1998 as the first from a multicellular organism.[8] At that moment, the genome was estimated to consist of approximately 97 Mb and to contain approximately 19,000 protein-coding genes. Further efforts have now completed the *C. elegans* genome sequence, indicating a 130-Mb genome containing 19,735 protein-coding genes and more than 1,300 noncoding RNA genes[10] (Table 6.1). The genome was also revealed to contain 88 genes encoding microRNAs (miRNAs), which represent 48 gene families.[11] Of these families, 46 are conserved in *C. briggsae,* and 22 families are conserved in humans.[11]

Pairwise comparison of the *C. elegans* genome with those of the bacteria *Escherichia coli*, the yeast *Saccharomyces cerevisiae*, and the human being *Homo sapiens* clearly showed that, as expected from evolutionary relationships, there were substantially more protein similarities found between *C. elegans* and *H. sapiens*. In fact, *C. elegans* and *H. sapiens* share highly conserved neurotransmitter receptors, neurotransmitter synthesis and release pathways, and heterotrimeric GTP-binding protein (G-protein)-coupled second-messenger pathways, although gap junction and chemosensory receptors have independent origin in vertebrates and nematodes.[12] Along with this similarity, the top 20 common protein domains that occur most frequently in the nematode genome are occupied by genes implicated in intracellular communication (the most frequent one was seven transmembrane chemoreceptor) or in transcriptional regulation. This strongly suggests that decoding of the invertebrate genome is critically important for understanding human genome and biology as well.[8,12]

Caenorhabditis briggsae diverged from common ancestors shared with *C. elegans* roughly 100 million years ago. They show similar outer morphology, have the same chromosome number, and occupy the same ecological niche. Decoding of the *C. briggsae* 104-Mb genome demonstrated the difference in genome size from that of *C. elegans* (100.3 Mb) is almost entirely due to repetitive sequence, which accounts for 22.4% of the *C. briggsae* genome, in contrast to 16.5% of the *C. elegans* genome.[13] Of approximately 19,500 protein-coding genes contained in both species, 12,200 have clear orthologs. On the other hand, approximately 800 genes were found only in *C. briggsae*. Comparison of genome sequences of the two closely related nematode species greatly improved the annotation of the *C. elegans* genome, and the comparison with the *C. briggsae* genome resulted in a finding of 1,300 new *C. elegans* genes. Comparison of the two *Caenorhabditis* genomes also shows dramatic differences in expansion of chemosensory genes[14] and for positive selection of members of the SRZ family (a distant relative of seven-pass receptor) of G-coupled receptors[15] between the two species.

6.2.2 GENOME OF A FRUIT FLY, *DROSOPHILA MELANOGASTER*

Drosophila melanogaster has over a 100-year history as a model organism of animal genetics. Due to its enormous contribution to our understanding of the biology of development, behavior, and evolution, the completion of the *D. melanogaster* genome was greatly anticipated. The *D. melanogaster* genome was accomplished in March 2000 as the second animal genome and was a landmark from technical and methodological viewpoints.[9] In this project, whole-genome shotgun sequencing was introduced by Craig Venter and his colleague, and the method, a combination of new capillary sequencing machines, very careful construction of clone libraries, and advanced software, succeeded for a large and complex genome of more than 100 Mb.

The *D. melanogaster* genome has about a 120-Mb euchromatic region, and about 13,600 protein-coding genes were predicted in this region. Thereafter, continuing efforts to complete the *D. melanogaster* genome have revised the genome several times to reach the object,[16] and now the genome predicts 14,461 protein-coding genes. Even in this mostly genomically advanced species, only 5,402 have known mutant alleles, and thousands of mutant alleles have yet to be identified among these DNA sequences. Most recent progress in the *D. melanogaster* gene annotation can be seen in the flybase (http://flybase.bio.indiana.edu).

Deciphering of the *D. melanogaster* genome also facilitated our understanding of transposable elements. The fly genome contains 6,013 transposable elements in 127 families. Analysis of the *D. melanogaster* genome also contributed to the discovery and understanding of small RNAs. Among them, miRNAs constitute nearly 1% of the annotated genes in the *D. melanogaster* genome. The complex heterochromatinic sequences of the telomeres and pericentromeric regions of chromosomes have also been analyzed in this genome. Much of the complex heterochromatin is composed of a graveyard of decaying, often nested, transposable elements with a sprinkling of protein-coding genes.[16]

In *D. melanogaster*, the large collection of inserted transposes used for gene disruption can now be mapped precisely to the genome sequence. About 65% of the genes of *D. melanogaster* have been disrupted by at least one transposon insertion.

The genomic sequences of an additional 12 species of *Drosophila* are now undergoing examination (http://rana.lbl.gov/drosophila/assemblies.html; Table 6.1), and the draft genome sequence of nine *Drosophila* species, including *D. pseudoobscura*, has been determined.[17] *Drosophila melanogaster* and *D. pseudoobscura* diverged from a common ancestor 25–55 million years ago. Comparison of the two *Drosophila* genomes suggests two important themes of genome divergence between these species of *Drosophila*: a pattern of repeat-meditated chromosomal rearrangement and high coadaptation in males and *cis*-regulatory sequences of both sexes. Although the vast majority of *Drosophila* genes have remained on the same chromosome arm, within each arm gene order has been extensively reshuffled (Figure 6.2), and a repetitive sequence is found in the *D. pseudoobscura* genome at many junctions between adjacent syntenic blocks. Of about 14,400 genes, 10,516 putative orthologs have been identified as a core gene set between the two species. Interestingly, genes expressed in the testes had higher amino acid sequence divergence than the genome-wide average, consistent with the rapid evolution of sex-specific proteins. The *cis*-regulatory sequences are more conserved than random and nearby sequences between the species, but the differences are slight, suggesting that the evolution of *cis*-regulatory elements is flexible. Comparisons of genome sequences of 22 *Drosophila* species could reveal much more

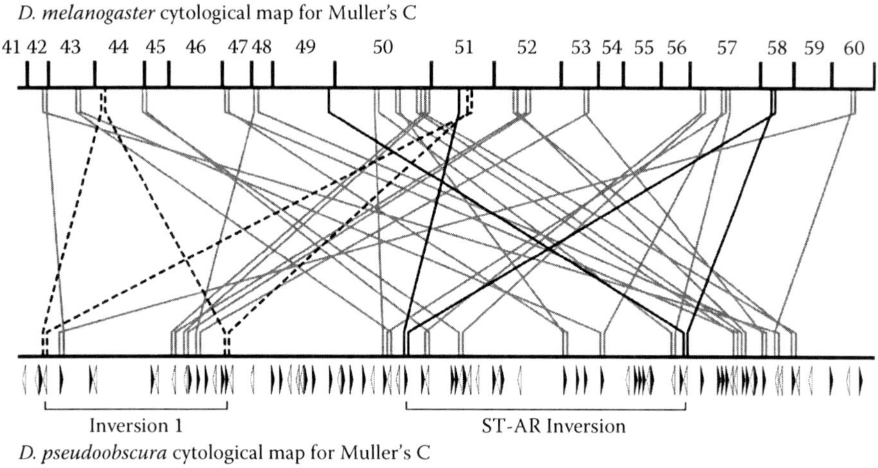

FIGURE 6.2 Rearrangement of conserved linkage groups between *D. melanogaster* and *D. pseudoobscura*. The thick horizontal lines represent the chromosomal maps of the *D. melanogaster* and *D. pseudoobscura* Mullar element C. Vertical lines drawn either down (*D. melanogaster*) or up (*D. pseudoobscura*) indicate conserved linkage groups. The location and orientation of 80 breakpoint motifs are indicated with open and filled triangles between breakpoint motifs will bring adjacent *D. melanogaster* genes together (dashed and gray lines). A second example that shows ectopic exchange between a pair of motifs for which only one breakpoint brings adjacent *D. melanogaster* genes together is indicated with black solid lines. (From Richards, S., et al., *Genome Res.* **15**, 1–18, 2005.)

definite answers for these questions and could greatly contribute to finding of conserved features, including *cis*-regulatory elements, small RNAs, and new exons.

6.2.3 Genome of a Mosquito, *Anopheles gambiae*

Malaria is a disease that afflicts more than 500 million people and causes over 1 million deaths each year. Malaria disease transmission is facilitated by mosquito vectors, and *Anopheles gambiae* is the principal carrier of the malaria parasite *Plasmodium falciparum*. Thus, the *A. gambiae* genome was sequenced in 2002. Tenfold shotgun sequence coverage was obtained from the PEST (pink eye standard) strain of *A. gambiae* and assembled into scaffolds that span 278 million bp.[18] There was substantial genetic variation within this strain. Analysis of the genome sequences revealed strong evidence for about 14,000 protein-coding transcripts. Prominent expansion in specific families of proteins likely involved in cell adhesion and immunity were noted. An EST analysis of genes regulated by blood feeding provided insights into the physiological adaptation of hematophagous insect.

In the same week of publication of the *A. gambiae* genome sequence, the sequence of the *P. falciparum* genome appeared.[19] The genomes of the two organisms along with that of the human provide a triad of critical genetic information relevant to all stages of the malaria transmission cycle and offer unprecedented opportunities to scientific examination of public health care and to create drugs against malaria.

6.2.4 Genome of a Silkworm, *Bombyx mori*

The silkworm *Bombyx mori* belongs to Lepidoptera insect order and was domesticated over the past 5,000 years because silk fibers are obtained from this animal. In addition, silkworms are a model for insect genetics, having mutants from genetically homogeneous inbred lines. *Bombyx mori* has 28 chromosomes. Its draft genome was publicized in 2004 by whole-genome shotgun sequencing of 5.9× coverage.[20] The genome is approximately 430 Mb, predicting 18,510 protein-coding genes. This genome size is 3.6 and 1.54 times larger than that of *D. melanogaster* and *A. gambiae*, respectively. This larger genome size may be explained by more protein-coding genes (compared to ~14,000 *Drosophila* genes) and larger genes as a result of the insertion of tranposable elements in introns.

6.2.5 Genome of a Honeybee, *Apis mellifera*

Honeybees belong to the insect order Hymenoptera, which includes 100,000 species of sawflies, wasps, ants, and bees. Hymenoptera exhibit haplodiploid sex determination, by which males arise from unfertilized haploid eggs, and females arise from fertilized diploid eggs. The transformation of an insect species from a solitary lifestyle to advanced colonial existence requires alternations in every system of the body coupled with sufficient plasticity in the traits prescribed by the genes to generate strong differences among the adult castes. These biological interests promoted the genome project of a honeybee, *Apis mellifera*.

The genome of *A. mellifera* is about 236 Mb in size, and sequences are distributed over 16 pairs of chromosomes.[21] Genome sequence analysis predicts 10,157 protein-coding genes. Compared with other sequenced insect genomes, the *A. mellifera* genome has high A + T and CpG contents (67% A + T in honeybee compared with 58% in *D. melanogaster* and 56% in *A. gambiae*). The genome lacks major transposon families, evolves more slowly, and is more similar to vertebrates for circadian rhythm, RNA interference, and DNA methylation genes, among other sequenced insect genomes.

The reading of the genome reveals that some of the genes have been modified from ancient precursors; namely, *A. mellifera* has more genes for odorant receptors, novel genes for nectar and pollen utilization, and fewer genes for innate immunity, detoxification enzymes, cuticle-forming proteins, and gustatory receptors, consistent with its ecology and social organization. For example, a cluster descended from a single progenitor gene that encoded a member of yellow protein family here prescribes the royal jelly used in caste determination and queen production. The honeybee has more genes encoding odorant receptors, mirroring the importance of pheromones in sensory communication during the various bee dances, as well as in distinguishing different castes and bees alien to the colony. On the other hand, the honeybee can get away with a simpler outer cuticle than the other insects, and so it has fewer genes encoding cuticle proteins, suggesting that their communal lifestyle contributes protection.

6.2.6 Genome of a Sea Urchin, *Strongylocentrotus purpuratus*

As shown in Figure 6.1, echinoderms are a group of deuterostomes, with hemichordates and chordates the two other groups of this animal superphyla. The genome of the sea urchin was sequenced primarily because of the remarkable usefulness of the echinoderm embryo as a research model system for modern molecular, evolutionary, and cell biology, especially disclosure of gene regulatory networks responsible for the construction of bilaterally organized embryo but a radial adult body plan.[22,23]

The DNA sequencing strategy combined whole-genome shotgun and bacterial artificial chromosome (BAC) sequences, and a scarcity of ESTs or complementary DNA (cDNA) information required for better understanding of transcriptomes and gene expression regulation was substantially covered by using custom tiling arrays covering the whole genome.[24] The *S. purpuratus* genome is 814 Mb in size, relatively large with high heterozygosity of the genome, and encodes about 23,000 genes.[25] Analysis suggests that there are many genes previously thought to be either vertebrate innovations or known only outside the deuterostomes, supporting the evolutionary context of echinoderms as one of the key transitional groups between invertebrates and vertebrates.

One of the triumphs of the sea urchin genome project was a follow-up of genome sequences by deeply characterized annotation of genes, especially genes involved in embryogenesis. Genes encoding transcription factors and cell-signaling molecules have been extensively annotated.[26] The high-resolution custom tiling arrays covering the whole genome were used to examine the complete repertoire of genes expressed during embryogenesis up to the late gastrula stage, demonstrating that at least 11,000–12,000 genes, including most of those encoding transcription factors and cell-signaling molecules, as well as some classes of general cytoskeletal and metabolic proteins, are expressed during early embryogenesis.

Comparative analysis of the sea urchin genome has broad implication for the primitive state of deuterostome host defense and the genetic underpinnings of the immunity of vertebrates.[27] The sea urchin has an unprecedented complexity of innate immune recognition receptors relative to other animal species yet characterized. These receptor genes include a vast repertoire of 222 Toll-like receptors, a superfamily of more than 200 NACHT (NTPase) domain-leucine-rich repeat proteins (similar to vertebrate nucleotide-binding and oligomerization domain [NOD] and NALP (a family of receptors with NACHT domain, leucine-rich repeat domain [LRR], and a pyrin domain [PYP]) proteins), and a large family of scavenge receptor cysteine-rich proteins. More typical numbers of genes encode other immune recognition factors. Homologs of important immune and hematopoietic regulators, many of which have previously been identified only from chordates, as well as genes that are critical in adaptive immaturity of jawed vertebrates, also are present. These results provide an evolutionary outgroup for chordates and yield insights into the evolution of deuterostomes.

6.2.7 Genome of an Ascidian, *Ciona intestinalis*

Ascidians are a major group of urochordates or tunicates, which are one of the chordate groups together with cephalochordates and vertebrates. They attract researchers in the field of developmental biology because their developing tadpole larvae represent one of the most simplified body plans of chordates[5] (Figure 6.1). Ascidians are also of evolutionary biology interest as a reference to analyze the origin and evolution of vertebrates.[5] *Ciona intestinalis* is now one of the model animals for developmental genomics.[28]

The draft genome of *C. intestinalis* has been read basically by the whole-genome shotgun method and BAC-end sequencing,[29] followed by detailed mapping of scaffold onto chromosomes using fluorescence in situ hybridization (FISH) of selected BAC clones.[30] The 160-Mb *C. intestinalis* genome is composed of about 117 Mb of nonrepetitive and euchromatic sequence. Protein-coding gene prediction based on the assembled genome sequences and a collection of over 480,000 ESTs suggests that the genome contains a total of 15,852 protein-coding genes.[29] Additional cDNA information (670,000 ESTs and 6,700 cDNA sequences in total, which are extraordinarily large in number in comparison to its genome size) has been used to improve the quality of the gene model set (http://ghost.zool.kyoto-u.ac.jp).[31]

The *Ciona* genome was the first example of genome sequencing of a "wild" animal since the sequenced *Ciona* individual was caught directly from the sea. In addition, the *C. intestinalis* genome is notably AT rich (65%) compared with the human genome. A high level of allelic polymorphism was found in the single individual used for determination of the genome sequence by the whole-genome shotgun method, namely, with 1.2% of the nucleotides differing between alleles (nearly 15-fold higher than in humans). Although these features made it more difficult to assemble the genome sequence appropriately, a high level of allelic polymorphism is useful for identification of conserved sequences associated with gene expression control (discussed below).

Comparison of the *Ciona* genome with the genomes of invertebrates and vertebrates revealed that approximately 62% of the genes are shared with metazoans, while 16% are chordate specific (e.g., genes encoding components of connexin and retinoic acid-related molecules), and 18% are specific to ascidians (e.g., cellulose synthase gene). In addition, the genome comparison revealed genes that are conserved in other animals but appear to be missing in urochordates.[29] For example, the *Hox* genes, which have clustered organization and collinearity between gene order within the cluster and a sequential pattern of expression during development, are broken in this animal. The *Ciona* genome lacks *Hox* 7, 8, and 9 genes, and the *Hox* cluster is grouped into two different chromosomes. This tendency of a type of shrinkage of the genome is more conspicuous in another order of tunicates, Appendicularia; the *Oikopleura dioica* genome is very compact (about 60 Mb) and has lost the clustering of Hox genes.[32,33]

Along with the genome project of *C. intestinalis*, it should be worth mentioning the mapping of genomic information onto chromosomes because chromosomal-level genome information is fundamental in every aspect of biology. Most animals with genomes so far decoded have well-characterized genetic background or strains representative to the species. On the other hand, advances in genomic technologies, especially the method of whole-genome shotgun, make it possible to read the genome sequences of various animals without genetic background. Among invertebrates for which decoded genomes were discussed above, the sea urchin and ascidian are included in this category. Due to increasing interest in species that occupy critical positions in consideration of animal evolution, it is easily expected that, in the near future, various pivotal animals will be targeted for genome projects. This situation raises one important problem of chromosomal localization or mapping of genome information. The use of FISH with BAC clones provides a powerful tool to bridge draft genome information and its chromosomal localization, as shown in the *C. intestinalis* genome. *Ciona intestinalis* has 14 pairs of chromosomes. The small size of the chromosomes (most pairs measuring less than 2 μm) and morphological polymorphisms made it difficult to perform precise karyotyping based on morphology alone. To overcome this difficulty, each chromosome was characterized by two-color FISH with representative BAC clones. Using these BACs as references, two-color FISH of 170 BAC clones succeeded in mapping approximately 65% of the deduced 117-Mb nonrepetitive sequences onto chromosomes.[30]

6.3 OVERALL COMPARISON OF INVERTEBRATE GENOMES

Since the genetic information is encoded in the genome, comparative analysis among sequenced genomes of invertebrates is expected to provide insights into the biologically most important question of how every animal species evolved and what kind of genomic changes are responsible for the speciation.[34] In other words, without genome sequences, truly meaningful comparisons between two or more species are impossible. For example, as discussed, decoding of the honeybee *A. mellifera* genome and its comparison with those of other insects with solitary lifestyle was aimed to explain how the honeybee created its eusociety system by altering genomic information.[16] In addition, as also discussed, the comparison of sequenced genomes between closely related species (e.g., between *C. elegans* and *C. briggsae* and between *D. melanogaster* and *D. pseudoobscura*) might demonstrate the genomic alternation associated with speciation.

On the other hand, comparison of sequenced genomes among evolutionarily distant animal groups is predicted to provide insight into the overall evolutionary scenario of invertebrates, that is, of metazoan phyla. As will be discussed, the sequenced genomes have been well utilized in molecular phylogenetic analyses of animals. Figure 6.3 shows a comparison of numbers of orthologous genes among the bilaterians. This analysis indicates that the sea urchin has more orthologs with the ascidian than the insect and nematode, supporting the grouping of deuterostomes. However, at the moment a real answer to the question has not been obtained, mainly due to difficulties or gaps between genetic information and biological phenomena. In other words, comparative genomics of invertebrates is a rather important subject of future genomics integrated with other field of biological sciences, including genetics, cell and developmental biology, evolutionary biology, and ecology. It should be emphasized here that more experimental data to understand molecular mechanisms of biological phenomena are inevitably necessary for better understanding of animal evolution through the comparative genomics.

Here, it should be worth mentioning that a natural outcome of accumulation of multiple genome sequences is comparative genomics. However, one of the difficulties in comparative genomics remains in the disunity of assembly and strategies of gene prediction or annotation among the genome projects. Researchers who would like to analyze the multiple genomes must know what kinds of materials and strategies are used for obtaining the data.

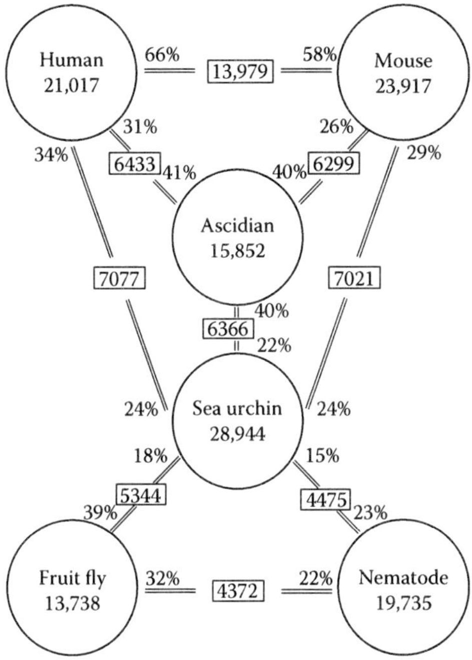

FIGURE 6.3 Orthologs among bilaterians. The number of 1:1 orthologs captured by BLAST alignments at a match value of $e = 1 \times 10^{-6}$ in comparisons of sequenced genomes among the bilaterian. The number of orthologs is indicated in the boxes along the arrows, and the total number of International Protein Index database sequences is shown under the species. (Modified from The Sea Urchin Genome Sequencing Consortium, *Science* **314**, 941–952, 2006.)

6.4 FUNDAMENTAL AND APPLIED PERSPECTIVE

The sequenced genomes of invertebrates have had vigorous impacts on every aspect of animal biology. Following are several examples of the fundamental and applied perspective of the sequenced invertebrate genomes.

6.4.1 Discovery of Novel Genes with Important Biological Function

The sequenced genomes together with cDNA and EST information provide a great opportunity to discover novel genes with yet-unknown function. One example is the discovery of a novel gene encoding voltage-sensor-containing phosphatase (VSP).[35] Usually, changes in membrane potential affect ion channels and transporter, which then alter intracellular chemical conditions. This gene was first found in *Ciona* (*Ci-VSP*) during the systematic genomic survey of ion channel genes using a comparative genomic approach. *Ci-VSP* encodes a protein that has a transmembrane voltage-sensing domain homologous to the S1–S4 segments of voltage-gated channels and a cytoplasmic domain similar to phosphatase and tensin homologs. Namely, this protein displays channel-like gating currents and directly translates changes in membrane potential into the turnover of phosphoinositides. Further characterization of the voltage-sensor domain (VSD) revealed that VSD is a voltage-gated proton channel.[36] Thus, the genome project and cDNA project have greatly helped the identification of novel genes with yet-unknown function, and such efforts may continue to find additional novel genes.

6.4.2 Contribution to Molecular Phylogenetic Analysis of Invertebrates

Molecules and sequenced genomes provide powerful tools to infer a phylogenetic relationship among living organisms. For example, molecular phylogenetic studies thus far have taught us that the unicellular animal most closely related to multicellular metazoans is the choanoflagellate,[2] and that protostomes are subgrouped into Ecdysozoa (e.g., nematodes and insects) and Lophotrochozoa (e.g., annelids and mollusks).[3] In addition, rare genomic changes also provide a good tool to infer phylogenetic relationships among invertebrates.[37]

A recent trend in this field is to analyze phylogenetic relationships using multiple, slowly evolving molecules, and only sequenced genomes provide information sufficient for these kinds of analyses. Delsuc et al.[38] examined the phylogenetic relationship among deuterostomes, using a phylogenetic data set of 146 nuclear genes (33,800 unambiguously aligned amino acids). Their result showed that tunicates (urochordates), not cephalochordates, are the closest living relatives of vertebrates. A following study with 35,000 homologous amino acids, including new data from a hemichordate (*Saccoglossus kowalevskii*) and *Xenoturbella* (a new phylum of deuterostomes) supported this view of earliest divergence of cephalochordates among chordate groups.[39]

To be expected, genomes of various animal groups that occupy a critical position among animal phylogeny will be sequenced in near future. This will provide a great opportunity to determine an in-depth scenario of animal evolution.

6.4.3 Polymorphism in Invertebrate Genomes and Conserved *cis*-Regulatory Sequences for Specific Gene Expression

As mentioned, the genomes of invertebrates, especially of wild-living animals such as sea urchins and tunicates, exhibit considerably high haplotype (or allelic) polymorphism. For example, sequence polymorphisms within individuals are remarkably 1.2% in *C. intestinalis* and 4.6% in *Ciona savignyi*, while the sea urchin *S. purpuratus* has about 4% haplotype polymorphism. Such a high grade of sequence polymorphism makes it troublesome to assemble genome sequences obtained by the whole-genome shotgun method into proper contigs and scaffolds, and thus the genome sequence of the sea urchin and ascidians are a mosaic combination of haplotype sequences. However, this type of polymorphism facilitates finding DNA sequences that are responsible for the regulation of spatiotemporal expression of genes, namely, noncoding DNA, which has regulatory functions that tend to be more highly conserved than other noncoding DNA, and sequence polymorphisms within individuals facilitate such studies to find conserved elements. For example, intraspecies sequence comparisons of individuals from different populations have been shown to be useful in finding conserved *cis*-regulatory sequences required for the specific expression of developmentally regulated genes.[40]

More frequently, a comparison is now carried out interspecifically. For example, comparison of *C. intestinalis* and *C. savignyi* genes and their 5′ upstream noncoding region clearly demonstrates the low level of conservation of noncoding versus coding regions and a higher level of noncoding conservation over the first 800 bp of 5′ flanking DNA. A direct test of this 5′ conserved region indicates that it contains an enhancer that recapitulates native expression. These methods have been used to identify a variety of tissue-specific enhancers in *Ciona*,[41] and a similar strategy of finding conserved *cis*-regulatory sequences has been applied to various invertebrates, including sea urchins. In sea urchins, sequences 5′ upstream of genes in *S. purpuratus* were compared with that of another species, *Lytechinus variegatus,* to find elements responsible for precise gene expression.[23]

6.4.4 Genome-wide Gene Regulatory Networks for Construction of Invertebrate Body Plans

One of the most spectacular phenomena in biology is the emergence of diverse animal shapes through embryogenesis, with each species specific and adapted over a long evolutionary history. The cellular and molecular mechanisms underlying this phenomenon have long been a hot topic of biological studies. Since 1980, there has been remarkable progress in identifying the regulatory genes and signaling pathways responsible for the development of a variety of tissues and organs in worms, flies, sea urchins, ascidians, and vertebrates. The best success has been obtained for the specification of endomesoderm in the pregastrular *S. purpuratus* embryo,[22] the dorsal-ventral patterning of the early *D. melanogaster* embryo,[42] the construction of the basic chordate body plan during early embryogenesis of *C. intestinalis*,[43,44] and the organization of the three germ layers of amphibians.[45]

Programs of animal development are encoded in the genome, and every gene is spatiotemporally regulated by this program. This program can be represented by gene regulatory networks, which constitute wiring diagrams of transcription factors and signaling molecules. Thus, animal evolutions would be best understood by comparisons of these networks rather than just comparisons of the genomes themselves. For this purpose, the gene regulatory networks must be analyzed genome-wide since animal development proceeds with the coordinated expression of all genes encoded in the genome. The *C. intestinalis* gene regulatory networks might be a good example to discuss. Taking advantage of both genomic DNA and cDNA/EST information, genes encoding transcription factors in the *Ciona* genome were intensively and comprehensively annotated, showing a total of 669 genes. Basically all transcription factor genes as well as all major signaling ligand and receptor genes were examined for their expression during embryogenesis to form tadpole-type larvae. As a result, it become evident that 76 regulatory genes are zygotically expressed in early embryos, at the time when naïve blastomeres are determined to follow specific cell fates. Systematic gene disruption assays provided more than 3,000 combinations of gene expression profiles responsible for constitution of a blueprint for the *Ciona* embryo, providing a foundation for understanding the evolutionary origins of the chordate body plan.[44] Although comparisons of the *Ciona* networks with those of other animals have not yet revealed significant conservations or divergences, this important question might be answered after networks in each species become known more precisely and comprehensively.

6.5 CONCLUSION AND PERSPECTIVE

An organism's genome contains all of its genetic information, and thus sequenced genomes provide the basis for the entire field of biological sciences. As shown in Figure 6.1, invertebrate groups subjected to genome sequencing to date are limited to nematodes, insects, a sea urchin, and an ascidian. As discussed in this review, each has a distinct reason why its genome should have been deciphered. Together with advances in the technologies in genomics, especially whole-genome shotgun sequencing and computational assembly methods, it is desired that the genomes of more invertebrates will be decoded in near future. For example, comparison of sequencing the genome of a unicellular choanoflagellate, *Monosiga* species,[46] and that of a sponge will provide insights into genomic changes responsible for multicellularity or molecular mechanisms involved in the origin of metazoans. The sequencing genome of a cnidarian sea anemone, *Nematostella vectensis*, might suggest genetic features of diploblast metazoans. In addition, the genome of a planarian, *Dugesia japonica*, and of some lophotrochozoans should be decoded at least in relation to the evolution of protostomes. Furthermore, the genome of a hemichordate, *Saccoglossus kowalevskii*, could provide clues about the determinates of deuterostomy, and that of a cephalochordate amphioxus, *Branchiostoma floridae*, will give further insight into the origin and evolution of chordates. The genome projects of these invertebrate groups are now under way, and we will be able to compare the sequenced genome in the near future.

The period of decoding of genomes coincided with the great advances in genomic technologies that have revolutionized our ability to study transcription, protein binding to specific DNA sequences, and genome variation at the molecular level. Especially, microarrays might open a new arena of genomic studies. Microarrays are used for expression profiling, targeted either to all known or predicted coding regions or against a whole-genome tiling path of high resolution. We can now map the binding sites of chromatin-associated proteins to the genome at high resolution using either DamID[47] or chromatin immunoprecipitation (ChIP).[48] Together with computational prediction, we can also conduct genome-scale surveys for polymorphisms using high-throughput polymerase chain reaction (PCR) strategies and effectively resequence other genomes of the same species using tiling paths of oligonucleotides. Taking advantage of characteristic features of each of the sequenced genomes, future studies of genomics will give us more fundamental and profound understanding of animal development, behavior, and evolution.

REFERENCES

1. Brusca, R. C. & Brusca, G. J. *Invertebrates* (Sinauer, Sunderland, MA, 2003).
2. Wainright, P. O., Hinkle, G., Sogin, M. L. & Stickel, S. K. Monophyletic origins of the metazoa: an evolutionary link with fungi. *Science* **260**, 340–342 (1993).
3. Aguinaldo, A. M. et al. Evidence for a clade of nematodes, arthropods and other moulting animals. *Nature* **387**, 489–493 (1997).

4. Delsuc, F., Brinkmann, H. & Philippe, H. Phylogenomics and the reconstruction of the tree of life. *Nat. Rev. Genet.* **6**, 361–375 (2005).
5. Satoh, N. The ascidian tadpole larva: comparative molecular development and genomics. *Nat. Rev. Genet.* **4**, 285–295 (2003).
6. Holland, P. W. H., Garcia-Fernàndez, J., Williams, N. A. & Sidow, A. Gene duplications and the origins of vertebrate development. *Development* **Suppl.**, 125–133 (1994).
7. Dehal, P. & Boore, J. L. Two rounds of whole genome duplication in the ancestral vertebrate. *PLoS Biol.* **3**, e314 (2005).
8. The *C. elegans* Sequencing Consortium. Genome sequence of the nematode *C. elegans*: A platform for investigating biology. *Science* **282**, 2012–2018 (1998).
9. Adams, M. D., Celniker, S. E., Holt, R. A. et al. The genome sequence of *Drosophila melanogaster. Science* **287**, 2185–2195 (2000).
10. Hillier, L. W. et al. Genomics in *C. elegans*: so many genes, such a little worm. *Genome Res.* **15**, 1651–1660 (2005).
11. Lim, L. P. et al. The microRNAs of *Caenorhabditis elegans. Genes Dev.* **17**, 991–1008 (2003).
12. Bargmann, C. I. Neurobiology of the *Caenorhabditis elegans* genome. *Science* **282**, 2028–2033 (1998).
13. Stein, L. D. et al. The genome sequence of *Caenorhabditis briggsae*: a platform for comparative genomics. *PLoS Biol.* **1**, 166–192 (2003).
14. Chen, N. et al. Identification of a nematode chemosensory gene family. *Proc. Natl. Acad. Sci. U. S. A.* **102**, 146–151 (2005).
15. Thomas, J. H., Kelley, J. L., Robertson, H. M., Ly, K. & Swanson, W. J. Adaptive evolution in the SRZ chemoreceptor families of *Caenorhabditis elegans* and *Caenorhabditis briggsae. Proc. Natl. Acad. Sci. U. S. A.* **102**, 4476–4481 (2005).
16. Ashburner, M. & Bergman, C. M. *Drosophila melanogaster*: a case study of a model genomic sequence and its consequences. *Genome Res.* **15**, 1661–1667 (2005).
17. Richards, S. et al. Comparative genome sequencing of *Drosophila pseudoobscura*: chromosomal, gene, and *cis*-element evolution. *Genome Res.* **15**, 1–18 (2005).
18. Holt, R. A. et al. The genome sequence of the malaria mosquito *Anopheles gambiae. Science* **298**, 129–149 (2002).
19. Gardner, M. J. et al. Genome sequence of the human malaria parasite *Plasmodium falciparum. Nature* **419**, 498–511 (2002).
20. Xia, Q. et al. A draft sequence for the genome of the domesticated silkworm (*Bombyx mori*). *Science* **306**, 1937–1940 (2004).
21. The Honeybee Genome Sequencing Consortium. Insights into social insects from the genome of the honeybee *Apis mellifera. Nature* **443**, 931–949 (2006).
22. Davidson, E. H. et al. A genomic regulatory network for development. *Science* **295**, 1669–1678 (2002).
23. Davidson, E. H. *The regulatory Genome: Gene Regulatory Networks in Development and Evolution* (Academic Press, New York, 2006).
24. Samanta, M. P. et al. The transcriptome of the sea urchin embryo. *Science* **314**, 960–962 (2006).
25. Sea Urchin Genome Sequencing Consortium. The genome of the sea urchin *Strongylocentrotus purpuratus. Science* **314**, 941–952 (2006).
26. Howard-Ashby, M. et al. Gene families encoding transcription factors expressed in early development of *Strongylocentrotus purpuratus. Dev. Biol.* **300**, 90–107 (2006).
27. Rast, J. P., Smith, L. C., Loza-Coll, M., Hibino, T. & Litman, G. W. Genomic insights into the immune system of the sea urchin. *Science* **314**, 952–956 (2006).
28. Satoh, N., Satou, Y., Davidson, B. & Levine, M. *Ciona intestinalis*: an emerging model for whole-genome analyses. *Trends Genet.* **19**, 376–381 (2003).

29. Dehal, P. et al. The draft genome of *Ciona intestinalis*: insights into chordate and vertebrate origins. *Science* **298**, 2157–2167 (2002).
30. Shoguchi, E. et al. Chromosomal mapping of 170 BAC clones in the ascidian *Ciona intestinalis*. *Genome Res.* **16**, 297–303 (2006).
31. Satou, Y., Kawashima, T., Shoguchi, E., Nakayama, A. & Satoh, N. An integrated database of the ascidian, *Ciona intestinalis*: towards functional genomics. *Zool. Sci.* **22**, 837–843 (2005).
32. Seo, H.-C. et al. Miniature genome in the marine chordate *Oikopleura dioica*. *Science* **294**, 2506 (2001).
33. Seo, H.-C. et al. Hox cluster disintegration with persistent anteroposterior order of expression in *Oikopleura dioica*. *Nature* **431**, 67–71 (2004).
34. Ureta-Vidal, A., Ettwiller, L. & Birney, E. Comparative genomics: genome-wide analysis in metazoan eukaryotes. *Nat. Rev. Genet.* **4**, 251–262 (2003).
35. Murata, Y., Iwasaki, H., Sasaki, M., Inaba, K. & Okamura, Y. Phosphoinositide phosphatase activity coupled to an intrinsic voltage sensor. *Nature* **435**, 1239–1243 (2005).
36. Sasaki, M., Takagi, M. & Okamura, Y. A voltage sensor-domain protein is a voltage-gated proton channel. *Science* **312**, 589–592 (2006).
37. Rokas, A. & Holland, P. W. H. Rare genomic changes as a tool for phylogenetics. *Trends Ecol. Evol.* **15**, 454–459 (2000).
38. Delsuc, F., Brinkmann, H., Chourrout, D. & Philippe, H. Tunicates and not cephalochordates are the closest living relatives of vertebrates. *Nature* **439**, 965–968 (2006).
39. Bourlat, S. J. et al. Deuterostome phylogeny reveals monophyletic chordates and the new phylum Xenoturbellida. *Nature* **444**, 85–88 (2006).
40. Boffelli, D. et al. Intraspecies sequence comparisons for annotating genomes. *Genome Res.* **14**, 2406–2411 (2004).
41. Johnson, D. S. et al. De novo discovery of a tissue-specific gene regulatory module in a chordate. *Genome Res.* **15**, 1315–1324 (2005).
42. Stathopoulos, A. & Levine, M. Genomic regulatory networks and animal development. *Dev. Cell* **9**, 449–462 (2005).
43. Imai, K. S., Hino, K., Yagi, K., Satoh, N. & Satou, Y. Gene expression profiles of transcription factors and signaling molecules in the ascidian embryo: towards a comprehensive understanding of gene networks. *Development* **131**, 4047–4058 (2004).
44. Imai, K. S., Levine, M., Satoh, N. & Satou, Y. Regulatory blueprint for a chordate embryo. *Science* **312**, 1183–1187 (2006).
45. Loose, M. & Patient, R. A genetic regulatory network for *Xenopus* mesendoderm formation. *Dev. Biol.* **271**, 467–478 (2004).
46. King, N. & Carroll, S. B. A receptor tyrosine kinase from choanoflagellates: molecular insights into early animal evolution. *Proc. Natl. Acad. Sci. U. S. A.* **98**, 15032–15037 (2001).
47. Orian, A. Chromatin profiling, DamID and the emerging landscape of gene expression. *Curr. Opin. Genet. Dev.* **16**, 157–164 (2006).
48. Bulyk, M. L. DNA microarray technologies for measuring protein–DNA interactions. *Curr. Opin. Biotechnol.* **17**, 422–430 (2006).
49. Carroll, S. B., Grenier, J. K. & Weatherbee, S. D. *From DNA to Diversity. Molecular Genetics and the Evolution of Animal Design* (Blackwell Science, Malden, MA, 2001).

7 Comparative Vertebrate Genomics

James W. Thomas

CONTENTS

ABSTRACT

With the application of whole-genome sequencing to an increasing number of vertebrates, comparative genomics has become an integral component of vertebrate genome analysis. In particular, comparative vertebrate genomics provides a unique and powerful perspective on how genomes are organized, what portions of the genome are functional, and what makes each species genetically distinct. This chapter provides an overview of the resources and fundamental principles of contemporary vertebrate genomics.

7.1 INTRODUCTION

Comparative genomics is a burgeoning field that leverages interspecies comparisons to gain insights into the function and evolution of the human and other vertebrate genomes. Spurred on by the advances in large-scale DNA sequencing technology, comparative genomic sequence analysis has become an integral and invaluable tool for elucidating the history and function of vertebrate genomes. This chapter is designed to provide a broad overview of the resources and fundamental principles that are the basis for contemporary studies in comparative vertebrate genomics.

7.2 VERTEBRATE PHYLOGENY AND GENOME SEQUENCING

The origin of all modern-day vertebrates dates back to 500–600 million years ago (MYA).[1] At present, there are an estimated approximately 50,000 species of vertebrates, which can be classified into four major groups (clades): jawless fishes, which include hagfish and lampreys; cartilaginous fishes, which include sharks and rays; bony fishes, which include all other fishes; and tetrapods, which include amphibians, birds, reptiles, and mammals.[2] From the point of view of humans, we share the closest evolutionary relationship with the chimpanzee, from which we diverged from a common ancestor about 5 MYA.[3] Our most distant evolutionary relationship within vertebrates is to the jawless fishes, with whom our most recent common ancestor dates back more than 500 MYA.[1]

In part due to sustained increases in worldwide DNA sequencing capacity initiated by the Human Genome Project, as well as the now-accepted power of comparative sequence analysis to interpret the sequence of the human genome, an ever-expanding set of vertebrates has been targeted for some level of whole-genome sequencing (Figure 7.1). As of June 2006, there were 50 vertebrates selected for whole-genome sequencing. Within this select group of species is a deep sampling of mammals (n = 40) and a limited sampling of other tetrapods (n = 4), bony fishes (n = 5), and jawless fishes (n = 1). The heavy bias toward mammalian genomes represents efforts to maximize the power of interspecies comparisons to identify putative functional elements in the human genome.[4,5] Indeed, most mammals targeted for whole-genome sequencing, such as the elephant, that are not experimental model systems have been selected primarily for the purpose of annotating the human genome. As a result, these genomes will only be whole-genome shotgun sequenced to a depth of about 2.5-fold coverage. Therefore, while providing a valuable comparison for annotating the human genome and an extensive sequence-based survey of these genomes, these efforts will not yield the type of stand-alone and high-quality assemblies associated with the human genome.[6,7]

Following the first publications describing the sequence of the human genome,[6,7] a series of articles describing several other vertebrate genome sequences, including fugu (marine puffer fish), mouse, rat, chicken, tetraodon (freshwater puffer fish), dog, and chimpanzee, have been published,[8–14] with many more expected in the future. In addition to published genomes, a hallmark of genomic sequencing projects has been the rapid release of data to the public prior to publication. Thus, for nearly all genome projects, even before a genome is assembled, the public at large has nearly immediate access to the trace and quality files of individual sequencing reads through the Trace Repository at the National Center for Biotechnology Information (NCBI) (http://www.ncbi.nlm.nih.gov/Traces/trace.cgi). Subsequently, the assembled and annotated sequences can be accessed via genome browsers, such as the University of California, Santa Cruz, Genome Browser (http://www.genome.ucsc.edu) and Ensembl (http://www.ensembl.org). The cumulative efforts to date to generate and analyze vertebrate genome sequences have provided the basis for unbiased and comprehensive genome-wide comparisons, which in turn are yielding highly detailed and accurate descriptions of the similarities and differences between

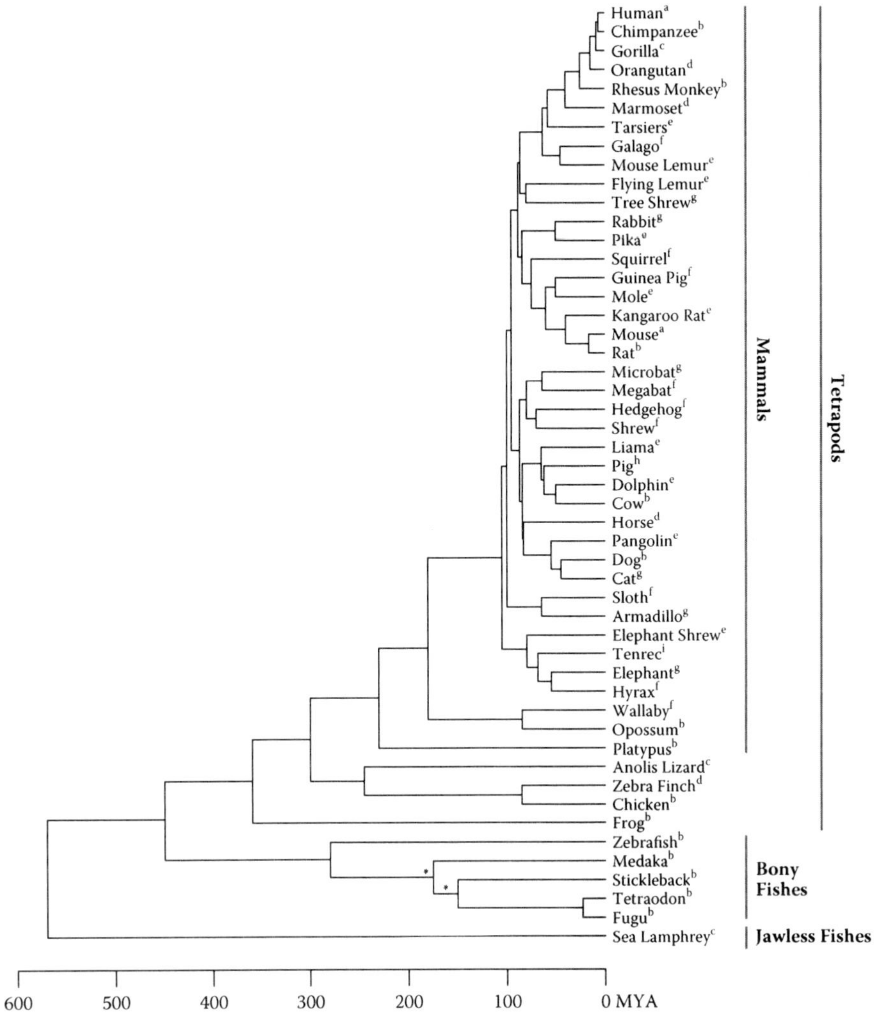

FIGURE 7.1 Phylogeny of the 50 vertebrates targeted for whole-genome sequencing. The evolutionary relationships and divergence times illustrated here were compiled from the literature.[1,87–99] Genome sequencing project status as of October 2006: [a]finished genome; [b]>5x whole-genome shotgun (WGS) assembly; [c]~6x WGS approved or in process; [d]~6x WGS complete; [e]~2x WGS approved or in process; [f]~2x WGS complete; [g]~2x WGS assembly complete and ~6x WGS approved or in process; [h]BAC-based sequencing and WGS; [i]~2x WGS assembly. MYA, million years ago; *, uncertain divergence time.

vertebrate genomes (see Sections 7.4 and 7.5). As more genomes are sequenced, it can be expected that our understanding of genomes, the functions encoded within them, and how they evolved will become both clearer and more complex.

7.3 VERTEBRATE BAC LIBRARIES: A RESOURCE FOR FUNCTIONAL GENOMICS

Prior to the ability to perform whole-genome shotgun sequencing and assemblies on large and complex genomes, a physical map based on genomic clones was a necessary template for sequencing a vertebrate genome.[15] Bacterial artificial chromosomes (BACs), which have proven to be highly stable and amenable to high-throughput mapping, emerged as the preferred large-insert genomic libraries of choice.[16] The typical vertebrate BAC library is comprised of clones with an average insert size of 100–200 kb, which in total represent about 10-fold redundancy of the target genome, and can be readily screened by hybridization-based methods.[16] At present, BAC libraries are available for a diverse collection of 91 vertebrates (Table 7.1).

Although clone-end read pairs from a combination of unmapped and randomly selected small-insert plasmids (~3–10 kb) and fosmids (~40 kb) are the primary substrates for most current whole-genome sequencing efforts, BAC libraries still have several key applications that complement and enhance whole-genome shotgun sequencing. At the whole-genome level, methods for generating BAC-based physical maps consisting of ordered and overlapping clones by restriction-enzyme fingerprint analysis of entire BAC libraries have been developed.[17,18] These BAC-based physical maps can be used to select a minimal tiling path of clones for sequencing,[19] and in conjunction with BAC-end sequencing, can be utilized to improve whole-genome shotgun assemblies[20] and to select clones from targeted regions of the genome for high-quality finished sequencing. Mapping BAC-end sequences onto whole-genome assemblies, which are commonly displayed in the genome browsers, also allows individual investigators a means to rapidly access genomic clones for their gene or region of interest without screening the library themselves. BAC clones are also the preferred probe substrate for fluorescence in situ hybridization (FISH) and therefore provide an important means by which a position in a whole-genome assembly can be translated to its corresponding physical location on a chromosome.[21]

Independent of whole-genome sequencing efforts, BAC libraries also provide the necessary reagents for targeted comparative mapping and sequencing of genes or regions of interest from multiple species,[22] and efficient methodologies and resources for the parallel construction of targeted BAC-based physical maps from diverse sets of vertebrate genomic libraries have been developed to support such projects.[23,24] BAC-based mapping and sequencing can therefore provide high-quality sequence in a greater diversity of species across targeted regions of the genome than can whole-genome shotgun sequencing. For example, this strategy is being used to generate comparative sequence data sets for projects such as ENCODE (Encyclopedia of DNA Elements),[25] the goal of which is to annotate all the functional elements in the human genome.

Finally, BAC clones represent an invaluable functional genomic resource. Because of their size, stability, and general availability, BAC clones are commonly used to make transgenic mice.[26] To support and broaden the application of BAC clones in transgenics and other functional assays, methods have been devised for engineering specific sequence modifications into BAC clones.[27] Such methods have greatly enhanced the capabilities for using BACs as experimental templates for

TABLE 7.1
Vertebrate BAC Libraries

Primates	Other Placental Mammals	Marsupials and Monotremes	Birds and Reptiles	Amphibians	Bony Fishes	Cartilaginous Fishes	Jawless Fishes
Baboon[a]	Armadillo[a]	Bandicoot[b]	Alligator[e]	*Xenopus laevis*[a]	Antarctic icefish[a]	Clearnose skate[f]	Hagfish[f]
Black lemur[a]	Cat[a]	Echidna[d]	California condor[a]	*Xenopus tropicalis*[a]	Antarctic toothfish[a]	Horn shark[f]	Sea lamprey[f]
Chimpanzee[a]	Chinese hamster[a]	Opossum (North American)[a]	Chicken[a]		Atlantic salmon[a]	Little skate[c]	
Colobus monkey[a]	Chinese muntjac[a]	Opossum (South American)[a]	Emu[e]		Bichir[f]	Nurse shark[d]	
Dusky titi[a]	Clouded leopard[a]	Platypus[c]	Garter snake[e]		Blind cavefish[g]	Spiny dogfish shark[c]	
Galago[a]	Cow[a]	Wallaby[d]	Gila monster[e]		Channel catfish[a]		
Gibbon[a]	Deer mouse[a]		Painted turtle[b]		Chinook salmon[a]		
Gorilla[a]	Dog[a]		Side-blotched lizard[b]		Coelecanth[f]		
Human[a]	Elephant[a]		Tuatara[b]		Fugu[h]		
Japanese macaque[a]	Ferret[a]		Turkey[a]		Haplochromine cichlid[f]		
Marmoset[a]	Guinea pig[a]		Zebra finch[c, d]		Lake Melawi zebra[g]		
Mouse lemur[a]	Hedgehog[a]				Medaka[i]		
Orangutan[a]	Horse[a]				Paddlefish[f]		
Owl monkey[a]	Horseshoe bat[a]				Platyfish[a]		

(*Continued*)

TABLE 7.1
Vertebrate BAC Libraries (Continued)

Primates	Other Placental Mammals	Marsupials and Monotremes	Birds and Reptiles	Amphibians	Bony Fishes	Cartilaginous Fishes	Jawless Fishes
Rhesus monkey[a]	Indian muntjac[a]				Rainbow trout[g]		
Ring-tailed lemur[a]	Little brown bat[a]				Southern puffer[f]		
Squirrel monkey[a]	Mouse[a]				Stickleback[a,f]		
Vervet monkey[a]	Mule deer[b]				Swordtail fish[a]		
	Pig[a]				Tetraodon[j]		
	Rabbit[a]				Tilapia[g]		
	Rat[a]				Yellowbelly rockcod[a]		
	Sheep[a]				Zebrafish[a]		
	Shrew[c]						
	Squirrel[a]						
	Tenrec[a]						

[a] BACPAC Resources (http://bacpac.chori.org/).
[b] Amplicon Express (http://www.genomex.com/).
[c] Clemson University Genomics Institute (https://www.genome.clemson.edu/).
[d] Arizona Genomics Institute (http://www.genome.arizona.edu/).
[e] Genome Project Solutions (http://genomeprojectsolutions.com).
[f] BRI (http://benaroyaresearch.org/investigators/amemiya_chris/).
[g] Hubbard Center for Genome Studies (http://hcgs.unh.edu/).
[h] Geneservice (http://www.geneservice.co.uk/home/).
[i] RZPD (http://www.rzpd.de/).
[j] Genoscope (http://www.cns.fr/externe/English/Projets/Projet_C/getDNA.html).

accurately assessing the effect of specific disease-causing mutations[28] or for identification and characterization of regulatory elements that specify when and where genes are expressed.[29] BAC clones also hold exceptional promise for the functional dissection of variation within a species. Specifically, as BAC clones represent a contiguous segment of DNA from a single chromosome, BACs can be used as templates to functionally compare alleles or haplotypes. In an analogous manner, BACs can also be used to directly compare the function of orthologous genes between species, which will be critical for experimentally interrogating and validating candidate genetic differences that underlie species-specific traits. Therefore, the application of BAC clones in experimental paradigms promises to be one avenue by which we can extend our description of genomes beyond mere DNA sequence.

7.4 VERTEBRATE GENOME EVOLUTION

Whole-genome comparisons between multiple species are increasing our knowledge of how genomes differ from one another, the mechanisms by which these differences have evolved, and the general rates at which small- and large-scale genomic changes occur. In this section, attention is focused on three fundamental properties of vertebrate genomes and how they are compared: (1) genome size, (2) gene content and structure, and (3) genome organization and comparative mapping.

7.4.1 Genome Size

Estimates and comparisons of genome size are some of the oldest and simplest methods in comparative genomics. More than 50 years ago, it was noted that the total amount of DNA within a genome varied considerably across species.[30] The observation that genomes of some primitive species, such as some fish and amphibians, were larger than the human genome presented a contradiction to the accepted theory that more complex species would have the most genes and thus the largest genomes.[31] This lack of correlation between species complexity and genome size was labeled the C-value paradox.[31] The discovery that in many genomes, including vertebrates, the vast majority of DNA did not code for proteins largely resolved the C-value paradox. However, the functional consequences and mechanisms by which the observed differences in genome size across species arose are still the subject of debate.[32–39]

Within vertebrates, genome size varies 300-fold, with the genomes of the puffer fish representing the smallest vertebrate genomes at approximately 0.4 Gb, while the largest genomes, such as the genome of the lungfish, can be upward of 120 Gb.[40,41] The human genome is on the order of approximately 2.88 Gb (not including heterochromatic regions) and is one of the largest vertebrate genomes sequenced to date (Table 7.2). Differences in genome size between vertebrates can be the result of polyploidy.[42] However, the gain and loss of DNA by insertions and deletions is likely to be more important in the divergence of vertebrate genome size.[43] Moreover, insertions and deletions are the primary molecular basis for sequence divergence between vertebrate genomes[10,14,22,44] and perhaps may even account for the majority of nucleotides that differ between humans.[45] For example, interspersed repetitive

TABLE 7.2
Vertebrate Genome Organization and Content[a]

	Human	Mouse	Chicken	*X. tropicalis*	Zebrafish	Tetraodon
Karyotype	2n = 46	2n = 40	2n = 78	2n = 20	2n = 50	2n = 42
Genome size (Gb)	2.88	2.57	1.05	1.36	1.63	0.34
Repetitive element content	48.8%	42.4%	9.9%	19.6%	48.1%	3.0%
Gene number[b]	23,732	24,438	18,632	18,473	21,503	28,005

[a] Statistics are based on whole-genome sequence assemblies: human (hg18), mouse (mm8), chicken (galGal2), *X. tropicalis* (xenTro2), zebrafish (Zv6), and tetraodon (tetNig1).

[b] Gene number refers to Ensembl (v39) annotated protein-coding genes, with the exception of tetraodon, which refers to annotation from Genoscope.[12]

element content, which reflects the portion of the genome derived from transposable element insertions, makes up anywhere from 3% to 50% of sequenced vertebrate genomes (Table 7.2).[6,12] Although high interspersed repetitive element content is generally correlated with large genome size, it should be noted that repetitive element content is not the sole factor that determines genome size. Consider the zebrafish genome, which at 1.63 Gb is much smaller than the human genome. Nonetheless, with approximately a 50% repetitive element content, the zebrafish genome is just as cluttered with repeats as our own genome (Table 7.2). It has been argued that population dynamics alone could lead to the variation in genome size among vertebrates, and that a "simple model incorporating random genetic drift and weak mutation pressure against intron-containing alleles"[46, p. 6118] is consistent with the evolution of intron number and size.[39,46] However, this theory has not been universally accepted,[47] and hypotheses related to nonneutral processes continue to be put forward to explain why some genomes are large and others relatively small.[37,38] Thus, while whole-genome sequencing efforts have provided a more precise picture of the size and composition of vertebrate genomes, fundamental questions regarding the importance and origin of genome size differences remain unanswered.

7.4.2 Gene Content and Structure

One of the most important outcomes of whole-genome sequencing projects is the accurate identification and annotation of genes. Although conceptually straightforward, truly complete and accurate gene annotation is an ongoing challenge, even in the completed human genome.[48] Current estimates of the number of genes within the human genome indicate that we have about 20,000–25,000 protein-coding genes (Table 7.2).[48] The number of protein-coding genes in other sequenced vertebrate genomes is estimated to range from 18,000[11] to about 28,000[12] (Table 7.2). The variability in estimated number of protein-coding genes between vertebrates in large part reflects the differences in the quality of whole-genome assemblies and availability of complementary DNAs (cDNA) and expressed sequence tags (ESTs) for a given species, both of which will strongly influence how accurate gene annotation is for a given genome. That said, it is likely that the estimate of about 20,000–25,000 protein-coding genes for the human genome will hold true for the typical vertebrate genome.

Although vertebrate genomes contain a similar number of protein-coding genes, comparisons of genes between species have made it apparent that no two genomes encode exactly the same set of genes. One factor that shaped the gene content of all vertebrate genomes was large-scale duplication(s) prior to the most recent common ancestor of vertebrates more than about 500 MYA, which likely included at least one and perhaps two whole-genome duplications.[49–51] Subsequently, an additional whole-genome duplication specific to the ray-finned fish lineage is hypothesized to have occurred approximately 300 MYA.[12,52,53] Despite the relatively recent whole-genome duplication in fish, estimates of the gene number within extant fishes are similar to those of other vertebrates, suggesting that massive gene loss must have occurred in ray-finned fish since that event. On a more recent timescale, it has been shown that segmental duplications have played a significant role in creating new genes in the human genome.[54]

The cumulative effect of the continuous gain and loss of genes is highlighted in a detailed comparison of gene content between humans and mice.[55] In this report, the authors found that while a mouse homolog could be identified for 90% of all human genes, only 65% of human genes have a simple 1:1 orthologous relationship with a single mouse gene, and for nearly 10% of all human genes there is no identifiable homolog in mouse. Differences in gene content have been hypothesized to be the underlying cause of biological differences between species,[56,57] and a number of genes that were recently lost in human evolution have been proposed as key genetic differences that distinguish us from chimpanzees and other apes.[58] For example, loss of the *MYH16* gene has been hypothesized to have been a key event facilitating the expansion of the human skull and brain size.[59] Thus, the gene content of vertebrate genomes is constantly being modified by the process of evolution.

At the level of individual genes, intron–exon structure tends to be highly conserved across vertebrates. For example, a large-scale comparison of human and mouse orthologs found that 92% of the orthologous gene pairs had identical intron–exon structures.[60] More specifically, it was shown that 98% of all constitutively spliced exons were conserved between humans and mice.[61] Such conservation of gene structure has been observed over even greater evolutionary distances as well, with few changes in gene structure observed among 12 diverse vertebrates, including mammals, chicken, and fish (see Thomas et al.[22]; personal observations, unpublished). Thus, since the average human gene is estimated to contain about 10 exons,[48] it is likely that the average vertebrate gene contains about 10 exons as well. There are, however, notable exceptions to the conservation of intron–exon structure. Only 28% of alternatively spliced exons present in minor-frequency transcripts were found to be conserved between humans and mice,[61] suggesting many of these exons have been either gained or lost since the most recent common ancestor between these species. In fact, it has been estimated that new exons were created in the mouse lineage at a minimum rate of about 81.3 exons/million years, and that most of the new mouse exons were derived from the exonization of unique intronic sequence.[62]

Interspersed repeats derived from transposable elements have also been a source for the evolution of intron–exon structure. For example, in one survey approximately 5% of alternatively spliced exons in the human genome contained sequence similar to *Alu* elements, which are a class of short interspersed nuclear elements (SINEs).[63] In addition, the use of a polyA signal and long terminal repeat (LTR) promoter encoded within the L1 class of long interspersed nuclear elements (LINEs) embedded within an intron has been shown to lead to "gene breakage."[64] Specifically, a novel 3′ truncated transcript can be generated by splicing in a L1 polyA signal, and a novel 5′ truncated transcript can be generated by initiation from the L1 LTR promoter that then includes the downstream exons of the preexisting gene.[64]

Finally, genes and their intron-exon structures can have complex and unexpected origins. In the case of the non-protein-coding gene *XIST*, which is critical for the initiation of X inactivation in placental mammals, it was hypothesized that pseudogenization of a protein-coding gene and subsequent recruitment of some of the degraded exons was at least in part responsible for the genesis and current intron-exon structure of this gene.[65] Thus, while the intron–exon structure of individual genes is a highly conserved feature among vertebrate genomes, as with genome size and gene content, it also can be the substrate for evolutionary innovations.

7.4.3 Genome Organization and Comparative Mapping

As was true of genome size, differences in the physical organization of vertebrate genomes at the level of chromosome number and size have been apparent for some time. In particular, the comparison of karyotypes across vertebrates has revealed a remarkable degree of variation in the number and size of chromosomes that are associated with an individual species (Table 7.2). Such differences reflect the cumulative effect of chromosomal rearrangements, such as chromosome fissions and fusions, translocations, inversions, and transpositions, that have occurred and, over time, become fixed in a given population. In fact, there appears to be substantial flexibility in terms of both the number and the size of chromosomes in a vertebrate karyotype. For example, the chicken and other bird genomes have more than 40 chromosomes, some of which are greater than 180 Mb, while others are less than 1 Mb.[11] Although in most instances visible changes in karyotypes between species accumulate at a relatively slow rate, there is precedent for rapid changes in genome organization. In the case of the Indian and Chinese muntjacs, although these species diverged less than 2 MYA and can produce viable hybrid offspring, their karyotypes are remarkably different, $2n = 6$ (or 7 in males) for the Indian versus $2n = 46$ for the Chinese muntjac.[66] This example also highlights the fact that while a pair of species may have very distinct karyotypes, the underlying genetic information encoded within their genomes can be quite similar.

As mentioned in Section 7.4.2, the majority of genes within any given vertebrate genome have orthologs or homologs in other vertebrate genomes. Thus, it is possible to compare the organization of vertebrate genomes on a gene-by-gene basis by comparing the physical linkage and order of orthologous genes between species. Such comparisons have a long history dating back to the early 20th century, when it was noted that two coat color mutations were linked in both mice and rats.[67] By the end of the 20th century, extensive comparative mapping between species, especially human and mouse, revealed that the rate at which gene linkage and order changed was slow enough such that large chromosomal segments that covered the majority of the genome could be identified in which gene content or gene linkage had been conserved between species.[68] The establishment of comparative maps between species therefore provides invaluable templates for (1) leveraging a highly detailed genome sequence or genetic maps from one species to predict the gene content or order along the chromosome in another species with a sparsely mapped genome and (2) reconstructing the series of chromosomal rearrangements that have led to the differences in genome organization between vertebrates.[69]

With the release of whole-genome sequence assemblies from a number of vertebrates, genome comparisons can now be done by genomic sequence alignments. Such detailed comparisons are providing a high-resolution picture of the extent of similarities and differences in genome organization between species that has both reinforced and modified the pre–genome sequencing era view of genome evolution. In particular, the concept that genomes can be subdivided into a finite number of relatively large blocks of chromosomal segments with conserved gene content or gene order clearly has held true. For example, comparisons of the human genome to the genomes of dog, mouse, and chicken have shown that each of these genomes is broken respectively into 371, 539, and 1,068 chromosomal segments with conserved gene content or order relative to the human genome.[11,13] Within placental mammals, the most conserved chromosome in terms of gene content and order is the X chromosome, which due to the functional constraints imposed by X inactivation, has remained intact in all eutherians.[69] On the other hand, evolutionary breakpoints, which mark the position of chromosomal rearrangements that have occurred over time, were previously thought to be randomly distributed across the genome.[68] However, it now appears that a small fraction of the mammalian genome is particularly susceptible to breakage, and that these chromosomal locations have been "reused" as evolutionary breakpoints independently during the past approximately 100 million years.[8,10–13,70,71] Another remarkable observation gleaned from genome era comparative mapping is that centromeres can emerge at a new position on a chromosome and disappear from the old location independent of a chromosomal rearrangement. This phenomenon, called *centromere repositioning*, has been demonstrated to have occurred in relatively recent timescales within groups as diverse as primates and birds.[72,73] Thus, comparative mapping provides a global view of how and when vertebrate genome organization evolved as well as an entry point for exploring new genomes.

7.5 COMPARATIVE GENOMIC SEQUENCE ANALYSIS

Large-scale genome sequencing projects for 50 vertebrates are currently at various stages of completion. The primary rationales that have driven the expansion of whole-genome sequencing efforts beyond the human genome are (1) developing a complete sequence catalog of the genomes of widely used or emerging genetic model organisms, such as mouse, rat, zebrafish, and stickleback, or important agricultural species, such as cow, pig, and chicken; and (2) enhancing the annotation of the human genome. In particular, the now broadly accepted concept that interspecies comparisons can be used to identify putative functional elements in the human genome is the primary impetus for the vast majority of vertebrate genome sequencing projects.

Perhaps the most important finding in early small-scale comparative genomic sequencing projects was that it was not uncommon to detect sequences outside of known protein-coding regions or untranslated regions (UTRs) that were highly conserved between species. While it was known and expected that protein-coding regions were highly conserved between humans and mice,[74] the detection of conserved noncoding elements was quite striking and suggested that comparative genomic sequencing could provide an unbiased and large-scale systematic method by which putative functional elements could be detected in the human genome.[75] Subsequent experimental studies that tested the ability of conserved sequences to regulate gene expression verified that many of these conserved elements were indeed functional.[76]

As a result, methods have been developed to generate whole-genome alignments between two or more species[77] that can then be scanned to identify the most-conserved elements in a genome. Although numerous methods have been developed to detect conserved elements,[78–82] in general each method incorporates a model by which some nucleotides are evolving freely without functional constraint at a neutral rate, while other nucleotides are evolving under functional constraint and thus at a rate slower than the neutral rate. Constrained sequences are said to be evolving under negative (purifying) selection, which means that changes within these sequences are deleterious to the organism. As a consequence, mutations in these constrained sequences are removed from the population by natural selection, resulting in a reduced number of observed differences between species than would otherwise be expected based on the rate at which random mutations occur over time. In its most extreme form, purifying selection and functional constraint have led to the absolute conservation of sequences up to 388 nucleotides in length between humans and chicken, which based on the neutral mutation rate would have been expected to accumulate more than 1 substitution/site.[83] Current estimates of the fraction of the human genome that is made up of conserved elements are on the order of about 5% of the genome.[79] Since only approximately 1.5% of the human genome codes for proteins, most of these conserved elements represent potential functional elements for which no specific function has been assigned.

The power to detect conserved elements depends on the species used in the comparison. For example, although it has been demonstrated that sequence comparisons between humans and fish are extremely effective for detecting functional conserved noncoding elements, such as enhancers,[84] only a subset of the sequences conserved among mammals is also conserved in more distantly related vertebrates (see Figure 7.2).[85] It has also been experimentally demonstrated that fish and mammals have distinct sets of conserved noncoding elements.[86] This result suggests that in each vertebrate lineage, including the human and other primate lineages,[80] old functional elements have been lost and new elements have emerged (see Figure 7.2). Therefore, when using comparative sequence analysis to identify putative functional elements, it is critical to ensure that the set of species used in the comparison is appropriate for the biological question

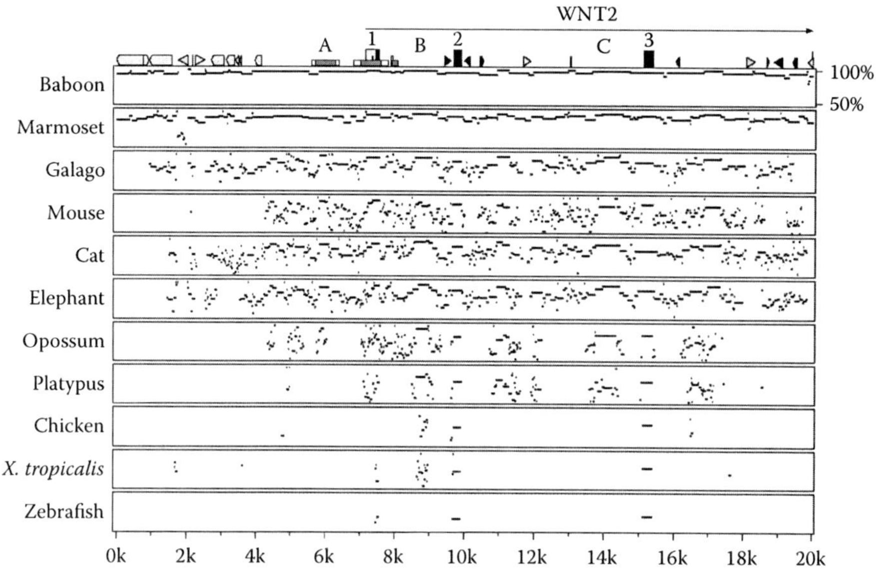

FIGURE 7.2 Comparison of vertebrate genomic sequence. Orthologous genomic sequence corresponding to a 20-kb portion of the *WNT2* locus on human chromosome 7 from 12 species was extracted from published whole-genome[6] and targeted BAC-based assemblies[22,82] and unpublished genome assemblies (*X. tropicalis*: xenTro2 and Zebrafish: Zv6) and aligned with MultiPipMaker[100] using the human sequence as the reference. *WNT2* exons[1–3] are represented by the numbered boxes (open = 5′ UTR and solid = protein-coding regions); short boxes represent CpG islands; and repetitive elements are indicated by the remaining symbols. The letters A, B, and C indicate the position of examples of non-protein-coding elements conserved in eutherians and marsupials, all tetrapods, and all mammals, respectively. Note that the *WNT2* protein-coding exons 2 and 3 are conserved in all species.

that is being asked. For example, if one is attempting to identify putative regulatory elements that modulate expression in the placenta, sequence comparisons to species outside placental mammals are likely to be of limited utility. Moreover, if one was seeking to identify the genetic basis of what makes humans unique, the most appropriate species to include in such a study would be our closest relatives, the great apes and other primates. Fortunately, with the expanding number of species targeted for whole-genome sequencing and the ability to use the extensive set of vertebrate BAC libraries for targeted comparative sequencing, there is an ever-increasing power to establish the optimal comparative genomic data set for the question at hand.

7.6 SUMMARY

Comparative vertebrate genomics is an expanding discipline that unites large-scale genomics with evolutionary biology toward the purpose of reconstructing the history of vertebrate genomes and elucidating the complete functional content encoded within our genome. Current and future resources, such as whole-genome sequences and BAC libraries, promise to support a wide range of applications. Future applications include the development of a better understanding of how human genetic susceptibility to certain diseases evolved, more accurate genetic and biological models of human disease, and ascertainment of all functional elements in the human genome by projects like ENCODE.[25] Comparative genomic resources can also be used to address more fundamental questions, like what are the key genetic determinants that make each species unique and how have they evolved. In conclusion, the explosion of genomic data in the past decade and remarkable discoveries that it has yielded are just the beginning of the genomic era and comparative vertebrate genomics.

REFERENCES

1. Kumar, S. & Hedges, S. B. A molecular timescale for vertebrate evolution. *Nature* **392**, 917–920 (1998).
2. Burnie, D. & Wilson, D. E. (Eds.). *Animal* (DK Publishing, New York, 2001).
3. Sarich, V. M. & Wilson, A. C. Immunological time scale for hominid evolution. *Science* **158**, 1200–1203 (1967).
4. Eddy, S. R. A model of the statistical power of comparative genome sequence analysis. *PLoS Biol* **3**, e10 (2005).
5. Margulies, E. H. et al. An initial strategy for the systematic identification of functional elements in the human genome by low-redundancy comparative sequencing. *Proc Natl Acad Sci, USA* **102**, 4795–800 (2005).
6. Lander, E. S. et al. Initial sequencing and analysis of the human genome. *Nature* **409**, 860–921 (2001).
7. Venter, J. C. et al. The sequence of the human genome. *Science* **291**, 1304–1351 (2001).
8. Waterston, R. H. et al. Initial sequencing and comparative analysis of the mouse genome. *Nature* **420**, 520–562 (2002).
9. Aparicio, S. et al. Whole-genome shotgun assembly and analysis of the genome of Fugu rubripes. *Science* **297**, 1301–1310 (2002).
10. Gibbs, R. A. et al. Genome sequence of the brown Norway rat yields insights into mammalian evolution. *Nature* **428**, 493–521 (2004).
11. Hillier, L. W. et al. Sequence and comparative analysis of the chicken genome provide unique perspectives on vertebrate evolution. *Nature* **432**, 695–716 (2004).
12. Jaillon, O. et al. Genome duplication in the teleost fish *Tetraodon nigroviridis* reveals the early vertebrate proto-karyotype. *Nature* **431**, 946–957 (2004).
13. Lindblad-Toh, K. et al. Genome sequence, comparative analysis and haplotype structure of the domestic dog. *Nature* **438**, 803–819 (2005).
14. Initial sequence of the chimpanzee genome and comparison with the human genome. *Nature* **437**, 69–87 (2005).
15. McPherson, J. D. Sequence ready — or not? *Genome Res* **7**, 1111–1113 (1997).
16. Dunham, I., Dewar, K., Kim, U.-J. & Ross, M. In: *Genome Analysis: A Laboratory Manual, Volume 3: Bacterial Cloning Systems* (Eds. Birren, B. et al.), pp. 1–86 (Cold Spring Harbor Laboratory Press, Cold Spring Harbor, NY, 1998).
17. Marra, M. A. et al. High throughput fingerprint analysis of large-insert clones. *Genome Res* **7**, 1072–1084 (1997).
18. Schein, J. et al. In: *Bacterial Artificial Chromosomes, Volume 1: Library Construction, Physical Mapping, and Sequencing* (Eds. Zhao, S. & Stodolsky, M.), pp. 143–156 (Humana Press, Totowa, NJ, 2004).

19. McPherson, J. D. et al. A physical map of the human genome. *Nature* **409**, 934–941 (2001).
20. Warren, R. L. et al. Physical map-assisted whole-genome shotgun sequence assemblies. *Genome Res* **16**, 768–775 (2006).
21. Kirsch, I. R. et al. A systematic, high-resolution linkage of the cytogenetic and physical maps of the human genome. *Nat Genet* **24**, 339–340 (2000).
22. Thomas, J. W. et al. Comparative analyses of multi-species sequences from targeted genomic regions. *Nature* **424**, 788–793 (2003).
23. Thomas, J. W. et al. Parallel construction of orthologous sequence-ready clone contig maps in multiple species. *Genome Res* **12**, 1277–1285 (2002).
24. Kellner, W. A., Sullivan, R. T., Carlson, B. H. & Thomas, J. W. Uprobe: a genome-wide universal probe resource for comparative physical mapping in vertebrates. *Genome Res* **15**, 166–173 (2005).
25. The ENCODE (Encyclopedia of DNA Elements) Project. *Science* **306**, 636–640 (2004).
26. Marshall, V. M., Allison, J., Templeton, T. & Foote, S. J. In: *Bacterial Artificial Chromosomes Volume 2: Functional Studies* (Eds. Zhao, S. & Stodolsky, M.), pp. 159–182 (Humana Press, Totowa, NJ, 2004).
27. Copeland, N. G., Jenkins, N. A. & Court, D. L. Recombineering: a powerful new tool for mouse functional genomics. *Nat Rev Genet* **2**, 769–779 (2001).
28. Yang, Y., Swaminathan, S., Martin, B. K. & Sharan, S. K. Aberrant splicing induced by missense mutations in BRCA1: clues from a humanized mouse model. *Hum Mol Genet* **12**, 2121–2131 (2003).
29. Mortlock, D. P., Guenther, C. & Kingsley, D. M. A general approach for identifying distant regulatory elements applied to the Gdf6 gene. *Genome Res* **13**, 2069–2081 (2003).
30. Mirsky, A. E. & Ris, H. The desoxyribonucleic acid content of animal cells and its evolutionary significance. *J Gen Physiol* **34**, 451–462 (1951).
31. Thomas, C. A. The genetic organization of chromosomes. *Annu Rev Genet* **5**, 237–256 (1971).
32. Cavalier-Smith, T. Nuclear volume control by nucleoskeletal DNA, selection for cell volume and cell growth rate, and the solution of the DNA C-value paradox. *J Cell Sci* **34**, 247–278 (1978).
33. Hughes, A. L. & Hughes, M. K. Small genomes for better flyers. *Nature* **377**, 391 (1995).
34. Castillo-Davis, C. I., Mekhedov, S. L., Hartl, D. L., Koonin, E. V. & Kondrashov, F. A. Selection for short introns in highly expressed genes. *Nat Genet* **31**, 415–418 (2002).
35. Petrov, D. A. Mutational equilibrium model of genome size evolution. *Theor Popul Biol* **61**, 531–544 (2002).
36. Vinogradov, A. E. Buffering: a possible passive-homeostasis role for redundant DNA. *J Theor Biol* **193**, 197–199 (1998).
37. Vinogradov, A. E. Evolution of genome size: multilevel selection, mutation bias or dynamical chaos? *Curr Opin Genet Dev* **14**, 620–626 (2004).
38. Vinogradov, A. E. "Genome design" model: evidence from conserved intronic sequence in human–mouse comparison. *Genome Res* **16**, 347–354 (2006).
39. Lynch, M. & Conery, J. S. The origins of genome complexity. *Science* **302**, 1401–1404 (2003).
40. Hinegardner, R. & Rosen, E. D. Cellular DNA content and the evolution of teleostean fishes. *Am Naturalist* **106**, 621–644 (1972).
41. Gregory, T. R. (2006). Animal Genome Size Database. Available at: http://genomesize.com.

42. Hirsch, N., Zimmerman, L. B. & Grainger, R. M. *Xenopus*, the next generation: *X. tropicalis* genetics and genomics. *Dev Dyn* **225**, 422–433 (2002).
43. Hartl, D. L. Molecular melodies in high and low C. *Nat Rev Genet* **1**, 145–149 (2000).
44. Britten, R. J., Rowen, L., Williams, J. & Cameron, R. A. Majority of divergence between closely related DNA samples is due to indels. *Proc Natl Acad Sci, USA* **100**, 4661–4665 (2003).
45. Freeman, J. L. et al. Copy number variation: new insights in genome diversity. *Genome Res* **16,** 949–961 (2006).
46. Lynch, M. Intron evolution as a population-genetic process. *Proc Natl Acad Sci, USA* **99**, 6118–6123 (2002).
47. Charlesworth, B. & Barton, N. Genome size: does bigger mean worse? *Curr Biol* **14**, R233–R235 (2004).
48. Finishing the euchromatic sequence of the human genome. *Nature* **431**, 931–945 (2004).
49. Gu, X., Wang, Y. & Gu, J. Age distribution of human gene families shows significant roles of both large- and small-scale duplications in vertebrate evolution. *Nat Genet* **31**, 205–209 (2002).
50. McLysaght, A., Hokamp, K. & Wolfe, K. H. Extensive genomic duplication during early chordate evolution. *Nat Genet* **31**, 200–204 (2002).
51. Friedman, R. & Hughes, A. L. Pattern and timing of gene duplication in animal genomes. *Genome Res* **11**, 1842–1847 (2001).
52. Vandepoele, K., De Vos, W., Taylor, J. S., Meyer, A. & Van de Peer, Y. Major events in the genome evolution of vertebrates: paranome age and size differ considerably between ray-finned fishes and land vertebrates. *Proc Natl Acad Sci, USA* **101**, 1638–1643 (2004).
53. Panopoulou, G. & Poustka, A. J. Timing and mechanism of ancient vertebrate genome duplications — the adventure of a hypothesis. *Trends Genet* **21**, 559–567 (2005).
54. Bailey, J. A. & Eichler, E. E. Primate segmental duplications: crucibles of evolution, diversity and disease. *Nat Rev Genet* **7**, 552–564 (2006).
55. Shiu, S. H., Byrnes, J. K., Pan, R., Zhang, P. & Li, W. H. Role of positive selection in the retention of duplicate genes in mammalian genomes. *Proc Natl Acad Sci, USA* **103**, 2232–2236 (2006).
56. Ohno, S. *Evolution by Gene Duplication* (Springer-Verlag, Berlin, 1970).
57. Olson, M. V. When less is more: gene loss as an engine of evolutionary change. *Am J Hum Genet* **64**, 18–23 (1999).
58. Wang, X., Grus, W. E. & Zhang, J. Gene losses during human origins. *PLoS Biol* **4**, e52 (2006).
59. Stedman, H. H. et al. Myosin gene mutation correlates with anatomical changes in the human lineage. *Nature* **428**, 415–418 (2004).
60. Yandell, M. et al. Large-scale trends in the evolution of gene structures within 11 animal genomes. *PLoS Comput Biol* **2**, e15 (2006).
61. Modrek, B. & Lee, C. J. Alternative splicing in the human, mouse and rat genomes is associated with an increased frequency of exon creation and/or loss. *Nat Genet* **34**, 177–180 (2003).
62. Wang, W. et al. Origin and evolution of new exons in rodents. *Genome Res* **15**, 1258–1264 (2005).
63. Sorek, R., Ast, G. & Graur, D. Alu-containing exons are alternatively spliced. *Genome Res* **12**, 1060–1067 (2002).
64. Wheelan, S. J., Aizawa, Y., Han, J. S. & Boeke, J. D. Gene-breaking: a new paradigm for human retrotransposon-mediated gene evolution. *Genome Res* **15**, 1073–1078 (2005).

65. Duret, L., Chureau, C., Samain, S., Weissenbach, J. & Avner, P. The Xist RNA gene evolved in eutherians by pseudogenization of a protein-coding gene. *Science* **312**, 1653–1655 (2006).
66. Wang, W. & Lan, H. Rapid and parallel chromosomal number reductions in muntjac deer inferred from mitochondrial DNA phylogeny. *Mol Biol Evol* **17**, 1326–1333 (2000).
67. Castle, W. E. *Studies of Heredity in Rabbits, Rats and Mice* (Carnegie Institute of Washington, DC, 1919).
68. Nadeau, J. H. & Taylor, B. A. Lengths of chromosomal segments conserved since divergence of man and mouse. *Proc Natl Acad Sci, USA* **81**, 814–818 (1984).
69. O'Brien, S. J. et al. The promise of comparative genomics in mammals. *Science* **286**, 458–481 (1999).
70. Pevzner, P. & Tesler, G. Human and mouse genomic sequences reveal extensive breakpoint reuse in mammalian evolution. *Proc Natl Acad Sci, USA* **100**, 7672–7677 (2003).
71. Murphy, W. J. et al. Dynamics of mammalian chromosome evolution inferred from multispecies comparative maps. *Science* **309**, 613–617 (2005).
72. Montefalcone, G., Tempesta, S., Rocchi, M. & Archidiacono, N. Centromere repositioning. *Genome Res* **9**, 1184–1188 (1999).
73. Kasai, F., Garcia, C., Arruga, M. V. & Ferguson-Smith, M. A. Chromosome homology between chicken (*Gallus gallus domesticus*) and the red-legged partridge (*Alectoris rufa*); evidence of the occurrence of a neocentromere during evolution. *Cytogenet Genome Res* **102**, 326–330 (2003).
74. Makalowski, W., Zhang, J. & Boguski, M. S. Comparative analysis of 1,196 orthologous mouse and human full-length mRNA and protein sequences. *Genome Res* **6**, 846–857 (1996).
75. Hardison, R. C., Oeltjen, J. & Miller, W. Long human-mouse sequence alignments reveal novel regulatory elements: a reason to sequence the mouse genome. *Genome Res* **7**, 959–966 (1997).
76. Gottgens, B. et al. Analysis of vertebrate *SCL* loci identifies conserved enhancers. *Nat Biotechnol* **18**, 181–186 (2000).
77. Blanchette, M. et al. Aligning multiple genomic sequences with the threaded blockset aligner. *Genome Res* **14**, 708–715 (2004).
78. Margulies, E. H., Blanchette, M., Haussler, D. & Green, E. D. Identification and characterization of multi-species conserved sequences. *Genome Res* **13**, 2507–2518 (2003).
79. Siepel, A. et al. Evolutionarily conserved elements in vertebrate, insect, worm, and yeast genomes. *Genome Res* **15**, 1034–1050 (2005).
80. Boffelli, D. et al. Phylogenetic shadowing of primate sequences to find functional regions of the human genome. *Science* **299**, 1391–1394 (2003).
81. Lunter, G., Ponting, C. P. & Hein, J. Genome-wide identification of human functional DNA using a neutral indel model. *PLoS Comput Biol* **2**, e5 (2006).
82. Cooper, G. M. et al. Distribution and intensity of constraint in mammalian genomic sequence. *Genome Res* **15**, 901–913 (2005).
83. Bejerano, G. et al. Ultraconserved elements in the human genome. *Science* **304**, 1321–1325 (2004).
84. Woolfe, A. et al. Highly conserved non-coding sequences are associated with vertebrate development. *PLoS Biol* **3**, e7 (2004).
85. Prabhakar, S. et al. Close sequence comparisons are sufficient to identify human cis-regulatory elements. *Genome Res* **16**, 855–863 (2006).
86. Fisher, S., Grice, E. A., Vinton, R. M., Bessling, S. L. & McCallion, A. S. Conservation of RET regulatory function from human to zebrafish without sequence similarity. *Science* **312**, 276–279 (2006).

87. Hedges, S. B. & Poling, L. L. A molecular phylogeny of reptiles. *Science* **283**, 998–1001 (1999).
88. van Tuinen, M. & Hedges, S. B. Calibration of avian molecular clocks. *Mol Biol Evol* **18**, 206–213 (2001).
89. Eizirik, E., Murphy, W. J. & O'Brien, S. J. Molecular dating and biogeography of the early placental mammal radiation. *J Hered* **92**, 212–219 (2001).
90. Murphy, W. J. et al. Resolution of the early placental mammal radiation using Bayesian phylogenetics. *Science* **294**, 2348–2351 (2001).
91. Lee, M. H., Shroff, R., Cooper, S. J. & Hope, R. Evolution and molecular characterization of a beta-globin gene from the Australian Echidna *Tachyglossus aculeatus* (Monotremata). *Mol Phylogenet Evol* **12**, 205–214 (1999).
92. Teeling, E. C. et al. A molecular phylogeny for bats illuminates biogeography and the fossil record. *Science* **307**, 580–584 (2005).
93. Steppan, S., Adkins, R. & Anderson, J. Phylogeny and divergence-date estimates of rapid radiations in muroid rodents based on multiple nuclear genes. *Syst Biol* **53**, 533–553 (2004).
94. Price, S. A., Bininda-Emonds, O. R. & Gittleman, J. L. A complete phylogeny of the whales, dolphins and even-toed hoofed mammals (Cetartiodactyla). *Biol Rev Camb Philos Soc* **80**, 445–473 (2005).
95. Delsuc, F., Vizcaino, S. F., & Douzery, E. J. Influence of tertiary paleoenvironmental changes on the diversification of South American mammals: a relaxed molecular clock study within xenarthrans. *BMC Evol Biol* **4**:11 (2004).
96. Drummond, A. J., Ho, S. Y., Phillips, M. J. & Rambaut, A. Relaxed phylogenetics and dating with confidence. *PLoS Biol* **4**, e88 (2006).
97. Kumazawa, Y., Yamaguchi, M. & Nishida, M. In: *The Biology of Biodiversity* (Ed. Kato, M.), pp. 35–52 (Springer-Verlag, Tokyo, 1999).
98. Crnogorac-Jurcevic, T., Brown, J. R., Lehrach, H. & Schalkwyk, L. C. *Tetraodon fluviatilis*, a new puffer fish model for genome studies. *Genomics* **41**, 177–184 (1997).
99. Goodman, M., Grossman, L. I. & Wildman, D. E. Moving primate genomics beyond the chimpanzee genome. *Trends Genet* **21**, 511–517 (2005).
100. Schwartz, S. et al. MultiPipMaker and supporting tools: alignments and analysis of multiple genomic DNA sequences. *Nucleic Acids Res* **31**, 3518–3524 (2003).

8 Gaining Insight into Human Population-Specific Selection Pressure

Michael R. Barnes

CONTENTS

ABSTRACT

The availability of large-scale catalogs of human genetic variation has stimulated many genome-wide scans for positive selection in human populations. Evidence for population-specific selective sweeps has now been found in many regions of the human genome, in genes known to be associated with diet, disease, and social development. However, detecting evidence of molecular selection may often be confounded by the influence of the underlying complex demographics of a population; the varying mutation and recombination rates in different populations; or the ascertainment schemes used to discover polymorphisms. Here, approaches to the analysis of selection in human populations are reviewed in the context of the available data, tools, and some of the key challenges to the interpretation of putative signals of selection in human populations.

> "I have called this principle, by which each slight variation, if useful, is preserved, by the term of Natural Selection."
>
> *On the Origin of Species* (1859), Charles Darwin

8.1 INTRODUCTION

Many of the chapters in this book have focused either directly or indirectly on the study of natural selection *between* species, helping to explain how an infinitesimal number of mutation events over billions of years led from single-cell organisms to the complexity of life today. In this context, study of the six million years of natural selection following the division of *Homo sapiens* and other primates[1] might superficially seem a little pedestrian; however, any closer examination of this period quickly uncovers the veritable evolutionary roller coaster that *Homo sapiens* has ridden in recent history. One could argue that the rapidity and complex nature of the events that led to the development of modern humans are largely unprecedented in the history of evolution. All the landmark events leading to the separation of humans from other ape species were accompanied by defined selective pressures, many which are reviewed here. These selective pressures were intensified as human culture became increasingly tribal by nature, leading to the isolation of populations and occasional population bottlenecks. As humans shifted from a hunter-gatherer lifestyle to a settled agricultural existence, diets changed, and population density increased and became more susceptible to epidemic outbreaks of disease. These events are all clearly evidenced in the genomes of extant populations. In fact, in this field of research, genome sequences may offer fascinating insights into human prehistory at which archeology could only hint. Using several recently generated genome-scale variation data sets, a number of groups have identified genomic regions with high levels of population differentiation, low levels of diversity, or unusually long stretches of DNA sequence showing very highly correlated alleles, a phenomenon known as *linkage disequilibrium* (LD).[2] These characteristics are all possible hallmarks of natural selection, but they could also be explained by other phenomena unrelated to selection. Validating the mark of selection in these regions will provide valuable insights on where and how selection acted during human evolution, with possible implications to health by identifying variants involved in common diseases. This is one of the major motivations for analysis of natural selection as there are already many examples for which disease alleles have been shown to confer a selective benefit to the carrier in particular environmental circumstances, hence balancing the deleterious effect of the disease, so-called balancing mutations (e.g., glucose-6-phosphate dehydrogenase [G6PD] alleles in malaria; see Section 8.3.2.1).

Today, we might like to convince ourselves that developed societies at least have raised themselves above all but the most severe forces of natural selection. However, selection still grips human evolution; we do not have to go back far in our own history to find striking examples of this — the current human immunodeficiency virus (HIV) pandemic is probably a case in point, with eerie echoes of simian immunodeficiency virus (SIV) pandemics in earlier primate evolution.[3] In this case, one could argue that natural selection on host immunity to HIV infection is acting largely unchecked across swaths of the human population in sub-Saharan Africa and Asia. Other pandemics, such as avian flu, threaten even more devastation. As we begin to gain a better understanding of these events, we not only learn about our own history, but also may gain valuable insights into the how the diversity of imprints of selection in human populations may have an impact on health and well-being.

This chapter cannot hope to offer an exhaustive review of the entire field of natural selection in human populations, so instead it is structured in six key sections. In Section 8.2, the key principles of natural selection within the human population are reviewed. In Section 8.3, some of the known examples of selection that have led to the propagation of disease alleles throughout human populations are examined. Section 8.4 starts to become technical by reviewing the principle methods that are used for the analysis of selection at a molecular level. In Section 8.5, the methods and results presented in some of the key publications that have carried out genome-wide screens for natural selection in human populations are reviewed and compared. Section 8.6 reviews some of the tools that can be used to help prioritize loci showing evidence of selection. The final section reviews some of the bioinformatics approaches that can be used to investigate the molecular basis of a putative signature of selection.

8.2 NATURAL SELECTION DURING HUMAN EVOLUTION

8.2.1 Natural Selection in the Context of the Human Diaspora

As any consideration of natural selection in humans quickly reveals, a grasp of anthropology is required as much as knowledge of genetics. Extensive human population migrations have occurred throughout history. These have led to both sustained periods of admixture and, in some cases, extended periods of population isolation, often leading to population bottlenecks. These conditions have combined to strongly favor the positive selection of certain advantage alleles in human populations. In this context, selection analysis can give us insights into our own evolution and the fundamental genetic differences that distinguish us from other apes. Selection generally manifests in either positive or negative forms. Positive selection is the evolutionary force that favors advantageous alleles within populations, while negative selection removes disadvantageous alleles. For example, positive selection has helped our immune systems evolve to deal with increasing human population densities and changing diets; it has also played a role in development of language and cognition — leading humans away from their hominid cousins.

8.2.2 Out of Africa

Most early human evolution is believed to have taken place in Africa. Mitochondrial DNA (mtDNA) analysis and later Y chromosome analysis of human populations have suggested that the so-called mitochondrial Eve, the most recent matrilinear common ancestor shared by all living human beings, is likely to have lived around 150,000–120,000 years ago, probably in the area of modern Ethiopia, Kenya, or Tanzania.[4] Many studies point to the probability that the human diaspora, as we know it, began around 100,000–80,000 years ago. Three main lines of humans began major migrations, leading to divergent population groups bearing the mitochondrial haplogroup L1 (mtDNA)/A (Y-DNA) colonizing Southern Africa, bearers of haplogroup L2 (mtDNA)/B (Y-DNA) settling Central and West Africa, while the bearers of haplogroup L3 remained in East Africa. Approximately 70,000 years ago, a part of the L3 bearers migrated into the Near East, spreading east to southern Asia and Australasia (~60,000 years ago), northwestward into Europe and eastward into

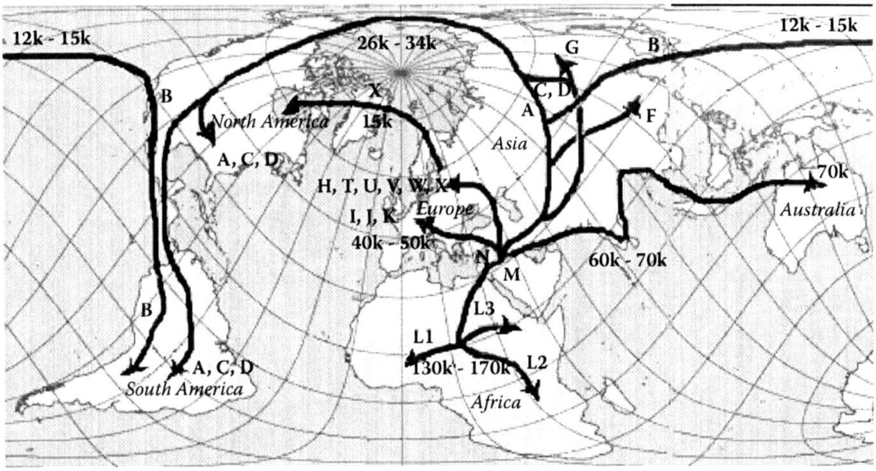

FIGURE 8.1 The human diaspora. A putative map of human migration based on mitochondrial DNA haplotypes.

Central Asia (~40,000 years ago), and further east to the Americas (~30,000 years ago)[4] (Figure 8.1). The full complexity of these human migrations and the ways that they are studied could be the subject of an entire chapter, but it is perhaps worth mentioning one final strand of evidence. Some time after the first mtDNA studies, the first genome-wide studies of LD presented compelling evidence to support the "out-of-Africa" theory. Gabriel et al.[5] showed radical differences in the extent of LD between African (L1 and L2) and Caucasian (L3) populations, supporting the demographic events that might be expected (e.g., bottlenecks and periods of population isolation) following the migration out of Africa.

8.2.3 Out of Africa in Context Today

These events in the recent history of man are the backdrop against which all analysis of human natural selection should take place. Considering that every human genome is unique, ideally we need to interpret each of the 12 billion copies currently populating our planet. Unfortunately, interpretation of individual genomes is impractical, so usually we seek to understand selection at the population level. This immediately creates a problem. Human populations have always been fluid and in many cases poorly defined; this issue applies doubly today as air travel, mass emigration, and interracial marriage have made the definition of ethnicity less and less precise. This makes the collection of ethnically homogeneous populations for the analysis of selection a very tall order indeed — a caveat that needs to be kept in mind at all times.

8.2.4 The HapMap

8.2.4.1 A Key Human Population Resource for Analysis of Selection

Arguably, the completion of the human genome did relatively little to inform on the full range of variation *between* human populations. Both public and private versions of the human genome were based on Caucasian individuals, the traditionally studied ethnic group in most biomedical research. However, recent advances in technology have led to the generation of some fundamental data sources that have enabled far-reaching analyses of the imprint of positive selection on different human population samples. The foremost among these resources is the HapMap,[6] which has yielded genotype and LD information on about four million single-nucleotide polymorphisms (SNPs) in four human population samples. As a relatively comprehensive sample of genetic variation in four population samples, the HapMap is an informative genome scan that is a more-than-adequate data set for the detection of the signatures of selection. In fact, by their nature, it might be expected that the majority of positively selected alleles would be present in the HapMap due to their increased (selected) allele frequency. The immediate objective of the HapMap was to support high-density genetic association analysis of human disease, but these data are already in use to address a diverse range of scientific issues,[7] ranging from disease gene discovery, regulation of expression, to the kinds of molecular evolutionary analysis that are reviewed here.

8.2.4.2 The HapMap Project: Background

The HapMap project was established in 2002 to study the LD relationships across the human genome in four different ethnic groups.[6] These included a panel of 30 trios from Yoruba, Nigeria (YRI); a panel of 30 CEPH (Centre d'Etude du Polymorphism Humain) trios from Utah residents with European ancestry (CEU); and a panel of 45 unrelated Japanese individuals from Tokyo (JPT) and 45 unrelated Han Chinese individuals from Beijing (CHB). It is worth noting that, by most genetic measures, the Japanese and Chinese populations are very similar, and so in many analyses they are combined as a single Asian population group (JPT&CHB). The sample sizes selected for each population are sufficient for the immediate purpose of the HapMap, that is, to characterize LD and haplotypes between common variants in these population samples. However, the sample sizes are not sufficient to be wholly representative of the specific "population" from which they were collected. So, the CHB sample is not representative of all Han Chinese, and it is even less representative of wider geographic populations from China. The degree of similarity between the HapMap samples and wider populations is one of the great challenges to the wider applicability of HapMap data.

The HapMap project has been run in three phases. HapMap phase I was completed in October 2005 and involved genotyping of about one million SNPs at an average spacing of 5 kb. Phase II HapMap provided a broader sampling of genomic variation. Using the same 269 samples, a further 2.9 million SNPs were genotyped, bringing the genome-wide total of polymorphic SNPs genotyped up to 3.9 million. Three years after the launch of the project, genotyping of 4.6 million SNPs is complete, and a number of tools are now offering an integrated view of LD across the human genome. A preliminary analysis of the phase I data set has been published,[8] but analysis of the HapMap is still ongoing. All the information produced by the HapMap project is freely available at the project Web site (http://www.hapmap.org).

As a sample of human genetic variation across populations, the HapMap variants are a fantastic tool for investigating the genetic diversity of humans. Although the HapMap sample size is modest, it is still highly informative considering the historically small size and shared ancestry of human populations. This offers a great opportunity to investigate selected variants that have historically affected human fitness, many of which are still segregating in populations today. In this chapter, how the HapMap and similar data sources can be used to study selection in human populations is examined. To illustrate this first, some of the principles underpinning analysis of selection and some of the methods that are in use to study selection at a molecular level are briefly reviewed.

8.3 NATURAL SELECTION, HUMAN HEALTH, AND DISEASE

8.3.1 Forces of Selection

When a new mutation that confers selective advantage arises in a population, it is likely to increase in frequency in the population by natural selection.[9] This event will also influence the standing variation neighboring the mutation, as the pattern of variation, in the individual in which the mutation arose, sweeps away other variations in the selected locus. This leads to a reduction in haplotype diversity, increased LD, and a skewed pattern of mostly low allele frequency variants in the selected locus — a chain of events known as a *selective sweep*[9] (Figure 8.2). Already, a number of signals of very strong and recent population-specific selection have been identified in human genes, many with an obvious impact on the differentiation of humans from other hominids. Taking a look at the selected genes that have been identified so far, clear themes emerge that highlight many of the key selective forces during human evolution. Signatures of selection have been identified in genes involved in immunity, reproduction, nervous system development, and sensory perception (Table 8.1). Researchers have used SNP genotype data to detect these signatures across the genome, using a variety of statistical measures, many of the results of which are fully available on the Web (reviewed in detail in Section 8.5).

8.3.2 Balancing Selection: The Double-Edged Sword of Evolution

It is obvious that selective events that have occurred during human evolution may have important implications today. This is because some selective advantages may carry with

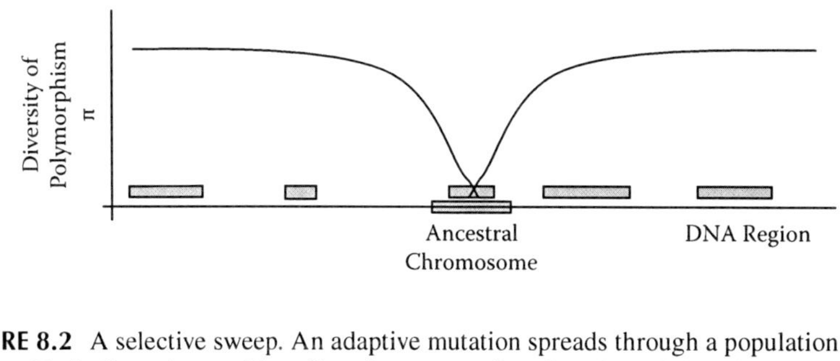

FIGURE 8.2 A selective sweep. An adaptive mutation spreads through a population toward fixation. Typically, polymorphism diversity surrounding the selected allele is reduced, forming a characteristic selective sweep signature.

TABLE 8.1
Signatures of Selection Identified in Human Genes

Gene	Putative Selective Pressure	Phenotype	Reference
AGT	Climate (thermoregulation)	Hypertension	Nakajima et al.[10]
CYP3A5	Climate (salt avidity)	Hypertension	Thompson et al.[11]
SLC24A5	Climate (UV exposure)	Skin pigmentation	Lamason et al.[12]
FY	Immunity (malaria)	Malaria resistance	Hamblin et al.[13]
G6PD	Immunity (malaria)	Malaria resistance	Kwiatkowski[14]
IL4 & IL13	Immunity (unknown)	Asthma	Sakagami et al.[15]
CASP12	Immunity (unknown)	CASP12 pseudogene protects against sepsis	Xue et al.[16]
CFTR	Immunity (cholera)	Cystic fibrosis	Gabriel et al.[17]
NAT2	Diet (agriculture)	Bladder cancer/adverse drug reactions	Patin et al.[18]
LCT	Diet (milk)	Lactose intolerance	Bersaglieri et al.[19]
TRPV6	Diet (milk)	Prostate cancer	Akey et al.[20]
MMP3	Diet (unknown)	Coronary heart disease	Rockman et al.[21]
ZAN	Reproduction	Reproductive success	Gasper & Swanson[22]
FOXP2	Social development	Language development	Enard et al.[23]

them severe disadvantages, which may only manifest after the allele has been widely selected into populations or perhaps when the conditions for a population change. This is one of the major reasons for studying positive selection. In some instances, positive selection can explain the unexpectedly high frequencies of disease alleles — the classic paradigm of *balancing selection*, by which an advantageous heterozygote allele of a mutation that is deleterious in the homozygous state is widely selected, conferring a heterozygote advantage but causing a disease in the homozygous state.

8.3.2.1 Infectious Disease as a Selective Force in Human Populations

Some of the best precedents for balancing selection events are related to enhanced resistance to infection and disease. Such events have accounted for some diseases that are widespread throughout human populations; a good example is cystic fibrosis, one of the most common Mendelian disorders (see entry OMIM (online Mendelian inheritance in man) *602421). Heterozygote mutant alleles of the cystic fibrosis transmembrane conductance regulator (CFTR) were believed to be selected for in humans by conferring greater resilience to typhoid infection[24]; unfortunately, in the homozygous state these alleles cause a highly debilitating illness.

Balancing selection events can also explain the extraordinarily high frequencies of some serious hemopathologies in sub-Saharan Africa. For example, low-activity G6PD alleles are common. Bienzle et al.[25] showed that these alleles conferred greater resistance to malaria, while subsequent studies showed that low-activity G6PD alleles were highly correlated with the prevalence of malaria.[14] This led to a typical balancing selection hypothesis that low-activity G6PD alleles may reduce risk from *Plasmodium* infection, hence explaining maintenance of alleles that otherwise cause quite serious hemopathologies.

This is just one of many malaria-resistance alleles that have arisen in Africa. Alleles causing both sickle cell anemia and α-thalassemia also occur at high frequencies in sub-Saharan Africa; individually, each is protective against severe malaria.[26] This illustrates the extraordinary evolutionary struggle between malaria and human populations. This has clearly led to a great deal of evolutionary selection in species, on host genes that contribute to resistance, and on parasite genes involved in the infection process and more recently drug resistance.[27]

Going back to our knowledge of human demographics, there is also evidence that much of this has happened recently in human history and certainly since humans started to migrate out of Africa. This is supported by haplotype analysis of *A*− and *Med* mutations at the G6PD locus. Tishkoff et al.[28] presented evidence to suggest that these alleles evolved independently and increased in frequency at a rate that is too rapid to be explained by random genetic drift alone. Application of a statistical model indicated that the *A*− allele arose within the past 3,840–11,760 years, and the *Med* allele arose within the past 1,600 to 6,640 years. These results directly support the hypothesis that malaria is only likely to have had a major impact on humans since the introduction of agriculture (within the past 10,000 years), providing a striking example of a signature of very recent selection in the human genome.

8.3.2.2 Diet as a Selective Force in Human Evolution: Lactase

In the majority of human populations, the ability to digest lactose in milk declines rapidly after weaning because of decreasing levels of the enzyme lactase-phlorizin hydrolase (LCT). However, some individuals maintain the ability to digest lactose into adulthood, so-called lactase persistence. The frequency of lactase persistence is high in northern European populations (>90% in Swedes and Danes) but decreases in frequency across southern Europe and the Middle East (50% in Spanish, French, and pastoralist Arab populations) and is low in nonpastoralist Asian and African populations (1% in Chinese, 5%–20% in West African agriculturalists). Notably, lactase persistence is common in pastoralist populations from Africa (90% in Tutsi, 50% in Fulani). Several studies have presented strong evidence to suggest that the LCT locus has been subjected to a recent strong selective sweep. This selective sweep is particularly evident in Caucasian populations; in fact, in some genome-wide studies the LCT locus is the most strongly selected locus in the human genome.[29] An explanation for this remarkably strongly selected trait may lie in the recent history of Caucasian populations. Lactase persistence is believed to have arisen soon after Caucasians entered Europe after the last Ice Age. As their culture shifted from a hunter-gatherer culture to an agricultural, more specifically dairy farming, culture, alleles conferring lactose tolerance afforded a major selective advantage.[19,30]

8.3.2.3 Where Does Selection Leave Us When Our Environment Changes?

The selective pressures on human populations in developed nations have changed radically in the last century. Generally, Western populations are sheltered from famine, severe drought, or the extremes of climate. This reversal of age-old selective pressures can create problems in itself. For example, individuals bearing alleles that help to store fat would better survive famines but would be more susceptible to obesity in a modern society. Thompson et al.[11] demonstrated this principle when they showed that that a high-expressing allele of the cytochrome p450 gene, CYP3A5, confers, by influencing salt and water retention, a selective advantage in equatorial populations who may experience water shortages. The allele showed an unusual geographic pattern significantly correlated with distance from the equator. In Western populations, however, the allele was identified as a major risk factor for salt-sensitive hypertension.

8.3.2.4 Psychiatric Diseases: The Selective Price of Intelligence?

The principles of balancing selection have led some researchers to hypothesize that the fierce recent selection pressures for so-called human-specific characteristics such as intelligence and language have also created new disease burdens on human populations, such as psychiatric diseases. One of the most interesting cases in point is in regard to schizophrenia, a disorder prevalent in all human populations and with a multifactorial but highly genetic etiology. A constant prevalence rate in the face of reduced fecundity have caused some to argue that an evolutionary advantage exists in unaffected relatives, while others have proposed that schizophrenia was essentially a by-product of the evolution of complex social cognition.[31] The latter argument seems more persuasive as paleoanthropological and comparative primate research suggests that hominids evolved complex cortical interconnectivity to regulate social cognition and the intellectual demands of group living.

Burns[31] suggested that the ontogenetic mechanism underlying this cerebral adaptation rendered the hominid brain vulnerable to genetic and environmental insults. Burns argued that the changes in genes regulating the timing of neurodevelopment occurred prior to the migration of man out of Africa, giving rise to the schizotypal spectrum that is observed in populations today. While some individuals within this spectrum may have exhibited unusual creativity or leadership, this phenotype was not necessarily adaptive in reproductive terms. However, because the disorder shared a common genetic basis with the evolving circuitry of the social brain, it persisted. Thus, Burns proposed that schizophrenia emerged as a costly trade-off in the evolution of complex social cognition. Others have suggested that shamanism and similar characteristics may have been "enhanced" by psychosis, ensuring the survival of these alleles in populations.[32]

8.4 STUDYING HUMAN NATURAL SELECTION AT A MOLECULAR LEVEL

8.4.1 The "Neutralist-Selectionist" Debate

Although Darwinian theory might appear to be widely accepted as the fundamental principle governing the evolution of species at a molecular level, other evolutionary forces are known to exist; naturally, these are constantly reexamined in the light of new molecular data like the HapMap. Perhaps the most widely considered is the neutral theory of molecular evolution, which is arguably complementary to natural selection. First proposed by Kimura[33] in the late 1960s, the neutral theory proposes that when the genomes of existing species are compared, the vast majority of variants are selectively neutral, with no impact on fitness and hence no natural selection. Instead, the neutral theory asserts that most evolutionary change is the result of genetic drift acting on neutral alleles. Through drift, these new alleles may become more common within the population. In most cases, they will subsequently be lost, but in rare cases they may become fixed. In this way, neutral substitutions accumulate, and genomes evolve. Following on from this, polymorphism within species and divergence between species are likely to be governed by the effective population size and neutral mutation rate, respectively.[34] Put simply, most variants can be assumed to have accumulated at the same rate as individuals with mutations are born. It has been widely argued that this latter mutation rate is predictable from the error rate of the highly conserved enzymes that carry out DNA replication. Thus, the neutral theory is the foundation of the "molecular clock" concept that is widely used in evolutionary biology as a measure of the time passed since a species diverged from a common ancestor. In terms of the analysis of molecular selection, the neutral theory is used as a "null model" for hypothesis testing — comparing the actual number of differences between two sequences and the number that the neutral theory predicts given the independently estimated divergence time. If the actual number of differences is much less than the prediction, then the null hypothesis has failed, and researchers may reasonably assume that selection has acted on the sequences in question.

Neutral theory and natural selection are still a subject of debate, although instead of arguing for the exclusive action of one or the other process, the debate tends to be focused on the relative percentages of alleles that are "neutral" versus "nonneutral" in any given genome. Neutral theory is also evolving with the concept of "near neutrality."[35,36] The nearly neutral theory states that genes are affected mostly by drift or mostly by selection, depending on the effective size of a breeding population. This theory is particularly relevant for the study of the evolution of human populations, a process that is in many cases defined by bottleneck events and isolated population histories.[37,38]

Large-scale catalogs of genetic variation and LD, such as those generated by the SNP consortium,[5] Perlegen,[39] and most recently the HapMap[8] have all stimulated reexamination of the theory and demographics of human evolution. Gabriel et al.[5] were among the first to use LD evidence to support the post-Ice Age bottleneck that Caucasian populations were believed to have endured.[37] It followed naturally that these same data sets would be employed to investigate regions showing evidence of positive selection. Ultimately, study of selection at a large-scale molecular level offers to clarify the roles of drift and selection, which have so occupied evolutionary biologists. Rather less esoterically, identification of the signatures of positive selection may also highlight regions of the genome that are functionally important. In Section 8.5, some of the best examples of these genome-wide analyses for positive selection are reviewed; however, before addressing the details, it is worth taking a brief overview of some of the statistical methods that are used to detect selection or, more precisely, deviation from the null hypothesis of neutrality.

8.4.2 Approaches for Detecting Evidence of Selection

This chapter is not intended as a comprehensive review of the statistical approaches that are used to test for the imprint of natural selection; there are several excellent reviews that explore this area in more detail.[40,41] In Section 8.4, the key principles that underpin the analysis of selection were reviewed. All essentially compare DNA or amino acid variation in populations or species and attempt to estimate the degree of divergence before evaluating against the neutral model. The power of these tests is generally established by performing simulations under a limited range of demographic models and parameters to estimate the threshold at which the neutral model can be rejected.[42] With this in mind, it quickly becomes clear why an understanding of population history is crucial for identifying the genes that are subject to selection. Table 8.2 reviews some of the most commonly used methods for the analysis of selection. To try to put some of these approaches into context, some of the most commonly used methods receive a closer look next.

8.4.2.1 Using Protein Sequences to Test for Selection between Species

In regions that code for proteins, the simplest and most commonly used measure of deviation from neutral evolution between species is the relative rate at which nonsynonymous (amino acid–substituting) and synonymous (silent) mutations are fixed in a population.[50] This is known as the Ka/Ks ratio (or sometimes the dN/dS ratio), with Ka the rate of nonsynonymous substitutions and Ks the rate of synonymous substitutions. For example, under neutrality Ka/Ks = 1. If an amino acid is subject to functional constraint, then deleterious substitutions are purged from the population (negative selection); in such a case, Ka/Ks < 1. If Ka/Ks > 1, then the assumption is that the protein is evolving at a faster rate than would be expected under the neutral theory. This suggests the protein is undergoing positive selection. Although this test is elegant in its simplicity, at the whole-protein level this is only likely to detect more extreme cases of selection; it also has a high potential false-discovery rate for selected sites.[43] More subtle and powerful methods have been developed more recently (e.g., PAML (phylogenetic analysis by maximum likelihood)[44,51]) that focus on detecting selection at the level of individual codons. These methods can be used to analyze a gene — codon by codon — using data from multiple species to pinpoint potential amino acids on which selection appears to be acting. The successful application of these methods is heavily dependent on assignment of appropriate species orthologs for analysis, ideally using data from a wide range of species. SPEED (Searchable Prototype Experimental Evolutionary Database) is a Web tool that was developed specifically to facilitate such analyses, presenting the user with precomputed ortholog alignments, which could then be analyzed by a number of methods.[52]. Other chapters in this book deal with PAML and other methods for the analysis of coding sequences, so this subject is not explored further here.

8.4.2.2 Exploring Signatures of Selection across the Genome

Although proteins are obvious targets of selection, protein-focused methods have limitations. For example, neutrality of synonymous mutations cannot always be assumed as synonymous mutations may affect splicing, messenger RNA (mRNA)

TABLE 8.2
Methods for Detecting Molecular Selection

Test	Type	Designed to Detect	Best Use	Caveats	Reference
Ka/Ks	Between sp. (synonymous vs. nonsynonymous)		Adaptive protein evolution; mutation/selection	High potential false discovery rate for selected sitese	Guindon et al.[43]
PAML	Between species (synonymous vs. nonsynonymous)	Adaptive evolution in coding regions	Adaptive protein evolution; mutation/selection		Yang[44]
HKA	Within vs. between species (two loci)	Differences in variation levels not accountable by constraints	Balancing selection; recent selective sweeps or other variation-reducing forces	High recombination rates may reduce effectiveness of tests	Hudson et al.[45]
McDonald Kreitman G	Within vs. between species (synonymous vs. nonsynonymous)	Adaptive evolution	Adaptive protein evolution; mutation/selection	Selection on codon usage can seriously jeopardize tests	McDonald[46]
Tajima's D	Within species	Skew in frequency spectrum	General-purpose test of frequency spectrum skew; hard sweeps	See Mu et al.[27] for situations in which the test performs poorly	Tajima[47]
Fu's F_s	Within species	Excess or rare alleles (one sided)	Population growth; genetic hitchhiking; background selection	May be best overall test for detecting genetic hitchhiking and population growth	Fu[42]
Fst	Within species	Measure of allelic variability within and between populations	Simple test to identify outlier variants	Oversimplified for genome-wide analysis	Wright[48]
iHS	Within species	Haplotype based	Soft sweeps		Voight et al.[29]
Number of haplotypes K	Within species	Haplotype based	Soft sweeps		Depaulis & Veuille[49]

stability, or binding by regulatory RNAs such as microRNAs.[53] Likewise, many functional elements are known to reside outside coding regions that are likely to be particularly relevant to the study of human evolution. In studies of human and chimp gene sequences, it quickly became evident that the rare amino acid changes explained few of the phenotypic differences between these hominid cousins. In a visionary 1975 article, considering the lack of sequence information at the time, King and Wilson[54] hypothesized that the main differences between chimps and humans would most likely be found in noncoding regulatory DNA. For this reason alone, it can be argued that selection needs to be evaluated on a genome-wide scale.

Robust confirmation of King and Wilson's hypothesis was not possible for more than 30 years, until Pollard et al.[55] compared human and chimp genome sequences to find DNA elements that show evidence of rapid evolution in the human lineage. They based their analysis on the rate of nucleotide substitution and identified 202 so-called human accelerated regions (HARs) that are evolving very slowly in vertebrates but have changed significantly in the human lineage. A Web page is available that summarizes the properties of the HARs (http://www.docpollard.com/HARs.html). Interestingly, as King and Wilson might have predicted, the HARs were mostly noncoding (66.3% intergenic, 31.7% intronic, with just 1.5% overlapping coding genes). This study highlights the dual roles of neutral theory and selection in the evolution of the human genome as it is evident that more than one evolutionary force is shaping these rapidly evolving regions. To characterize each region, Pollard et al.[55] used the presence and extent of selective sweeps and likelihood ratio tests (LRTs) to detect substitution bias in HARs. The LRT statistic for a region was defined as the ratio of the likelihood of the model with acceleration on the human branch to the model without human acceleration. The significance of the LRT statistics was assessed against a genome-wide null model.[55] The top five most accelerated HARs (HAR1–5) were further evaluated for evidence of selective sweeps using a Hudson–Kreitman–Aguadé (HKA) test on genotype data (see Table 8.2). No evidence of departures from neutrality was identified in three of the top five. However, HAR1 and HAR2 showed significant departures from the neutral model ($p < 1.0e^{-4}$ and $p = 6.0e^{-4}$, respectively), suggesting the action of positive selection in the two most accelerated regions.

8.4.2.3 Using Genotype Data to Test for Selection between and within Species

Although amino acid and DNA sequences can be used for the analysis of selection, analysis of large within-species population samples calls for a different approach. Data resources like the HapMap create an opportunity to review evidence of selection in individual genes in the context of genotypes, which are effectively a genome-wide sampling of variation. This allows for a subjective evaluation of the difference between a putative selected locus and the background variation across the genome in a large number of samples. Polymorphism data can be used to distinguish different evolutionary signatures; for example, it can be used to distinguish between mutation bias and fixation bias.[56] As already seen, an accurate assessment of substitution rate variation is essential to enable construction of a neutral model against which accelerated evolution can be evaluated. Again, the key principle underlying the analysis of genotypes to detect positive selection is the same as it is with conventional sequence data, that is, the relationship between the age and frequency of an allele in a population. If selection is occurring neutrally, then higher-frequency alleles would be expected to be older than lower-frequency alleles due to genetic drift toward fixation.[33] Where a more recently arisen allele confers an advantage, it may undergo positive selection, leading to a rapid increase in population frequency as carriers of the allele are preferentially selected. Tishkoff et al.[30] provided a textbook example of such an analysis in their investigation of the positive selection of lactose tolerance alleles, examined previously in Section 8.3.2.2.

8.4.2.4 Using LD to Detect Selection

While genotypes can be used in an unlinked form to detect selection, LD data offer some distinct advantages over unlinked genotypes in the detection of selection. The principle underpinning the use of LD for detection of selection is simple, as illustrated in Figure 8.3.

A range of tests is available that use LD and haplotypes to detect selection; all are variations on similar themes (Table 8.2). The long-range haplotype (LRH) test examines the relationship between allele frequency and the extent of LD.[57] Positive selection is expected to accelerate the frequency of an advantageous allele faster than recombination can break down LD at the selected haplotype. To capture the hallmark of positive selection (an allele that has greater long-range LD than would be expected given its frequency in the population), the LRH test begins by selecting a

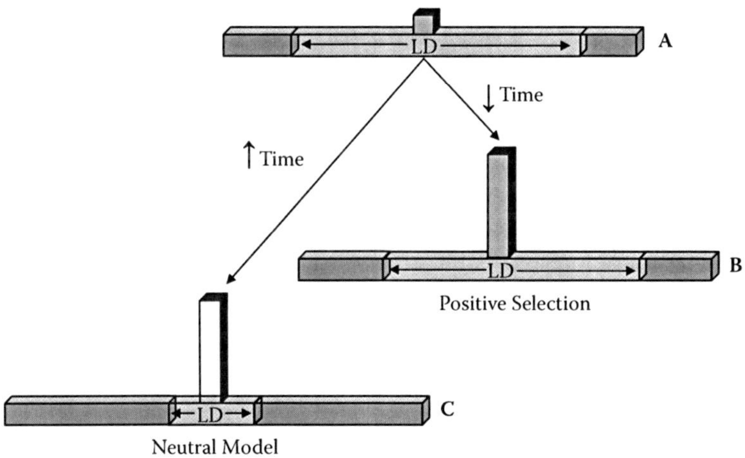

FIGURE 8.3 Using linkage disequilibrium information to detect selection. A new allele enters the population (indicated by the height of the vertical bar in Figure 8.3a) on a background haplotype that is characterized by long-range LD between the allele and the linked markers. In the case of positive selection (Figure 8.3b), the selected allele increases in frequency faster than local recombination can reduce the range of LD between the allele and the linked markers. In the case of neutral evolution (Figure 8.3c), the frequency of the allele increases slowly as a result of genetic drift, and local recombination reduces the range of the LD between the allele and the linked markers.

core haplotype. The relative decay in LD is assessed for flanking markers by calculating extended haplotype homozygosity (EHH),[58] defined as the probability that two randomly chosen chromosomes carrying the core SNP or haplotype are identical by descent. For each core, haplotype homozygosity is initially 1, and as distance increases, it decays to 0. Positive selection is formally tested by finding core haplotypes that have elevated EHH relative to other core haplotypes at the locus conditional on haplotype frequency. By focusing on relative levels of EHH in each region, the various core haplotypes control for local rates of recombination.

One of the advantages of haplotype methods for the detection of selection is that they allow estimation of the age of an allele, taking a genealogical view of LD. This makes it possible to uncover historical patterns of recombination that reflect the age of an allele.[59] A haplotype-sharing method particularly suited for this task is DHS, a method that estimates the decay of ancestral haplotype sharing.[60] The term *ancestral haplotype* refers to the original configuration of linked variants that were present in the ancestral chromosome carrying the selected allele. The length of the ancestral haplotype retained shortens, as seen in Figure 8.3. Methods such as DHS test for deviations from the expected level of preservation of the ancestral haplotype (in terms of genetic distance) as a reciprocal of the time in generations back to the most recent common ancestor (TMRCA) of the allele.

Toomajian et al.[60] clearly summarized the steps required to use DHS to detect deviation from the neutral model of evolution. First, haplotype data are collected from a population using markers that flank a region of interest (e.g., a gene or exon). The haplotypes are sorted by the alleles carried at the region of interest. The ages of the alleles are estimated using DHS from the observed decay of haplotype sharing at the flanking markers within the haplotype set and compared to the background LD and allele frequencies found at the marker loci on the remaining haplotypes. The frequencies of the alleles are then compared against the neutral model to identify alleles estimated to be young but at unexpectedly high frequencies. These simulations model uncertainty in the geneology of alleles and provide an appropriate statistical comparison for the observed alleles. In a final step, the ages of observed alleles need to be compared with the distribution of ages for simulated alleles at the same frequency produced under different demographic models. Alleles that are younger than the vast majority or all of the simulated alleles are unlikely to occur by chance under the neutral models and are indicative of a possible selection event.

8.4.3 Deviations from Classical Models of Selection

The classical model of a selective sweep that we might want to detect assumes that the beneficial allele arose on a single occasion by mutation, a so-called hard sweep.[61] However, this assumption may not always hold. For example, an advantageous substitution might originate by several independent but identical mutation events on different haplotype backgrounds. As one can imagine, this throws the proverbial spanner in the works for many classical methods of detecting selective sweep signatures. Pennings and Hermisson[62] coined the term *soft sweep* to describe this phenomenon and evaluated the power of different analytical tools and coalescent simulations to detect soft-sweep signatures. In this valuable study, they showed that soft sweeps tended to be characterized by strong LD. This has obvious implications for the tests used to detect soft-sweep signatures, suggesting that existing LD-based tests (such as iHS [integrated haplotype score],[29] DHS,[60] and K[49]) might have increased power to detect soft sweeps. Pennings and Hermisson confirmed this, showing that LD-based tests actually performed better on soft sweeps than classical sweeps, particularly when a sweep was recent and had not yet reached fixation.

This may be a very important concept to consider during analysis of selection and may go some way toward explaining the difference in the performance and lack of overlap between the results of different selection tests. It also underlines the need for considering the results of different selection tests across a region of interest to take account of hard and soft sweeps in addition to the age of a selection event, something explored further in Section 8.5.

8.4.4 The Role of Demographics and Other Mutational Events in Molecular Evolution

As seen in the study of HARs by Pollard et al.,[55] positive selection is not the exclusive evolutionary force resulting in accelerated sequence evolution. Demographic events, such as population subdivisions and rapid changes in population size, can lead to the accelerated fixation of segregating alleles.[40] Unlike the local effects of natural selection, demographic events affect all genes in a genome. While this creates obvious problems for the analysis of selection, the genome-wide nature of demographic effects makes a genome-wide correction possible.

Stajich and Hahn[63] used publicly available data from 151 loci sequenced in both European and African American populations to distinguish the effects of demography and selection. Their analyses confirmed that demographics can account for a large proportion of the frequency of genomic variation. For example, they showed that African American populations show both a higher level of nucleotide diversity and more negative values of Tajima's D than European populations. These observations could be explained using relatively simple coalescent models of population admixture and bottleneck, respectively. However, even working within such a framework, they were still able to demonstrate deviations from neutral expectations at a number of loci and in many regions of low recombination. They concluded that these results were consistent with the combined effects of population bottlenecks and repeated selective sweeps during the human migration out of Africa, in agreement with previous reports.[64]

The nature of certain mutational events can also confound the analysis of selection by any method. Studies of guanine (G) and cytosine (C) nucleotide-enriched genomic regions, known as (GC)-isochores, have highlighted another selectively neutral evolutionary process with a potentially important influence on nucleotide evolution. Biased gene conversion (BGC) is a mechanism caused by the mutagenic effects of recombination[65] combined with the preference in recombination-associated DNA repair toward strong (GC) versus weak (adenine and thymine [AT]) nucleotide pairs at non-Watson-Crick heterozygous sites in heteroduplex DNA during crossover in meiosis. BGC results in an increased probability of fixation of G and C alleles despite beginning with random mutations. Recent studies have shown that increasing the GC content of transcribed sequences may increase their expression level, which in some cases may offer a selective advantage.[66] So, this is another example for which tests seeking evidence of selection against the neutral model may be confounded by BGC or alleles fixed by demographic factors.

8.4.5 Investigating the Link among Selection, Sequence Conservation, and Linkage Disequilibrium

One striking observation made during studies of genome-wide LD was that exonic regions were often associated with strong LD in human populations. For example, Hinds et al.[39] observed significant overrepresentation of genic SNPs in extended LD regions, while Tsunoda et al.[67] found that LD was significantly stronger between exonic variants within a gene compared with intronic or intergenic SNPs.

Kato et al.[68] used HapMap data to rigorously evaluate these observations in an evolutionary context. They hypothesized that LD might be stronger in regions conserved among species than in nonconserved regions since regions exposed to natural selection would tend to be conserved. To evaluate this, they examined LD in regions conserved between the human and mouse genomes. Their results were somewhat unexpected. They observed that LD was significantly weaker in conserved regions than in nonconserved regions. To try to explain this observation, they looked for sequence features that might distort the relationship between LD and conserved regions. Interestingly, they found that interspersed repeats were associated with the tendency toward weak LD in conserved regions. After removing the effect of repetitive elements, they found that, as originally expected, a high degree of sequence conservation was indeed strongly associated with high LD in coding regions but not in noncoding regions. Combining these observations, they concluded that negative selection may act on the polymorphic patterns of coding sequences but may not influence the patterns of functional units such as regulatory elements present in noncoding regions. They suggested that the action of negative selection on coding sequences might be due to the constraint of maintaining a functional protein product across multiple exons compared to the relative lack of restraint required to maintain a regulatory element as an individually isolated unit.

8.5 EVALUATING SELECTION IN HUMAN POPULATIONS USING GENOME-WIDE SCREENS

8.5.1 A Genome-wide Approach to the Analysis of Selection

The hypothesis-free approach of genome-wide selection analysis is an attractive alternative to the candidate gene approach for the detection of selection. The advantages of going genome wide are obvious: We know that function is not limited to gene regions, so why test only these regions? Genome-wide scans can be performed using either SNPs or microsatellites. Each exhibits different rates of return to mutation-drift equilibrium because SNP and microsatellite mutation rates differ by several orders of magnitude.[69] Thus, comparison of patterns of SNP and microsatellite polymorphism can be expected to provide valuable information about the timing of selection events. The most appropriate form of variation for use in selection analysis may be dependent on the age of the event under study. For example, microsatellites generally show a higher mutability. Wiehe[69] suggested that microsatellite-based studies would be most appropriate for detecting selective sweeps that were both strong and recent (e.g., during the Neolithic era). By contrast, SNP variation may be more appropriate for detecting relatively ancient sweeps as mutation rates for SNPs are generally at least four orders of magnitude lower.[70]

8.5.2 A Review of Published Genome-wide Studies of Selection

Early (pre-HapMap) attempts to detect selection using pairwise LD data were of limited success due to the limited information available and a tendency to oversimplify analysis by assuming independence between linked alleles. Sabeti et al.[58] performed one of the first robust LD-based studies of positive selection, although they did not use a genome-wide LD data set. Their method improved on earlier methods by allowing the addition of loci at increasing distances. They were also able to distinguish recent from ancient events that had already reached fixation. To do this, they used the relationship between haplotype frequency and the extent of LD associated with haplotypes to determine if and when positive selection might have occurred. Voight et al.[29] later extended this method using phase I HapMap data to identify human variants under directional selection that have not yet reached fixation.

The HapMap[8] has made LD data widely available; this has led to a flurry of genome-wide studies of selection based on the same data set.[8,29,71–73] These studies are summarized and compared in Table 8.3.

8.5.2.1 Selection Data Available Online

Among the studies summarized in Table 8.3, two studies have published their data online. These are valuable resources that allow the user to rapidly query, using precomputed analysis, a gene or region of interest for evidence of selection. Voight et al.[29] scanned phase I HapMap SNP data in the CEU, YRI, and JPT&CHB populations using the haplotype-based test iHS and found evidence of recent positive selection in all three population samples. They identified nominally significant

TABLE 8.3
A Comparison of Published Genome-wide Scans for Positive Selection

Study	Wang et al.[71]	Voight et al.[29]	Carlson et al.[73]	Altshuler et al.[8]	Nielsen et al.[72]	Bustamante et al.[74]
Data set	Perlegen & HapMap	HapMap	Perlegen	HapMap	Chimp vs. human genomes	Chimp vs. human genomes
Data type	Genotypes	Genotypes	Genotypes	Genotypes	Protein coding	Protein coding
Positive selected genes	1,799	455	176	19 (926 SNPs)	56	304
Methods applied	LD decay	iHS	Tajima's D	LRH	HKA/PAML	McDonald-Kreitman G

evidence of positive selection in at least one population in 2,532 genes. The results of these analyses are available to query using a stand-alone application, Haplotter (http://hg-wen.uchicago.edu/selection/). This allows the user to query an SNP, gene, or genomic region. One of the most valuable features of the Haplotter tool is that it allows the user to compare iHS scores against other measures of selection across a region, including measures of Tajima's D,[47] H,[75] and FST.[48] In Figure 8.4, the output from Haplotter across the lactase (LCT) locus is shown; this serves to illustrate the different performances of these methods across one of the strongest signals of recent selection in the human genome.

Another source of selection data online comes from an earlier study by Carlson et al.[73] They used the 1.5 million SNPs described by Hinds et al.[39] to carry out a Tajima's D analysis in three populations. This analysis is available in the Tajima's D track in the University of California, Santa Cruz (UCSC), genome browser (Table 8.4).

8.5.2.2 Investigating Overlap between Genome-wide Studies of Selection

Biswas and Akey[2] attempted to summarize the pairwise overlap in positively selected genes identified by the genome-wide scans reviewed in Table 8.3. In total, the various scans identified several thousand loci with putative signatures of positive selection; many of these encompassed large regions, including 2,316 genes in total. As most loci contained multiple genes, further analysis of each locus is required to attempt to identify the selected allele in the gene in question and to exclude the other genes. In many cases, resolution of a selective sweep signature to an allele in a gene may be difficult indeed as the selected allele may not be directly localized in the gene or even the gene region (in the case of *cis*-regulatory elements, which may be at great distances from the gene).[76] The size of the selected region is likely to depend on the age of the selection event; the more recent the event, the larger the expected locus will be. The lactase (LCT) locus is a great case in point of many of these problems. First, the selective sweep signature spans over 1 Mb of sequence, encompassing five large genes. Second, the putative selected alleles are not located within the lactase gene as might be expected but are instead localized in the intron of the neighboring gene MCM6.[30] Section 8.7 takes a close look at the lactase locus as an example of the steps needed to follow-up a signature of selection.

Perhaps unsurprising considering the differing properties of tests for selection, Biswas and Akey[2] found only a modest overlap between genes identified by the various published genome scans for selection. They found the greatest number of significantly associated genes shared between the studies of Voight et al.[29] and Wang et al.[77] In this case, both studies shared 27% of the significant genes, as might be expected as both used extended regions of LD to detect selection. Interestingly, although 27% of the significant genes from Carlson et al.,[73] who used Tajima D (a test of frequency skew), overlap with Wang et al.,[77] only 8% overlap with Voight et al.[29] This may not necessarily be due to false-positive signals but may instead reflect the difference in the ability of the Tajima D and LD-based methods to detect different age selection events.

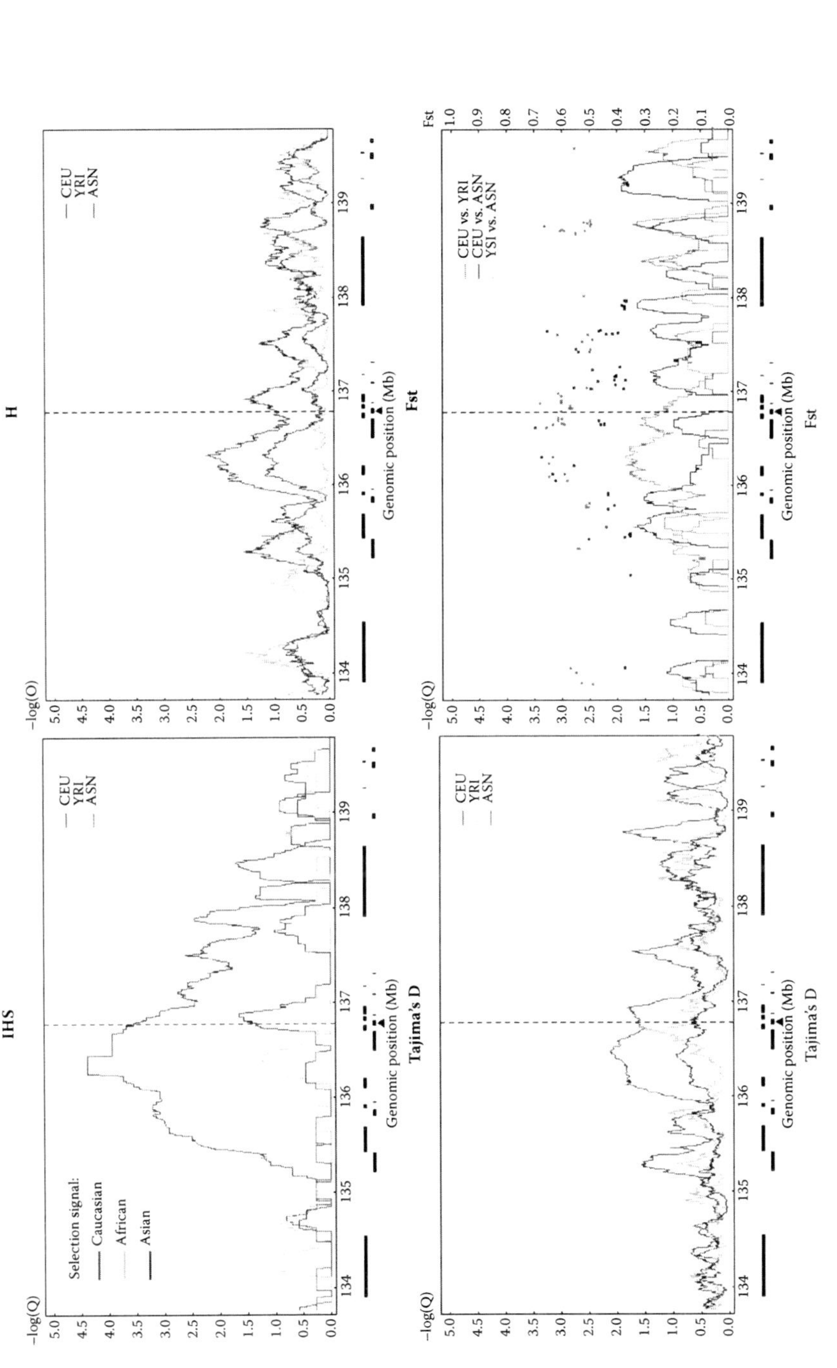

FIGURE 8.4 (See color figure in the insert following page 48.) Haplotter output across the LCT locus. Results of four different molecular selection analysis methods (iHS, H, Tajima's D, Fst) are presented across the LCT locus.

TABLE 8.4
Tools for Analysis of Human Population-Specific Selection

Tool	URL
Software for Mapping and Analysis of Selection	
Pritchard lab tools	http://pritch.bsd.uchicago.edu/software.html
Popgen analysis tools	http://www.biology.lsu.edu/general/software.html
BIOPERL popgen WIKI	http://www.bioperl.org/wiki/HOWTO:PopGen
Detecting Selective Sweep Signatures	
Haplotter	http://hg-wen.uchicago.edu/selection/
Sweep	http://www.broad.mit.edu/mpg/sweep/
Variscan	http://www.ub.es/softevol/variscan/
Detecting Signatures of Mammalian Selection	
SPEED	http://bioinfobase.umkc.edu/speed/
PAML	http://abacus.gene.ucl.ac.uk/software/paml.html
Genome Visualization	
UCSC Genome Browser	http://genome.ucsc.edu
ENSEMBL	http://www.ensembl.org
LocusView	http://www.broad.mit.edu/mpg/locusview/
NCBI MapViewer	http://www.ncbi.nlm.nih.gov/mapview/map_search.cgi/
LD and Haplotype Data	
HapMap Web site	http://www.hapmap.org
HapMap Genome Browser	http://www.hapmap.org/cgi-perl/gbrowse/gbrowse/
HapMart	http://hapmart.hapmap.org/BioMart/martview
Integrated Genome-scale Data Annotation Tools	
DAVID	http://http://david.abcc.ncifcrf.gov/
GSEA	http://www.broad.mit.edu/gsea/
GEPAS	http://gepas.bioinfo.cipf.es/cgi-bin/anno
GFINDer	http://www.medinfopoli.polimi.it/GFINDer/
L2L	http://depts.washington.edu/l2l/
Specialist Gene Ontology (GO) Analysis	
GO tools	http://www.geneontology.org/GO.tools.shtml
Gene Ontology Tree	http://bioinfo.vanderbilt.edu/gotm/
Building Biological Rationale	
Stanford SOURCE	http://source.stanford.edu
OMIM	http://www.ncbi.nlm.nih.gov/entrez/query.fcgi?db=OMIM
UniProt	http://www.uniprot.org
Functional Analysis of Variation	
FastSNP	http://fastsnp.ibms.sinica.edu.tw/fastSNP/index.htm
PupaSNP	http://pupasnp.bioinfo.cnio.es

8.5.3 Caveats of the Genome-wide Approach

All genome-wide analysis approaches, such as association analysis or expression analysis, carry a burden of false-positive associations (type I error) due to multiple testing. Genome scans for signatures of selective sweeps are no exception to this rule; indeed, the problem may be compounded in some cases by other factors, such as ascertainment bias among the polymorphisms tested. The HapMap SNP ascertainment strategy has generated some debate. Phase I and II HapMap SNPs were prioritized for analysis primarily on the basis of prior validation; failing this, they were also considered validated if they matched a variant in chimpanzee sequence data.[8] This means that the phase I, and to a lesser extent the phase II, HapMap data sets show significant ascertainment bias toward ancestral (generally common) alleles.[78] The impact of this is complex and dependent on the specific analysis undertaken. In theory, it is possible to correct analyses for the ascertainment scheme used to select SNPs,[73,79] but in some cases such corrections are at best approximate. This is a major issue in the interpretation of the results of scans for selection.

Considering the problems of multiple testing, ascertainment bias, and the existence of demographic events that mimic selective sweeps, it really is difficult to completely exclude false-positive signals. However, there are ways to limit them. In a microsatellite-based study, Wiehe et al.[80] showed that the analysis of flanking markers drastically reduced the number of false positives among the candidate regions identified in a genome-wide survey of unlinked loci. However, in some severe population bottleneck scenarios, they found genomic signatures that were very similar to those produced by a selective sweep. They concluded that, in such worst-case scenarios, the power of microsatellite methods remained high, but the false-positive rate reaches values close to 50%. With this in mind, they concluded that selective sweeps may be hard to identify even if multiple linked loci are analyzed.

Aside from the problems of type I error, there are many other challenges, such as the demographic effects and mutation effects discussed, which could potentially confound signals of selection. Ultimately, like most other genomic data, signals of selection need to be considered alongside other information that might support a selective event in the genomic region in question. Such supporting information might include evidence of functionality for a selected allele or a rationale in a selection event for a selected gene. Methods for pulling together other supporting evidence for selection are addressed in the following sections.

8.6 PRIORITIZING GENES TO INVESTIGATE SIGNALS OF NATURAL SELECTION

8.6.1 Following Up a Signal of Selection at Gene Level

The flurry of studies of selection stimulated by the HapMap have raised the standard for reporting evidence of selection, calling for robust experimental evidence to provide a molecular or functional basis by which selection is likely to act. This is a necessary requirement due to the high level of false-positive associations that the genome-wide approach generates. However, it is simply not possible to perform follow-up experiments on every gene when thousands of genes are implicated.

Preliminary associations need to be prioritized using in silico analysis methods to determine the appropriate experiments to test a functional hypothesis. Genes need to be prioritized based on their likely function and involvement in known pathways, possibly leading to some rationale for selection (e.g., involvement in key processes such as immunity). In this section, some of the best methods available on the Web for analysis of large-scale gene-based data sets are reviewed.

8.6.2 Functional Annotation of Genome-Scale Data Sets

There are currently many public efforts that focus on the functional annotation of genes and proteins; Entrez Gene, UniProt, and OMIM (Table 8.4.) are notable examples of tools that are leading the field in this area. However, most of these tools can only be queried on a gene-by-gene basis, making them unsuitable for analysis of genome-scale gene sets, such as those generated during genome-wide scans of selection. Microarray analysis of gene expression is a mature area of research with similar analysis requirements to genome-wide scans for selection; both deal with highly multidimensional data on a genome scale, and both have issues of multiple testing, generating many thousands of results, with a large burden of false positives. There are no tools specifically developed to deal with the output of genome scans for selection; fortunately, there are a number of tools that focus on similar issues in the microarray domain that are more than adequate for our needs (see Table 8.4 and Verducci et al.[81] for a review).

One of the most versatile tools for functional annotation of large gene sets is the Database for Annotation, Visualization, and Integrated Discovery (DAVID)[82] (http://david.abcc.ncifcrf.gov/). DAVID provides a suite of data-mining tools that systematically combine functionally descriptive gene annotation based on gene ontology[83] (GO), *Kyoto Encyclopedia of Genes and Genomes* (KEGG) (http://www.genome.jp/kegg/), BioCarta (http://www.biocarta.com), and other pathway tools with intuitive graphical displays. The tool provides exploratory visualizations of functional categories, pathways, and GO terms that are enriched at statistically significant levels in the data set. Tools such as DAVID can be used in two distinct ways; first, they can be used to simply expedite the process of functional annotation and analysis of a list of genes for further analysis, or they can be used as a means to attempt to identify genes that are significantly enriched in specific pathways or functional classes.

The controlled vocabulary of GO provides a structured language that can be applied to the functions of genes and proteins in all organisms, with up-to-date knowledge of gene function added as it continues to accumulate and evolve.[83,84] The GO module in DAVID offers the opportunity to evaluate the distribution of submitted genes across three general types of classification: biological process (GOTERM_BP), cellular component (GOTERM_CC), and molecular function (GOTERM_MF). These are divided further into five levels of annotation of increasing specificity of term coverage. These differing levels can be useful for modifying the threshold of inclusion for selection of genes for follow-up based on biological rationale. For example, given a list of several hundred genes, one might want to identify genes that might be selected during the development of cognition in humans. In this case, the level 3 biological process term "nervous system development" is of particular interest. Evaluation of the GO annotations in DAVID quickly identifies a number of genes

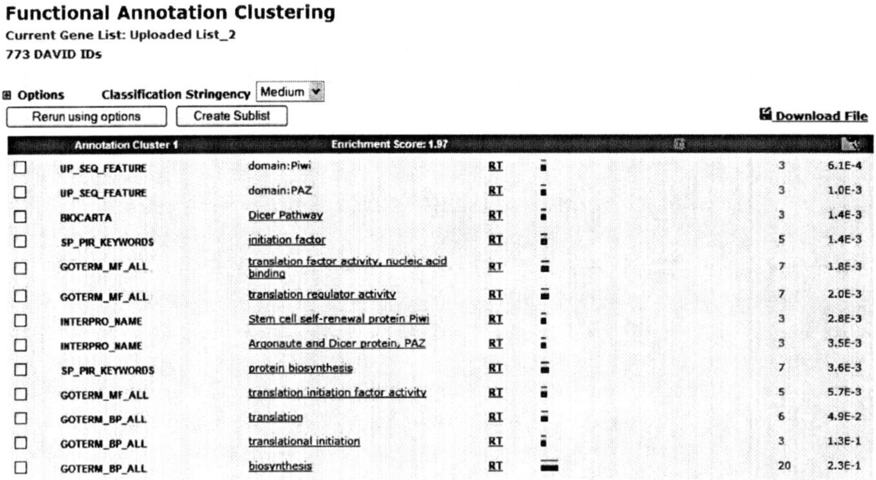

Functional Annotation Clustering
Current Gene List: Uploaded List_2
773 DAVID IDs

Options Classification Stringency Medium
Rerun using options Create Sublist Download File

Annotation Cluster 1	Enrichment Score: 1.97			
UP_SEQ_FEATURE	domain:Piwi	RT	3	6.1E-4
UP_SEQ_FEATURE	domain:PAZ	RT	3	1.0E-3
BIOCARTA	Dicer Pathway	RT	3	1.4E-3
SP_PIR_KEYWORDS	initiation factor	RT	5	1.4E-3
GOTERM_MF_ALL	translation factor activity, nucleic acid binding	RT	7	1.8E-3
GOTERM_MF_ALL	translation regulator activity	RT	7	2.0E-3
INTERPRO_NAME	Stem cell self-renewal protein Piwi	RT	3	2.8E-3
INTERPRO_NAME	Argonaute and Dicer protein, PAZ	RT	3	3.5E-3
SP_PIR_KEYWORDS	protein biosynthesis	RT	7	3.6E-3
GOTERM_MF_ALL	translation initiation factor activity	RT	5	5.7E-3
GOTERM_BP_ALL	translation	RT	6	4.9E-2
GOTERM_BP_ALL	translational initiation	RT	3	1.3E-1
GOTERM_BP_ALL	biosynthesis	RT	20	2.3E-1

FIGURE 8.5 Functional annotation of selected genes using DAVID. Genes showing statistically significant enrichment in specific pathways or Gene Ontology terms are highlighted and assigned a p value.

that are involved in processes that might be highly relevant to cognitive development in humans; these are summarized by a tabular visualization (Figure 8.5).

8.6.3 Using Pathway Tools

DAVID also annotates highly characterized pathways contained in KEGG, Biocarta, and a selection of other databases. While GO is based mainly on functional inference by homology, the information in these databases is based on experimental evidence and can be valuable for placing a gene in a validated pathway context. The amount of data is sometimes limited but generally of very high quality. Looking at the disease tab, in the OMIM_phenotype section two genes, DTNBP1 and APOL2, are linked to schizophrenia. In each case, if the user follows the hyperlinked terms, detailed information is returned that can rapidly put a gene into full biological context.

Annotation is a critical first step to move from a long list of possibly selected genes to a short list of genes worthy of detailed analysis. However, in a genome-wide study, even the narrowest definition for pathways of interest (e.g., cognitive development) are likely to generate lengthy lists of plausible genes. The next step calls for more focus on a gene and locus level to try to sort the real signatures of selection from the false. The next section reviews some possible approaches to achieve this.

8.7 FOLLOWING UP INDIVIDUAL SIGNALS OF POSITIVE SELECTION

8.7.1 Take a Second Statistical Opinion

Before committing costly laboratory resources or even in silico resources to the further analysis of a candidate selected gene, it is probably worth reviewing the locus in the light of a range of different tests for selection. As described in Section 8.4, different tests of selection have different power to detect selection events based on the age and nature of the event. Different methods can build confidence or cast light on the age of the selection event. For example, as described earlier, LD-based methods have more power to detect soft sweeps than frequency skew-based methods. The easiest way to review this kind of information without rerunning the analysis is to use Haplotter (Table 8.4). As seen in Figure 8.4, Haplotter plots several different measures of selection across a given locus; this makes it relatively easy to compare a range of tests.

Just as different tests can detect different events, the same principle applies to the type of marker. As discussed, highly mutable markers, such as microsatellites, are more suited to the detection of recent selection events than less-mutable markers, such as SNPs. In converse, less-mutable SNPs are more suited to the detection of ancient events.

8.7.2 Placing Signatures of Selection into a Genomic Context

Understanding the wider genomic context of a region containing a selective sweep signature is also an important next step toward an understanding of the molecular basis of the event that led to selection. Variants that may either be directly selected or "hitchhiking" with selected alleles need to be reviewed in the wider context of LD and haplotype information across a genomic locus. The UCSC genome browser[85] and the HapMap genome browser[86] are key tools to achieve this; both integrate HapMap LD and selection data with other genomic information.

Viewed in a genome-integrated form, in the UCSC or HapMap genome browser selection signals can also be reviewed in the context of the physical nature of the genome, which may be relevant. It is important to know about any physical features that might influence the evolution of a region. For example, structural variation may have a functional impact.[87] Information on recombination rates may also be important as recombination hot spots and cold spots might bias tests for natural selection. HapMap LD data itself can also provide information on functional relationships among genes, variants, and regulatory elements by highlighting selectively constrained relationships between variants (e.g., between groups of genes or a gene and *cis*-regulatory elements).[88]

Although the UCSC and HapMap genome browsers have many similarities, each contains distinct information and data interpretation, so it usually pays to consult both viewers. The UCSC genome browser has one great advantage over the HapMap genome browser as it allows visualization of LD across regions of greater than 1 Mb or even whole chromosomes. This robust LD visualization really makes the UCSC browser an exceptional tool for integrated LD/genomic analysis.[7]

8.7.3 Identifying Candidate Selected Alleles

Narrowing a selective sweep signal to the putative allele undergoing selection is a process fraught with difficulties. First, the actual selected allele may not be present in the available data. The location of the allele can also be a source of problems. One should not assume that a selected allele will be located in the gene undergoing selection. The lactase gene LCT provides an excellent example of the complexity that may often exist. An LD and haplotype analysis of Finnish pedigrees with lactase persistence identified two SNPs associated with the lactase persistence trait located 14 kb and 22 kb upstream of LCT, respectively, within introns 9 and 13 of the adjacent MCM6 gene.[30] These alleles were 100% and 97% associated with lactase persistence, respectively. Although these alleles could simply be in LD with an unknown regulatory mutation, several additional lines of evidence, including mRNA transcription studies and reporter gene assays driven by the LCT promoter in vitro, suggest that these are SNPs located in a *cis*-acting regulator of LCT transcription in Europeans.[30]

The HapMap genome browser can help in the search for selected alleles by allowing the user to visually review allele frequencies in all populations across a region showing selection by using the population-specific SNP frequency pie charts. If a selective sweep signature is restricted to an individual population, then the selected allele should show a significantly higher frequency in the selected population. Similar analysis can also be completed using tools such as HapMart, a data mining tool on the HapMap Web site. HapMart can be used to export allele frequency data in bulk to evaluate population-specific differences.

8.7.4 Functional Analysis of Putative Selected Variants

One of the most convincing pieces of in silico evidence that can be used to support the case for selection is function. It follows logically that if an allele is subject to selection, it will modify function. In the case of negative selection, it might be expected to be deleterious; in the case of positive selection, it would be expected to be advantageous — although the reverse might apply depending on the role of the gene in the selected trait. Proving function using in silico methods might sound relatively straightforward, but variation can have an impact on almost any biological process; hence, the scope of analysis required is immense. Much of the precedent in the area of functional analysis of variation has focused on the most obvious variation: nonsynonymous changes in genes. Alterations in amino acid sequences have been identified in a great number of diseases, particularly those that show Mendelian inheritance. This may reflect the severity of many Mendelian phenotypes, but this is probably not due to an increased likelihood that coding variation changes function but rather a bias in analysis that focuses in functional terms on the low-hanging fruit — the coding variation. Nonsynonymous variants may have an impact on protein folding, active sites, protein–protein interactions, protein solubility, or stability. But, the effects of DNA polymorphism are by no means restricted to coding regions. Variants in regulatory regions may alter the consensus of transcription factor-binding sites or promoter elements; variants in the untranslated region (UTR) of mRNA may alter mRNA stability or microRNA regulation[89]; variants in the introns and silent variants in exons may alter splicing efficiency.[90] Many of these noncoding changes may have an almost imperceptibly subtle impact on phenotype, but they may still be subject to strong selection as the subtlest alterations can nonetheless lead to major phenotypic effects in combination with other factors, such as lifestyle, environment, or disease.

8.7.5 Functional Analysis of Variants

Approaches for evaluating the potential functional effects of genetic variation are almost limitless, but there are only a few tools designed specifically for this task.

Instead, almost any bioinformatics tool that makes a prediction based on a DNA, RNA, or protein sequence can be commandeered to analyze polymorphisms — simply by analyzing both alleles of a variant and looking for an alteration in predicted outcome by the tool (many such tools are listed in Table 8.4). Polymorphisms can also be evaluated at a more fundamental level by looking at physical considerations of the properties of genes and proteins, or they can be evaluated in the context of a variant within a family of homologous or orthologous genes or proteins. Mooney[91] presented an excellent overview of some of the bioinformatics approaches to analyze the function of putative selected alleles.

8.7.6 Taking a Signature of Selection into the Lab

No matter how exhaustive any in silico analysis of function might be, the final proof of a hypothesis usually lies in the lab. Appropriate experimental evidence to support a signature of selection might involve a combination of sequence analysis with biochemical assays of recombinant proteins. For example, Zhang et al.[92] demonstrated how positive selection and relaxation of negative selection shaped the functional divergence of duplicated genes of a digestive enzyme (ribonuclease [RNase]) in colobine monkeys. Based on these experiments, Zhang et al. were able to attribute the selective force to an earlier change in diet. Other methods to prove a functional hypothesis can usually be found by judicious review of the literature; naturally, an experiment with a precedent is the best guarantee of success.

8.8 CONCLUSION: REPAYING THE DEBT OF BEING HUMAN

It is hoped that this chapter has shown that data previously the reserve of evolutionary biologists are now available in the public domain for all researchers. This offers an exciting opportunity to add selection into the general gamut of analysis methods for molecular genetic research. The field of molecular selection analysis is moving fast. This is a credit to researchers in the field; they have made something quite extraordinary accessible but never ordinary. We know that evolution shaped humanity, but it is clear that there was a cost — you could say that we are paying for this with some of the unique diseases that make us human. It is quite a debt, but hopefully the advances over the last few years will help us to start making the repayments.

REFERENCES

1. Chen, F.C. & Li, W.H. Genomic divergences between humans and other hominoids and the effective population size of the common ancestor of humans and chimpanzees. *Am J Hum Genet* **68**, 444–456 (2001).
2. Biswas, S. & Akey, J.M. Genomic insights into positive selection. *Trends Genet* **22**,437–446 (2006).
3. de Groot, N.G. et al. Evidence for an ancient selective sweep in the MHC class I gene repertoire of chimpanzees. *Proc Natl Acad Sci U S A* **99**, 11748–11753 (2002).
4. Ingman, M., Kaessmann, H., Paabo, S. & Gyllensten, U. Mitochondrial genome variation and the origin of modern humans. *Nature* **408**, 708–713 (2000).
5. Gabriel, S.B. et al. The structure of haplotype blocks in the human genome. *Science* **296**, 2225–2229 (2002).

6. The International HapMap Consortium. The International HapMap Project. *Nature* **426**, 789–796 (2003).
7. Barnes, M.R. Navigating the HapMap. *Brief Bioinform* **7**, 211–24 (2006).
8. Altshuler, D., Brooks, L.D., Chakravarti, A. et al. A haplotype map of the human genome. *Nature* **437**, 1299–1320 (2005).
9. Barton, N.H. The effect of hitch-hiking on neutral genealogies. *Genet Res* **72**, 123–133 (1998).
10. Nakajima, T. et al. Natural selection and population history in the human angiotensinogen gene (AGT), 736 complete AGT sequences in chromosomes from around the world. *Am J Hum Genet* **74**, 898–916 (2004).
11. Thompson, E.E. et al. CYP3A variation and the evolution of salt-sensitivity variants. *Am J Hum Genet* **75**, 1059–1069 (2004).
12. Lamason, R.L. et al. SLC24A5, a putative cation exchanger, affects pigmentation in zebrafish and humans. *Science* **310**, 1782–1786 (2005).
13. Hamblin, M.T., Thompson, E.E. & Di Rienzo, A. Complex signatures of natural selection at the Duffy blood group locus. *Am J Hum Genet* **70**, 369–383 (2002).
14. Kwiatkowski, D.P. How malaria has affected the human genome and what human genetics can teach us about malaria. *Am J Hum Genet* **77**, 171–192 (2005).
15. Sakagami, T. et al. Local adaptation and population differentiation at the interleukin 13 and interleukin 4 loci. *Genes Immun* **5**, 389–397 (2004).
16. Xue, Y. et al. Spread of an inactive form of caspase-12 in humans is due to recent positive selection. *Am J Hum Genet* **78**, 659–670 (2006).
17. Gabriel, S.E. et al. Cystic fibrosis heterozygote resistance to cholera toxin in the cystic fibrosis mouse model. *Science* **266**, 107–109 (1994).
18. Patin, E. et al. Deciphering the ancient and complex evolutionary history of human arylamine *N*-acetyltransferase genes. *Am J Hum Genet* **78**(3), 423–436 (2006).
19. Bersaglieri, T. et al. Genetic signatures of strong recent positive selection at the lactase gene. *Am J Hum Genet* **74**, 1111–1120 (2004).
20. Akey, J.M., Swanson, W.J., Madeoy, J., Eberle, M. & Shriver, M.D. TRPV6 exhibits unusual patterns of polymorphism and divergence in worldwide populations. *Hum Mol Genet* **13**, 2106–2113 (2006).
21. Rockman, M.V. et al. Positive selection on MMP3 regulation has shaped heart disease risk. *Curr Biol* **14**, 1531–1539 (2004).
22. Gasper, J. & Swanson, W.J. Molecular population genetics of the gene encoding the human fertilization protein zonadhesin reveals rapid adaptive evolution. *Am J Hum Genet* **79**, 820–830 (2006).
23. Enard, W. et al. Molecular evolution of FOXP2, a gene involved in speech and language. *Nature* **418**, 869–872 (2002).
24. Pier, G.B. et al. *Salmonella typhi* uses CFTR to enter intestinal epithelial cells. *Nature* **393**, 79–82 (1998).
25. Bienzle, U., Ayeni, O., Lucas, A.O. & Luzzatto, L. Glucose-6-phosphate dehydrogenase and malaria, greater resistance of females heterozygous for enzyme deficiency and of males with non-deficient variant. *Lancet* **1**, 107–110 (1972).
26. Williams, T.N. et al. Negative epistasis between the malaria-protective effects of alpha(+)-thalassemia and the sickle cell trait. *Nature Genet* **37**, 1253–1257 (2005).
27. Mu, J. et al. Recombination hotspots and population structure in *Plasmodium falciparum*. *PLoS Biol* **3**, e335 (2005).
28. Tishkoff, S.A. et al. Haplotype diversity and linkage disequilibrium at human G6PD, recent origin of alleles that confer malarial resistance. *Science* **293**, 455–462 (2001).
29. Voight, B.F., Kudaravalli, S., Wen, X. & Pritchard, J.K. A map of recent positive selection in the human genome. *PLoS Biol* **4**, e72 (2006).

30. Tishkoff, S.A. et al. Convergent adaptation of human lactase persistence in Africa and Europe. *Nat Genet* **39**, 31–40 (2007).
31. Burns, J.K. An evolutionary theory of schizophrenia: cortical connectivity, metarepresentation and the social brain. *Behav Brain Sci* **27**, 831–855 (2004).
32. Polimeni, J. & Reiss, J.P. How shamanism and group selection may reveal the origins of schizophrenia. *Med Hypotheses* **58**, 244–248 (2002).
33. Kimura, M. *The Neutral Theory of Molecular Evolution*, Cambridge University Press (1983).
34. Ohta, T. Slightly deleterious mutant substitutions in evolution. *Nature* **246**, 96–98 (1973).
35. Ohta, T. The nearly neutral theory of molecular evolution. *Annu Rev Ecol Syst* **23**, 263–286 (1992).
36. Ohta, T. Near-neutrality in evolution of genes and gene regulation. *Proc Natl Acad Sci, USA* **99**, 16134–16137 (2002).
37. Stringer, C.B. & Andrews, P. Genetic and fossil evidence for the origin of modern humans. *Science* **239**, 1263–1268 (1988).
38. Ambrose, S.H. Late Pleistocene human population bottlenecks, volcanic winter and differentiation of modern humans. *J Hum Evol* **34**, 623–651 (1998).
39. Hinds, D.A. et al. Whole-genome patterns of common DNA variation in three human populations. *Science* **307**, 1072–1079 (2005).
40. Kreitman, M. Methods to detect selection in populations with applications to the human. *Annu Rev Genomics Hum Genet* **1**, 539–559 (2000).
41. Bamshad, M. & Wooding, S.P. Signatures of natural selection in the human genome. *Nat Rev Genet* **4**, 99–111 (2003).
42. Fu, Y.X. Statistical tests of neutrality of mutations against population growth, hitchhiking and background selection. *Genetics* **147**, 915–925 (1997).
43. Guindon, S., Black, M. & Rodrigo, A. Control of the false discovery rate applied to the detection of positively selected amino acid sites. *Mol Biol Evol* **23**, 919–926 (2006).
44. Yang, Z. PAML: a program package for phylogenetic analysis by maximum likelihood. *Comput Appl Biosci* **13**, 555–556 (1997).
45. Hudson, R.R., Kreitman, M. & Aguade, M. A test of neutral molecular evolution based on nucleotide data. *Genetics* **116**, 153–159 (1987).
46. McDonald, J.H. Detecting non-neutral heterogeneity across a region of DNA sequence in the ratio of polymorphism to divergence. *Mol Biol Evol* **13**, 253–260 (1996).
47. Tajima, F. Simple methods for testing the molecular evolutionary clock hypothesis. *Genetics* **135**, 599–607 (1993).
48. Wright, S. *Evolution and the Genetics of Populations: The Theory of Gene Frequencies. Volume 2: The Theory of Gene Frequencies*, University of Chicago Press (1969).
49. Depaulis, F. & Veuille, M. Neutrality tests based on the distribution of haplotypes under an infinite-sites model. *Mol Biol Evol* **15**, 1788–1790 (1998).
50. Nekrutenko, A., Makova, K.D. & Li, W.H. The K(A)/K(S) ratio test for assessing the protein-coding potential of genomic regions: an empirical and simulation study. *Genome Res* **12**, 198–202 (2002).
51. Yang, Z. & Bielawski, J.P. Statistical methods for detecting molecular adaptation. *Trends Ecol Evol* **15**, 496–503 (2000).
52. Vallender, E.J., Paschall, J.E., Malcom, C.M., Lahn, B.T. & Wyckoff, G.J. SPEED: a molecular-evolution-based database of mammalian orthologous groups. *Bioinformatics* **22**, 2835–2837 (2006).
53. Chamary, J.V., Parmley, J.L. & Hurst, L.D. Hearing silence: non-neutral evolution at synonymous sites in mammals. *Nat Rev Genet* **7**, 98–108 (2006).

54. King, M.C. & Wilson, A.C. Evolution at two levels in humans and chimpanzees. *Science* **188**, 107–116 (1975).
55. Pollard, K.S. et al. Forces shaping the fastest evolving regions in the human genome. *PLoS Genet* **2**, e168 (2006).
56. Meunier, J. & Duret, L. Recombination drives the evolution of GC content in the human genome. *Mol Biol Evol* **21**, 984–990 (2004).
57. Zhang, C. et al. A whole genome long-range haplotype (WGLRH) test for detecting imprints of positive selection in human populations. *Bioinformatics* **22**, 2122–2128 (2006).
58. Sabeti, P.C. et al. Detecting recent positive selection in the human genome from haplotype structure. *Nature* **419**, 832–837 (2002).
59. Nordborg, M. & Tavare, S. Linkage disequilibrium: what history has to tell us. *Trends Genet* **18**, 83–90 (2002).
60. Toomajian, C., Ajioka, R.S., Jorde, L.B., Kushner, J.P. & Kreitman, M. A method for detecting recent selection in the human genome from allele age estimates. *Genetics* **165**, 287–297 (2003).
61. Hermisson, J. & Pennings, P.S. Soft sweeps: molecular population genetics of adaptation from standing genetic variation. *Genetics* **169**, 2335–2352 (2005).
62. Pennings, P.S. & Hermisson, J. Soft Sweeps III: the signature of positive selection from recurrent mutation. *PLoS Genet* **2**, e186 (2006).
63. Stajich, J.E. & Hahn, M.W. Disentangling the effects of demography and selection in human history. *Mol Biol Evol* **22**, 63–73 (2005).
64. Kayser, M., Brauer, S. & Stoneking, M. A genome scan to detect candidate regions influenced by local natural selection in human populations. *Mol Biol Evol* **20**, 893–900 (2003).
65. Duret, L. Evolution of synonymous codon usage in metazoans. *Curr Opin Genet Dev* **12**, 640–649 (2002).
66. Kudla, G., Lipinski, L., Caffin, F., Helwak, A. & Zylicz, M. High guanine and cytosine content increases mRNA levels in mammalian cells. *PLoS Biol* **4**, e180 (2006).
67. Tsunoda, T. et al. Variation of gene-based SNPs and linkage disequilibrium patterns in the human genome. *Hum Mol Genet* **13**, 1623–1632 (2004).
68. Kato, M. et al. Linkage disequilibrium of evolutionarily conserved regions in the human genome. *BMC Genomics* **7**, 326 (2006).
69. Wiehe, T. The effect of selective sweeps on the variance of the allele distribution of a linked multiallele locus: hitchhiking of microsatellites. *Theor Popul Biol* **53**, 272–283 (1998).
70. Nachman, M.W. & Crowell, S.L. Estimate of the mutation rate per nucleotide in humans. *Genetics* **156**, 297–304 (2000).
71. Wang, E.T., Kodama, G., Baldi, P. & Moyzis, R.K. Global landscape of recent inferred Darwinian selection for *Homo sapiens*. *Proc Natl Acad Sci U S A* **103**, 135–140 (2006).
72. Nielsen, R. et al. Genomic scans for selective sweeps using SNP data. *Genome Res* **15**, 1566–1575 (2005).
73. Carlson, C.S. et al. Genomic regions exhibiting positive selection identified from dense genotype data. *Genome Res* **15**, 1553–1565 (2005).
74. Bustamante, C.D. et al. Natural selection on protein-coding genes in the human genome. *Nature* **437**, 1153–1157, (2005).
75. Fay, J.C. & Wu, C.I. Hitchhiking under positive Darwinian selection. *Genetics* **155**, 1405–1413 (2000).
76. Stranger, B.E. et al. Genome-wide associations of gene expression variation in humans. *PLoS Genet* **1**, e78 (2005).

77. Wang, X., Grus, W.E., & Zhang, J. Gene losses during human origins. *PLoS Biol* **4**, e52 (2006).
78. Clark, A.G., Hubisz, M.J., Bustamante, C.D., Williamson, S.H. & Nielsen, R. Ascertainment bias in studies of human genome-wide polymorphism. *Genome Res* **15**, 1496–1502 (2005).
79. Nielsen, R., Hubisz, M.J. & Clark, A.G. Reconstituting the frequency spectrum of ascertained single-nucleotide polymorphism data. *Genetics* **168**, 2373–2382 (2004).
80. Wiehe, T., Nolte, V., Zivkovic, D. & Schlotterer, C. Identification of selective sweeps using a dynamically adjusted number of linked microsatellites. *Genetics* **175**, 207–218 (2007).
81. Verducci, J.S. et al. Microarray analysis of gene expression: considerations in data mining and statistical treatment. *Physiol Genomics* **25**, 355–363 (2006).
82. Dennis, G., Jr. et al. DAVID: Database for Annotation, Visualization and Integrated Discovery. *Genome Biol* **4**, P3 (2003).
83. Ashburner, M. et al. Gene Ontology: tool for the unification of biology. The Gene Ontology Consortium. *Nat Genet* **25**, 25–29, 2000.
84. Lomax, J. Get ready to GO! A biologist's guide to the Gene Ontology. *Brief Bioinform* **6**, 298–304 (2005).
85. Kent, W.J. et al. The Human Genome Browser at UCSC. *Genome Res* **12**, 996–1006 (2002).
86. Thorisson, G.A. et al. The International HapMap Project Web site. *Genome Res* **15**, 1592–1593 (2005).
87. McCarroll, S.A. et al. Common deletion polymorphisms in the human genome. *Nat Genet* **38**, 86–92 (2006).
88. Petkov, P.M. et al. Evidence of a large-scale functional organization of mammalian chromosomes. *PLoS Genet* **1**, e33 (2005).
89. Abelson, J.F. et al. Sequence variants in SLITRK1 are associated with Tourette's syndrome. *Science* **5746**, 317–320 (2005).
90. Kimchi-Sarfaty, C. et al. A "silent" polymorphism in the MDR1 gene changes substrate specificity. *Science* **315**, 525–528 (2007).
91. Mooney, S. Bioinformatics approaches and resources for single nucleotide polymorphism functional analysis. *Brief Bioinform* **6**, 44–56 (2005).
92. Zhang, J., Zhang, Y.P. & Rosenberg, H.F. Adaptive evolution of a duplicated pancreatic ribonuclease gene in a leaf-eating monkey. *Nat Genet* **30**, 411–415 (2002).

Part II

Applied Research in Comparative Genomics

9 Comparative Genomics in Drug Discovery

James R. Brown

CONTENTS

ABSTRACT

Drug discovery is a multistage process designed to rapidly progress the most promising candidate therapies while minimizing loss due to project attrition. Any technological or scientific discipline that can further either or both of these goals is an important addition to pharmaceutical research and development. Comparative genomics approaches have shown practical benefits in the validation of disease–gene relationships as well as establishing a better understanding of drug–target interaction effects. In this review, these various applications of comparative genomics in the pharmaceutical industry are discussed, and specific examples concerning the development of targeted kinase therapeutics are given.

9.1 INTRODUCTION

Drug discovery is one of the most challenging areas of scientific endeavor. Delivering a marketable drug can take decades from the time of initial gene target association with disease to the final approval by government regulatory agencies (Figure 9.1). Historically important drugs, such as penicillin and statin, were mostly discovered by screening compounds against whole cells or animal models and then looking for specific phenotypes. The mechanism of action on a molecular target was not obvious and often not fully determined until years after the drug's clinical deployment. This lack of genomic knowledge hindered the development of new therapeutics since

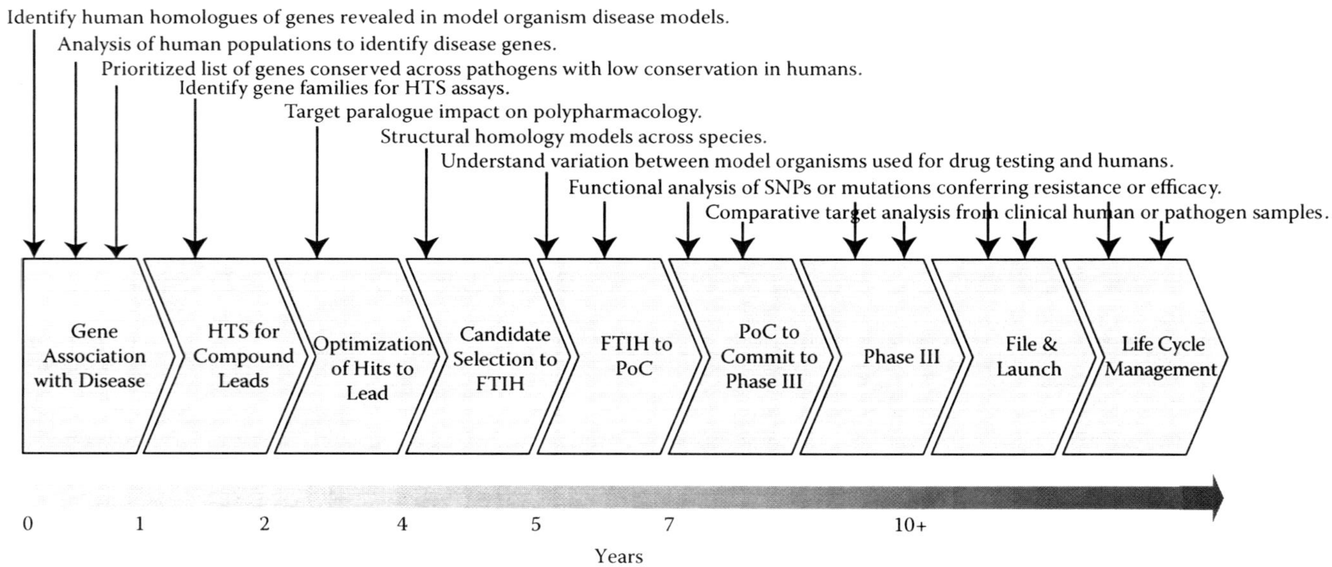

FIGURE 9.1 Schematic diagram of the drug discovery process. The time frame, in years, is approximate since the speed of progression, particularly in last-stage clinical phases, can be highly affected by factors other than drug–target interactions such as the time frame for patient recruitment into clinical trials, drug compound manufacturing issues, and the complexities of government regulatory decisions. Above the drug discovery pathway are some generalized comparative genomic approaches; the arrows indicate those stages at which they potentially make the greatest impact. HTS, high-throughput compound screening; FTIH, first time in human; POC, proof of concept.

relatively few diseases could be attacked. With the advent of genomics, the pharmaceutical industry seemed poised for a revolution in which unlocking the secrets of the human and pathogen genomes offered replacement of older, "low-hanging fruit" with baskets full of bountiful and more profitable targets.

Yet, nearly two decades into the genomics revolution (starting with expressed sequence tags [ESTs] and other fragmental views of human and bacterial genomes available before the completion of the human genome), pharmaceutical industry growth in terms of approved new drugs is still stumbling. From 1994 to 2005, the number of approved new molecular entities (NMEs), including both small molecules and biologicals, has declined by about 20%, although total investment in research and development (R&D) has risen across the pharmaceutical sector (CME International 2006). A number of hindering factors have been at play, including changing regulatory conditions and higher hurdles for safety compliance. Funding for innovative but costly R&D in established pharmaceutical companies is under intense pressure as revenues from older blockbuster drugs (i.e., annual sales more than $1 billion) erode from loss of patent protection and the consequential emergence of cheaper generic products.

However, the industrywide trend of reduced R&D productivity is in no small part due to the fundamental challenge of finding the right targets for a particular disease. Most diseases have complex genetic underpinnings that are still not adequately understood. Modulation of a target gene associated with a disease pathway could have detrimental effects because the gene product also fulfills an essential biochemical function in a different pathway — so-called drug pleiotrophy.[1] Compounds can also have off-target effects, which mean nonspecific activity against similar targets in the same or different pathways. Finally, resistance mechanisms in the form of alternative pathways or drug efflux transports can subvert the effects of any small molecule compound.

Industrial drug discovery is an incremental process designed to rapidly progress the most promising molecules yet control the financial risk involved with those that inevitably fail. Early preclinical stage studies are carefully designed to mitigate risks associated with both the compound and the target prior to further commitment of resources to costly clinical trial phases. However, unknown genetic variability among individuals means that long-term efficacy or liability for most drugs is often unknown until large populations of patients have been treated. Drug projects, broadly meant here to include small molecules as well as the biological agent vaccines, peptides, and antibodies, have brutally high attrition rates with a small percentage of initiated programs successfully progressing from target validation to government regulatory approval. Some areas are more challenging than others; for example, anticancer drugs have a failure rate nearly three times that of neurological or cardiovascular drugs in phases from candidate compound selection to clinical development.[2]

Therefore, the twofold challenges for controlling costs in R&D are ensuring success of late-stage efforts and moving attrition or termination decisions to the earliest phases, so-called "fast to fail." Perhaps overly hyped in its infancy, genomics has disappointed some in that there has not been an exponential leap in new pharmaceutical agents. However, the melding of biomedical research and genomics has been more evolutionary, rather than revolutionary, with genomics slowly proving its worth as drug discovery strives toward the right balance of high early-stage attrition and low late-phase failure rates.

9.2 THE DRUG DISCOVERY PATHWAY

The conventional pathway for drug discovery and development, whether of a chemical compound or biological agent, involves multiple, sequential steps beginning with initial gene-to-disease associations and ending with the registration and product management of the drug (Figure 9.1). Although the nomenclature might differ among organizations, the overall process is broadly comparable across the pharmaceutical industry. Alternative approaches to discovering new molecules using chemical genomics or genetics methods serve potentially to expand the universe of druggable targets yet must travel the same road to clinical development.[3] Taking into consideration the additional early years of fundamental academic research establishing the gene–disease association, it can often take longer than a decade before a new drug appears on the market.

Increasingly, comparative genomics is finding application throughout the drug discovery process as a valuable tool in helping to mitigate risk and promote success at various stages. The expansion of comparative genomics analyses finding utility in drug discovery is broad, ranging from identification of human homologs for model organism disease-linked genes to the functional analysis of resistance mutations and polymorphisms detected in the clinic (Figure 9.1). The rest of this review elaborates on some specific examples, and subsequent chapters discuss other roles of comparative genomics in biomedical and pharmaceutical research.

9.3 TARGET DISCOVERY AND VALIDATION

The initial step in the drug discovery process is the uncovering of a target association with a disease. As discussed by Barnes in chapter 8, human disease genetics focusing on the analysis of variation between diseased and normal human cohorts has been a powerful tool for revealing gene–disease associations. However, genetic linkage to the disease does not necessarily mean that the particular gene is a causative or maintenance factor for that disease. Also, many genes that are linked to some disease etiology are refractory to pharmaceutical approaches. The actual number of available drug targets in the human genome has been an area of intense investigation and speculation. Several well-known protein families are highly pursued because they can be modulated by small-molecule interactions and considered as tractable targets. In particular, G protein-coupled receptors (GPCRs) comprise the largest single target group, with interactions to nearly 40% of known drugs (see chapter 15 by Foord for an in-depth review). Other protein families include kinases, ion channels, and nuclear receptors, as well as pathogen-specific targets such as bacterial penicillin-binding proteins and human immunodeficiency virus (HIV) reverse transcriptase. Recent estimates suggest that approved drug substances with known mode of action (i.e., the compound is proven to modulate a particular protein and cause a disease response) affect as few as 324 molecular targets, of which 266 are human-derived proteins, with the remainder targets of viruses, bacteria, fungi, or other pathogens.[4] However, the universe of potential drug targets is much larger, with over 700 GPCRs alone in the human genome. But, among individual GPCRs, chemical tractability can range from very high to negligible, and other drug-tractable protein families have similar variances. Thus, establishing disease associations and tractability as well as the phenotypic effects of either activating (agonistic) or deactivating (antagonistic) a particular molecular target is a key concern in the initial target identification stage.

Model organisms, with their decoded genomic sequences and advanced molecular biology tools, are becoming increasingly important for discovering and validating drug targets. Targets can be validated in vivo using the arsenal of sophisticated molecular biological tools, such as RNA interference (RNAi) and gene knockouts.[5] Selective inhibitors can also be used as tool compounds to modulate the intended target for phenotypic effects. The nematode *Caenorhabditis elegans* is a particularly powerful platform for target identification. The small size and short life cycle of this organism make it suitable for large-scale phenotypic screening of genome-wide RNAi experiments.[6] In addition, some gene conservation across species allows for the rescue of *C. elegans* knockout mutants by transplanted human genes. For example, expression of human presenilin-1, a gene associated by mutations with early-onset familial Alzheimer's disease, rescued the neuronal deficiencies of *C. elegans* sel-12 presenilin mutants.[7,8]

The fruit fly *Drosophila melanogaster* is also a useful model species for studying many disorders, such as diseases of aging, including sleep and organ-specific aging effects.[9,10] For example, aging experiments in yeast, nematodes, fruit fly, and most recently, mice have extended the validation of the conserved class III histone deacetylase SirT1 as a potential regulator of life span that has been shown to be modulation amiable by small pharmacological molecules.[11] As mammals, rodent models for specific diseases, such as cancer[12] and neurodegenerative diseases,[13] are important in the preclinical stages of target discovery and target validation. One biotech company has developed high-throughput gene knockouts and phenotypic screens in the mouse as a platform for new target discovery.[14]

As discussed in subsequent chapters, comparative genomic analysis has a particular purpose in the discovery of new anti-infective drugs as well as in the life cycle management of approved drugs combating increasingly resistant viral and bacterial pathogens. In antimicrobial drug discovery, comparative genomics has been used to develop prioritized lists of potential novel targets. Some of the oldest drugs in clinical use are antibiotics, such as penicillin derivatives. However, the rapid spread of drug-resistant bacteria is driving an unmet medical need for new classes of antibiotics that can overcome "superbugs" like methicillin-resistant *Staphylococcus aureus* (MRSA).[15] The largest antibiotic market is for broad-spectrum agents that can kill a wide variety of gram-positive and gram-negative bacteria.

Several years ago, GlaxoSmithKline (GSK) as well as other pharmaceutical companies initiated genomics-based approaches for discovering novel antibiotic targets. GSK used comparative genomics to identify genes that were widely conserved among the genomes of key pathogens from both Gram positives (*S. aureus*, *Streptococcus pneumoniae*) and Gram negatives (*Haemophilus influenzae*).[16] Using gene-targeted knockout technology, the essentiality of genes was determined by in vitro culture and in vivo animal infection models.[17] Over 300 genes were determined to be putative targets, and 70 extensive high-throughput screening campaigns were launched.[18]

Despite this large-scale effort, the number of tractable broad-spectrum antimicrobial targets was low, mainly due to the high sequence diversity of bacterial genes and the poor chemical diversity of industrial compound libraries with respect to inhibiting bacterial enzymes. Nonetheless, because of bacterial species diversity, comparative genomic analysis has continued relevance in understanding the natural variation of potential antibiotic targets, particularly in isolates of clinical pathogens. Recent revival of antibacterial drug discovery efforts focusing on a narrower species spectrum combined with better diagnostics will be even more reliant on comparative bacterial genomics for target and biomarker identification.[15]

9.4 GENE ORTHOLOGY AND PARALOGY

No molecular entity is ever completely validated as a drug target until it has been proven actually to modulate a specific target that results in some tangible clinical benefit to the patient. Thus, each stage in the drug discovery process is designed to increase confidence in the validity of a target as well as establish the efficacy and safety of the intended modulator. Since preclinical in vivo testing can only be conducted on human cell lines and given the expense of clinical trials, model organism-oriented experiments are highly critical at each phase of the drug development process. A key challenge for comparative genomics is the interpretation and transference of results from model organism studies to humans. In this respect, molecular evolutionary concepts and methodologies have an increasingly important role in drug discovery.

Since the majority of drug targets belong to large, multigene families, a clear understanding of the homology relationships of a particular drug target between model organism species and humans is critical. However, it is well known that the gene complement even between closely related organisms can be highly variable. The genomes of eukaryotic species have highly variable complements of key drug targets. For example, genome-wide surveys of the eukaryotic protein kinases, the so-called kinome, reveal that the mouse has 510 orthologs[19] to the 518 putative human kinases.[20] *Drosophila* has only 239 kinases, while the sea urchin has 353 kinases[21] — a contrast that might reflect the divergence of signaling pathways in the regulation of protostome versus deutrostome development. More kinase families are in common between humans and sea urchins as opposed to humans and fruit fly. Despite its body plan simplicity, *C. elegans* has 454 kinases, nearly double the complement of *Drosophila*. However, the fruit fly has a better representation of homologous genes relative to the human kinome than the nematode, which suggests that numerous kinases in the worm evolved from lineage-specific expansions.[22]

Although not always fully appreciated, most steps in drug discovery are based on the assumption of evolutionary equivalence across multiple species. Yet, highly similar or homologous genes can have a variety of evolutionary relationships. Traditionally, orthologous genes are those that evolved by direct descent and hence show greater similarity between rather than within species. In contrast, paralogous genes emerged from ancestral gene duplications and tend to show greater similarity within a species. However, orthologs can have a one-to-many relationship if the gene duplicated in one species but not another or a many-to-many relationship if the gene duplicated in an earlier ancestor to both species. Additional gene homology nomenclature has been proposed in which the former situation is now called *inparalogs* and the latter termed *outparalogs*.[23]

A practical example is the evolutionary relationships of Aurora kinases (Figure 9.2). As key regulators of mitotic chromosome segregation, the Aurora family of serine/

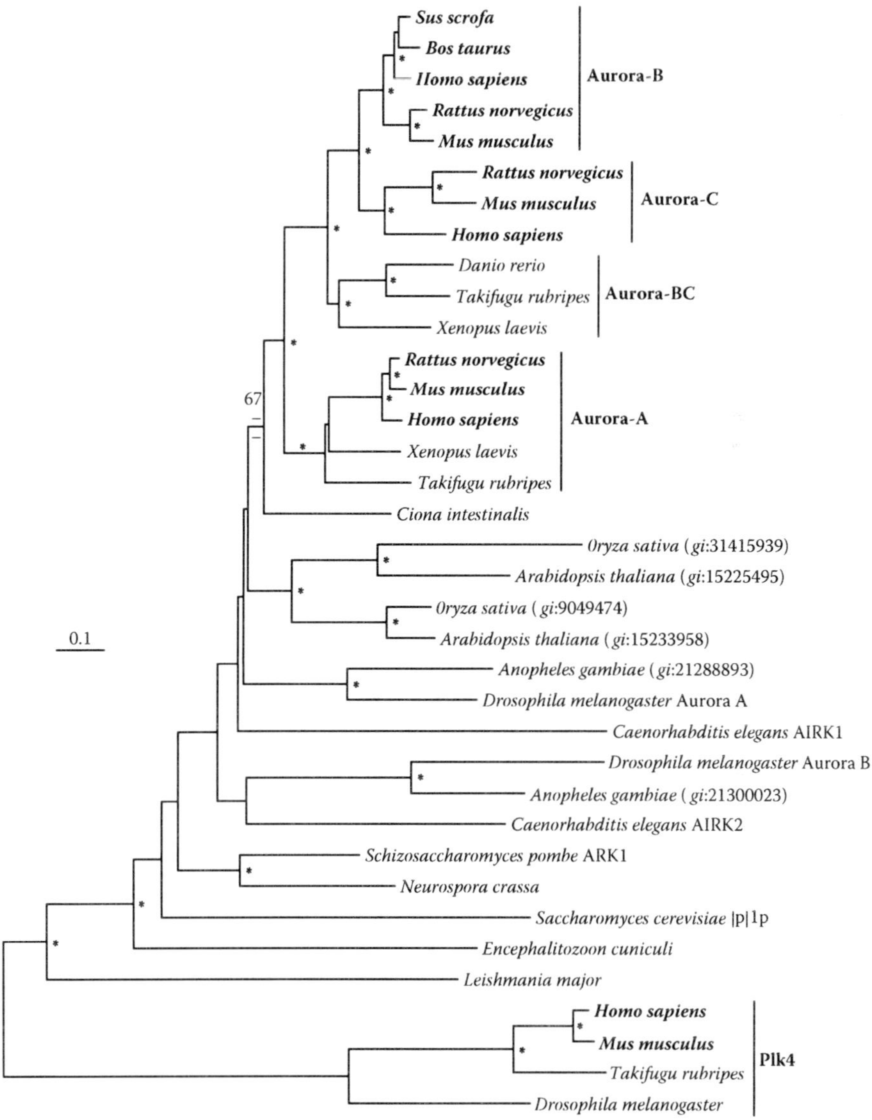

FIGURE 9.2 Neighbor-joining phylogenetic tree of Aurora kinases rooted by polo-like kinase 4 (PLK4) outgroup. Mammalian species names are in bold font, and major clusters of Aurora-A, Aurora-B, and Aurora-C kinases are indicated. Plant sequences are identified by their Genbank accession number. This adapted tree by the author[27] is based on pairwise distances between amino acid sequences using the programs NEIGHBOR and PROTDIST (Dayhoff option) of the PHYLIP 3.6 package.[58] Asterisks (*) indicate those nodes supported 70% or greater of 1,000 random bootstrap replicates. Scale bar represents 0.1 expected amino acid residue substitutions per site.

threonine kinases plays an important role in cell division.[24] Abnormalities in Aurora kinases have been strongly linked with cancer, which has led to the development of new classes of anticancer drugs that specifically target the Aurora adenosine triphosphate (ATP)-binding domain.[25,26] From an evolutionary perspective, the species distribution of the Aurora kinase family is intriguing. Mammals uniquely have three Aurora kinases: Aurora-A, Aurora-B, and Aurora-C, which appear to have arisen from a prechordate, possibly urochordate, ancestor as represented in the tree by the tunicate *Ciona intestinalis*.[27]

Interestingly, all other species suffice with one or two Aurora genes. Cold-blooded vertebrates have a direct Aurora-A ortholog to mammalian versions but only a single ortholog to Aurora-B and Aurora-C, termed here Aurora-BC. Therefore, mammalian Aurora-B and Aurora-C are considered inparalogs relative to cold-blooded Aurora-BC since they were derived from a mammalian-specific gene duplication. The functional significance of Aurora-C is poorly understood, although it does associate with the mitotic complex and is highly expressed in rapidly growing tissues such as testis.[28,29] The relationship of invertebrate Aurora-A and Aurora-B kinases (represented in Figure 9.2 by nematodes and insects) to vertebrate counterparts is ancestral and homologous. However, the phylogeny clearly shows that fruit fly and nematode Aurora-A and Aurora-B genes appear to have arisen from an invertebrate-specific gene duplication event, and that neither are orthologous to the similarly named counterparts in mammals.

The Aurora phylogeny is informative to drug discovery in two ways. First, it provides a context for the transference of knowledge from model organism studies to human cellular biology. While all metazoan Aurora kinases have similar roles in mitosis, it would be incorrect to infer from Aurora-A or Aurora-B kinase manipulations in the model invertebrates *Drosophila* or *C. elegans* the precise functioning of similarly named Aurora kinases in mammals. Second, the vast majority of small-molecule inhibitors of kinase activation bind to the ATP-binding pocket. Structure- and sequence-based comparisons of the 26 amino acids lining the ATP-binding site reveal that mammalian Aurora-B and Aurora-C have complete identity, while Aurora-A has three variant residues.[27] From a pharmacological perspective, the potential phenotypic effects of dual inhibition of Aurora-B and Aurora-C should be taken into consideration.

Orthologous and paralogous relationships among sequences are best determined using phylogenetic reconstruction. However, such tree building can be both computational and labor intensive, particularly if there are large numbers of genes or species to be analyzed. Identification of homologs using reciprocal-best-BLAST (Basic Local Alignment Search Tool) hits (RBH) is a common bioinformatics shortcut when dealing with genome-wide collections of genes.[30] Briefly, the concept is as follows: Hypothetical gene A in the species 1 is orthologous to gene B in species 2 if BLAST searches using either gene against the other species genome pulls in its counterpart as the top hit with the most significant E-value. There are Web resources available that have precomputed genome-wide ortholog identification based on RBH methodology, such as the Clusters of Orthologous Groups (COG) database[31,32]; The Institute for Genomic Research (TIGR) EGO database[33]; and INPARANOID, which has a separate subsection on orthologous disease genes called OrthoDisease.[34,35]

However, such scoring is prone to errors, and the ranking BLAST similarities are often not compatible with phylogenetic relationships.[36] For example, Kamath et al. determined RNAi mutant phenotypes in a genome-wide scan for 1,722 genes in *C. elegans*, of which 33 genes were stated to be homologous to human disease genes according to BLAST searches.[37] However, phylogenetic analysis revealed that only 5 of the 33 genes have confirmed orthologous relationships between human and nematode (personal unpublished data). Alternative and more sensitive methods to RBH have been proposed for large-scale ortholog and paralog predictions.[38] While careful phylogenetic analysis is time consuming, the evolutionary relationships are predicted with greater confidence than by BLAST homology alone, and such scientific rigor is worth the investment when functional conservation between putative drug target genes of model organisms and humans is a critical factor.

But, confirmed orthologous genes can still have alternative functions in different species. Pharmacogenetic studies have revealed several examples among the cytochrome P450 (CYP) genes, a large multigene family of drug-metabolizing enzymes. Polymorphisms in sequence and copy number of CYP genes have been linked to patient heterogeneity in treatment effects.[39] About 20%–25% of clinically used drugs are believed to be metabolized by one particular CYP gene, CYP2D6. Multiple allelic variants of CYP2D6, including gene duplications as well as missense mutations and defective splice variants, have been linked to changes in enzyme activity among different racial and population groups.[40] Rodents and humans show remarkable differences in the CYP2D loci. Humans have a single active, and highly polymorphic, CYP2D6 gene along with two pseudogenes, CYP2D7 and CYP2D8, while the mouse has nine CYP2D genes encoding fully functional enzymes. The diversification of CYP2D genes in the rodent compared to humans could be an adaptation in mice to digest a broad vegetarian diet since the CYP2D6 enzyme has an affinity for plant toxins such as alkaloids. It has been suggested that the detoxification benefits of CYP2D for ingested plant material would be strongly selected for in rodents, while the narrowing of human diet because of agriculture could have led to more relaxed selection on these loci. Interestingly, mutations leading to either increased expression or improved catalytic activity of CYP2D homologs in insects also appear to confirm increased resistance to toxic insecticides.

Another example of where species differences in cytochrome P450 account for changes in metabolic function is the CYP2A family. The rat isoform CYP2A1 expressed in the liver is considerably diverged from the orthologs in human, CYP2A6, and mouse, CYP2A4, as well as from a second rat paralog expressed in the lung.[1] This sequence divergence corresponds to the severe hepatotoxic effects of coumarin metabolism specific to the rat.

9.5 EVOLUTIONARY CONTEXT FOR CANCER MUTATIONS

While comparative genomics is applied to a wide variety of therapeutic areas, it is especially relevant to cancer, which is widely viewed as a genetic disease. Genetic abnormalities are hallmarks of tumor cell lines, which can be assigned to two broad categories: loss-of-function or gain-of-function mutations.[41] Loss of function for genes acting as tumor suppressors can occur by gene deletions and epigenetic silencing as well as inactivating mutations in the gene itself, which are called *intragenic mutations*. Gain of function can result from gene translocations, gene amplifications, and activating intragenic mutations. Different technologies are used to detect these different types of cancer mutations at a genomic level, such as array comparative genomic hybridizations (aCGHs) for establishing the presence of chromosomal aberrations and DNA methylation-specific arrays for detecting epigenomic configurations, both of which are discussed elsewhere in this book (Buys et al. in chapter 13 and Kuo et al. in chapter 14, respectively). Intragenic mutations are detected by a more conventional DNA resequencing approach. The occurrence of point mutations in cancer is highly variable and dependent on both the gene and the tumor type. The largest public source of cancer intragenic mutation data is the Catalogue of Somatic Mutations in Cancer (COSMIC) database (http://www.sanger.ac.uk/genetics/CGP/cosmic/) maintained by the Sanger Centre. The latest release (April 4, 2007) has records on 43,021 mutations in 2,671 genes across 204,457 tumors.

Understanding the effects of intragenic variants and assigning their causative role in tumorigenesis is not straightforward. In fact, there can be four plausible explanations for any sequence variant seen in a cancer gene. First, the variant could be a known germ-line single-nucleotide polymorphism (SNP) indicative of a particular population or race. Known SNPs are easy to identify from comparisons to SNP repositories such as the dbSNP of the National Center for Biotechnology Information (NCBI). Although not necessarily tumorigenic, an SNP might mark a susceptibility loci for cancer,[42] such as the pattern of SNPs in *N*-acetyltransferase (NAT) 1 and 2, enzymes important for the metabolism of carcinogenic aromatic and heterocyclic amines that have been associated with the certain types of cancer.[43,44] Second, the variant could be a novel or private germ-line SNP. These are impossible to differentiate from somatic mutations unless there is a corresponding nontumor or germ-line tissue sample available from the same individual. The third possibility is that the intragenic mutation is specific to the somatic tumor tissue, but it is unrelated to the advancement of tumorigenesis. Tumors can have defective DNA repair machinery; thus, an overall elevated mutation rate in cancer cells relative to those of normal tissue is often seen. Some mutations can be "passengers" or a mere consequence of the tumor's hypermutable state. Finally, the mutation can be a somatic, tumor cell-specific variant that is responsible for initiating or sustaining tumorigenesis. These "driver" mutations are the ones of principal interest for understanding cancer biology.

Computational methods for distinguishing between passenger and driver mutations are inexact and, at best, deliver proximate hypotheses. Several studies have reported on large-scale resequencing of several hundred to thousands of genes from multiple tumor types to catalog cancer-associated mutations. Sjoblom et al. sequenced 13,023 genes in 11 colorectal and 11 breast cell lines and found 1,307 validated nucleotide changes in 1,149 genes.[45] Using a statistical method to determine if a particular gene had a higher mutation rate than background, they identified 189 genes that were mutated at a significant frequency.[45] The distribution of mutations within these genes suggested some clustering in a specific protein domain, and 31 changes were stated to have occurred in evolutionarily conserved positions.

Another study by Greenman et al. resequenced 518 protein kinase genes in 210 primary tumors and cell lines.[46] Protein kinases play critical roles in various cell-signaling pathways known to regulate tumor cell proliferation and are widely viewed as a key class of anticancer targets.[47,48] Importantly, protein kinases are the targets of clinically approved small-molecule inhibitors such as the drug imatinib (Gleevec), which inactivates the kinase fusion BCR-ABL found in chronic myeloid leukemia, and trastuzumab (Herceptin), a monoclonal antibody targeting HER2 (ErbB2) kinase, which is overexpressed in many breast cancers. Assuming strong positive selection, Greenman and coworkers suggested that driver mutations could occur if the observed ratio of nonsynonymous to synonymous substitutions was significantly greater than 1.0 as compared to chance. Of the 921 base substitutions in their primary screen, 763 were estimated to be passenger mutations. They estimated a total of 158 driver mutations among 119 genes across 66 or about one-third of their samples. Several putative driver mutations occurred in the protein kinase P loop and activation domains, which might affect kinase function, but many others were located outside the kinase domain. Interestingly, there were few overlapping mutations in kinases found between these two studies, which might be indicative of the genomic diversity and heterogeneity of human cancers.[41] New studies, such as the proposed Cancer Genome Atlas (http://cancergenome.nih.gov/index.asp) funded by the U.S. National Cancer Institute and the National Human Genome Research Institute, seek to expand the available cancer mutation data by resequencing many more genes from a greatly expanded tumor collection.

Further knowledge might be gained by comparing mutations relative to orthologs in other species as well as paralogs of related kinases. As an example, the gene for phosphatidylinositol-3-kinase α (PIK3CA) peptide is highly mutated in colon, brain, and gastric cancers, where apparent gain-of-function mutations confer increased activity for this lipid kinase.[49,50] PIK3CA, also known as p110α, belongs to a family of 10 phosphatidylinositol-3- and -4-kinases, all involved in lipid second-message processing for various cellular pathways.[51] Phylogenetic analysis shows that PIK3CA is most closely related to three other class I kinases: PIK3C-β (PIK3CB), PIK3C-δ (PIK3CD), and PIK3C-γ (PIK3CG) (Figure 9.3). All four kinases are found throughout mammals and cold-blooded vertebrates, while invertebrates have only a single PIK3C-like kinase as well as PIK3C3.

Alignment of a consensus sequence of nonsynonymous cancer mutations reported in the COSMIC database with normal human PIK3CA as well as orthologs from mammals and human paralogs for PIK3CB, PIK3CD, and PIK3CG are shown in Figure 9.4. Several mutations occur in regions of PIK3CA that are conserved throughout mammalian isoforms. At least seven mutations, while nonconserved among PIK3CA orthologs, are conservative changes matching residues in one or more of the three other corresponding human PIK3C paralogs. According to the COSMIC database, one of the most frequent variants observed in cancer is H1047R in the terminal end of the kinase domain, which is also a potentially activating or gain-of-function mutation. The variants H1047L and, more rarely, H0147Y have also been recovered from clinical tumor samples.

Gymnopoulos et al. measured the oncogenic potential of the 15 most common PIK3CA mutations found in tumors by introducing retroviral expression vectors with each of the variants into avian cells and measuring their individual efficiencies for tumorigenic transformation.[52] Their functional assays confirmed that the mutation H1047R strongly conferred oncogenic potency, while moderate and weak potency was induced by the variants H1047L and H1047Y, respectively. Interestingly, H1047R corresponds with R1047 found in the normal human paralog PIK3CG,

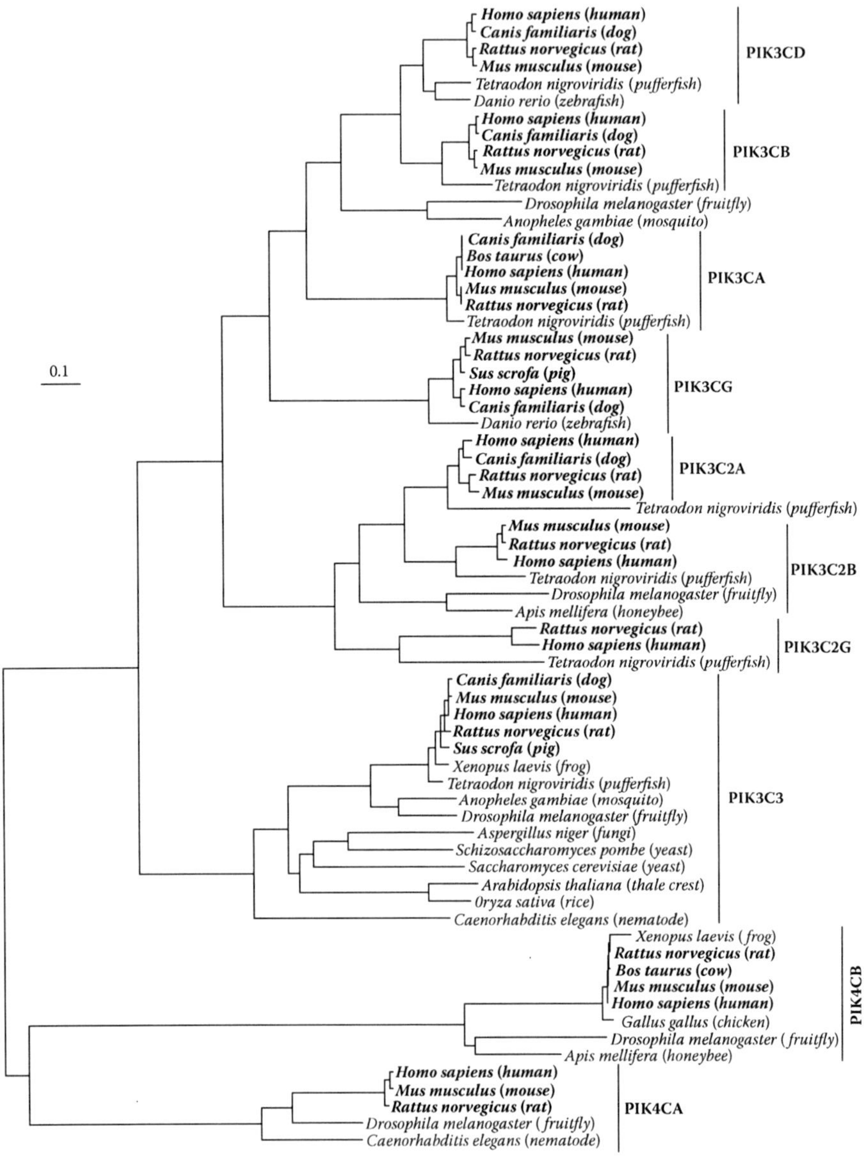

FIGURE 9.3 Neighbor-joining phylogenetic tree of phosphatidylinositol 3,4-kinases. Mammalian species names are in bold font. Gene groupings are the PIK3C kinases of PIK3C-α (PIK3CA), PIK3C-β (PIK3CB), PIK3C-δ (PIK3CD), and PIK3C-γ (PIK3CG). Also included are the kinases PIK3C2-α (PIK3C2A), PIK3C2-β (PIK3C2B), and PIK3C2-γ (PIK3C2G), PIK3C3 as well as PIK4C-α (PIK4CA) and PIK4C-β (PIK4CB). Tree construction methods are described for Figure 9.2, except no bootstrap values are shown.

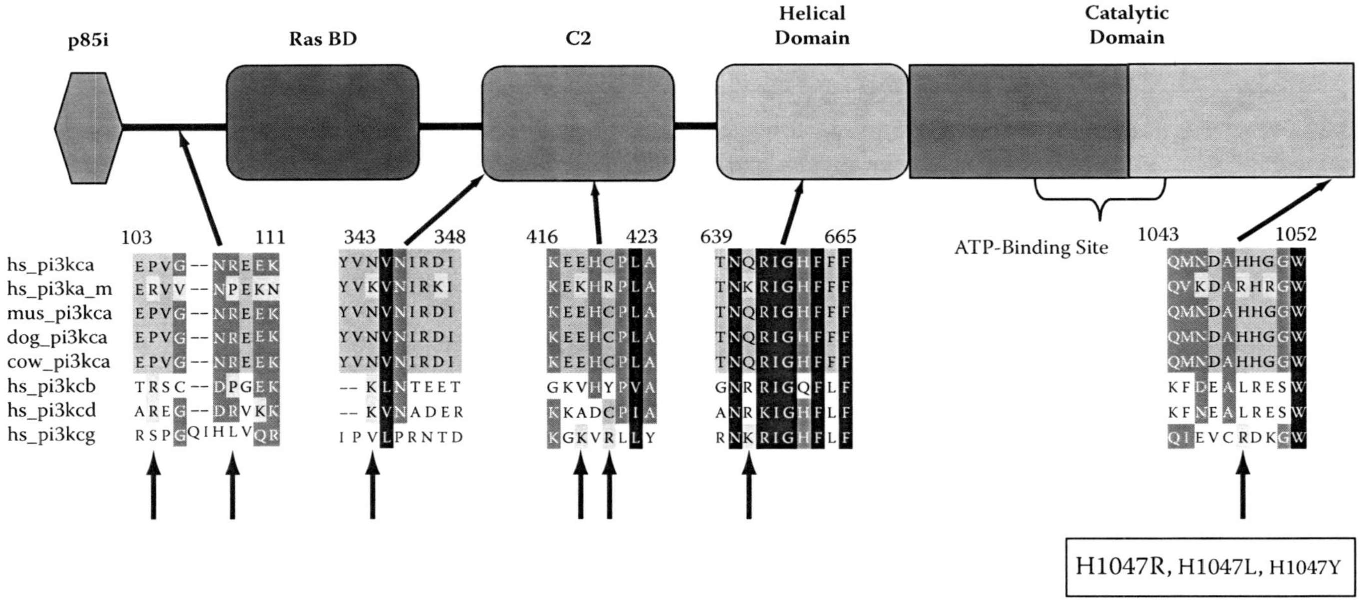

FIGURE 9.4 Occurrence of some missense cancer mutations in PIK3CA gene relative to orthologous and paralogous PI3K kinases. PI3KCA sequences are from human (hs_PI3Kca), mouse (mus_PI3Kca), dog (dog_PI3Kca), cow (cow_PI3kca), and chicken (chick_pi3K) as well as human PI3K paralogs PIK3CB (hs_PI3Kb), PIK3CD (hs_PI3Kd), and PIK3CG (hs_PI3Kg). Shown in the second row is a composite cancer mutant human PI3KCA (hs_PI3Ka_m) with amino acid substitutions (mutations) mapped as reported by Samuels et al.[49] and the Sanger COSMIC database. Regions of the alignments are shown where a cancer missense mutation is identical to an amino acid occurring in normal (wild-type) human paralogs. Numbers indicate coordinates in normal human PIK3CA. Arrows at the bottom of the alignment point to those specific changes across paralogs. Note that for H1047, three different amino acid substitutions have been observed, and font size of label indicates the relative high (large font) to low (small font) oncogenic potency of each type.[52] Structural domains were taken from the alignment of PI3K kinases to the PI3K C-γ structure reported by Walker et al.[59] and are not drawn to scale.

while H1047L corresponds to L1047 in wild-type paralogs PIK3CB and PIK3CD (Figure 9.4). H1047Y, rarely found in tumors and appearing to convey much weaker oncogenic potency than the other two mutations, is not found in any PIK3C family kinase. The correspondence of certain cancer mutations in PIK3CA to those found in normal paralogs suggests that selection pressures might limit the range of acceptable changes. Moreover, such mutations could potentially shift functionality of the protein toward that of a closely related paralog, perhaps converging on substrates, regulatory mechanisms, or protein interactions. Mapping H1047R/L mutations onto a structural model of PIK3CA by Gymnopoulos and coworkers suggests that these changes are located near the hinge region of the activation loop and could serve to increase catalytic activity. Given the importance of mutated kinases in tumor cell viability and their increased exploitation as cancer drug targets, better insights into delineating between passenger and driver mutations might be gained through broader sequence comparisons across different species as well as related protein family members.

9.6 GENOMICS AND POLYPHARMACOLOGY

Medicinal chemistry has always been the core of the pharmaceutical industry; thus, analysis approaches to combined chemistry and genomics data are highly synergistic for drug discovery. Understanding the relationships between the target gene and other potential binding partners can assist in the improvement of compound structure–activity relationships (SARs) and rational drug design. Since nearly all pharmaceutical targets belong to large, multigene families, drugs can have varying ranges of *specificity* (the degree of focused effect on a target) and *spectrum* (effects beyond the intended target due to interaction with similar or paralogous proteins). Comparative genomics plays an important role in identifying potential proteins for counterscreening in high-throughput screens, focusing compound target optimization, and suggesting potential off-target effects on related proteins (Figure 9.1).

Early postgenomic viewpoints that a drug needed to have high specificity for a single target are now tempered by the desirability for controlled sets of multiple target interactions for some therapeutic indications — a drug characteristic known as *polypharmacology*. Multiple target interactions can lead to a more effective drug because shunt pathways or resistance mechanisms can be countered. Structure-based design of promiscuous compounds are being applied to HIV-1 antiviral and anti-cancer therapeutics as a strategy to overcome multidrug resistance.[53]

Prediction of promiscuous compound interactions on the basis of target amino acid sequence alone has been attempted with mixed results for protein kinases. The human "kinome" is comprised of 518 kinases that share varying levels of homology across the core kinase domain, which ranges between approximately 250 and 350 amino acids in length depending on the kinase.[20] However, most kinase inhibitors depend on interactions with the 30–40 residues lining the ATP-binding pocket, and even there, as few as two amino acid changes can determine inhibitor specificity.[54]

Fabian et al. screened 20 kinase inhibitors against a panel of 119 kinases using an ATP-binding competition assay that determined the effectiveness of compounds to out compete ATP down to concentrations of less than 1 μM ($K_d < 1$ μM).[55] The resultant inhibitor assay data were overlaid on a previously published human kinome phylogenetic tree derived from sequence alignments of the kinase domains.[20] In many cases, the compounds bound to kinases that appeared to be poorly related by sequence, such as the compound BIRB-796, which bound kinases from two disparate groups, the serine/threonine kinase p38 and the tyrosine kinase ABL. Conversely, other compounds showed very fine-scale discrimination between nearly identical kinases. An obvious explanation for the diversity of interactions is that the key compound discriminating factors could be limited to very few residues in the crucial ATP-binding site, which as a small component of the overall kinase domain, would not have greatly influenced the overall phylogenetic tree topology. In addition, proteins with low sequence homology can have significant three-dimensional structural similarity, which would also result in similar small-molecule interactions with the protein. Reconstructing phylogenetic trees of kinases based on only the key residues of the ATP-binding pocket can improve the overall predictability of compound interactions from tree topologies, although there can still be significant off-target effects unaccountable by sequence homology (personal unpublished data).

Knight et al. published a similar study of 13 inhibitors targeting the entire PIK3C family of lipid kinases and included assays for several more distantly related lipids as well as protein kinases.[56] Their study did not include a phylogeny but rather used separate principal component analysis (PCA) plots to compare the statistical space of target similarity versus compound–target inhibition values. A phylogenetic perspective of these data based on an alignment of kinase domains is shown here in Figure 9.5, where the IC_{50} (median inhibition concentration) values for

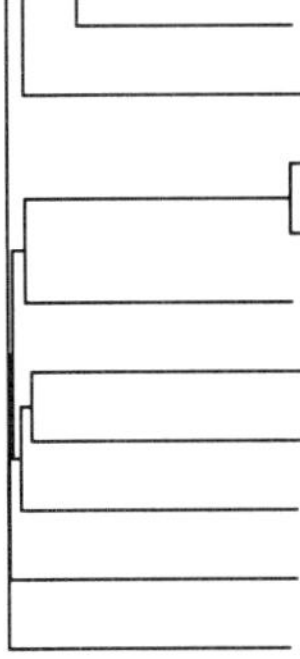

	PIK23	TGX115	AMA37	PIK39	IC87114	TGX286	PIK75	PIK90	PIK93	PIK108	PI103	PIK124	KU55399
PIK3CD	0.097	0.63	22	0.18	0.13	1	0.51	0.058	0.12	0.26	0.048	0.34	0.72
PIK3CB	42	0.13	3.7	11	16	0.12	1.3	0.35	0.59	0.057	0.088	1.1	1.2
PIK3CA	>200	61	32	>200	>200	4.5	0.0058	0.011	0.039	2.6	0.008	0.023	3.3
PIK3CG	50	100	100	17	61	10	0.076	0.018	0.016	4.1	0.15	0.054	9.9
PIK3C2B	100	50	>100	100	>100	100	1	0.064	0.14	20	0.026	0.37	ND
PIK3C2A	>100	>100	>100	>100	>100	>100	10	0.047	16	100	1	0.14	ND
PIK3C2G	>100	100	50	100	>100	ND	ND	ND	ND	ND	ND	ND	ND
PIK3C3	50	5.2	>100	>100	>100	3.1	2.8	0.83	0.32	5	2.3	10	10
PIK4CA1	>100	>100	>100	>100	>100	>100	>100	>100	>100	>100	>100	>100	>100
PIK4CA2	>100	>100	>100	>100	>100	>100	>100	0.83	1.1	50	>100	>100	>100
PIK4CB	>100	>100	>100	>100	>100	>100	50	3.1	0.019	>100	50	>100	>100
ATM	>100	20	ND	>100	>100	>100	2.3	0.61	0.49	35	0.92	3.9	0.005
ATR	>100	>100	>100	>100	>100	>100	21	15	17	>100	0.85	2	20
PRKDC	>100	1.2	0.27	>100	>100	50	0.002	0.013	0.064	0.12	0.002	1.5	10
FRAP1	>100	>100	>100	>100	>100	>100	1	1.05	1.38	10	0.02	9	20
SMG-1													

$IC_{50} \leq 1$ $1 < IC_{50} < 10$ $10 < IC_{50} < 32$ $IC_{50} > 32$

FIGURE 9.5 Phylogenetic tree of PIK3/4 and related protein kinases with IC_{50} values from tested inhibitors as reported by Knight et al.[56] Compound names are in column headers, while the rows are tested kinases aligned with their branching order in the phylogenetic tree. IC_{50} values are shaded according to potency, with smaller values representing more effective inhibitors of kinase activity. The tree was constructed using the neighbor-joining method as described for Figure 9.2.

each kinase reported by Knight and coworkers are aligned with terminal nodes in the tree. (Extensive homology searches of GenBank did not reveal further kinases that would have been intermediate branches in the tree, other than SMG-1, which is included as an outgroup but was not tested by Knight et al.[55]) Several inhibitors are highly specific to particular PIK3C kinases, such as the compound PIK23, which at low concentrations inhibits only PIK3C-δ kinase. Other compounds, such as PIK75, PIK90, PIK93, and PI103, show a pharmacological range primarily limited to the PIK types but also inhibit one or more distantly related kinases (ATM, ATR, PRKDC, and FRAP1). Further molecular modeling and testing with additional compound chemotypes might help illuminate the particular binding interactions that are involved in compound specificities. Moreover, this type of phylogenetic visualization, which incorporates data on both target homology and compound activities, can be very useful for guiding further medicinal chemistry efforts.

9.7 CONCLUSION

There are many other important applications of comparative genomics to drug discovery, some of which are covered elsewhere in this book. The plethora of genomic sequence data is driving new therapeutic approaches to the treatment of pathogen infection diseases such as acquired immunodeficiency syndrome (AIDS), malaria, tuberculosis, and drug-resistant bacteria. Drug toxicity profiling is now incorporating comparative genomic analysis of the genomes, transcriptomes, and proteomes of drug-testing organisms, such as mouse, rat, and dog. Early target discovery was advanced through the use of a few model organisms with well-established genetics, such as yeast, *C. elegans*, *Drosophila*, and mouse. However, new technologies such as RNAi have unshackled biologists from use of traditional experimental species to study disease and now allow for the genetic manipulation of practically any species provided there is sufficient genomic DNA sequence. As further human genomes are sequenced, comparative analysis will become important for understanding individual patient variance in drug efficacy and adverse events. Finally, new modalities for therapeutic intervention could emerge in the coming decades as we learn more about the role of nonprotein elements in the disease progression, such as microRNAs and other noncoding RNAs.[57]

Multidisciplinary approaches that merge bioinformatics, evolutionary biology, and molecular biology to exploit multispecies genomic data for the benefit of enhanced pharmacology are playing an increasing role in drug discovery. The complexities of these technologies and data sets as well as the breadth of disease treatment opportunities are also driving major structural changes in the pharmaceutical industry. It is no longer possible for any single organization to proficiently encompass all these capabilities in-house. Thus, new R&D paradigms are emerging where large pharmaceutical companies, with their expertise in later phase drug development, are seeking closer, highly integrated partnerships with innovative biotechnology companies to invigorate and revolutionize their drug discovery pipelines.

ACKNOWLEDGMENTS

This work was supported by Informatics, Molecular Discovery Research, GlaxoSmithKline. I thank Aaron Mackey, Heather A. Madsen, and Joanna Betts for some useful discussions and references.

REFERENCES

1. Searls, D.B. Pharmacophylogenomics: genes, evolution and drug targets. *Nat. Rev. Drug Discov.* **2**, 613–623 (2003).
2. Kamb, A., Wee, S. & Lengauer, C. Why is cancer drug discovery so difficult? *Nat. Rev. Drug Discov.* **6**, 115–120 (2007).
3. Lipinski, C. & Hopkins, A. Navigating chemical space for biology and medicine. *Nature* **432**, 855–861 (2004).
4. Overington, J.P., Al Lazikani, B. & Hopkins, A.L. How many drug targets are there? *Nat. Rev. Drug Discov.* **5**, 993–996 (2006).
5. Kramer, R. & Cohen, D. Functional genomics to new drug targets. *Nat. Rev. Drug Discov.* **3**, 965–972 (2004).
6. Kaletta, T. & Hengartner, M.O. Finding function in novel targets: *C. elegans* as a model organism. *Nat. Rev. Drug Discov.* **5**, 387–398 (2006).
7. Wittenburg, N. et al. Presenilin is required for proper morphology and function of neurons in *C. elegans. Nature* **406**, 306–309 (2000).
8. Levitan, D. et al. Assessment of normal and mutant human presenilin function in *Caenorhabditis elegans. Proc. Natl. Acad. Sci. U. S. A.* **93**, 14940–14944 (1996).
9. Lim, H.Y., Bodmer, R. & Perrin, L. *Drosophila* aging 2005/06. *Exp. Gerontol.* **41**, 1213–1216 (2006).
10. Jafari, M., Long, A.D., Mueller, L.D. & Rose, M.R. The pharmacology of ageing in *Drosophila. Curr. Drug Targets.* **7**, 1479–1483 (2006).
11. Porcu, M. & Chiarugi, A. The emerging therapeutic potential of sirtuin-interacting drugs: from cell death to lifespan extension. *Trends Pharmacol. Sci.* **26**, 94–103 (2005).
12. Sharpless, N.E. & DePinho, R.A. The mighty mouse: genetically engineered mouse models in cancer drug development. *Nat. Rev. Drug Discov.* **5**, 741–754 (2006).
13. Van Dam, D. & De Deyn, P.P. Drug discovery in dementia: the role of rodent models. *Nat. Rev. Drug Discov.* **5**, 956–970 (2006).
14. Zambrowicz, B.P. & Sands, A.T. Knockouts model the 100 best-selling drugs — will they model the next 100? *Nat. Rev. Drug Discov.* **2**, 38–51 (2003).
15. Kresse, H., Belsey, M.J. & Rovini, H. The antibacterial drugs market. *Nat. Rev. Drug Discov.* **6**, 19–20 (2007).
16. Brown, J.R. & Warren, P.V. Antibiotic discovery: is it in the genes? *Drug Discov. Today* **3**, 564–566 (1998).
17. Payne, D.J., Gwynn, M.N., Holmes, D.J. & Rosenberg, M. Genomic approaches to antibacterial discovery. *Methods Mol. Biol.* **266**, 231–259 (2004).
18. Payne, D.J., Gwynn, M.N., Holmes, D.J. & Pompliano, D.L. Drugs for bad bugs: confronting the challenges of antibacterial discovery. *Nat. Rev. Drug Discov.* **6**, 29–40 (2007).
19. Caenepeel, S., Charydczak, G., Sudarsanam, S., Hunter, T. & Manning, G. The mouse kinome: discovery and comparative genomics of all mouse protein kinases. *Proc. Natl. Acad. Sci. U. S. A.* **101**, 11707–11712 (2004).
20. Manning, G., Whyte, D.B., Martinez, R., Hunter, T. & Sudarsanam, S. The protein kinase complement of the human genome. *Science* **298**, 1912–1934 (2002).

21. Bradham, C.A. et al. The sea urchin kinome: a first look. *Dev. Biol.* **300**, 180–193 (2006).
22. Manning, G., Plowman, G.D., Hunter, T. & Sudarsanam, S. Evolution of protein kinase signaling from yeast to man. *Trends Biochem. Sci.* **27**, 514–520 (2002).
23. Sonnhammer, E.L. & Koonin, E.V. Orthology, paralogy and proposed classification for paralog subtypes. *Trends Genet.* **18**, 619–620 (2002).
24. Carmena, M. & Earnshaw, W.C. The cellular geography of aurora kinases. *Nat. Rev. Mol. Cell Biol.* **4**, 842–854 (2003).
25. Mahadevan, D., Bearss, D.J. & Vankayalapati, H. Structure-based design of novel anti-cancer agents targeting aurora kinases. *Curr. Med. Chem. Anticancer Agents* **3**, 25–34 (2003).
26. Warner, S.L. et al. Identification of a lead small-molecule inhibitor of the Aurora kinases using a structure-assisted, fragment-based approach. *Mol. Cancer Ther.* **5**, 1764–1773 (2006).
27. Brown, J.R., Koretke, K.K., Birkeland, M.L., Sanseau, P. & Patrick, D.R. Evolutionary relationships of Aurora kinases: implications for model organism studies and the development of anti-cancer drugs. *BMC. Evol. Biol.* **4**, 39 (2004).
28. Bernard, M., Sanseau, P., Henry, C., Couturier, A. & Prigent, C. Cloning of STK13, a third human protein kinase related to *Drosophila aurora* and budding yeast Ipl1 that maps on chromosome 19q13.3-ter. *Genomics* **53**, 406–409 (1998).
29. Kimura, M., Matsuda, Y., Yoshioka, T. & Okano, Y. Cell cycle-dependent expression and centrosome localization of a third human aurora/Ipl1-related protein kinase, AIK3. *J. Biol. Chem.* **274**, 7334–7340 (1999).
30. Altschul, S.F. et al. Gapped BLAST and PSI-BLAST: a new generation of protein database search programs. *Nucleic Acids Res.* **25**, 3389–3402 (1997).
31. Tatusov, R.L. et al. The COG database: an updated version includes eukaryotes. *BMC Bioinformatics.* **4**, 41 (2003).
32. Tatusov, R.L., Galperin, M.Y., Natale, D.A. & Koonin, E.V. The COG database: a tool for genome-scale analysis of protein functions and evolution. *Nucleic Acids Res.* **28**, 33–36 (2000).
33. Lee, Y. et al. Cross-referencing eukaryotic genomes: TIGR Orthologous Gene Alignments (TOGA). *Genome Res.* **12**, 493–502 (2002).
34. O'Brien, K.P., Westerlund, I. & Sonnhammer, E.L. OrthoDisease: a database of human disease orthologs. *Hum. Mutat.* **24**, 112–119 (2004).
35. O'Brien, K.P., Remm, M. & Sonnhammer, E.L. Inparanoid: a comprehensive database of eukaryotic orthologs. *Nucleic Acids Res.* **33**, D476–D480 (2005).
36. Koski, L.B. & Golding, G.B. The closest BLAST hit is often not the nearest neighbor. *J. Mol. Evol.* **52**, 540–542 (2001).
37. Kamath, R.S. et al. Systematic functional analysis of the *Caenorhabditis elegans* genome using RNAi. *Nature* **421**, 231–237 (2003).
38. Fulton, D.L. et al. Improving the specificity of high-throughput ortholog prediction. *BMC Bioinformatics.* **7**, 270 (2006).
39. Goldstein, D.B., Need, A.C., Singh, R. & Sisodiya, S.M. Potential genetic causes of heterogeneity of treatment effects. *Am. J. Med.* **120**, S21–S25 (2007).
40. Ingelman-Sundberg, M. Genetic polymorphisms of cytochrome P450 2D6 (CYP2D6): clinical consequences, evolutionary aspects and functional diversity. *Pharmacogenomics J.* **5**, 6–13 (2005).
41. Haber, D.A. & Settleman, J. Cancer: drivers and passengers. *Nature* **446**, 145–146 (2007).
42. Erichsen, H.C. & Chanock, S.J. SNPs in cancer research and treatment. *Br. J. Cancer* **90**, 747–751 (2004).

43. Hein, D.W. Molecular genetics and function of NAT1 and NAT2: role in aromatic amine metabolism and carcinogenesis. *Mutat. Res.* **506–507**, 65–77 (2002).
44. Morton, L.M. et al. Genetic variation in *N*-acetyltransferase 1 (NAT1) and 2 (NAT2) and risk of non-Hodgkin lymphoma. *Pharmacogenet. Genomics* **16**, 537–545 (2006).
45. Sjoblom, T. et al. The consensus coding sequences of human breast and colorectal cancers. *Science* **314**, 268–274 (2006).
46. Greenman, C. et al. Patterns of somatic mutation in human cancer genomes. *Nature* **446**, 153–158 (2007).
47. Dancey, J. & Sausville, E.A. Issues and progress with protein kinase inhibitors for cancer treatment. *Nat. Rev. Drug Discov.* **2**, 296–313 (2003).
48. Cohen, P. Protein kinases — the major drug targets of the 21st century? *Nat. Rev. Drug Discov.* **1**, 309–315 (2002).
49. Samuels, Y. et al. High frequency of mutations of the PIK3CA gene in human cancers. *Science* **304**, 554 (2004).
50. Ikenoue, T. et al. Functional analysis of PIK3CA gene mutations in human colorectal cancer. *Cancer Res.* **65**, 4562–4567 (2005).
51. Vanhaesebroeck, B., Leevers, S.J., Panayotou, G. & Waterfield, M.D. Phosphoinositide 3-kinases: a conserved family of signal transducers. *Trends Biochem. Sci.* **22**, 267–272 (1997).
52. Gymnopoulos, M., Elsliger, M.A. & Vogt, P.K. Rare cancer-specific mutations in PIK3CA show gain of function. *Proc. Natl. Acad. Sci. U. S. A.* **104**, 5569–5574 (2007).
53. Hopkins, A.L., Mason, J.S. & Overington, J.P. Can we rationally design promiscuous drugs? *Curr. Opin. Struct. Biol.* **16**, 127–136 (2006).
54. Cohen, M.S., Zhang, C., Shokat, K.M. & Taunton, J. Structural bioinformatics-based design of selective, irreversible kinase inhibitors. *Science* **308**, 1318–1321 (2005).
55. Fabian, M.A. et al. A small molecule-kinase interaction map for clinical kinase inhibitors. *Nat. Biotechnol.* **23,** 329–336 (2005).
56. Knight, Z.A. et al. A pharmacological map of the PI3-K family defines a role for p110alpha in insulin signaling. *Cell* **125**, 733–747 (2006).
57. Esquela-Kerscher, A. & Slack, F.J. Oncomirs — microRNAs with a role in cancer. *Nat. Rev. Cancer* **6**, 259–269 (2006).
58. Felsentein, J. PHYLIP (Phylogenetic Inference Package). Version 3.6. University of Washington, Seattle, 2000.
59. Walker, E.H., Perisic, O., Ried, C., Stephens, L. & Williams, R.L. Structural insights into phosphoinositide 3-kinase catalysis and signalling. *Nature* **402**, 313–320 (1999).

10 Comparative Genomics and the Development of Novel Antimicrobials

Diarmaid Hughes

CONTENTS

ABSTRACT

Genomics has opened up the previously mysterious world of microbiology to the possibility of a systematic analysis. A comparative genomics approach is now an integral part of efforts to identify novel broad-spectrum targets for new antimicrobial drugs. However, genomics is also revealing high levels of diversity within bacterial species and significant horizontal transfer of resistance elements through the gene pool. Together, these problematic factors suggest that the number of novel, essential, and susceptible broad-spectrum drug targets is very small. As a consequence, successful development of novel classes of antimicrobials may in the longer term increasingly be tied with the economics of exploiting narrow-spectrum targets. Comparative genomics in its broad sense, including comparative proteomics and structural biology, may be of practical use in this important area if it can lead to a more accurate determination of which candidate drugs should be taken into the expensive stage of clinical trials.

10.1 INTRODUCTION: THE NEED FOR NEW ANTIMICROBIALS

Effective antimicrobial therapies have been an important medical tool in controlling infections for a little more than six decades. During that period, approximately two dozen chemically different classes of antimicrobial drugs were developed and introduced successfully to the market. These drugs target essential cellular processes, including bacterial cell wall synthesis (β-lactams, cephalosporins, monobactams, carbapenems, bacitracin, glycopeptides, isoniazid); DNA replication (quinolones); RNA transcription (rifampicin); protein synthesis (macrolides, tetracyclines, chloramphenicol, aminoglycosides, lincomycin, oxazolidinones, fusidic acid, mupirocin, etc.); cell membrane integrity (polymixins, gramicidin); and folic acid synthesis (trimethoprim, sulfonamides). The overwhelming bulk of antimicrobials sold today for human medicine are modifications of a few chemical classes that were discovered or initially marketed between the 1940s and the 1960s: β-lactams, cephalosporins, macrolides, tetracyclines, and quinolones. More recently, the pipeline of new classes of antimicrobial drugs has slowed to a trickle. Unfortunately, the slowdown in development of novel antimicrobials is coinciding with a continuing increase in the prevalence of resistance in most countries. The most recent (2004) European Antimicrobial Resistance Surveillance System (EARSS) report[1] found on average the following: 24% of *Staphylococcus aureus* were methicillin-resistant *S. aureus* (MRSA); for *Streptococcus pneumoniae*, 9% and 16% were nonsusceptible to penicillin and erythromycin, respectively; and for *Escherichia coli*, 48% and 14% were resistant to aminopenicillins and fluoroquinolones, respectively. The figures are much worse for some countries, with Spain, for example, having resistance levels of 25% or higher for each of the above drug–bacteria combinations. The resistance problem is a worldwide phenomenon, and of particular worry is the rise in the frequency of multidrug-resistant tuberculosis in many developing countries.[2] As a consequence of resistance, infections associated with a high level of morbidity and mortality become increasingly difficult to treat effectively.

There are several reasons for the slowdown in development of novel antimicrobials. Beginning in the 1960s, there was the perception that the existing antimicrobial agents were sufficient to solve the problems caused by bacterial infections. This was exemplified in the well-publicized statement by the U.S. surgeon general in 1967 that it was "time to close the book on infectious disease" and shift attention (and dollars) to the new dimension of health: chronic diseases.[3] This was in line with a perception in the big pharmaceutical companies that it was more profitable to invest research money in developing drugs to treat chronic conditions such as arthritis and depression.[4–6] The saturation of the market by existing antimicrobial drugs strengthened the economic argument, as did the availability of generic compounds for some of the largest-selling drugs and the enormous costs of the clinical trials that were required to bring new drugs to the market. In addition, there have emerged increasing political pressures to reduce the unnecessary consumption of antibacterial agents[7] because this is regarded as a major driving force for the increasing prevalence of antibiotic-resistant bacteria globally.[8,9] The argument is that restrictive use may extend the useful life of a drug by halting or slowing the rise in resistance, although both theoretical and experimental analysis suggest that this is unlikely in most cases to reverse existing levels of resistance.[10,11] Restrictive use may be highly relevant for novel classes of antimicrobials, for which resistance, or linkage to resistance, does not preexist. However, restricting sales and consumption further exacerbates the economic issues in drug development, and creates a dilemma between encouraging investment in antimicrobial development and preserving the usefulness of current and new drugs. Resolving this dilemma will probably require working out new policy agreements, between government regulatory agencies and pharmaceutical companies, that succeed in combining profitability with long-term public health requirements.[4,12]

There is general agreement that the worsening antibiotic resistance problem necessitates some action if we are to avoid a serious public health threat in the near future.[4,5,12,13] This problem comes at a time when societies face the additional threats of emerging and reemerging infections and of bioterrorism and when there is a growing appreciation of infectious disease as a possible cause of chronic disease.[3] Among the proposed actions are the development of new antimicrobial vaccines, the exploration of the utility of phage therapy, and not surprisingly, the development of novel classes of antimicrobial drugs.[12] A large part of the initial stages in the research and development of novel antimicrobials, in the continued absence of renewed interest by big pharma, will probably be carried out by relatively small pharmaceutical and biotechnological companies,[14] and almost all of it will include, or be based on, the concepts of comparative genomics.

10.2 WHAT CAN COMPARATIVE GENOMICS DO FOR ANTIMICROBIALS?

The principles of using comparative genomics as an integral part of an approach to antimicrobial drug development are simple and straightforward. The first step is to identify a novel drug target. This is broadly defined as a bacterial structure (DNA, RNA, protein, lipid, etc.) that is essential, at least in relevant environments, and has not previously been used as an antimicrobial drug target. The drug interaction with the target should cause bacterial death or severe growth inhibition. The drug target should be widely conserved among bacteria to ensure a broad spectrum of activity, in particular against organisms such as staphylococci, streptococci, pneumococci, enterococci, pseudomonas, and mycobacteria, for which mortality and resistance are currently most problematic. The desire for a broad spectrum of activity is partly driven by economics but also by the empirical nature of most diagnosis, although this could change with the development of new rapid diagnostic technologies.[15–20] Thus, notwithstanding the opposing concerns of those who wish to restrict antibiotic use and employ more narrow-spectrum drugs,[7] the pharmaceutical industry is more likely to focus on drug targets conserved across many bacterial groups. The chosen drug target should also be absent in humans, with the aim to reduce the risk of drug failure due to toxicity in clinical trials. One can question the wisdom of excluding targets with human counterparts given that one of the best antimicrobial targets, the ribosome, is highly conserved in bacteria and humans.

Within these parameters, comparative genomics is essentially the process of sifting through, and comparing, bacterial and human genome sequences with the aim of picking out widespread, conserved, essential, and uniquely bacterial genes or genetic pathways for more detailed analysis. In the early days (only a few years ago), this initial phase required in-house genomic sequencing of target organisms. Now, there is a huge and rapidly growing amount of freely available genome sequence data to support and drive the comparative genomics approach.[21] The development of advanced bioinformatics methods that facilitate whole-genome analyses has also progressed rapidly,[22–27] and an increasing integration with a systems biology approach[28–31] promises to enhance the value of the raw sequence information for identifying useful drug targets.

Candidate target genes must be validated, typically by genetic inactivation, to confirm that they are essential in relevant environments. Methods for target validation by gene inactivation include transposon mutagenesis,[32,33] targeted allelic exchange,[34] and expression of antisense RNA.[35] In addition, it is usually important to determine that the target is essential in vivo, and one of the most useful techniques to address this issue is signature-tagged mutagenesis.[36,37] Identifying targets at key nodes in metabolic or regulatory networks, where the effects of drug binding are pleiotropic and therefore difficult to compensate, should be one benefit of this highly informed genomics approach.

The systems biology approach is itself the integrative analytical branch of transcriptomics and proteomics research[31,38] that facilitate high-throughput evaluation of the gene expression profile across the whole genome in a variety of environments.[39–41] Another important approach that has advanced in step with genomics analysis is the ability to rapidly solve the three-dimensional structures of potential or actual drug targets. Structural genomics is already an important tool in guiding the rational modification of the chemical structure of drug candidates to optimize their abilities to interact and inhibit target molecules in specific bacteria.[42,43] In the future, structure-guided design of antimicrobial drugs, *ab initio*, might also become a feasible approach to the creation of drugs specific for rationally chosen targets.[44]

Thus, comparative genomics information includes (1) in silico comparisons that allow correlations to be made between genotype and phenotype and the initial identification of candidate target genes; (2) target validation methodologies that address the essentiality of the target candidates; and (3) transcriptomic and proteomic analyses that provide insights into gene expression, in relation to virulence,[45] to the presence of antimicrobials,[46–49] and to genetic alterations associated with antimicrobial resistance.[49,50] Transcriptomic and proteomic analysis can also inform about the mechanism of action of drug candidates. This information is useful in screening drug candidates to identify those that most likely have novel targets, novel mechanisms of action, or multiple targets and for which there is less likelihood of preexisting resistance. Finally, it should be noted that bacteriophage have coevolved with bacteria and have developed a variety of effective means of killing or otherwise inhibiting bacterial growth. Bacteriophage genomics analysis has been used to identify potential antibacterial targets in, for example, *S. aureus*.[51]

The comparative genomics approach is radically different from the traditional approach to finding new antimicrobials. Traditionally, the starting point in the search for a new antimicrobial drug had been either a chemical compound library or a microorganism extract library. These libraries were assayed for growth inhibitory activity against a panel of interesting bacteria. A positive outcome would be the identification of chemicals or extracts that inhibited the growth of some, or all, of the panel of bacteria. This approach, while yielding positive hits, had several major drawbacks. First, we should consider the nature of the libraries, chemical and biological, that are used in the screening process. A biological extract library gives access potentially to the full range of natural molecules resulting from four billion years of biological evolution. The major drawback, however, is that some of these molecules are already known and in use as drugs. Thus, screening a biological library for growth inhibitory activity will yield known drugs such as chloramphenicol, tetracycline, β-lactams, and the like, and these have to be screened away before any novel molecules can be identified.

A chemical library avoids this problem because it can be designed not to contain any known drug structures and also to contain structures that do not exist in living organisms. The major drawback of a chemical library is that it is more limited in variety compared to a biological extract library. Using the traditional approach to drug discovery, there is a further drawback that is common to both types of library, namely, the target of the drug hit is initially unknown. Ignorance of the target means that it is more difficult to interpret the significance of the activity spectrum of the drug or to predict whether it might have a toxic effect in humans. This is a serious problem because hundreds of hits may be found in a traditional screening, many with only weak inhibitory effects. It is not possible to decide in any rational way which ones, if any, might make the best drugs (after suitable modifications) without spending a large amount of time on a program to identify their targets. This limitation was of course well known even before the genomics era.

The way to counter the problem was to decide on a target (e.g., the ribosome or the cell wall) or a specific step associated with the target (e.g., protein elongation on the ribosome) at the beginning and design a biochemical or genetic assay that facilitated compound screening directed to the chosen target. A recent example of the successful application of this approach has identified hits from a biological extract library that are specific for inhibition of translation initiation.[52–54] Another recent success came from screening a library of 250,000 commercially available compounds against *S. aureus* RNA polymerase holoenzyme in a functional assay. This yielded a small molecule (2-ureidothiophene-3-carboxylate) that has been used successfully as the basis for the development of a set of potent inhibitors with good antibacterial activities, including against rifampicin-resistant *S. aureus*.[55] In another example of the identification of novel antimicrobials from a chemical library against an established target, a set of small molecules targeting the interaction between RNA polymerase and sigma factor have been reported.[56] In this case, comparative genomics was used to establish the conservation of the protein–protein interface across a wide spectrum of bacteria before the screening process was begun.

In the pregenomic era, target-based screening could only be directed against targets that were already known to be conserved, such as the ribosome, the cell wall, DNA synthesis, and so on. This approach is not without value because these targets have been validated by the discovery of many active antimicrobials, and as illustrated by the examples, it is still possible to discover new drugs for old targets. However, the great advance that genomics has brought is the possibility to gain access to a complete catalog of genetic and physiological information on bacteria that can form the basis for rational choices of novel targets that have not previously been exploited in drug discovery programs. This is where the comparative genomics approach potentially provides a big boost to the process of novel drug discovery. The libraries to be screened may be the same (chemical or biological), but by beginning with the definition of a novel validated target, it is possible in principle to ensure that any hits that emerge will be novel, at least in terms of action, and unique to bacteria.

10.3 LIMITATIONS AND POTENTIAL OF COMPARATIVE GENOMICS

One of the obvious advantages of genome sequencing and comparative genomics as an approach to developing novel antimicrobials is that it provides lists of candidate genes common to the infectious organisms of interest. In mid-2007, there were in the public domain 523 completely sequenced bacterial genomes, and sequencing of 1300 was ongoing.[21] The expectation that genome sequencing would reveal a wealth of diversity within the microbial world and facilitate a rational classification of bacteria in terms of their phylogenetic relationships is being realized. What was largely unexpected was the diversity that genome sequencing would reveal within bacterial species, even allowing for problems in defining a species concept for bacteria.[57] For example, *E. coli* K-12, the gold standard organism for microbiology, and its enterohemorrhagic relative *E. coli* O157:H7, differ in gene content by 30%.[58] A three-way comparison of *E. coli* MG1655 K-12, O157:H7, and the uropathogen CFT073 showed that they have only 39% of their combined (nonredundant) set of proteins in common.[59] The pathogenic *E. coli* genomes are as different from each other as each pathogen is from the benign K-12 strain. Thus, without fairly extensive genomic sequencing and comparison, it cannot be assumed that all varieties of an important group of infectious bacteria carry the gene coding for a particular novel target. Genetic diversity at both the inter- and intraspecies level, assessed by DNA microarrays and genomic comparisons, appears to set tight limits on the number of widely conserved targets for broad-spectrum antimicrobials.[39,60] More comparative genomic analysis based on the much larger number of genomes now available is needed to quantify the actual limitations on target selection associated with the inverse relationship between the number of conserved targets and the spectrum of bacteria diversity.

Comparative genomics may also provide valuable information on the potential for resistance development against novel antimicrobials by increasing understanding of how horizontally transferable resistance elements move through the gene pool. Thus, genomic comparisons are revealing that the sources of genetic diversity within and between bacterial species are several. These include divergent evolution of specific gene sets[61]; genome rearrangements, often mediated by insertion sequence (IS) elements[62,63]; and horizontal gene transfers (HGTs).[63–65]

Horizontal gene transfer in particular means that bacterial phylogenies are better represented by a network of vertically and horizontally transferred genes rather than as a single tree.[66,67] Part of the significance of bacterial evolution by HGT is that mechanisms of resistance to antimicrobial agents, and novel virulence genes, can potentially travel across large genetic distances by a small number of HGT events.[67] This poses a dilemma for the development of antimicrobials. HGT makes available an almost limitless number of potential sources of resistance mechanisms. The potential problems associated with HGT of resistance mechanisms are currently difficult to quantify. One of the benefits of continuing basic studies in comparative genomics will be to provide more detailed information on the rates of HGT. In particular, it will be of great interest to know whether HGT is essentially random (akin to Brownian motion of genes in a gene pool) or whether it tends to follow particular paths (akin to main routes in a gene network).

The concept of the pan-genome has been proposed to describe the amount of the total global genome that might be available to, or associated with, a particular bacterial species, and this also shows great variation.[57] Thus, the total number of genes associated with *Streptococcus agalactiae* appears to be unlimited,[57,68] whereas for *Bacillus anthracis*, the pan-genome may be limited to only four genome sequences.[57] It is predicted that species that colonize multiple environments and have multiple ways of exchanging genetic information, such as streptococci, meningococci, salmonellae, and *E. coli*, will have relatively open pan-genomes in contrast to those that live in isolated niches such as *B. anthracis, Mycobacterium tuberculosis*, and *Chlamydia trachomatis*. Quantitative and qualitative information on the pan-genome of medically important bacterial species will facilitate improved risk assessment for the acquisition of resistance by HGT and will assist the prospective evaluation of novel antimicrobial agents.

Gathering information on HGT rates and preferred transfer pathways requires that we learn much more of the true diversity of the microbial world. It has been estimated that more than 99% of all bacteria are unculturable in the laboratory.[69] Attempts to access this vast pool of genetic information are occurring based on the development of metagenomic technologies to sequence and assemble genomes independently of the ability to culture organisms.[70,71] However, to fully exploit the power of genomics, methods to culture the unculturable need to be developed, and efforts in the area are meeting with some success.[72]

10.4 PROSPECTS FOR ANTIMICROBIAL DEVELOPMENT

The comparative genomics approach to drug discovery is essentially a "target first" approach coupled to the possibility of making a rational choice from all possible targets. Although it has been around for only a decade, there are already reviews suggesting that genomics is regarded by some as a disappointment for not yielding a bonanza of novel antimicrobials.[6,73] In part, this reflects the overly optimistic expectations associated with a new field of research. In part, it reflects the apparent reality emerging from comparative genomics studies that the number of universally conserved and essential novel targets is actually quite small. In addition, with the benefit of a decade of experience, there is now a greater appreciation that a target-based genomics approach to drug discovery requires the successful development and integration of a host of new technologies and methodologies, as discussed in this chapter. The genome sequences themselves are only the basic raw materials, and there are now many more of them available to examine and compare than there were even a few years ago. Between the pessimism and the hyperbole about genomic approaches, there are actually some novel drug targets and associated drug candidates that are in the process of evaluation.

10.4.1 Aminoacyl-tRNA Synthetases

Synthetases belong to one of the traditional antimicrobial target classes, the translation machinery, and so cannot be claimed as an example of the success of genomics in identifying novel essential targets. However, genomic comparisons and associated genetic validation studies have been useful in showing that aminoacyl-tRNA (transfer RNA) synthetases as a group are widely conserved essential bacterial enzymes. In addition, genomics-driven structural analysis of synthetases from different bacteria has been critical in directing the modification of inhibitors to achieve improved activity or broader spectrum. Isoleucyl-tRNA synthetase is the target of mupirocin, a small molecule with good antistaphylococcal activity.[74] The success of mupirocin as an antimicrobial makes other members of the tRNA synthetase family attractive targets for drug discovery programs.

The approach taken to find an inhibitor of prolyl-tRNA synthetase is especially interesting. A specific peptide that bound to the synthetase was initially selected in vitro.[75] Expression of the peptide in vivo was shown to rescue an animal model from a lethal infection, validating the synthetase, and more specifically the peptide-binding site, as a good target for inhibition. A small-molecule library was then screened for hits that could displace the peptide from the synthetase as a way to obtain new drug leads.[75]

This approach has since been used in the discovery of lead compounds that target several other tRNA synthetases.[76–78] Over the past several years, there has been a significant investment in finding compounds that target each of the tRNA synthetases, resulting in the identification of a series of small molecules with antimicrobial activity.[79,80] One of the hopes in this field is that structural conservation of catalytic residues between related synthetases might lead to the development of multienzyme inhibitors. This could be advantageous in terms of associating major fitness costs to resistance and that might restrict resistance development. However, problems with poor in vivo and whole-cell activity are holding up the development of these leads into clinically useful drugs.

In addition, extensive HGT of aminoacyl-tRNA synthetases has also frustrated development of drugs against this class of targets.[81,82] Thus, an inhibitor of methionyl-tRNA synthetase encountered a small but significant population of resistant *S. pneumoniae* strains isolated from clinical samples.[83] The mode of resistance was shown to be due to a second copy of the MetRS gene that was acquired via HGT from a species related to *B. anthracis* and also harboring two methionyl-tRNA synthetase (MetRS) genes.[84] The second MetRS gene is more similar to archael or eukaryotic orthologs and hence refractory to the inhibitor. Ancient and more recent HGT could be problematic across aminoacyl-tRNA synthetases.

10.4.2 Peptide Deformylase

Formylation of the initiator methionine in protein synthesis occurs in most bacteria.[85] When translation is complete, the formyl group is removed by the enzyme peptide deformylase (PDF).[86] Genetic knockout experiments showed that PDF is an essential enzyme and thus a potential target for antimicrobial action.[87] Although PDF was identified as a potential target for antimicrobials in the pregenomic era, it was genomic comparisons, and genomics-driven structural comparisons, that subsequently showed that it was a near-universal bacterial gene with highly conserved motifs.[88,89] PDF is a metalloenzyme, and its activity is inhibited by divalent metal ion inhibitors.[90] A natural inhibitor of PDF, actinonin, and several synthetic inhibitors act by having a structure resembling the enzyme substrate coupled to a metal ion chelator.[89] Synthetic inhibitors of PDF created by Ocsient Pharmaceuticals and by Novartis Pharmaceuticals have good in vitro and in vivo activities and have progressed to phase I clinical trials.[89,91,92]

10.4.3 Fatty Acid Biosynthesis

Type II fatty acid synthesis as a potential target for antimicrobial development was established prior to the genomics era, and several inhibitors were known, including triclosan and isoniazid.[93,94] Genomics has contributed to the interest in these targets mainly by providing information on the conservation of genes in the pathway in pathogenic bacteria and by supporting the structural analysis of each of the enzymes in the pathway.[95] The small molecule platensimycin was identified from a biological extract library as a potent inhibitor of FabF, an enzyme involved in fatty acid biosynthesis with broad-spectrum activity and good in vivo efficacy.[96] No other drugs targeting FabF are used clinically, and platensimycin shows no cross resistance.[96] Platensimycin has good activity against MRSA, vancomycin-intermediate staphlococcus (VISA), and vancomycin-resistant enterococci (VRE) but has not yet entered clinical trials.

10.4.4 Cofactor Biosynthesis Enzymes

A comparative genomics analysis identified cofactor biosynthetic pathways as potential broad-spectrum drug targets.[32] Using a non-genomics-directed approach, screening compounds from different chemical series for whole-cell growth inhibition of *Mycobacterium smegmatis*, a novel antimycobacterial was discovered.[97] The drug is a diarylquinoline (DARQ) and targets the proton pump of adenosine triphosphate (ATP) synthase. Chemical optimization has led to DARQs with potent activity in vitro and in vivo against drug-sensitive and drug-resistant *M. tuberculosis*.[97] The drug is very specific for mycobacteria and shows no cross resistance to other antituberculosis drugs.

10.4.5 Bacteriophage Genomics

Comparative genomics of bacteriophage, particularly learning how specific bacteriophage proteins inhibit bacterial growth, is a promising path to novel bacterial targets.[98,99] One target that has been identified and validated by this approach in *S. aureus* is DnaI, a protein that is required for primosome assembly and is essential during the initiation of DNA replication.[51] A small-molecule library (125,000 compounds from commercially available libraries) was screened for inhibitors of the interaction between the phage protein and DnaI, resulting in the identification of 36 hits, of which 11 compounds had whole-cell activity with a minimum inhibitory concentration (MIC) of 16 μg/ml or less.

10.5 CONCLUSIONS AND THE NEAR FUTURE

The need for comparative genomics as a tool to identify novel broad-spectrum targets for antimicrobial drug development is not going to last forever. Soon, if not already, we will have access to all relevant genome sequences for the major bacterial infections. How many broad-spectrum targets will emerge from this analysis and how many will be druggable remains to be seen. Viewed pessimistically, it appears that the number of essential, broad-spectrum drug targets will be small, much fewer than 100, and that most of these may belong to pathways already targeted by existing antimicrobials.[23,100] Viewed optimistically, the structural and functional complexity of most currently used targets and pathways shows that most can be independently targeted by several structurally different and non-cross-reacting small molecules. The same may be true for the new targets discovered through comparative genomics. Thus, the real number of structural targets should be greater than the number of protein complexes or pathways that are validated. Indeed, there are several antimicrobial drugs currently in development that belong to novel structural classes but are directed to specific parts of traditional targets, such as the cell wall, RNA polymerase, folic acid pathway, and so on.[101–103]

The reality of antimicrobial drug development today, as illustrated by the short review of those targets and drugs now in development, suggests that genomics has not yet revealed any novel drug target for which an inhibitor has been found and that is exciting and promising enough to tempt the big pharmaceutical companies into a development program. What genomics has undoubtedly done is to open up the previously mysterious world of microbiology to the possibility of a systematic analysis. If in the end the conclusion should be that there is far more variation among bacteria than earlier expected, then at least we can approach the problem of infection control with that knowledge as a base. It may be that we already know of and utilize most of the broad-spectrum drug targets, and that the future development of novel classes of antimicrobials will increasingly be tied with narrow-spectrum targets.

The emphasis in antimicrobial drug discovery will shift downstream in the development process. The next bottleneck will be to develop high-throughput assays for each of the interesting targets to use in drug-screening programs. The drug candidates themselves are another obvious development bottleneck. The chemical libraries, although large, are inevitably limited in terms of chemical structures, and this may be the cause of a failure to discover a drug that can inhibit a particular target. The current alternative, to use biological extract libraries, has two advantages: (1) the number of molecules assayed is possibly much greater; and (2) more importantly, they will certainly contain molecules designed by evolution to interact with the chosen target. This should in theory greatly increase the probability of finding inhibitor molecules. A third alternative is to analyze the structure of the chosen target and then design and chemically synthesize a small inhibitory molecule. This approach is in its infancy, but the rapid advances made in structural biology and drug design hold out the promise that this may eventually become the method of choice in drug discovery.

There is one other critical and economically important bottleneck in the drug discovery process, namely, the high failure rate during clinical trials. If potential drugs could be screened more effectively at an early stage in the discovery process to filter out more of those that would later show toxicity or other undesirable side effects in clinical trials, then it could contribute to a massive reduction in the overall costs of development. This in turn would radically alter the economics of developing narrow-spectrum antimicrobials, with the double knock on benefits that many more drug targets could then be exploited, and because the drugs would be narrow spectrum, the selection pressure for resistance development would be that much smaller. There is reason to hope that comparative genomics in its broad sense, including comparative proteomics and structural biology, will be of practical use in this important area, more accurately determining which candidate drug molecules should be taken into the expensive stage of clinical trials.

ACKNOWLEDGMENTS

I acknowledge support for my research from the Swedish Research Council (Vetenskapsrådet) and the European Union Sixth Framework Programme (LSHM-CT-2005-518152).

REFERENCES

1. EARSS Annual Report 2004. Available at: http://www.rivm.nl/earss/.
2. Okeke, I. N. et al. Antimicrobial resistance in developing countries. Part I: recent trends and current status. *Lancet Infect Dis* **5**, 481–493 (2005).
3. Fauci, A. S. Infectious diseases: considerations for the 21st century. *Clin Infect Dis* **32**, 675–685 (2001).
4. Projan, S. J. Why is big pharma getting out of antibacterial drug discovery? *Curr Opin Microbiol* **6**, 427–430 (2003).
5. Projan, S. J. & Shlaes, D. M. Antibacterial drug discovery: is it all downhill from here? *Clin Microbiol Infect* **10 Suppl 4**, 18–22 (2004).
6. Shlaes, D. M. The abandonment of antibacterials: why and wherefore? *Curr Opin Pharmacol* **3**, 470–473 (2003).
7. Goossens, H. et al. National campaigns to improve antibiotic use. *Eur J Clin Pharmacol* **62**, 373–379 (2006).
8. Austin, D. J., Kristinsson, K. G. & Anderson, R. M. The relationship between the volume of antimicrobial consumption in human communities and the frequency of resistance. *Proc Natl Acad Sci USA* **96**, 1152–1156 (1999).
9. Seppala, H. et al. The effect of changes in the consumption of macrolide antibiotics on erythromycin resistance in group A streptococci in Finland. Finnish Study Group for Antimicrobial Resistance. *N Engl J Med* **337**, 441–446 (1997).
10. Andersson, D. I. Persistence of antibiotic resistant bacteria. *Curr Opin Microbiol* **6**, 452–456 (2003).
11. Levin, B. R., Perrot, V. & Walker, N. Compensatory mutations, antibiotic resistance and the population genetics of adaptive evolution in bacteria. *Genetics* **154**, 985–997 (2000).

12. Hughes, D. Exploiting genomics, genetics and chemistry to combat antibiotic resistance. *Nat Rev Genet* **4**, 432–441 (2003).
13. Overbye, K. M. & Barrett, J. F. Antibiotics: where did we go wrong? *Drug Discov Today* **10**, 45–52 (2005).
14. Barrett, J. F. Can biotech deliver new antibiotics? *Curr Opin Microbiol* **8**, 498–503 (2005).
15. Sanguinetti, M. et al. Use of microelectronic array technology for rapid identification of clinically relevant mycobacteria. *J Clin Microbiol* **43**, 6189–6193 (2005).
16. Peters, R. P., van Agtmael, M. A., Danner, S. A., Savelkoul, P. H. & Vandenbroucke-Grauls, C. M. New developments in the diagnosis of bloodstream infections. *Lancet Infect Dis* **4**, 751–760 (2004).
17. Peters, R. P. et al. Faster identification of pathogens in positive blood cultures by fluorescence in situ hybridization in routine practice. *J Clin Microbiol* **44**, 119–123 (2006).
18. Poppert, S. et al. Rapid diagnosis of bacterial meningitis by real-time PCR and fluorescence in situ hybridization. *J Clin Microbiol* **43**, 3390–3397 (2005).
19. Honest, H., Sharma, S. & Khan, K. S. Rapid tests for group B streptococcus colonization in laboring women: a systematic review. *Pediatrics* **117**, 1055–1066 (2006).
20. Eigner, U., Weizenegger, M., Fahr, A. M. & Witte, W. Evaluation of a rapid direct assay for identification of bacteria and the mec A and van genes from positive-testing blood cultures. *J Clin Microbiol* **43**, 5256–5262 (2005).
21. Gold (Genomes Online Database) Available at: http://www.genomesonline.org.
22. Yoon, S. H. et al. A computational approach for identifying pathogenicity islands in prokaryotic genomes. *BMC Bioinformatics* **6**, 184 (2005).
23. Anishetty, S., Pulimi, M. & Pennathur, G. Potential drug targets in *Mycobacterium tuberculosis* through metabolic pathway analysis. *Comput Biol Chem* **29**, 368–378 (2005).
24. Chen, T., Abbey, K., Deng, W. J. & Cheng, M. C. The bioinformatics resource for oral pathogens. *Nucleic Acids Res* **33**, W734–W740 (2005).
25. Raskin, D. M., Seshadri, R., Pukatzki, S. U. & Mekalanos, J. J. Bacterial genomics and pathogen evolution. *Cell* **124**, 703–714 (2006).
26. Bansal, A. K. Bioinformatics in microbial biotechnology — a mini review. *Microb Cell Fact* **4**, 19 (2005).
27. Dieterich, G., Karst, U., Fischer, E., Wehland, J. & Jansch, L. LEGER: knowledge database and visualization tool for comparative genomics of pathogenic and non-pathogenic *Listeria* species. *Nucleic Acids Res* **34**, D402–D406 (2006).
28. Watson, M. ProGenExpress: visualization of quantitative data on prokaryotic genomes. *BMC Bioinformatics* **6**, 98 (2005).
29. Kell, D. B. et al. Metabolic footprinting and systems biology: the medium is the message. *Nat Rev Microbiol* **3**, 557–565 (2005).
30. Mori, H. From the sequence to cell modeling: comprehensive functional genomics in *Escherichia coli*. *J Biochem Mol Biol* **37**, 83–92 (2004).
31. Gerdes, S. Y. et al. Experimental determination and system level analysis of essential genes in *Escherichia coli* MG1655. *J Bacteriol* **185**, 5673–5684 (2003).
32. Gerdes, S. Y. et al. From genetic footprinting to antimicrobial drug targets: examples in cofactor biosynthetic pathways. *J Bacteriol* **184**, 4555–4572 (2002).
33. Akerley, B. J. et al. A genome-scale analysis for identification of genes required for growth or survival of *Haemophilus influenzae*. *Proc Natl Acad Sci USA* **99**, 966–971 (2002).
34. Thanassi, J. A., Hartman-Neumann, S. L., Dougherty, T. J., Dougherty, B. A. & Pucci, M. J. Identification of 113 conserved essential genes using a high-throughput gene disruption system in *Streptococcus pneumoniae*. *Nucleic Acids Res* **30**, 3152–3162 (2002).

35. Ji, Y. et al. Identification of critical staphylococcal genes using conditional phenotypes generated by antisense RNA. *Science* **293**, 2266–2269 (2001).
36. Hensel, M. et al. Simultaneous identification of bacterial virulence genes by negative selection. *Science* **269**, 400–403 (1995).
37. Mecsas, J. Use of signature-tagged mutagenesis in pathogenesis studies. *Curr Opin Microbiol* **5**, 33–37 (2002).
38. Brotz-Oesterhelt, H., Bandow, J. E. & Labischinski, H. Bacterial proteomics and its role in antibacterial drug discovery. *Mass Spectrom Rev* **24**, 549–565 (2005).
39. Dorrell, N., Hinchliffe, S. J. & Wren, B. W. Comparative phylogenomics of pathogenic bacteria by microarray analysis. *Curr Opin Microbiol* **8**, 620–626 (2005).
40. Alberts, R. et al. Combining microarrays and genetic analysis. *Brief Bioinform* **6**, 135–145 (2005).
41. Lindsay, J. A. et al. Microarrays reveal that each of the 10 dominant lineages of *Staphylococcus aureus* has a unique combination of surface-associated and regulatory genes. *J Bacteriol* **188**, 669–676 (2006).
42. Schmid, M. B. Crystallizing new approaches for antimicrobial drug discovery. *Biochem Pharmacol* **71**, 1048–1056 (2006).
43. Barker, J. J. Antibacterial drug discovery and structure-based design. *Drug Discov Today* **11**, 391–404 (2006).
44. Banfi, E. et al. Antifungal and antimycobacterial activity of new imidazole and triazole derivatives. A combined experimental and computational approach. *J Antimicrob Chemother* **58,** 76–84, (2006).
45. Liautard, J. P., Jubier-Maurin, V., Boigegrain, R. A. & Kohler, S. Antimicrobials: targeting virulence genes necessary for intracellular multiplication. *Trends Microbiol* **14**, 109–113 (2006).
46. Goh, E. B. et al. Transcriptional modulation of bacterial gene expression by subinhibitory concentrations of antibiotics. *Proc Natl Acad Sci USA* **99**, 17025–17030 (2002).
47. Tsui, W. H. et al. Dual effects of MLS antibiotics: transcriptional modulation and interactions on the ribosome. *Chem Biol* **11**, 1307–1316 (2004).
48. Yim, G., Wang, H. H. & Davies, J. The truth about antibiotics. *Int J Med Microbiol* **296**, 163–170 (2006).
49. Aakra, A. et al. Transcriptional response of *Enterococcus faecalis* V583 to erythromycin. *Antimicrob Agents Chemother* **49**, 2246–2259 (2005).
50. Marrer, E., Satoh, A. T., Johnson, M. M., Piddock, L. J. & Page, M. G. Global transcriptome analysis of the responses of a fluoroquinolone-resistant *Streptococcus pneumoniae* mutant and its parent to ciprofloxacin. *Antimicrob Agents Chemother* **50**, 269–278 (2006).
51. Liu, J. et al. Antimicrobial drug discovery through bacteriophage genomics. *Nat Biotechnol* **22**, 185–191 (2004).
52. Brandi, L. et al. Specific, efficient, and selective inhibition of prokaryotic translation initiation by a novel peptide antibiotic. *Proc Natl Acad Sci USA* **103**, 39–44 (2006).
53. Brandi, L. et al. Novel tetrapeptide inhibitors of bacterial protein synthesis produced by a *Streptomyces* sp. *Biochemistry* **45**, 3692–3702 (2006).
54. Brandi, L., et al. Characterization of GE82832, a peptide inhibitor of translocation interacting with bacterial 30S ribosomal subunits. *RNA* **12**, 1262–1270 (2006).
55. Arhin, F. et al. A new class of small molecule RNA polymerase inhibitors with activity against rifampicin-resistant *Staphylococcus aureus*. *Bioorg Med Chem* **14**, 5812–5832 (2006).
56. Andre, E. et al. Novel synthetic molecules targeting the bacterial RNA polymerase assembly. *J Antimicrob Chemother* **57**, 245–251 (2006).

57. Medini, D., Donati, C., Tettelin, H., Masignani, V. & Rappuoli, R. The microbial pan-genome. *Curr Opin Genet Dev* **15**, 589–594 (2005).
58. Perna, N. T. et al. Genome sequence of enterohaemorrhagic *Escherichia coli* O157: H7. *Nature* **409**, 529–533 (2001).
59. Welch, R. A., et al. Extensive mosaic structure revealed by the complete genome sequence of uropathogenic *Escherichia coli*. *Proc Natl Acad Sci USA* **99**, 17020–17024 (2002).
60. Santos, S. R. & Ochman, H. Identification and phylogenetic sorting of bacterial lineages with universally conserved genes and proteins. *Environ Microbiol* **6**, 754–759 (2004).
61. Kim, H. S. et al. Bacterial genome adaptation to niches: divergence of the potential virulence genes in three *Burkholderia* species of different survival strategies. *BMC Genomics* **6**, 174 (2005).
62. Nierman, W. C. et al. Structural flexibility in the *Burkholderia mallei* genome. *Proc Natl Acad Sci USA* **101**, 14246–14251 (2004).
63. Holden, M. T. et al. Genomic plasticity of the causative agent of melioidosis, *Burkholderia pseudomallei*. *Proc Natl Acad Sci USA* **101**, 14240–14245 (2004).
64. Fitzgerald, J. R. et al. Genome diversification in *Staphylococcus aureus*: molecular evolution of a highly variable chromosomal region encoding the staphylococcal exotoxin-like family of proteins. *Infect Immun* **71**, 2827–2838 (2003).
65. Gill, S. R. et al. Insights on evolution of virulence and resistance from the complete genome analysis of an early methicillin-resistant *Staphylococcus aureus* strain and a biofilm-producing methicillin-resistant *Staphylococcus epidermidis* strain. *J Bacteriol* **187**, 2426–2438 (2005).
66. Doolittle, W. F. Phylogenetic classification and the universal tree. *Science* **284**, 2124–2129 (1999).
67. Kunin, V., Goldovsky, L., Darzentas, N., & Ouzounis, C. A. The net of life: reconstructing the microbial phylogenetic network. *Genome Res* **15**, 954–959 (2005).
68. Tettelin, H. et al. Genome analysis of multiple pathogenic isolates of *Streptococcus agalactiae*: implications for the microbial "pan-genome." *Proc Natl Acad Sci USA* **102**, 13950–13955 (2005).
69. Schloss, P. D. & Handelsman, J. Metagenomics for studying unculturable microorganisms: cutting the Gordian knot. *Genome Biol* **6**, 229 (2005).
70. Venter, J. C. et al. Environmental genome shotgun sequencing of the Sargasso Sea. *Science* **304**, 66–74 (2004).
71. Ley, R. E., et al. Unexpected diversity and complexity of the guerrero negro hypersaline microbial mat. *Appl Environ Microbiol* **72**, 3685–3695 (2006).
72. Tyson, G. W. & Banfield, J. F. Cultivating the uncultivated: a community genomics perspective. *Trends Microbiol* **13**, 411–415 (2005).
73. Projan, S. J. New (and not so new) antibacterial targets — from where and when will the novel drugs come? *Curr Opin Pharmacol* **2**, 513–522 (2002).
74. Sutherland, R. et al. Antibacterial activity of mupirocin (pseudomonic acid), a new antibiotic for topical use. *Antimicrob Agents Chemother* **27**, 495–498 (1985).
75. Tao, J. et al. Drug target validation: lethal infection blocked by inducible peptide. *Proc Natl Acad Sci USA* **97**, 783–786 (2000).
76. Brown, M. J. et al. Rational design of femtomolar inhibitors of isoleucyl tRNA synthetase from a binding model for pseudomonic acid-A. *Biochemistry* **39**, 6003–6011 (2000).
77. Jarvest, R. L. et al. Potent synthetic inhibitors of tyrosyl tRNA synthetase derived from C-pyranosyl analogues of SB-219383. *Bioorg Med Chem Lett* **11**, 715–718 (2001).

78. Stefanska, A. L., Fulston, M., Houge-Frydrych, C. S., Jones, J. J. & Warr, S. R. A potent seryl tRNA synthetase inhibitor SB-217452 isolated from a *Streptomyces* species. *J Antibiot (Tokyo)* **53**, 1346–1353 (2000).
79. Kim, S., Lee, S. W., Choi, E. C. & Choi, S. Y. Aminoacyl-tRNA synthetases and their inhibitors as a novel family of antibiotics. *Appl Microbiol Biotechnol* **61**, 278–288 (2003).
80. Hurdle, J. G., O'Neill, A. J. & Chopra, I. Prospects for aminoacyl-tRNA synthetase inhibitors as new antimicrobial agents. *Antimicrob Agents Chemother* **49**, 4821–4833 (2005).
81. Brown, J. R. et al. Horizontal transfer of drug resistant aminoacyl-tRNA synthetases of anthrax and Gram-positive pathogens. *EMBO Rep.* **4**, 692–698 (2003).
82. Gentry, D. R. et al. Variable sensitivity to bacterial methionyl-tRNA synthetase inhibitors reveals sub populations of Streptococcus pneumoniae with two distinct methionyl tRNA synthetase genes. *Antimicrobial Agents Chemother.* **47**, 1784–1789 (2003).
83. Gentry, D. R. et al. Variable sensitivity to bacterial methionyl-tRNA synthetase inhibitors reveals subpopulations of *Streptococcus pneumoniae* with two distinct methionyl-tRNA synthetase genes. *Antimicrob Agents Chemother* **47**, 1784–1789 (2003).
84. Brown, J. R. et al. Horizontal transfer of drug-resistant aminoacyl-transfer-RNA synthetases of anthrax and gram-positive pathogens. *EMBO Rep* **4**, 692–698 (2003).
85. Newton, D. T., Creuzenet, C. & Mangroo, D. Formylation is not essential for initiation of protein synthesis in all eubacteria. *J Biol Chem* **274**, 22143–22146 (1999).
86. Adams, J. M. On the release of the formyl group from nascent protein. *J Mol Biol* **33**, 571–589 (1968).
87. Mazel, D., Pochet, S. & Marliere, P. Genetic characterization of polypeptide deformylase, a distinctive enzyme of eubacterial translation. *EMBO J* **13**, 914–923 (1994).
88. Giglione, C., Pierre, M. & Meinnel, T. Peptide deformylase as a target for new generation, broad spectrum antimicrobial agents. *Mol Microbiol* **36**, 1197–1205 (2000).
89. Yuan, Z. & White, R. J. The evolution of peptide deformylase as a target: contribution of biochemistry, genetics and genomics. *Biochem Pharmacol* **71**, 1042–1047 (2006).
90. Rajagopalan, P. T., Datta, A. & Pei, D. Purification, characterization, and inhibition of peptide deformylase from *Escherichia coli*. *Biochemistry* **36**, 13910–13918 (1997).
91. Watters, A. A. et al. Antimicrobial activity of a novel peptide deformylase inhibitor, LBM415, tested against respiratory tract and cutaneous infection pathogens: a global surveillance report (2003–2004). *J Antimicrob Chemother* **57**, 914–923 (2006).
92. Ramanathan-Girish, S. et al. Pharmacokinetics in animals and humans of a first-in-class peptide deformylase inhibitor. *Antimicrob Agents Chemother* **48**, 4835–4842 (2004).
93. Campbell, J. W. & Cronan, J. E., Jr. Bacterial fatty acid biosynthesis: targets for antibacterial drug discovery. *Annu Rev Microbiol* **55**, 305–332 (2001).
94. Heath, R. J. & Rock, C. O. Fatty acid biosynthesis as a target for novel antibacterials. *Curr Opin Investig Drugs* **5**, 146–153 (2004).
95. Zhang, Y. M., White, S. W. & Rock, C. O. Inhibiting bacterial fatty acid synthesis. *J Biol Chem* **281**, 17541–17544 (2006).
96. Wang, J. et al. Platensimycin is a selective FabF inhibitor with potent antibiotic properties. *Nature* **441**, 358–361 (2006).
97. Andries, K. et al. A diarylquinoline drug active on the ATP synthase of *Mycobacterium tuberculosis*. *Science* **307**, 223–227 (2005).

98. Kwan, T., Liu, J., Dubow, M., Gros, P. & Pelletier, J. Comparative genomic analysis of 18 *Pseudomonas aeruginosa* bacteriophages. *J Bacteriol* **188**, 1184–1187 (2006).
99. Kwan, T., Liu, J., DuBow, M., Gros, P. & Pelletier, J. The complete genomes and proteomes of 27 *Staphylococcus aureus* bacteriophages. *Proc Natl Acad Sci USA* **102**, 5174–5179 (2005).
100. Becker, D. et al. Robust *Salmonella* metabolism limits possibilities for new antimicrobials. *Nature* **440**, 303–307 (2006).
101. Appelbaum, P. C. & Jacobs, M. R. Recently approved and investigational antibiotics for treatment of severe infections caused by gram-positive bacteria. *Curr Opin Microbiol* **8**, 510–517 (2005).
102. Butler, M. S. & Buss, A. D. Natural products — the future scaffolds for novel antibiotics? *Biochem Pharmacol* **71**, 919–929 (2006).
103. Mariani, R. et al. Antibiotics GE23077, novel inhibitors of bacterial RNA polymerase. Part 3: chemical derivatization. *Bioorg Med Chem Lett* **15**, 3748–3752 (2005).

11 Comparative Genomics and the Development of Antimalarial and Antiparasitic Therapeutics

Emilio F. Merino, Steven A. Sullivan, and Jane M. Carlton

CONTENTS

ABSTRACT

We are in the midst of a transformation in the study of eukaryotic parasites, a transformation sparked by the vast amounts of genome sequence data becoming available for many of the species in this diverse group. In this review, we summarize the current state of parasite genomics, provide details concerning the available drug and vaccine therapies for the diseases caused by these parasites, and describe the roles comparative genomics is playing in the design of new drugs and vaccines against them. These roles include the identification of various metabolic pathways or proteins that might serve as therapeutic targets by virtue of their presence in the parasite but absence in humans; elucidation of the causes of drug resistance and antibiotic sensitivity; identification of genes expressed in a stage-specific fashion; and detection of potential antigens for vaccine development. The future is bright for comparative genomic analysis of parasites, and the development of several public–private partnerships that foster collaborations among scientists in academia, big pharmaceutical companies, and the public sector provide new hope for the development of the next generation of antiparasitic therapeutics.

11.1 INTRODUCTION

Parasitology, the study of eukaryotic parasites, has undergone a revolution in recent years with the availability of vast amounts of genome sequence data from many of the species that make up this eclectic grouping. Two major groups of parasites, protists and helminths, account for most of the human suffering and agricultural loss caused by pathogenic eukaryotes, and in many cases the available antiparasitic therapeutics, (i.e., drugs and vaccines) are woefully inadequate or becoming obsolete as parasite species develop resistance. Genome sequence data, in particular comparative genome sequence analysis, thus provide an alternative for development of novel therapeutics through the identification of species-specific proteins, metabolic pathways, and parasite-specific molecular mechanisms.

In this chapter, we first describe the current status of parasite genomics and the development of antiparasitic therapeutics. We then provide specific examples of how comparative genomics is used to identify novel drugs and vaccines for the treatment and prophylaxis of several important diseases, such as malaria, East Coast fever, amebiasis, and filariasis. This chapter is not meant to be exhaustive but rather to illustrate some of the first steps taken to harness the power of comparative genomics in the discovery, design, and application of therapies for diseases. The relative "tree-of-life" positions of the parasitic organisms discussed in this review are shown in Figure 11.1.

11.2 THE CURRENT STATUS OF PARASITE GENOMICS

Several billion people suffer from infection by parasitic protists and helminths at any given moment. The diseases they cause are frequently referred to as "neglected" due to their prevalence in developing countries, where poor sanitation and lack of access to clean water enhance disease transmission and vector proliferation. Eukaryotic parasites also ravage agricultural livestock, compounding their negative economic effects on the livelihoods of endemic country people. The initial momentum for the sequencing of many of these infectious disease pathogens came from scientists within the affected communities of both developed and developing countries. Several of the consortiums they formed, such as the International Malaria Genome Sequencing Project Consortium created in the mid-1990s, drove the introductory phase of network formation, genome

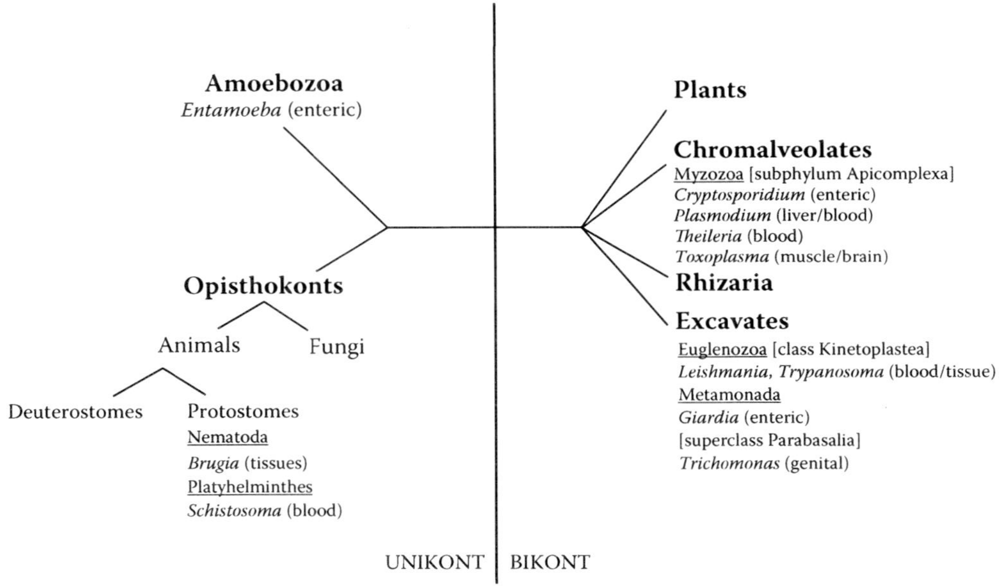

FIGURE 11.1 Parasites in the context of a tree of eukaryotes. Recent reconstructions of the global phylogeny of eukaryotes have divided them broadly into unikonts (cells with a single flagellum) and bikonts (cells with two flagella) and further into six "supergroups" (reviewed in Keeling[108]). Parasite species discussed at length in this review are shown according to their respective supergroups (bold) and phyla (underscore); additional taxonomic levels are added for clarity or to elucidate commonly occurring groupings in the literature (e.g., parabasalids, kinetoplastids). Sites of parasite residence within the host are shown in parentheses. Branch lengths do not reflect evolutionary distances.

mapping, and resource building. Generating funds for a sequencing effort was the principal aim of the consortiums, but a secondary component for many was the building of expertise and collaborative North–South and South–South networks for molecular biology, genomics, and associated bioinformatics research.[1] This led to international workshops (often within endemic countries) to promote technology transfer and foster scientific exchange and the development of biological reagent repositories such as the Malaria Research and Reference Reagent Resource.[2] Subsequently, parasite species identified by funding agencies as representing a serious threat to human health were targeted for genome sequencing funding (see, e.g., the National Institute of Allergy and Infectious Diseases [NIAID] Blue Ribbon Panel on Genomics report at http://www.niaid.nih.gov/dmid/genomes/ribbon.htm and more recently the NIAID Microbial Sequencing Centers' initiative at http://www.niaid.nih.gov/dmid/genomes/mscs/default.htm).

Of the unicellular taxa, the phylum Apicomplexa contains many disease-causing organisms. Malaria is caused by parasites of the apicomplexan genus *Plasmodium*. More than 200 *Plasmodium* species are known to exist that cause varying degrees of morbidity and mortality in different hosts, such as mammals, birds, and reptiles. Four species (*P. falciparum*, *P. vivax*, *P. malaria*, and *P. ovale*) cause human malaria, although cases of human infection by the monkey parasite *Plasmodium knowlesi* in Malaysia have recently been reported.[3] There are between 300 million and 500 million human malaria cases and about 2–3 million malaria deaths per year, mostly of African children.[4]

Several genome sequencing projects of different *Plasmodium* species have been published (Table 11.1), including the complete sequence of the most deadly human malaria *P. falciparum*,[5] and partial coverage of the laboratory rodent parasites *P. yoelii yoelii*,[6] *P. berghei*,[7] and *P. chabaudi*.[7] Genome sequencing of *P. vivax*, the most geographically widespread human malaria parasite,[8] as well as the closely related model monkey malaria species *P. knowlesi* and several other *Plasmodium* species, are in progress. Other apicomplexan genome sequencing projects either completed or under way include several *Cryptosporidium* species that are common waterborne agents of diarrhea[9,10]; several species of tick-borne hemoparasites that give rise to diseases of livestock (e.g., *Theileria*)[11,12]; and a number of genotypes of *Toxoplasma*, the causative agent of congenital toxoplasmosis (Table 11.1).

Sequencing projects of several luminal parasite genomes are in varying degrees of completion. *Entamoeba histolytica*[13] is the causative agent of amoebiasis and is a significant source of morbidity and mortality in developing countries, causing an estimated 40,000–100,000 deaths yearly. *Giardia lamblia*,[14,15] which infects the small intestines of human and other mammalian hosts, is one of the most common causes of gastrointestinal disorders. *Trichomonas vaginalis*[16] causes one of the most common nonviral sexually transmitted diseases, responsible for about 170 million new cases yearly worldwide.

Genome sequences and analyses of three trypanosomatid genomes *Trypanosoma cruzi*, *Trypanosoma brucei*, and *Leishmania major*, the "tri-Tryps,"have been published.[17–19] *Trypanosoma cruzi* causes Chagas disease and is transmitted by several kinds of reduviid, blood-sucking insects; *T. brucei* is transmitted by tsetse flies and causes human sleeping sickness; and *L. major* causes cutaneous leishmaniasis, one of

TABLE 11.1
Current Status of Parasitic Protist and Helminth Whole-Genome Sequencing Projects

Parasite	Host/Disease	Genome Size (Mb)	No. of Genes	Project Status	Sequencing Center	Genome Project Web Site or Reference
Protists: Apicomplexa						
Babesia bovis	Bovine/babesiosis	~9	N/A	In progress	TIGR	http://www.tigr.org/tdb/e2k1/bba1
Cryptosporidium hominis	Human/cryptosporidiosis	9	3,994	Published	UMN/VCU	10
Cryptosporidium parvum	Human/cryptosporidiosis	9	3,807	Published	UMN/VCU	9
Cryptosporidium muris	Human/cryptosporidiosis	N/A	N/A	In progress	TIGR	http://msc.tigr.org/status.shtml
Eimeria tenella	Avian/coccidiosis	60	N/A	In progress	WTSI	http://www.sanger.ac.uk/Projects/E_tenella
Plasmodium berghei	Rodent/malaria	23	5,864	Published	WTSI	7
Plasmodium chabaudi	Rodent/malaria	23	5,698	Published	WTSI	7
Plasmodium falciparum 3D7	Human/malaria	24	5,268	Published	TIGR/WTSI/SU	5
Plasmodium falciparum Ghana	Human/malaria	24	N/A	In progress	WTSI	http://www.sanger.ac.uk/Projects/P_falciparum
Plasmodium falciparum IT	Human/malaria	24	N/A	In progress	WTSI	http://www.sanger.ac.uk/Projects/P_falciparum
Plasmodium falciparum Dd2	Human/malaria	24	N/A	Complete	BI	http://www.broad.mit.edu/annotation/genome/plasmodium_falciparum_spp/MultiHome.html
Plasmodium falciparum HB3	Human/malaria	24	N/A	Complete	BI	http://www.broad.mit.edu/annotation/genome/plasmodium_falciparum_spp/MultiHome.html
Plasmodium gallinaceum	Avian/malaria	N/A	N/A	In progress	WTSI	http://www.sanger.ac.uk/Projects/P_gallinaceum

(*Continued*)

TABLE 11.1
Current Status of Parasitic Protist and Helminth Whole-Genome Sequencing Projects (Continued)

Parasite	Host/Disease	Genome Size (Mb)	No. of Genes	Project Status	Sequencing Center	Genome Project Web Site or Reference
Plasmodium knowlesi	Nonhuman primate/malaria	25	N/A	Complete	WTSI	http://www.sanger.ac.uk/Projects/P_knowlesi
Plasmodium reichenowi	Nonhuman primate/malaria	N/A	N/A	In progress	WTSI	http://www.sanger.ac.uk/Projects/P_reichenowi
Plasmodium vivax	Human/malaria	26	5,433	Complete	TIGR	http://www.tigr.org/tdb/e2k1/pva1
Plasmodium yoelii yoelii	Rodent/malaria	23	5,878	Published	TIGR	6
Theileria annulata	Bovine/tropical theileriosis	8.5	3,792	Published	WTSI	12
Theileria parva	Bovine/East Coast fever	8.5	4,035	Published	TIGR	11
Toxoplasma gondii type I	Human/toxoplasmosis	~65	N/A	Complete	TIGR/WTSI	http://www.tigr.org/tdb/e2k1/tga1http://www.sanger.ac.uk/Projects/T_gondii
Toxoplasma gondii type III	Human/toxoplasmosis	~65	N/A	In progress	TIGR	http://msc.tigr.org/t_gondii/toxoplasma_gondii_type_iii/index.shtml
Protists: Kinetoplastida						
Leishmania braziliensis	Human/leishmaniasis	~34	N/A	In progress	WTSI	http://www.sanger.ac.uk/Projects/L_braziliensis
Leishmania infantum	Human/leishmaniasis	~34	N/A	In progress	WTSI	http://www.sanger.ac.uk/Projects/L_infantum
Leishmania major	Human/leishmaniasis	~34	8,272	Published	EULEISH/SBRI/WTSI	19
Trypanosoma brucei	Human/African sleeping sickness	35	9,068	Published	TIGR/WTSI	17
Trypanosoma congolense	Human/trypanosomiasis	35	N/A	In progress	WTSI	http://www.sanger.ac.uk/Projects/T_congolense

Trypanosoma cruzi	Human/Chagas disease	44	~12,000	Published	TIGR/SBRI/KI	18
Trypanosoma vivax	Bovine/ trypanosomiasis	35	N/A	In progress	WTSI	http://www.sanger.ac.uk/Projects/T_vivax
Protists: Luminal						
Entamoaeba histolytica	Human/amebiasis	24	9,938	Published	TIGR/WTSI	13
Entamoeba invadens	Reptile/amebiasis	20	N/A	In progress	TIGR/WTSI	http://www.sanger.ac.uk/Projects/E_invadens/ http://msc.tigr.org/entamoeba/entamoeba_invadens
Entamoeba dispar	Human/nonpathogenic	N/A	N/A	In progress	TIGR	http://msc.tigr.org/entamoeba/entamoeba_dispar
Giardia lamblia	Human/giardiasis	12	N/A	Complete	MBL	http://www.mbl.edu/Giardia
Trichomonas vaginalis	Human/trichomoniasis	160	~60,000	Complete	TIGR	http://www.tigr.org/tdb/e2k1/tvg/
Helminths: Platyhelminths						
Echinococcus multilocularis	Human/hydatid disease	150	N/A	In progress	WTSI	http://www.sanger.ac.uk/Projects/Helminths
Nippostrongylus brasiliensis	Rodent/ nippo-strongyloidiasis	N/A	N/A	In progress	WTSI	http://www.sanger.ac.uk/Projects/Helminths
Schistosoma mansoni	Human/schistosomiasis	270	N/A	In progress	TIGR/WTSI	http://www.tigr.org/tdb/e2k1/sma1/ http://www.sanger.ac.uk/Projects/S_mansoni
Helminths: Nematoda						
Ancylostoma duodenale	Human/hookworm disease	N/A	N/A	Planned	WTSI	http://www.sanger.ac.uk/Projects/Helminths
Ascaris lumbricoides	Human/ascariasis	230	N/A	In progress	WTSI	http://www.sanger.ac.uk/Projects/Helminths
Brugia malayi	Human/lymphatic filariasis	100	N/A	In progress	TIGR/WTSI/UE	http://www.tigr.org/tdb/e2k1/bma1

(Continued)

TABLE 11.1
Current Status of Parasitic Protist and Helminth Whole-Genome Sequencing Projects (Continued)

Parasite	Host/Disease	Genome Size (Mb)	No. of Genes	Project Status	Sequencing Center	Genome Project Web Site or Reference
Haemonchus contortus	Ovine/hemonchosis	60	N/A	In progress	WTSI	http://www.sanger.ac.uk/Projects/H_contortus
Heterorhabditis bacteriophora	Insect/biocontrol of soil-dwelling insects	N/A	N/A	In progress	GSC	http://genome.wustl.edu/genome_group_index.cgi
Onchocerca volvulus	Human/river blindness	150	N/A	In progress	WTSI	http://www.sanger.ac.uk/Projects/Helminths
Strongyloides ratti	Rodent/strongyloidiasis	N/A	N/A	In progress	WTSI	http://www.sanger.ac.uk/Projects/Helminths
Trichinella spiralis	Porcine, human/trichinosis	N/A	N/A	In progress	GSC	http://genome.wustl.edu/genome_group_index.cgi
Trichuris muris	Rodent/trichuriasis	96	N/A	In progress	WTSI	http://www.sanger.ac.uk/Projects/Helminths

Note: EST and genome survey sequencing projects are not shown. BI, Broad Institute; EULEISH, European *Leishmania major* Friedlin Genome Sequencing Consortium; GSC, Genome Sequencing Center, Washington University, St Louis; KI, Karolinska Institute; N/A, no data available; SBRI, Seattle Biomedical Research Institute; SU, Stanford University; TIGR, The Institute for Genomic Research; UE, University of Edinburgh; UMN, University of Minnesota; VCU, Virginia Commonwealth University; WTSI, Wellcome Trust Sanger Institute.

the three types of leishmaniasis (cutaneous, mucocutaneous, and visceral) transmitted by sand flies.

Production of whole-genome sequence data and analysis of parasitic helminths lags behind that of the protist species, although the published sequence of the free-living nematode *Caenorhabditis elegans* genome in 1998 was one of the signal achievements of genomic science.[20] Targets of ongoing genome sequencing projects of human-infective helminths include several nematode (e.g., *Brugia malayi* and *Trichinella spiralis*) and three platyhelminth (*Schistosoma mansoni, Nippostrongylus brasiliensis*, and *Echinococcus multilocularis*) species (Table 11.1). Of these, the *B. malayi*[21] and *S. mansoni*[22] projects are the most advanced. *Brugia malayi* is the principal cause (along with *Wuchereria bancrofti*) of lymphatic filariasis, which afflicts about 120 million people worldwide, a third of whom show disfigurement due to swelling of the lymph system in the legs and groin. Four *Schistosoma* species cause schistosomiasis or bilharzia, a major cause of morbidity in tropical areas such as Africa, South America, and Southeast Asia. The *B. malayi* and *Schistosoma* genomes are expected to be completed in 2007. In addition, more than 30 expressed sequence tag (EST) and mitochondrial genome sequencing projects are ongoing for a variety of helminth species that infect humans, animals, and plants.[23]

Table 11.1 is an attempt at a comprehensive list of eukaryotic parasite genome sequencing projects as of mid-2006. The reader is also referred to reviews that detail the current status of several of these genome projects[24,25] and, as many projects hinge on the vagaries of funding, to the Web sites of the sequencing centers themselves. Many of the genome sequencing centers have made their sequence data available in advance of final publication to support and "jump-start" research. Sequence databases such as the Wellcome Trust Sanger Institute's GeneDB,[26] The Institute for Genomic Research's (TIGR's) database SYBTIGR linked to individual project Web pages, and species-specific databases such as the ApiDB suite of databases PlasmoDB, ToxoDB, and CryptoDB,[27] have provided researchers with access to the preliminary sequence data. In many instances, genome data release has been accompanied by a data policy outlining the pitfalls associated with draft sequence data (which is error prone and may contain contaminating sequences) and outlining the sequencing center's plans for final gene prediction, annotation, and publication.

One of the most exciting prospects arising from the flood of genome sequence data is the opportunity to do comparative genomics — the analysis and comparison of genomes within or between different species or strains. Through comparative genomics, we hope to gain a better understanding of how species have evolved and to determine the function of genes, proteins, and noncoding regions of the genome. Comparative genomics encompasses analysis of relative genome composition, chromosome organization, conservation of gene synteny, gene orthology and paralogy, species-specific genes, and evolution of the genomes compared. As such, it is a powerful tool for identifying the differences between pathogen and host and elucidating gaps in a parasite's armor that may be exploited for control or intervention methods. Comparative genomics of eukaryotic parasites is still a young science, and as will be evident in the coming sections, its use in the development of antiparasitic therapeutics has yet to be exploited fully.[28]

11.3 THE CURRENT STATUS OF ANTIPARASITIC DRUG AND VACCINE RESEARCH AND DEVELOPMENT

At the end of the last millennium, drug research and development (R&D) for neglected parasitic diseases was at an all time low, with only 13 of the 1,393 new drugs marketed during the last 25 years being for the cure of tropical diseases.[29] The lack of interest shown by the pharmaceutical industry is undoubtedly one of the reasons for this poor record, stemming from the high costs associated with R&D for diseases for which normal market incentives do not exist. This has had devastating effects: There are no vaccines available for many tropical diseases, and existing drugs are either inadequate or toxic and increasingly fail due to resistance. The available diagnostic tests for some of these diseases are equally deficient, with many techniques being invasive, nonpredictive, or utilizing poor biomarkers. What follows is a brief overview of the drugs and vaccines currently available for the parasitic diseases of humans discussed in this review.

The arsenal of antimalarial drugs, classically consisting of chloroquine, quinine, and artemisinin, has grown modestly over the past 20 years, mostly due to the generation of drug combinations that have extended the life of old drugs (e.g., LapDap, a combination of the antifolate drugs chlorproguanil and dapsone). However, resistance has developed to almost all antimalarial drugs,[30] adding urgency to the development of new leads.[31] The relatively new nitrothiazolide antiprotozoal agent nitazoxanide (2-acetyloxy-*N*-benzamide) is the only currently approved drug for treating cryptosporidial diarrhea, while spiramycin is used to treat acute toxoplasmosis in pregnant women, the healthy human population at primary risk of this apixomplexan disease. Current drugs of choice for treatment of infection by the luminal parasites *E. histolytica, T. vaginalis*, and *G. lamblia* include metranidazole, tinidazole, and other 5-nitroimidazole derivatives, although resistance is an emerging problem.[32] Current chemotherapy for the human trypanosomiases relies on only six drugs (pentamidine, miltefosine, suramin, melarsoprol, eflornithine, benznidazole), five of which were developed more than 30 years ago.[33] The toxicity and poor efficacy of these drugs and the emergence of drug-resistant trypanosomes have spurred recent progress in identifying novel therapeutic compounds.[34] Regarding helminths, the most effective therapeutics against parasitic nematodes such as *B. malayi* are the benzimidazoles and pyrantel and ivermectin. Praziquantel is the only commercially available treatment for infection by the blood flukes *S. mansoni* and *Schistosoma japonicum*, but it requires repeated treatments in endemic areas and does not prevent reinfection.[35] Moreover, while not yet a problem commonly associated with helminth-caused diseases, drug resistance could become an issue in their treatment, based on observations in the field.[36]

Vaccines are an alternative to drug treatment of infectious diseases. A limited number of commercially available vaccines based on live parasites are used successfully and extensively against several eukaryotic parasitic diseases of livestock (e.g., coccidiosis in poultry[37] and toxoplasmosis in sheep[38]). However, the number of human parasites for which a vaccine is currently in development is pitifully small (Table 11.2). Encouragingly, more than 20 different malaria vaccine candidates are under study, as both epidemiological and experimental data support the feasibility

TABLE 11.2
Development Status of Various Parasitic Disease Vaccines

Disease	Vaccine Name/Antigen	Pharmaceutical Company or Research Group	Stage of Development
Malaria	RTS,S/AS02A	GSK/WRAIR/MVI	Phase IIb
	CSP	Dictagen/Lausanne University	Phase Ib
	ICC-1132	Apovia/MVI	Phase II
Preerythrocytic stage	DNA vaccines	US Navy/Vical	Phase I
	CSP-LSA-1	Oxford Univ/Oxxon/MVI	Phase Ib
	TRAP + multiepitope string	Crucell/GSK/WRAIR/NIAID	Phase Ia
	CSP	Oxford University; NYU	Preclinical
	LSA-3	Pasteur Institute/WRAIR/GSK	Phase Ia
	LSA-1, SALSA, other liver-stage antigens	Hawaii Biotech; Epimmune	Preclinical
Blood stage	MSP1	GSK/WRAIR/MVI	Phase Ib/II
		NIAID; Hawaii Biotech; AECOM; University of Maryland	Preclinical to phase I
	MSP1, MSP2, RESA	Queensland Medical Research Institute/WEHRI	Phase II
	AMA1	MVDU; NIAID	Phase Ib
	MSP1 AMA1	Second Military University/Wanxing Pharmaceuticals/WHO	Phase I
	MSP3	Pasteur Institute/AMANET/EMVI	Phase Ib
	GLURP	EMVI/SSI	Phase I
	MSP3-GLURP	EMVI/SSI	Phase I
	MSP4, MSP5	Monash	Preclinical

(Continued)

TABLE 11.2
Development Status of Various Parasitic Disease Vaccines (Continued)

Disease	Vaccine Name/Antigen	Pharmaceutical Company or Research Group	Stage of Development
	SE36	Osaka University/Biken	Phase I
	Other blood-stage antigens (EBA-175, RAP-2, EMP-1)	Various groups	Preclinical
Sexual stage	PfS25 (yeast)	NIH	Phase I
	PvS25 and other sexual-stage antigens	NIH	Preclinical
	Live attenuated/drug-sensitive strains	Various laboratories	Preclinical
	Killed promastigotes	Razi Institute	Phase II
Leishmaniasis	LeIF/LmSTI-1/TSA subunit vaccine	IDRI/Corixa	Phase I/Ib
	DNA vaccines	Various laboratories	Preclinical
Hookworm disease	ASP2 subunit vaccine	HHVI	Phase I
Schistosomiasis	*S. haematobium* 28-kDa GST subunit vaccine	IPL	Phase II
	S. mansoni paramyosin + TPI multiepitope	Bachem/USAID/SVDP	Preclinical
	S. mansoni Sm14	FioCruz	Preclinical

Source: Adapted from the World Health Organization's Initiative for Vaccine Research, http://www.who.int/vaccine_research/documents/en/Status_Table.pdf.

Notes: AECOM, Albert Einstein College of Medicine; AMANET, African Malaria Network Trust; EMVI, European Malaria Vaccine Initiative; GSK, GlaxoSmithKline Biologicals; HHVI, Human Hookworm Vaccine Initiative; IDRI, Infectious Disease Research Institute; IPL, Pasteur Institute of Lille; MVI, Malaria Vaccine Initiative; NIAID, National Institute of Allergy and Infectious Diseases; NIH, National Institutes of Health; NYU, New York University; SSI, Statens Serum Institut; SVDP, Schistosomiasis Vaccine Development Programme; USAID, U.S. Agency for International Development; WEHRI, Walter and Eliza Hall Institute of Medical Research; WHO, World Health Organization; WRAIR, Walter Reed Army Institute of Research.

of such a vaccine; immunity to malaria is known to be acquired by adults from malaria-endemic regions,[39] and humans have been immunized against malaria using irradiated sporozoites, the infective stage from mosquito salivary glands.[40,41] Indeed, promising evidence of the effectiveness of antisporozoite vaccine against *P. falciparum* malaria in children has emerged from a trial in Mozambique (reviewed in Alonso[42]). The use of comparative genomics to develop safe, effective, and affordable vaccines that provide sustained protection against parasite diseases, however, is still in its nascent stages.

11.4 COMPARATIVE GENOMICS OF MALARIA PARASITES AND DRUG AND VACCINE DESIGN

The organisms that cause malaria are obligate, intracellular parasites that have a complex life cycle in two hosts, mosquito and man. Sporozoites inoculated into the vertebrate host through the bite of a female mosquito travel to the liver, where they invade hepatocytes and undergo successive rounds of mitotic replication to generate liver schizonts. Merozoites released from mature liver schizonts enter the bloodstream, where they invade erythrocytes and develop into trophozoite and erythrocytic schizont forms. The schizonts rupture at maturity and release merozoites into the bloodstream, which can invade further erythrocytes, completing the asexual cycle. Some merozoite-infected red blood cells may develop into gametocytes, the sexual stage of the parasite. When these are taken up in the blood meal of a mosquito, male and female gametes from the gametocytes are generated, which then fuse to form ookinetes. These cross the wall of the mosquito midgut and form sporozoite-filled oocysts on the midgut surface. When the oocysts burst, sporozoites migrate to the mosquito salivary glands, ready to be transmitted during the mosquito's next bite, and the life cycle is repeated.

With the publication of several *Plasmodium* genome sequencing projects and functional genomics studies in the past few years, comparative genomics of malaria parasites has become an important field in malaria research (see Carlton, Silva, and Hall[43] and Hall and Carlton[44] for review). The first whole-genome comparison established that *P. falciparum* and *P. yoelii yoelii* genomes have many similarities.[6] Both are haploid and about 23 Mb in size, distributed among 14 linear chromosomes that range in size from 500 kb to over 3 Mb. Of the approximately 5,500 predicted genes, between 60% and 70% are orthologs, found in extensive regions of synteny. Species-specific genes are localized to subtelomeric regions of the chromosomes, and many of these are involved in specialized mechanisms of invasion and pathogenesis. Subsequent comparative analyses for several other *Plasmodium* species provided further evidence of the conserved nature of chromosome-internal *Plasmodium* genes.[43]

The availability of genome sequences from the malaria parasite projects has undoubtedly facilitated discovery of novel antimalarial drug targets. One of the best-known examples came from bioinformatic screening of the *P. falciparum* genome, which identified a distinctive eukaryotic pathway for isoprenoid biosynthesis (Figure 11.2). Isoprenoids, found in several important membrane components such as sterols and ubiquinone, are synthesized via the mevalonate pathway in mammals and fungi, whereas algae, plants, and some bacteria employ the 1-deoxy-d-xylulose-5-phosphate

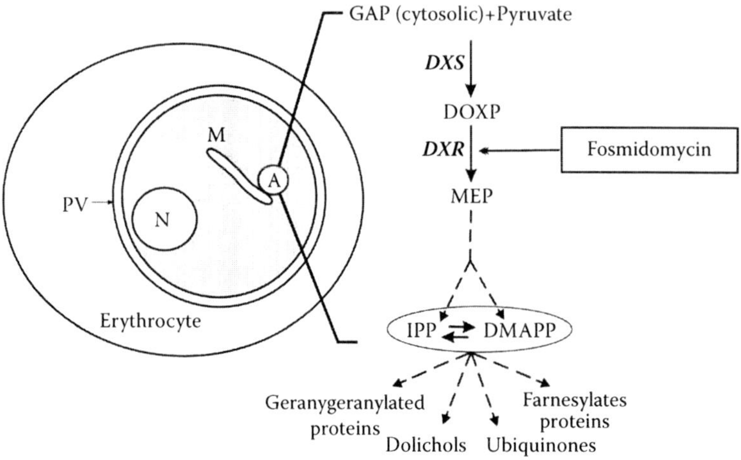

FIGURE 11.2 Schematic representation of the isoprenoid biosynthesis pathway in *P. falciparum*, indicating the step inhibited by fosmidomycin. The parasite is located within a parasitophorous vacuole (PV) inside the erythrocyte. The pathway is localized to an apicomplexan-specific organelle, the apicoplast (A). N, nucleus; M, mitochondrion; GAP, glyceraldehyde-3-phosphate; DOXP, 1-deoxy-d-xylulose-5-phosphate; *DXS*, DOXP synthase; *DXR*, DOXP reductoisomerase; MEP, 2*C*-methyl-d-erythritol-4-phosphate; IPP, isopentenyl diphosphate; DMAPP, dimethylallyl diphosphate. Broken arrow indicates other steps in the pathway omitted for space constraints.

(DOXP) pathway. Noting that antimalarials based on the mevalonate pathway had failed, Jomaa et al.[45] used bacterial DOXP pathway enzyme sequences to identify DOXP synthase and DOXP reductoisomerase genes in screens of the *P. falciparum* sequence data, and demonstrated that the pathway is critical for the parasite since in vitro cultures of *P. falciparum* were inhibited by treatment with the antibiotic fosmidomycin and its derivative FR-900089. These potential antimalarial drugs proved extremely effective against in vivo rodent malaria, resulting in total cure after eight days of oral treatment,[45] and fosmidomycin was used to treat malaria successfully in a clinical study.[46,47]

Another good example of the use of bioinformatics approaches to identify drug targets essential for parasite growth is the identification of several genes of the type II fatty acid biosynthesis pathway from the *P. falciparum* sequence.[48] This metabolic pathway occurs in plants and bacteria but is absent in mammals. In vitro activity against *P. falciparum* was demonstrated for the triclosan inhibitor of one enzyme of the pathway, enoyl-acyl-carrier protein (enoyl-ACP) reductase (FabI).[49] Orthologs of FabI have been identified in rodent *Plasmodium* species,[6,49] enabling testing of the efficacy of the drug in vivo.

Both the type II fatty acid biosynthesis pathway and DOXP pathway occur in an unusual organelle, the apicoplast,[50] which is peculiar to members of the apicomplexan phylum. This relict plastid, a nonphotosynthetic homolog of the chloroplasts of plants, synthesizes iron sulfur clusters and heme as well as fatty acids and isoprenoid precursors. Plastids are derived from the endosymbiosis of cyanobacteria, which means that many of the plastid-encoded proteins are bacterial in nature and different from their mammalian homologs. Moreover, in malaria parasites and the majority of other apicomplexans (although not in *Cryptosporidium*, which appears to lack the organelle), the apicoplast is indispensable, making it an attractive target for antiparasitic drugs. Apicoplasts not only contain their own genome and gene expression machinery but also import proteins encoded by nuclear genes. These nuclear genes originated from the endosymbiont genome but relocated to the nuclear genome by a process of intracellular gene relocation. Analysis of reconstructed metabolic pathways in the organelle has identified several other potential targets for drug development in addition to those outlined above,[50] illustrating how the unique biology of the apicoplast has been central to the identification of several novel drug targets. Postgenomic drug targets for malaria were the subject of a review,[51] which provides a more comprehensive description of the current set of candidates, particularly their weighting toward metabolic pathways.

Analysis of hourly changes in the *P. falciparum* transcriptome during the intraerythrocytic developmental cycle[52] exemplifies the use of comparative expression data to identify new vaccine targets. At least 60% of the genome was found to be transcriptionally active during the cycle, exhibiting "just-in-time" expression by which any given gene is induced just once per cycle and only when required. Approximately 260 ORFs (open reading frames) whose expression profiles tracked those of the seven best-known vaccine candidates in *Plasmodium* were identified; of those, 189 were of unknown function, representing new potential vaccine targets.[52] Another example was provided by Kappe, Matuschewski, and colleagues, who compared the transcriptome of rodent *Plasmodium* salivary sporozoites to those of oocyst sporozoites by suppression subtractive complementary DNA hybridization to identify novel infective (salivary) sporozoite transcripts.[53,54] One of the genes thus identified in *P. berghei* as upregulated in infective sporozoites (UIS3) was experimentally targeted for disruption, and immunization with the resulting UIS3-deficient sporozoites conferred complete protection against infectious sporozoite challenge in the rodent malaria model.[55] Using comparative genomics, they identified a UIS3 ortholog in the *P. falciparum* genome sequence, and studies are ongoing to use this to generate a genetically attenuated whole-organism malaria vaccine.

Recent malaria vaccine work has been predicated on the view that multiantigen vaccines will be needed to induce high protective immunity against the parasite[56] since clinical trials conducted with vaccines based on single antigens have been unsatisfactory. Doolan et al.[57] used the power of comparative genomics and proteomics to identify potential new *P. falciparum* antigens. Mass spectra of sporozoite peptide sequences, generated during a *P. falciparum* proteomics project,[58] were scanned against *P. falciparum* and host genomic databases to identify potential sporozoite-specific gene products. Amino acid sequences of 27 candidates were then scanned with human leukocyte antigen supertype algorithms to generate a list of probable epitopes from each protein. Finally, the predicted epitopes were tested for their ability to induce immune responses in blood cells from individuals immunized with radiation-attenuated sporozoites. In this fashion, 16 new antigenic proteins were experimentally identified, several of which were more antigenic than previously well-characterized antigens, such as CSP (circumsporozoite protein).

Vaccine development in the more prevalent malaria species *P. vivax* is far less advanced than for *P. falciparum,* as neither a long-term in vitro culture system nor an irradiated sporozoite vaccine model is available. However, Wang and colleagues[59] developed a high-throughput method of antigen identification that exploits the newly available *P. vivax* genome sequence data[8] and comparative genomics with *P. falciparum*. In endemic regions, *P. vivax*–exposed individuals who lack the DARC (Duffy antigen/receptor for chemokines) receptor do not develop blood-stage infections because DARC is the receptor used by *P. vivax* to invade erythrocytes. Hypothesizing that exposure to the parasite nevertheless elicits an immune response specific to pre–blood stages in these individuals, they compared the immune response to *P. vivax* antigens in exposed versus nonexposed DARC-positive and DARC-negative individuals. The authors selected five known antigens (CSP, SSP2 [sporozoite surface protein 2], MSP1 [merozoite surface protein 1], AMA1 [apical membrane protein 1], and DBP [Duffy binding protein]) and 18 candidate *P. vivax* proteins from the draft genome sequence for evaluation based on their homology to *P. falciparum* proteins established to be expressed during the sporozoite stage. They found that both of the known sporozoite-stage antigens (CSP and SSP2) and three of the candidate sporozoite-specific proteins were antigenic only in exposed individuals lacking DARC, demonstrating the potential of the model for developing new *P. vivax* vaccine candidates.

11.5 COMPARATIVE GENOMICS OF OTHER APICOMPLEXANS AND DRUG AND VACCINE DESIGN

The availability of *Cryptosporidium* and *Toxoplasma* genome sequences along with those of *Plasmodium* species has allowed creation of an apicomplexan comparative genomics database (ApiDB) that has been used to identify commonalities and differences between these organisms. Moreover, inherent characteristics of *Cryptosporidium* and *Toxoplasma* provide opportunities for genomics-based research into apicomplexans that are not available using *Plasmodium. Cryptosporidium* genomes are small and relatively lacking in introns, making them among the easiest apicomplexan genomes to analyze in silico, while the experimental tractability of *Toxosplasma gondii* far exceeds that of *Plasmodium* and *Cryptosporidium* species, fostering its use as a model for in vivo research in Apicomplexa (see review in Kim and Weiss[60]).

Comparative genomics and in vivo testing of apicomplexan genes have complemented each other and helped identify potential therapeutic targets. For example, comparative genomics has demonstrated that apicomplexan genomes to date lack de novo purine synthesis genes, relying instead on salvage pathways, and that *Cryptosporidium* in particular relies on adenosine salvage,[62] which requires the enzymes adenosine kinase (AK) and inosine monophosphate dehydrogenase (IMPDH). The experimental advantages of two apicomplexans were combined when *Cryptosporidium parvum* DNA fragments were transfected into a *T. gondii* mutant with a crippled salvage pathway and were able to complement the mutation, with the *C. parvum* IMPDH gene proving to be the rescuer.[63] Comparative genomics also showed that *Cryptosporidium* lacks genes for de novo synthesis of pyrimidine that are present in all other apicomplexan genomes studied to date. Instead, it contains genes for pyrimidine salvage enzymes,[62] including a gene for thymidine kinase, the target of the antiviral drug gancyclovir.[61] Apicomplexan pathways for purine and pyrimidine salvage show signs of having originated in bacteria, rendering several of their enzymes either unique or sufficiently distant enough from any human homologs to make them promising targets for parasite-specific drug therapies. Indeed, recent work shows that *Cryptosporidium* IMPDH is inhibited by the drugs mycophenolic acid and ribavarin[62] (drugs approved by the Food and Drug Administration), while 4-nitro-6-benzylthioinosine, a compound that demonstrates therapeutic promise against *T. gondii*, also inhibits *Cryptosporidium* AK.[63] Comparative genomics also revealed apicomplexan amino acid metabolic pathways that are absent in humans, making them promising potential targets for therapeutics. These include the conversion of aspartate to lysine in *Toxoplasma* and the metabolism of serine to tryptophan in *Cryptosporidium*.[9,64]

Calcium is an important second messenger, controlling processes such as motility, secretion, and differentiation in apicomplexan parasites. A comparative genomic analysis of *T. gondii* and *Cryptosporidium* and *Plasmodium* species was carried out to identify all the major calcium pathways in Apicomplexa.[65] Comparative and phylogenetic analyses of genes related to calcium metabolism revealed conserved pathways and more importantly from a drug development standpoint, several interesting differences from animal model organisms, such as plant-like pathways for calcium release channels and calcium-dependent kinases. Conceivably, the *T. gondii* system could be used experimentally to validate the functions of the genes involved in this pathway.

An example of the use of comparative genomics for antiapicomplexan vaccine development is provided by analysis of the genome of *Theileria parva*, the agent of bovine East Coast fever. Genes predicted to contain a secretory signal were identified from the *T. parva* genome sequence[11] and used to transfect bovine antigen-presenting cells. Transfected antigen-presenting cells were then subject to immunoassays with cytotoxic T lymphocytes (CTLs) from immune cattle resolving a challenge infection. Five candidate vaccine antigens that are targets of major histocompatibility complex (MHC) class I–restricted CD8+ from immune cattle were identified, and subsequent experiments showed that immunization of cattle with these antigens induced CTL responses that correlated with survival from a lethal parasite challenge.[66] Thus, these results provide a foundation for developing a CTL-targeted anti–East Coast fever subunit vaccine. Furthermore, orthologs of these antigens were identified in *Theileria annulata, C. parvum*, and *P. falciparum*, thus providing potential vaccine antigen candidates for other apicomplexan parasites.

11.6 COMPARATIVE GENOMICS OF LUMINAL PARASITES AND DRUG AND VACCINE DESIGN

To date, three kinds of parasitic luminal protist have been the focus of whole-genome sequencing projects: the diplomonad *G. lamblia*, the parabasalid *T. vaginalis*, and several species of the amoebid *Entamoeba*. Although historically these organisms were studied together due to perceived shared characteristics, such as the lack of mitochondria, the genomes and biology of these species are now understood to be widely different. Indeed, the term *amitochondriate* once used to lump them together is misleading since the species are now known to contain mitochondrial-derived proteins and organelles (hydrogenosomes and mitosomes).[67] Relatively little in the way of comprehensive comparative genomic analysis exists for these organisms, and scant progress in genomics-based drug discovery has occurred since sequencing was completed, although this is expected to change over the next decade.

Formally published in 2005, the *E. histolytica* genome contains about 10,000 predicted genes, a third of which have no identifiable homologs.[13] Sequence mining using bioinformatic tools has been the main mode of drug target identification, as exemplified by the sulfur metabolism pathway. Prior to the genome project, cysteine synthesis enzymes of the sulfur assimilation pathway previously thought to be exclusive to plants, fungi, and bacteria had been identified, suggesting sulfur metabolism as a possible target for new antiamebic drug therapies (reviewed in Nozaki[68]). Subsequent searches of the *E. histolytica* genome for sulfur metabolism genes revealed an absence of typical eukaryotic pathways for neutralizing toxic sulfur-containing amino acids[68,69] and two isotypes of methionine γ-lyase (MGL). These MGLs were apparently derived from archaeal lateral gene transfer and shown to be expressed in vivo and to catalyze degradation of sulfur-containing amino acids in vitro. Most promisingly, a methionine analog trifluoromethionine (TFMET), with catabolism that yields a protein cross-linker, was found to have a cytotoxic effect on *E. histolytica* trophozoites that is mediated by MGL.

The *E. histolytica* genome contains evidence of considerable gene loss, including loss of genes for folate and fatty acid metabolism and for synthesis of purines, pyrimidines, and most amino acids. In particular, the absence of genes for the biosynthesis of isoprenoids and the sphingolipid head group aminoethyphosphonate has led to speculation that novel pathways for biosynthesis of these membrane components could serve as drug targets.[13] Unusual or novel pathways have also been predicted for energy metabolism and pyrimidine synthesis based on further analysis of the *E. histolytica* "metabolome," although no therapeutic targets have been explicitly proposed.[70] There has been intense focus in both the pre- and postgenomic eras on entamoebic virulence factors such as the cell-adhesion lectin GalGalNAc, cysteine proteinases (CPs) that degrade the extracellular matrix, and pore-forming peptides (amoebapores) that insert into the host cells and cause cytolysis. Analysis of the draft genome identified new homologs of all three groups.[13] Expression profiling of *E. histolytica* trophozoites found upregulation of select CP and amoebapore genes after binding to collagen[71] and after intestinal colonization.[72] Although the *E. histolytica* genome appears to lack typical cystatin-like CP inhibitors, a homolog of the novel *T. cruzi* CP inhibitor chagasin was identified in a screen of the genome sequence data. A synthetic hexapeptide based on a conserved chagasin motif was able to inhibit protease activity in a trophozoite extract, suggesting such peptides as promising candidates for development of antiamebic drugs.[73]

Meanwhile, broader genomic comparisons have generated a growing list of "genes of interest," though their relevance to drug discovery is currently speculative. Expression profiling of virulent versus nonvirulent strains of *Entamoeba* identified several dozen transcripts and retrotranspons preferentially expressed in virulent strains. While some of these have been ascribed potential roles in stress response and virulence (e.g., CP5 and CP1, periredoxin), most are hypotheticals and have undetermined roles.[74–77] Intriguingly, transfecting trophozoites with a plasmid containing a segment of an *E. histolytica* SINE (short interspersed element) retrotransposon found upstream of the amoebapore-A gene completely silenced transcription of that gene in the transfected line, even after the plasmid was removed by antibiotic selection. Moreover, additional genes (specifically CP5 and the light subunit of Gal-lectin) could be targeted for shutdown in the altered trophozoites by subsequent transfection with a SINE/gene construct. In all three cases, virulence was substantially reduced, opening a new avenue for *E. histolytica* vaccine development using attenuated amoebae.[78,79]

The *G. lamblia* genome has been completed but was unpublished as of November 2006. An early survey of the genome[80] indicated that about 150 of the approximately 6,000 coding genes encode variant-specific proteins (VSPs), which confer protease resistance and exhibit antigenic variation,[81] making VSP genes an attractive subject for studies of parasite survival in the host. Subsequently, the *G. lamblia* genome has been mined for genes for cyst wall proteins,[82] RNA interference (RNAi) pathway components,[83,84] type II DNA topoisomerase,[85] and cathepsin-like proteases,[86] all of which could be relevant to development of drug therapies for giardiasis.

11.7 COMPARATIVE GENOMICS OF TRYPANOSOMATID PARASITES AND DRUG AND VACCINE DESIGN

The *T. brucei, T. cruzi,* and *L. major* (together referred to as the tri-Tryps) genomes share many general characteristics, including about 6,200 orthologs arranged in long syntenic blocks, nonsyntenic subtelomeric regions containing species-specific genes, polycistronic transcription, and chromosomal GC-bias and AT-skew.[87] Comparative mining of the genome sequence data of all three species has identified several possible novel drug targets, for example, the pathway for generation of aminoethylphosphonate, a molecule that attaches parasite surface glycoproteins (involved in immune evasion, attachment, or invasion) via their glycosylphosphatidylinositol (GPI) anchors. The pathway is found exclusively in *T. cruzi*, and components of it represent novel drug targets because of their absence in humans.[17]

Rresults highly relevant to drug discovery were obtained from a proteomic analysis of the *T. brucei* flagellum.88 The proteomic data were screened against genome sequence data of flagellated and nonflagellated eukaryotes to elucidate flagellar evolution and identify trypanosome-specific flagellar proteins. Of 331 proteins tested, a small fraction had homologs in nonflagellated species, while 208 proved to be trypanosomatid specific. RNAi studies showed that flagellar function is essential in the bloodstream trypanosome, suggesting that impairment of this function may provide a new opportunity for selective intervention.[88]

Another study of interest used mining of the *T. cruzi* genomic and EST sequence databases to identify novel secreted or membrane-associated GPI proteins as potential vaccine candidates.[89] Such proteins are expected to be abundantly expressed in the infective and intracellular stages of this parasite and thus to be recognized as antigenic targets by the immune system. Eight candidates selected from the screen induced antibodies when used to immunize mice; the majority of the antibodies were trypanolytic, validating the sequence-mining strategy for identifying potential vaccine candidates in *T. cruzi*.

Similarly, in a screen of the *L. major* genome sequence, approximately 100 genes expressed in the amastigote stage (the nonmotile form in the mammalian host) were tested in a mouse footpad assay for antigens that would provide some measure of protection against the severe clinical outcome. Fourteen antigens were identified that showed some protection against virulent *L. major* in susceptible mice, providing a potential source of antigens for immune screening of T cells from *Leishmania*-infected mice and as multiantigen cocktails in trials on other mammals, including humans.[90]

11.8 COMPARATIVE GENOMICS OF PARASITIC HELMINTHS AND DRUG AND VACCINE DESIGN

As there are no vaccines available for parasitic helminths, there is much hope that genomic discoveries will broaden the range of antihelminthic therapeutics, although comparative genomics of helmiths is still in its infancy. Complete, annotated genomes are available only for *Caenorhabditis* species, with the model organism *C. elegans* usually serving as the reference helminth genome for comparative genomics. A whole-genome comparison of *C. elegans* to *B. malayi* has revealed overall conservation of gene synteny but a high rate of intrachromosomal rearrangement.[91] A survey of EST libraries from 28 parasitic and 2 free-living nematode genomes identified over 4,000 genes unique to *B. malayi*, in concordance with an earlier genome survey project that found approximately 20% of *B. malayi* putative coding sequences (~3,600, assuming a gene complement of 18,000) to be unique to the species.[91,92] Indeed, the multinematode EST survey found that, on average, 27% of the putative genes of each species were unique to it, indicating remarkable genomic diversity among nematodes. This finding, along with the high rate of intrachromosomal rearrangement observed between nematode genomes, has provoked concern that *C. elegans* may be a less-than-optimal model genome for understanding nematode parasitism.[91,93] At the same time, genomic diversity holds out the possibility of very specific drug targeting of nematode species and suggests that there is a substantial pool of potential filariasis drug targets to be mined from the *B. malayi* genome in particular. Once these have been identified, techniques are in place to analyze their function. The species has proved tractable to RNAi[94] as well as heterologous gene expression,[95] although high-throughput techniques required to test multiple drug candidates are far from perfected. Interestingly, the sequenced genome of the *B. malayi* bacterial endosymbiont *Wolbachia*[91,96] metabolically complements that of its host, containing genes that *B. malayi* genome lacks for biosynthesis of flavins, haem, nucleotides, and glutathione. Antirickettsial antibiotics such as tetracycline, rifampicin, and chloramphenicol that clear the *Wolbachia* endosymbiont also target its nematode host, suggesting the *Wolbachia* genome may be a rich resource for antifilariasis drug discovery.[97]

Sequenced genomes of the African blood fluke *S. mansoni* and its Asian counterpart *S. japonicum* are about two or three times larger than the *C. elegans* genome (reviewed in Brindley[98]). The two *Schistosoma* transcriptomes are estimated to each comprise about 14,000 genes,[99–101] of which approximately 50% are estimated to be schistosome specific — and thus perhaps also parasitism related.[102,103] Moreover, about 400 *S. japonicum* genes identified through transcriptome analysis as having significant similarity to mammalian genes were localized to the host–parasite interface (i.e., tegument and eggshell). Among these were numerous cytoskeletal, extracellular matrix, and receptor-like genes that might be involved in immune system evasion via host antigen mimicry, as well as homologs of molecules (e.g., immunophilin) that might be involved in modulating the host immune system.[103] A proteomic survey of the tegument found 43 tegument-specific proteins, more than a quarter of which were unique to schistosomes.[104] Together with the approximately 1,300 other *S. japonicum* ESTs identified as being *Schistosoma* specific,[103] the proteins listed above constitute a substantial pool of potential drug therapy targets for schistosomiasis. Transcriptome-wide comparisons have also shed light on the refractory nature of *Schistosoma* species to drugs and vaccines by identifying several multidrug resistance genes (e.g., efflux transporters) as well as paralogs of previously investigated proteins (e.g., cathepsin B) whose ineffectiveness as vaccine targets that might thus be due to functional redundancy in the genome.[100,101]

11.9 SUMMARY

The recent completion of the genome sequences for a wide variety of parasites that cause some of the most severe diseases of humans has led to increased optimism that genomic approaches are the panacea for which drug and vaccine development has been waiting. There is no question that the availability of these sequences has accelerated basic research into the biology of many of these organisms. The accessibility of sequence data from different strains of the same species, from different species of the same genus, and from related but nonpathogenic species has also allowed for the development of comparative genomic analysis and the development of novel comparative bioinformatic tools.

However, translation of this work into the identification of new drug targets and vaccine candidates using high-throughput discovery pipelines has yet to be achieved, most likely for several reasons. The first is that extensive gene expression data provided by analysis of the transcriptome and proteome of parasites is only just being gathered for many of the parasites that have been sequenced. Gene expression data are required to identify genes that are expressed in the stages to which drugs need to be targeted and provide important data on the RNA and protein composition of cells and how this may change in response to the effects of a drug or vaccine. Mapping of protein interactions and modeling of cellular networks will also aid in this endeavor, providing a systems biology approach to identification of drug and vaccine candidates.[105] Second, the genome sequences of parasites have been found to contain a large number of hypothetical genes of unknown function (in some instances, as many as 60% of the identified genes), indicating that we still do know the full range of metabolic pathways and structural and housekeeping activities of many parasite species. Finally, the pharmaceutical industry itself has low interest in developing novel therapeutics for parasitic diseases, which occur predominantly in developing countries, due to the high cost and low returns. The formation of public–private partnerships[29] that foster collaborations among scientists in academia, big pharmaceutical companies, and the public sector; provision of economic incentives[106]; and alternative financial options[107] provide new hope that a change may be on the horizon.

REFERENCES

1. Degrave, W. M., Melville, S., Ivens, A. & Aslett, M. Parasite genome initiatives. *Int J Parasitol* 31, 532–536 (2001).
2. Adams, J. H., Wu, Y. & Fairfield, A. Malaria Research and Reference Reagent Resource Center. *Parasitol Today* 16, 89 (2000).
3. Singh, B. et al. A large focus of naturally acquired *Plasmodium knowlesi* infections in human beings. *Lancet* 363, 1017–1024 (2004).
4. Snow, R. W., Guerra, C. A., Noor, A. M., Myint, H. Y. & Hay, S. I. The global distribution of clinical episodes of *Plasmodium falciparum* malaria. *Nature* 434, 214–217 (2005).
5. Gardner, M. J. et al. Genome sequence of the human malaria parasite *Plasmodium falciparum*. *Nature* 419, 498–511 (2002).
6. Carlton, J. M. et al. Genome sequence and comparative analysis of the model rodent malaria parasite *Plasmodium yoelii yoelii*. *Nature* 419, 512–519 (2002).
7. Hall, N. et al. A comprehensive survey of the *Plasmodium* life cycle by genomic, transcriptomic, and proteomic analyses. *Science* 307, 82–86 (2005).
8. Carlton, J. The *Plasmodium vivax* genome sequencing project. *Trends Parasitol* 19, 227–231 (2003).
9. Abrahamsen, M. S. et al. Complete genome sequence of the apicomplexan, *Cryptosporidium parvum*. *Science* 304, 441–445 (2004).
10. Xu, P. et al. The genome of *Cryptosporidium hominis*. *Nature* 431, 1107–1112 (2004).
11. Gardner, M. J. et al. Genome sequence of *Theileria parva*, a bovine pathogen that transforms lymphocytes. *Science* 309, 134–137 (2005).
12. Pain, A. et al. Genome of the host-cell transforming parasite *Theileria annulata* compared with *T. parva*. *Science* 309, 131–133 (2005).
13. Loftus, B. et al. The genome of the protist parasite *Entamoeba histolytica*. *Nature* 433, 865–868 (2005).
14. Adam, R. D. The *Giardia lamblia* genome. *Int J Parasitol* 30, 475–484 (2000).
15. McArthur, A. G. et al. The *Giardia* genome project database. *FEMS Microbiol Lett* 189, 271–273 (2000).
16. Carlton, J. M. et al. Draft genome sequence of the sexually-transmitted pathogen *Trichomonas vaginalis*. *Science* 315, 207–212 (2007).
17. Berriman, M. et al. The genome of the African trypanosome *Trypanosoma brucei*. *Science* 309, 416–422 (2005).
18. El-Sayed, N. M. et al. The genome sequence of *Trypanosoma cruzi*, etiologic agent of Chagas disease. *Science* 309, 409–415 (2005).
19. Ivens, A. C. et al. The genome of the kinetoplastid parasite, *Leishmania major*. *Science* 309, 436–442 (2005).
20. Consortium, C. E. S. Genome sequence of the nematode *C. elegans*: a platform for investigating biology. *Science* 282, 2012–2018 (1998).
21. Ghedin, E., Wang, S., Foster, J. M., & Slatko, B. E. First sequenced genome of a parasitic nematode. *Trends Parasitol* 20, 151–153 (2004).
22. El-Sayed, N. M., Bartholomeu, D., Ivens, A., Johnston, D. A., & LoVerde, P. T. Advances in schistosome genomics. *Trends Parasitol* 20, 154–157 (2004).

23. Foster, J. M., Zhang, Y., Kumar, S., & Carlow, C. K. Mining nematode genome data for novel drug targets. *Trends Parasitol* 21, 101–104 (2005).
24. Coppel, R. L. & Black, C. G. Parasite genomes. *Int J Parasitol* 35, 465–479 (2005).
25. Worthey, E. A. & Myler, P. J. Protozoan genomes: gene identification and annotation. *Int J Parasitol* 35, 495–512 (2005).
26. Aslett, M. et al. Integration of tools and resources for display and analysis of genomic data for protozoan parasites. *Int J Parasitol* 35, 481–493 (2005).
27. Aurrecoechea, C. et al. ApiDB: Integrated resources for the apicomplexan bioinformatics resource center. *Neucleic Acids Res* 35, 427–430 (2007).
28. Cowman, A. F. & Crabb, B. S. Functional genomics: identifying drug targets for parasitic diseases. *Trends Parasito*l 19, 538–543 (2003).
29. Croft, S. L. Public–private partnership: from there to here. *Trans R Soc Trop Med Hyg* 99 Suppl 1, S9–S14 (2005).
30. Hyde, J. E. Drug-resistant malaria. *Trends Parasitol* 21, 494–498 (2005).
31. Bathurst, I. & Hentschel, C. Medicines for malaria venture: sustaining antimalarial drug development. *Trends Parasitol* 22, 301–307 (2006).
32. Upcroft, P. & Upcroft, J. A. Drug targets and mechanisms of resistance in the anaerobic protozoa. *Clin Microbiol Rev* 14, 150–164 (2001).
33. Croft, S. L., Barrett, M. P., & Urbina, J. A. Chemotherapy of trypanosomiases and leishmaniasis. *Trends Parasitol* 21, 508–512 (2005).
34. Steverding, D. & Tyler, K. M. Novel antitrypanosomal agents. *Expert Opin Investig Drugs* 14, 939–955 (2005).
35. Ribeiro-Dos-Santos, G., Verjovski-Almeida, S., & Leite, L. C. Schistosomiasis — a century searching for chemotherapeutic drugs. *Parasitol Res* 99, 505–521 (2006).
36. Fenwick, A., Rollinson, D., & Southgate, V. Implementation of human schistosomiasis control: challenges and prospects. *Adv Parasitol* 61, 567–622 (2006).
37. Chapman, H. D. et al. Sustainable coccidiosis control in poultry production: the role of live vaccines. *Int J Parasitol* 32, 617–629 (2002).
38. Buxton, D. & Innes, E. A. A commercial vaccine for ovine toxoplasmosis. *Parasitology* 110 Suppl, S11–S16 (1995).
39. Gupta, S. & Day, K. P. A theoretical framework for the immunoepidemiology of *Plasmodium falciparum* malaria. *Parasite Immunol* 16, 361–370 (1994).
40. Nussenzweig, R. S., Vanderberg, J., Most, H., & Orton, C. Protective immunity produced by the injection of x-irradiated sporozoites of *Plasmodium berghei. Nature* 216, 160–162 (1967).
41. Clyde, D. F., Most, H., McCarthy, V. C., & Vanderberg, J. P. Immunization of man against sporozite-induced falciparum malaria. *Am J Med Sci* 266, 169–177 (1973).
42. Alonso, P. L. Malaria: deploying a candidate vaccine (RTS,S/AS02A) for an old scourge of humankind. *Int Microbiol* 9, 83–93 (2006).
43. Carlton, J., Silva, J., & Hall, N. The genome of model malaria parasites, and comparative genomics. *Curr Issues Mol Biol* 7, 23–37 (2005).
44. Hall, N. & Carlton, J. Comparative genomics of malaria parasites. *Curr Opin Genet Dev* 15, 609–613 (2005).
45. Jomaa, H. et al. Inhibitors of the nonmevalonate pathway of isoprenoid biosynthesis as antimalarial drugs. *Science* 285, 1573–1576 (1999).
46. Missinou, M. A. et al. Fosmidomycin for malaria. *Lancet* 360, 1941–1942 (2002).
47. Borrmann, S. et al. Fosmidomycin-clindamycin for the treatment of *Plasmodium falciparum* malaria. *J Infect Dis* 190, 1534–1540 (2004).
48. Waller, R. F. et al. Nuclear-encoded proteins target to the plastid in *Toxoplasma gondii* and *Plasmodium falciparum. Proc Natl Acad Sci USA* 95, 12352–12357 (1998).

49. Surolia, N. & Surolia, A. Triclosan offers protection against blood stages of malaria by inhibiting enoyl-ACP reductase of *Plasmodium falciparum*. *Nat Med* 7, 167–173 (2001).
50. Ralph, S. A. et al. Tropical infectious diseases: metabolic maps and functions of the *Plasmodium falciparum* apicoplast. *Nat Rev Microbiol* 2, 203–216 (2004).
51. Yeh, I. & Altman, R. B. Drug targets for *Plasmodium falciparum*: a post-genomic review/survey. *Mini Rev Med Chem* 6, 177–202 (2006).
52. Bozdech, Z. et al. The transcriptome of the intraerythrocytic developmental cycle of *Plasmodium falciparum*. *PLoS Biol* 1, E5 (2003).
53. Matuschewski, K. et al. Infectivity-associated changes in the transcriptional repertoire of the malaria parasite sporozoite stage. *J Biol Chem* 277, 41948–41953 (2002).
54. Kaiser, K., Matuschewski, K., Camargo, N., Ross, J., & Kappe, S. H. Differential transcriptome profiling identifies *Plasmodium* genes encoding pre-erythrocytic stage-specific proteins. *Mol Microbiol* 51, 1221–1232 (2004).
55. Mueller, A. K., Labaied, M., Kappe, S. H., & Matuschewski, K. Genetically modified *Plasmodium* parasites as a protective experimental malaria vaccine. *Nature* 433, 164–167 (2005).
56. Doolan, D. L. et al. Utilization of genomic sequence information to develop malaria vaccines. *J Exp Biol* 206, 3789–3802 (2003).
57. Doolan, D. L. et al. Identification of *Plasmodium falciparum* antigens by antigenic analysis of genomic and proteomic data. *Proc Natl Acad Sci USA* 100, 9952–9957 (2003).
58. Florens, L. et al. A proteomic view of the *Plasmodium falciparum* life cycle. *Nature* 419, 520–526 (2002).
59. Wang, R. et al. Immune responses to *Plasmodium vivax* pre-erythrocytic stage antigens in naturally exposed Duffy-negative humans: a potential model for identification of liver-stage antigens. *Eur J Immunol* 35, 1859–1868 (2005).
60. Kim, K. & Weiss, L. M. *Toxoplasma gondii*: the model apicomplexan. *Int J Parasitol* 34, 423–432 (2004).
61. Striepen, B. & Kissinger, J. C. Genomics meets transgenics in search of the elusive *Cryptosporidium* drug target. *Trends Parasitol* 20, 355–358 (2004).
62. Umejiego, N. N., Li, C., Riera, T., Hedstrom, L. & Striepen, B. *Cryptosporidium parvum* IMP dehydrogenase: identification of functional, structural, and dynamic properties that can be exploited for drug design. *J Biol Chem* 279, 40320–40327 (2004).
63. Galazka, J., Striepen, B. & Ullman, B. Adenosine kinase from *Cryptosporidium parvum*. *Mol Biochem Parasitol* 149, 223–230 (2006).
64. Chaudhary, K. & Roos, D. S. Protozoan genomics for drug discovery. *Nat Biotechnol* 23, 1089–1091 (2005).
65. Nagamune, K. & Sibley, L. D. Comparative genomic and phylogenetic analyses of calcium ATPases and calcium-regulated proteins in the apicomplexa. *Mol Biol Evol* 23, 1613–1627 (2006).
66. Graham, S. P. et al. *Theileria parva* candidate vaccine antigens recognized by immune bovine cytotoxic T lymphocytes. *Proc Natl Acad Sci USA* 103, 3286–3291 (2006).
67. Embley, T. M. & Martin, W. Eukaryotic evolution, changes and challenges. *Nature* 440, 623–630 (2006).
68. Nozaki, T., Ali, V. & Tokoro, M. Sulfur-containing amino acid metabolism in parasitic protozoa. *Adv Parasitol* 60, 1–99 (2005).

69. Tokoro, M., Asai, T., Kobayashi, S., Takeuchi, T. & Nozaki, T. Identification and characterization of two isoenzymes of methionine gamma-lyase from *Entamoeba histolytica*: a key enzyme of sulfur-amino acid degradation in an anaerobic parasitic protist that lacks forward and reverse trans-sulfuration pathways. *J Biol Chem* 278, 42717–42727 (2003).
70. Anderson, I. J. & Loftus, B. J. *Entamoeba histolytica*: observations on metabolism based on the genome sequence. *Exp Parasitol* 110, 173–177 (2005).
71. Debnath, A., Das, P., Sajid, M. & McKerrow, J. H. Identification of genomic responses to collagen binding by trophozoites of *Entamoeba histolytica*. *J Infect Dis* 190, 448–457 (2004).
72. Gilchrist, C. A. et al. Impact of intestinal colonization and invasion on the *Entamoeba histolytica* transcriptome. *Mol Biochem Parasitol* 147, 163–176 (2006).
73. Riekenberg, S., Witjes, B., Saric, M., Bruchhaus, I. & Scholze, H. Identification of EhICP1, a chagasin-like cysteine protease inhibitor of *Entamoeba histolytica*. *FEBS Lett* 579, 1573–1578 (2005).
74. Ackers, J. P. & Mirelman, D. Progress in research on *Entamoeba histolytica* pathogenesis. *Curr Opin Microbiol* 9, 367–373 (2006).
75. Bruchhaus, I., Loftus, B. J., Hall, N. & Tannich, E. The intestinal protozoan parasite *Entamoeba histolytica* contains 20 cysteine protease genes, of which only a small subset is expressed during in vitro cultivation. *Eukaryot Cell* 2, 501–509 (2003).
76. MacFarlane, R. C. & Singh, U. Identification of differentially expressed genes in virulent and nonvirulent *Entamoeba* species: potential implications for amebic pathogenesis. *Infect Immun* 74, 340–351 (2006).
77. Shah, P. H. et al. Comparative genomic hybridizations of *Entamoeba* strains reveal unique genetic fingerprints that correlate with virulence. *Eukaryot Cell* 4, 504–515 (2005).
78. Bracha, R., Nuchamowitz, Y., Anbar, M. & Mirelman, D. Transcriptional silencing of multiple genes in trophozoites of *Entamoeba histolytica*. *PLoS Pathog* 2, e48 (2006).
79. Mirelman, D., Anbar, M., Nuchamowitz, Y. & Bracha, R. Epigenetic silencing of gene expression in *Entamoeba histolytica*. *Arch Med Res* 37, 226–233 (2006).
80. Smith, M. W., Aley, S. B., Sogin, M., Gillin, F. D. & Evans, G. A. Sequence survey of the *Giardia lamblia* genome. *Mol Biochem Parasitol* 95, 267–280 (1998).
81. Nash, T. E. Surface antigenic variation in *Giardia lamblia*. *Mol Microbiol* 45, 585–590 (2002).
82. Sun, C. H., McCaffery, J. M., Reiner, D. S. & Gillin, F. D. Mining the *Giardia lamblia* genome for new cyst wall proteins. *J Biol Chem* 278, 21701–21708 (2003).
83. Ullu, E., Lujan, H. D. & Tschudi, C. Small sense and antisense RNAs derived from a telomeric retroposon family in *Giardia intestinalis*. *Eukaryot Cell* 4, 1155–1157 (2005).
84. Ullu, E., Tschudi, C. & Chakraborty, T. RNA interference in protozoan parasites. *Cell Microbiol* 6, 509–519 (2004).
85. He, D., Wen, J. F., Chen, W. Q., Lu, S. Q. & Xin de, D. Identification, characteristic and phylogenetic analysis of type II DNA topoisomerase gene in *Giardia lamblia*. *Cell Res* 15, 474–482 (2005).
86. Dubois, K. N., Abodeely, M., Sajid, M., Engel, J. C. & McKerrow, J. H. *Giardia lamblia* cysteine proteases. *Parasitol Res* 99, 313–316 (2006).
87. El-Sayed, N. M. et al. Comparative genomics of trypanosomatid parasitic protozoa. *Science* 309, 404–409 (2005).
88. Broadhead, R. et al. Flagellar motility is required for the viability of the bloodstream trypanosome. *Nature* 440, 224–227 (2006).

89. Bhatia, V., Sinha, M., Luxon, B. & Garg, N. Utility of the *Trypanosoma cruzi* sequence database for identification of potential vaccine candidates by in silico and in vitro screening. *Infect Immun* 72, 6245–6254 (2004).
90. Stober, C. B. et al. From genome to vaccines for leishmaniasis: screening 100 novel vaccine candidates against murine *Leishmania major* infection. *Vaccine* 24, 2602–2616 (2006).
91. Guiliano, D. B. et al. Conservation of long-range synteny and microsynteny between the genomes of two distantly related nematodes. *Genome Biol* 3, RESEARCH0057 (2002).
92. Parkinson, J. et al. A transcriptomic analysis of the phylum Nematoda. *Nat Genet* 36, 1259–1267 (2004).
93. Viney, M. E. The biology and genomics of Strongyloides. *Med Microbiol Immunol (Berl)* 195, 49–54 (2006).
94. Aboobaker, A. A. & Blaxter, M. L. Use of RNA interference to investigate gene function in the human filarial nematode parasite *Brugia malayi*. *Mol Biochem Parasitol* 129, 41–51 (2003).
95. Gomez-Escobar, N. et al. Heterologous expression of the filarial nematode alt gene products reveals their potential to inhibit immune function. *BMC Biol* 3, 8 (2005).
96. Foster, J. et al. The *Wolbachia* genome of *Brugia malayi*: endosymbiont evolution within a human pathogenic nematode. *PLoS Biol* 3, e121 (2005).
97. Rao, R. U. Endosymbiotic *Wolbachia* of parasitic filarial nematodes as drug targets. *Indian J Med Res* 122, 199–204 (2005).
98. Brindley, P. J. The molecular biology of schistosomes. *Trends Parasitol* 21, 533–536 (2005).
99. Hu, W., Brindley, P. J., McManus, D. P., Feng, Z. & Han, Z. G. Schistosome transcriptomes: new insights into the parasite and schistosomiasis. *Trends Mol Med* 10, 217–225 (2004).
100. Hu, W. et al. Evolutionary and biomedical implications of a *Schistosoma japonicum* complementary DNA resource. *Nat Genet* 35, 139–147 (2003).
101. Verjovski-Almeida, S. et al. Transcriptome analysis of the acoelomate human parasite *Schistosoma mansoni*. *Nat Genet* 35, 148–157 (2003).
102. Hoffmann, K. F. & Dunne, D. W. Characterization of the *Schistosoma* transcriptome opens up the world of helminth genomics. *Genome Biol* 5, 203 (2003).
103. Liu, F. et al. New perspectives on host–parasite interplay by comparative transcriptomic and proteomic analyses of *Schistosoma japonicum*. *PLoS Pathog* 2, e29 (2006).
104. van Balkom, B. W. et al. Mass spectrometric analysis of the *Schistosoma mansoni* tegumental sub-proteome. *J Proteome Res* 4, 958–966 (2005).
105. Winzeler, E. A. Applied systems biology and malaria. *Nat Rev Microbiol* 4, 145–151 (2006).
106. Fehr, A., Thurmann, P. & Razum, O. Editorial: drug development for neglected diseases: a public health challenge. *Trop Med Int Health* 11, 1335–1338 (2006).
107. Brogan, D. & Mossialos, E. Applying the concepts of financial options to stimulate vaccine development. *Nat Rev Drug Discov* 5, 641–647 (2006).
108. Keeling, P. J. et al. The tree of eukaryotes. *Trends Ecol Evol* 20, 670–676 (2005).

12 Comparative Genomics in AIDS Research

Philippe Lemey, Koen Deforche, and Anne-Mieke Vandamme

CONTENTS

ABSTRACT

In this chapter, we provide a basic introduction to human immunodeficiency virus (HIV) biology and evolution and highlight many applications of comparative genomics. The wealth of available HIV sequence data has been used to investigate the epidemic history, HIV transmission dynamics, and within-host evolution of the virus. Because of the clinical impact, the main focus of within-host evolutionary studies has been the development of resistance to antiviral drug treatment. Therefore, our discussion on HIV comparative genomics concludes with a particular emphasis on data-mining techniques to investigate drug resistance.

12.1 INTRODUCTION

The acquired immunodeficiency syndrome (AIDS) epidemic is among the most devastating global epidemics in human history. According to the 2006 report from the UNAIDS organization (Joint United Nations Program on HIV/AIDS), the number of people who were living with the human immunodeficiency virus (HIV) worldwide in 2005 was estimated at around 39 million and still increases at an alarming rate (http://www.unaids.org). Despite tremendous research effort, HIV has been elusive to control, and its rapidly mutating genome remains a challenge for the development of both vaccines and antiviral drugs.

Shortly after the AIDS epidemic had been recognized in the United States,[1] the causative agent was identified as a complex retrovirus.[2] Because two other human retroviruses had just been isolated, the human T-cell lymphotropic virus types 1 and 2 (HTLV-1 and HTLV-2),[3,4] many essential tools to characterize retroviruses were already available at the time of HIV discovery.[5] Originally called lymphadenopathy-associated virus (LAV) or HTLV-3, the virus was renamed the human immunodeficiency virus in 1986 because it was shown to belong to the lentiviruses rather than oncoviruses.[6,7] Because of major research interest, the relatively short genome of HIV was quickly deciphered. Not surprisingly, genetic studies of HIV have rapidly moved beyond many standard research questions in comparative genomics, like gene finding and the identification of regulatory regions. The main focus has now shifted toward elucidating the evolutionary and population genetic processes that shape HIV diversity and how such knowledge can be used in an epidemiological context or in the struggle against HIV infection. However, the underlying evolutionary principles and computational aspects in tackling such problems have remained the same. Compared to organisms for which comparative genomics is now widely applied, there is a different dimensionality to the available HIV sequence data. On the one hand, the HIV genome size is rather restricted (approximately 9.6 kb). On the other hand, a massive amount of sequences have been obtained at different population levels (both within and among human hosts) and from their simian counterparts in different primate hosts.

Comparative genomics can also assist in characterizing host cell factors that interact with HIV, which could reveal new targets for drug intervention. Retroviruses are intimately associated with the host cell machinery, and many molecular interactions have not been fully unraveled (for a review of currently known interactions for HIV-1, see Trkola[8]). Two such examples in relationship to the HIV life cycle are discussed. Although this research arises from molecular studies of viral replication, the comparative genomic approaches to identify and characterize cellular factors apply to the host.

HIV sequence data have been accumulating at staggering rates, making the immunodeficiency viruses the most data-rich group of organisms for evolutionary analyses.[9] Several advances in polymerase chain reaction (PCR) and sequencing technology have stimulated the determination of HIV complete genomes[10]; about

800 complete genome sequences are now available at the Los Alamos HIV database,[11] a specialized and highly annotated database for HIV sequence data (http://www.hiv.lanl.gov/). In this chapter, we introduce the fundamentals of HIV biology relevant to therapeutic intervention and virus–host interactions and discuss how computational approaches can be used to study viral evolution and epidemiology, with special reference to vaccine development and antiviral drug resistance.

12.2 HIV PRIMER

12.2.1 HIV BIOLOGY

The HIV genome consists of two positive, single-stranded RNA molecules, which are approximately 9.6 kb long (Figure 12.1A). The diploid genome is embedded in a protein capsid (CA) together with viral enzymes required for HIV replication. A matrix (MA) composed of viral protein p17 surrounds the CA and is in turn enclosed by the envelope. The envelope is formed by a cell-derived lipid bilayer and is associated with the viral glycoproteins gp120 and gp41.

The HIV genome is flanked by two long terminal repeats (LTRs) and contains nine open reading frames, with three major genes encoding structural proteins: *gag*, *pol,* and *env* (Figure 12.1B). The *gag* region codes for the internal nonglycosylated proteins: CA, MA, and nucleocapsid (NC). The three products encoded by the *pol* gene are protease (PRO), reverse transcriptase (RT), and integrase (IN). The *env* gene product is a polyprotein (gp160) that is cleaved into the transmembrane (TM) (gp41) and surface (SU) (gp120) components, which are linked together by disulfide bonds. In addition to the structural proteins, complex retroviruses possess genes encoding regulatory and accessory proteins. The functions of these proteins are, among others, to stimulate and regulate viral transcription and to modulate the host cell machinery favoring the virus replication cycle (reviewed in Coffin[12] ; Luciw[13]; Frankel and Young[14]; Turner and Summers[15]; Cann and Chen[16]; and Coffin[17]).

Primarily, HIV infects T lymphocytes, and the first step in the replication cycle requires the attachment of the parental virus to a specific receptor on the host cell surface (Figure 12.2). The CD4 molecule has been characterized as the main cellular receptor for HIV.[18] This binding induces conformational changes in the SU glycoprotein gp120, exposing other regions that can bind to chemokine (C-C motif) receptor 5 (CCR5) and chemokine (C-X-C motif) receptor 4 (CXCR4). Coreceptor binding induces further conformational changes in the TM gp41, eventually triggering the fusion of the viral envelope to the cell membrane.

After delivery of the viral core to the cytoplasm and disassembly of MA and CA proteins (*uncoating*); (Figure 12.2), reverse transcription generates a double-stranded DNA copy of the RNA genome. The viral DNA is then transported into the nucleus and integrated into chromosomal DNA. The integrated provirus can now be transcribed by cellular RNA polymerase II. Part of the synthesized RNA copies is processed into messenger RNAs, which will be translated into viral proteins in the cytoplasm. Other RNA copies become full-length progeny virion RNA. The regulatory proteins Tat and Rev upregulate transcription and promote the translocation of unspliced or single-spliced transcripts to the cytoplasm. Finally, the virion core is

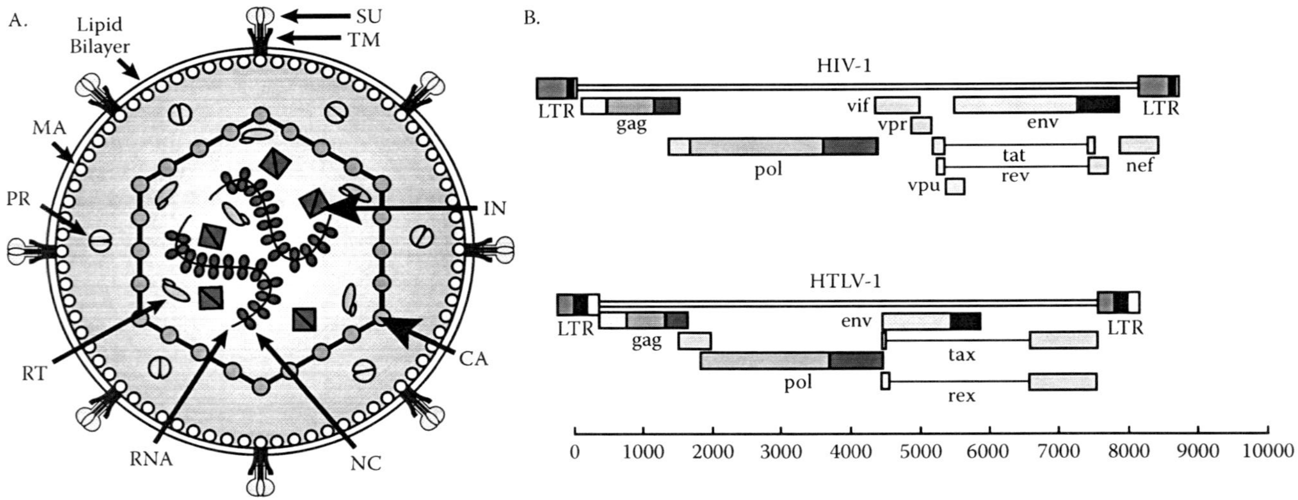

FIGURE 12.1 (See color figure in the insert following page 48.) (A) Schematic cross section through a retroviral particle. CA, capsid; IN, integrase; MA, matrix; NC, nucleocapsid; PR, protease; RT, reverse transcriptase; SU, surface unit; TM, transmembrane. (B) Schematic organization of the HIV genome. As a comparison, the genome of another complex retrovirus, HTLV-1, is depicted. The color codes in the genomes correspond to the encoded proteins in the particle. (Adapted from Voght, P. K., in *Retroviruses,* Eds. Coffin, J.M., Hughes, S.H., & Varmus, H.E., Cold Spring Harbor Press, New York, 1997.)

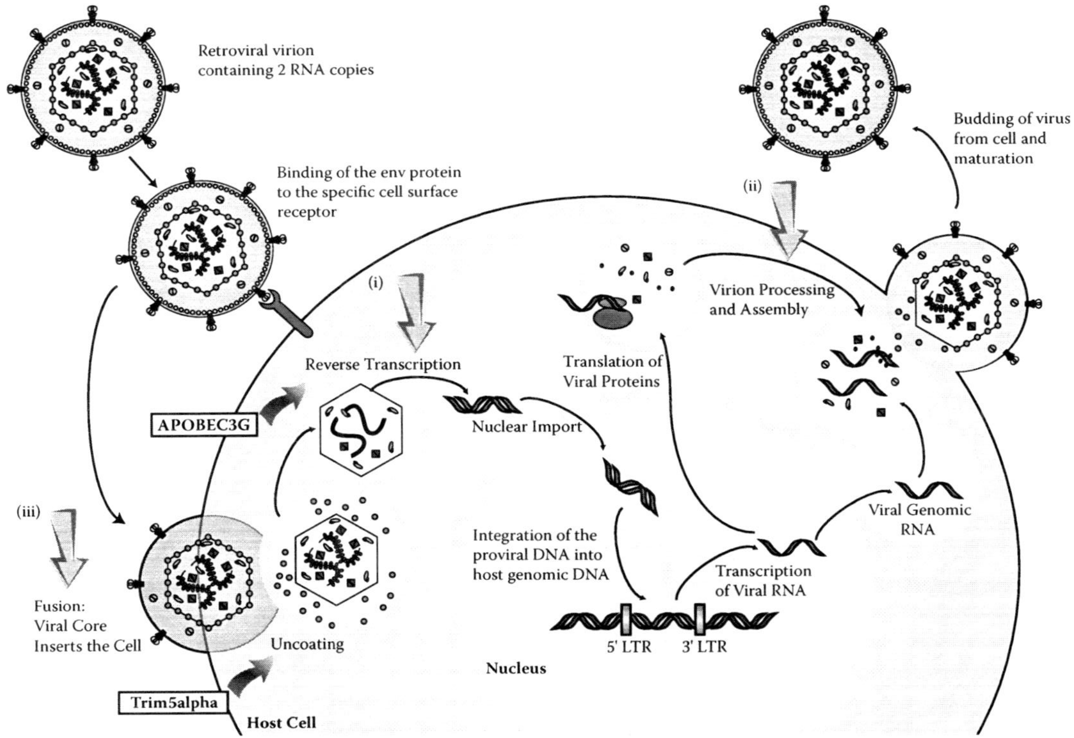

FIGURE 12.2 The retroviral replication cycle. The three different steps in the replication process targeted by currently available antivirals are indicated with vertical arrows: (i) reverse transcription, (ii) virion processing and assembly, and (iii) fusion. The interaction of Trim5α with the capsid and the uncoating process and the action of APOBEC3G during the reverse transcription process are indicated with arrows in the cell. (Adapted from Rambaut, A., et al., *Nat. Rev. Genet.* 5, 52–61, 2004.)

assembled at the plasma membrane and progeny virus is released by a process of budding and subsequent maturation into infectious virus (reviewed in Coffin[12]; Luciw[13]; Frankel and Young[14]; Turner and Summers[15]; Cann and Chen[16]; and Coffin[17]).

12.2.2 HIV Genetic Variability

Immunodeficiency viruses are among the most genetically diverse pathogens.[19] The rapid evolution of HIV can be attributed to a combination of high mutation rates (~3×10^{-5} substitutions/site/generation) due to the lack of RT proofreading activity,[20] short generation times (~2.6 days),[21,22] and enormous virion production (~10^{10} to 10^{12} new virions each day).[23] In addition, HIV genomes are subject to a great deal of recombination because the RT frequently alternates between the two RNA molecules as templates for complementary DNA synthesis. The frequency of template crossover has been estimated as between 7 and 30 events per replication round.[24] Therefore, copackaging of two distinct RNA molecules in a single virion, due to co- or superinfection with different viral variants infecting the same cells, will undoubtedly lead to the generation of progeny with mosaic genomes during the next replication cycle. In addition, HIV proteins have a high plasticity; for example, about 49 natural polymorphisms and 20 drug resistance–associated mutations are known in the 99 amino acid viral PRO. The rapid rate of genetic change represents an enormous evolutionary potential for HIV: A significant amount of nucleotide substitutions are usually accumulated over a time span of months or years. Therefore, both within hosts and between hosts the virus is considered as a measurably evolving population,[25] and phylogenetic as well as population genetic models have been developed to incorporate this temporal aspect.[26–29]

12.2.3 Drug Targets and Viral Drug Resistance

The HIV inhibitors currently used in clinical practice interfere with three different steps in the replication process (indicated in Figure 12.2). First, nucleoside RT inhibitors (NRTIs) target the RT-catalyzed transcription of the viral RNA genome to a DNA copy by mimicking the structure of nucleoside bases and thus competing with the natural substrates for binding to RT. Due to their modifications, incorporation of NRTI products into newly synthesized viral DNA results in DNA chain termination. Nonnucleoside RT inhibitors (NNRTIs) inhibit the same process by allosteric binding close to the active site of the enzyme, thereby inhibiting the HIV-1 RT activity. Next, protease inhibitors (PIs) inhibit the PRO-mediated cleavage of immature viral proteins into new enzymatic and structural HIV proteins by binding to the active site of PRO. Finally, more recently, peptides blocking the fusion of the virus with the host cell have been developed that bind competitively to a substructure of the gp41 undergoing conformational changes during the fusion process. New agents in existing drug classes (e.g., TMC125 and TMC278; see Pauwels[30]) and in new drug classes (e.g., coreceptor inhibitors and IN inhibitors) have reached the clinical testing phase and offer the hope for broader therapeutic options in the near future.

Because currently available antiretrovirals will not eradicate HIV, therapeutic intervention is aimed at durably inhibiting viral replication to reduce HIV load to levels below the limits of detection, to prevent ongoing host cell destruction, and to allow for immune restoration to some degree. Treatment should have a high genetic barrier to resistance, which quantifies the "evolutionary difficulty" for the virus to become resistant. To this purpose, combinations of drugs are used, also referred to as highly active antiretroviral therapy (HAART), which effectively increase the potency and the genetic barrier to resistance. In addition, recent drugs such as lopinavir or darunavir are designed specifically with a high genetic barrier to resistance, requiring multiple substitutions to become ineffective. When the virus has a "wild-type" genome that is susceptible to all drugs, which is the case for the majority of patients before start of treatment, most HAART drug combinations will reach the objective of reducing the viral load to undetectable levels. However, fluctuations in plasma levels of the drugs, in many cases caused by nonperfect adherence of the patient to drug intake, may allow the virus to replicate in an environment with strong selective pressure. Treatment failure is still common and usually associated with emergence of resistance.

12.3 UNDERSTANDING AND TARGETING WITH VIRUS–HOST INTERACTIONS

While NRTIs, NNRTIs, and PIs result from classical drug development, which concentrates on the inhibition of viral enzymes, fusion inhibitors were designed to intervene with specific virus–host interactions. Insights into interactions between virus and host proteins that promote or suppress steps in the HIV life cycle can further stimulate drug discovery, and this might be assisted by comparative genomics. One such example is the retrovirus restriction factor Trim5α, which blocks HIV-1 infection in simian cells.[31] This cytoplasmic restriction factor is known to bind to the HIV CA protein (Figure 12.2), thereby successfully disrupting the ordered process of viral uncoating and reverse transcription in Old World monkeys.[32,33] Human Trim5α, however, does not restrict HIV-1 infection, and this difference in susceptibility was attributed to species-specific CA binding.[32] Evolutionary analyses provided strong evidence for ancient positive selection in the primate TRIM5α gene,[34–36] which is interpreted as the molecular signal for adaptation to recognize viruses with new CA variants.[37] Interestingly, the Trim5α gene regions exhibiting the strongest signal for positive selection coincided with those identified as essential for biochemical CA recognition.[33,37] Comparative genomics can also shed light on the functional importance of other isoforms of this restriction factor. For example, an unexpectedly high frequency of a deleterious mutation in all Trim5 isoforms has been reported in the human population, implying that a function other than retroviral immune surveillance is probably not essential.[38] Recently, it has been shown that human Trim5α protects against infection by *Pan troglodytes endogenous retrovirus* (PtERV1), an endogenous retrovirus that is absent in humans.[39] This immune defense mechanism was probably an evolutionary advantage in humans, but unfortunately, it also seems to have increased our cells' susceptibility to HIV infection.[39]

Evolutionary genomics approaches have also been used to characterize other host factors involved in viral–host genetic conflicts. A particular interest has been shown in APOBEC3G (apolepoprotein B mRNA editing enzyme, catalytic polypeptide-like 3G), which belongs to a family of enzymes that edits RNA/DNA by deaminating cytosine to yield uracil.[39,40] This protein is packaged into the virions and performs its detrimental editing during the reverse transcription process (Figure 12.2), resulting in hypermutated and thus frequently damaged viral DNA. The protein encoded by the HIV-1 accessory *vif* gene can counteract APOBEC3G by promoting its degradation in the ubiquitin–proteasome pathway before its incorporation in the viral particles.[41] As expected from a long-standing genetic conflict with viral proteins, there is a clear molecular footprint of positive selection during primate evolution in the APOBEC3G gene.[42,43] Although APOBEC3G adaptive evolution appears to have occurred proteinwide,[42] a particular cluster of positively selected sites was recently revealed in the Vif-interaction domain.[44] Interestingly, the *vif* gene appears to be conserved between all primate and most nonprimate lentiviruses. It has now been shown that more members of the APOBEC3 family exert potent activity against Vif-deficient HIV-1, like APOBEC3F,[45] against or Vif-deficient simian immunodeficiency viruses (SIVs), like APOBEC3B and APOBEC3C,[46] and it has been suggested that an HIV-1 Vif-resistant mutant APOBEC3G could provide a gene therapy approach to combat HIV-1 infection.[47]

12.4 MOLECULAR EPIDEMIOLOGICAL TECHNIQUES

12.4.1 THE ORIGIN AND EPIDEMIC HISTORY OF HIV

Molecular methods have become invaluable tools to investigate important questions about the epidemiology and transmission patterns of infectious diseases. By focusing on the etiological agent, they complement traditional epidemiological studies that primarily concentrate on the host.[48] Phylogenetic inference of the viral evolutionary history plays a central role in molecular epidemiology, and many methods for phylogenetic analyses have been developed. These methods and models of molecular evolution are extensively reviewed elsewhere.[49–51]

AIDS can be caused by two types of HIV, HIV-1 and HIV-2, which have a genetic similarity of about 40%. Phylogenetic analyses have clearly demonstrated that the sources of HIV-1 and HIV-2 are SIVs that infect different African primates (Figure 12.3).[52] Three separate cross-species transmissions from chimpanzees have introduced distinct HIV-1 lineages in the human population, denoted M, N, and O.[53,54] HIV-1 group M is responsible for the worldwide pandemic and has radiated into nine

FIGURE 12.3 (Opposite; see also color figure in the insert following page 48.) Evolutionary history of the primate lentiviruses. The viral lineages infecting human hosts are indicated with red branches. The phylogenetic tree was reconstructed using Bayesian inference as implemented in MrBayes[120]; an alignment of 55 partial *pol* amino acid sequences was used, and the clustering was generally well supported by posterior probability values (full details are available from the authors on request). The magenta/green arrows indicate the branches along which a significant loss of Nef-mediated TCR-CD3 downmodulation has occurred.[81] (*Pan troglodytes*: photo by Hans-Georg Michna; *Cercopithecus neglectus*: photo by Aaron Logan, licensed under Creative Commons Attribution 1.0 License; *Cercopithecus albogularis*: photo by Eva Hejda, licensed under Creative Commons Attribution ShareAlike 2.0 Germany; *Cercopithecus mona:* photo from www.zoo.lyon.fr; *Cercocebus torquatus*: photo by Mike Kaplan; *Colobus guereza*: photo by Duncan Wright, licensed under the GNU Free Documentation License; *Mandrillus sphinx*: photo by Malene Thyssen, licensed under Creative Commons Attribution ShareAlike 2.5; *Cercopithecus cephus*: licensed under Creative Commons Attribution ShareAlike 2.5.)

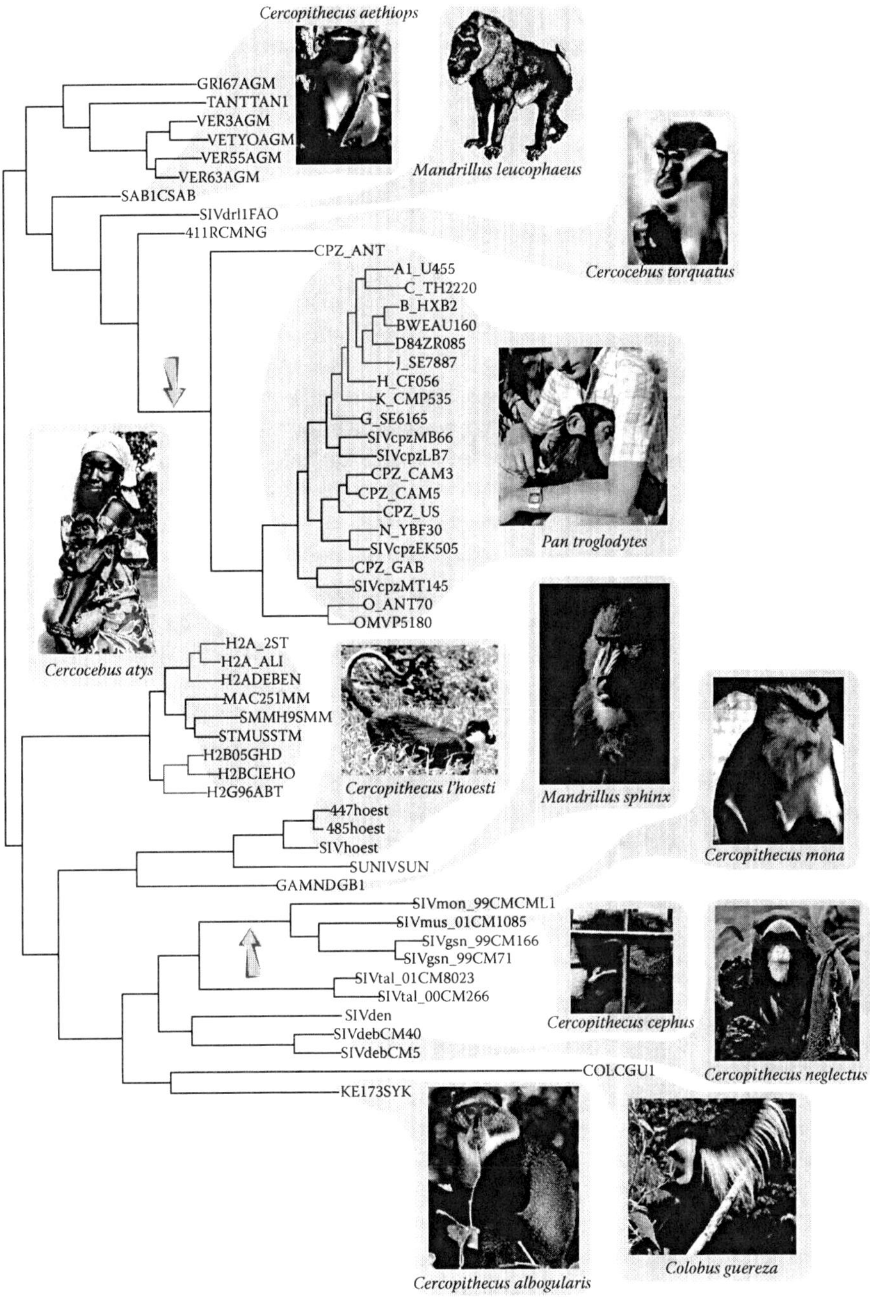
Cercopithecus aethiops
Mandrillus leucophaeus
Cercocebus torquatus
GRI67AGM
TANTTAN1
VER3AGM
VETYOAGM
VER55AGM
VER63AGM
SAB1CSAB
SIVdrl1FAO
411RCMNG
CPZ_ANT
A1_U455
C_TH2220
B_HXB2
BWEAU160
D84ZR085
J_SE7887
H_CF056
K_CMP535
G_SE6165
SIVcpzMB66
SIVcpzLB7
CPZ_CAM3
CPZ_CAM5
CPZ_US
N_YBF30
SIVcpzEK505
CPZ_GAB
SIVcpzMT145
O_ANT70
OMVP5180
Pan troglodytes
Cercocebus atys
H2A_2ST
H2A_ALI
H2ADEBEN
MAC251MM
SMMH9SMM
STMUSSTM
H2B05GHD
H2BCIEHO
H2G96ABT
Cercopithecus l'hoesti
Mandrillus sphinx
Cercopithecus mona
447hoest
485hoest
SIVhoest
SUNIVSUN
GAMNDGB1
SIVmon_99CMCML1
SIVmus_01CM1085
SIVgsn_99CM166
SIVgsn_99CM71
SIVtal_01CM8023
SIVtal_00CM266
SIVden
SIVdebCM40
SIVdebCM5
COLCGU1
KE173SYK
Cercopithecus cephus
Cercopithecus neglectus
Cercopithecus albogularis
Colobus guereza

roughly equidistant subtypes (A–D, F–H, J, and K). Using sequences sampled over time and assuming that the rate of evolution has remained fairly constant throughout the evolutionary history (the *molecular clock hypothesis*), it has become feasible to estimate timescales for viral epidemics. HIV-1 group M radiation originated in central Africa and has been dated back to around 1930 (1915–1941).[55–57]

The HIV-1 group M subtypes are unevenly distributed worldwide, and their phylogenetic structure appears to have resulted from founder effects and incomplete sampling.[58] For example, subtype B is the most prevalent strain in industrialized countries (North America, western Europe, and Australia), and has been introduced from Haiti into high-risk groups in the United States, allowing for an explosive viral spread during the 1970s.[59] At present, there is still an association between subtype B infections and HIV transmission through homosexual sex and injecting drug use. The overwhelming majority of HIV infections in the developing world stem from heterosexual transmission[60,61] and, to a lesser extent, perinatal transmission (http://www.unaids.org). The epidemic history of HIV variants and the impact of transmission dynamics on viral spread have been increasingly studied using population genetic techniques. More particularly, *coalescent theory*, modeling how changes in population size over time influence the shape of HIV phylogenies,[62,63] has become a popular application in molecular epidemiology. For example, coalescent analyses have characterized the viral epidemic spread of HIV-1 in central Africa,[64] HIV-2 in west Africa,[65] and the impact of high-risk groups on the early epidemic spread of HIV-1 subtype B in the United States[59] (for a review of these studies, see Lemey, Rambaut, and Pybus[66]).

Mosaic HIV gene sequences were identified relatively late in the pandemic,[67–69] but the full impact of recombination on global HIV diversity became apparent when complete genome sequencing was performed on a larger scale.[70] To date, a large number of circulating recombinant forms (CRFs), which have spread to some extent, and unique recombinant forms (URFs) have been identified; both forms are now part of the complex and dynamic epidemic (for the role of CRFs in the global epidemic, see McCutchan[11] and Peeters, Toure-Kane, and Nkengasong[71]). Although the detection of recombinant sequences has been aided by the development of many different bioinformatics tools (for an overview, see http://bioinf.man.ac.uk/recombination/programs.shtml), it still remains a challenging problem. The CRFs, which are the result of coinfection or superinfection of two genetically distinct strains, can be characterized relatively easy using phylogenetic-based methods (e.g., Simplot[72]). The inference, however, critically depends on the correct a priori assignment of "pure" lineages.[73] Detecting recombination within hosts harboring a more genetically homogeneous population is far more difficult and often requires a population genetic approach to quantify the rate of recombination.[74–76]

The broad range of African primates infected with SIV emphasizes the importance of studying primate evolution to identify host factors interacting with the viral replication (see Section 12.3). HIV comparative genomics, on the other hand, can complement host studies and unravel the role of viral adaptation in viral–host genetic conflicts. It has been reported that SIVs infecting chimpanzees have a methionine or leucine at residue 29 in the p17 MA protein, and this has been substituted by an arginine in the ancestral sequences of distinct viral lineages infecting human hosts

(HIV-1 groups M, O, and N).[77] Since these HIV-1 and SIV chimpanzee (SIVcpz) lineages are phylogenetically interspersed (Figure 12.3), such homoplasic polymorphism strongly suggests viral adaptation to the human host[77]; the functional importance of this adaptation remains to be elucidated.

Viral genetic differences might also be responsible for differences in pathogenicity among HIV/SIV infections. In contrast to HIV-1 and HIV-2 infections of humans, natural SIV infections are usually not pathogenic for their primary hosts (reviewed in Hirsch[78]). In turn, HIV-2 is known to be less transmissible and less pathogenic than HIV-1 group M.[79,80] Because T-cell activation appears to be a consistent difference between pathogenic and nonpathogenic lentiviral infections, and the accessory protein Nef has been implicated in immune activation, Schindler et al. in 2006 performed a functional characterization of Nef from an evolutionary perspective. They clearly showed that most SIV Nefs downmodulate T-cell receptor (TCR) CD3, thereby protecting against activation-induced cell death (AICD).[81] The fact that AICD might be more important than CTL killing in depleting the T-cell pool[82] highlights the protective role of TCR-CD3 downmodulation in SIV infection. The chimpanzee precursor of HIV-1, SIVcpz, and three *Cercopithecus* viruses (SIVgsn [SIV infecting greater spot-nosed monkeys], SIVmus [SIV infecting mustached monkeys], and SIVmon [SIV infecting mona monkeys]) are, however, remarkable exceptions. *Nef* alleles from these viral lineages fail to downmodulate TCR-CD3 and to inhibit cell death and may thus be key determinants in AIDS pathogenesis.[81] The SIV phylogenetic relationships indicate that the protective role of Nef represents a characteristic feature of long-standing virus–host interactions, which has been lost independently on the branch leading to the SIVgsn/SIVmus/SIVmon clade and after the recombination event that generated the simian precursor of HIV-1 (Figure 12.3).[81] Because TCR-CD3 downmodulation also correlated with $CD4^+$ T-cell depletion in SIV infecting sooty mangabyes (SIVsmm),[81] a similar phenomenon might be important for differences in pathogenicity among HIV-1 fast-, moderate-, and slow-progressing patients.

12.4.2 HIV Vaccine Design

Because of the ever-expanding HIV genetic diversity, it is not surprising that the virus is a difficult target for vaccine development. The ultimate goal for an effective vaccine is to elicit a potent immune response capable of preventing HIV infection or controlling disease. Both humoral and cell-mediated immune responses are mounted and sustained during natural infection. Unfortunately, the rapid generation of viruses that can escape immune recognition will eventually lead to $CD4^+$ T-cell depletion and clinical progression to AIDS. In addition, the pool of resting memory $CD4^+$ T cells that carry integrated proviral genomes represents a stable reservoir for latent HIV infection hidden from immune surveillance. Latent reservoirs are also the cause of viral persistence long after initiation of therapy.[83] Although the initial hope for HIV vaccine design was based on neutralizing antibodies, their ability to control viral replication might have been overestimated. Neutralizing antibodies predominantly target the hypervariable loops in the Env gp120 and rarely recognize the concealed receptor-binding sites.[84] Moreover, HIV rapidly escapes neutralizing antibodies, leaving a clear trace of adaptive evolution in the *env* gene sequences.[85] Phase III clinical trials of vaccines that largely elicit antibody responses have generally been disappointing.[86,87] This setback has shifted the focus toward cytotoxic T cell (CTL) responses, which may play a more important protective role in HIV infection. Evidence has shown that partial control of HIV replication in vivo is temporally associated with the appearance of CTL responses,[88] and that the rate of disease progression is strongly dependent on human leukocyte antigen (HLA) class I alleles.[89,90] By stimulating T lymphocytes that can identify and kill HIV-infected cells, vaccines inducing cellular responses will not prevent infection, but it is hoped that they will limit viral replication and delay disease progression.[91]

Irrespective of which immune response is induced, it is essential to maximize immunogen antigenic similarity to viruses likely to be encountered by the population at risk.[92] Therefore, molecular epidemiological surveys play an important role in tracing the geographic circulation of viral diversity. If circulating strains are chosen as vaccine candidates, however, then the degree of dissimilarity to other strains might still be too large to conserve key epitopes.[92] Population-level phylogenies for HIV typically exhibit starlike or approximately starlike tree topologies (Figure 12.4), as expected from exponentially growing populations. Consequently, the expected genetic distance between any two sampled HIV sequences is about twice the mean distance of the tips to the root of the tree (Figure 12.4). Computational analyses can be used to minimize the expected genetic distance between contemporary and candidate vaccine strains by inferring "centralized immunogens."[93]

A simple approach would be to employ a consensus sequence of strains sampled from the population. Consensus sequences, however, may be subject to sampling bias and link polymorphisms to combinations that are not found in natural strains.[92] Therefore, ancestral sequences have been proposed as more appropriate candidate vaccines.[92,93] Phylogenetic inference of ancestral protein sequences can be performed using maximum parsimony, maximum likelihood, and Bayesian methods and should lead to immunogens that elicit, on average, more cross-reactive immune responses than immunogens from contemporary strains.[94] The codon usage of centralized genes can be optimized to enhance protein expression in vivo and in vitro (e.g., see Andre[95] and Gao et al.[96]), and the resulting constructs are now increasingly evaluated in animal models (e.g., see Doria-Rose et al.,[92] Kothe et al.,[94] and Gao et al.[97]). Although centralized *env* genes were shown to express functional envelope glycoproteins, the breadth and potency of neutralizing antibody responses were sometimes limited,[92,97] and the advantage of ancestral sequences over consensus sequences or even over contemporary strains are not necessarily substantiated.[94] More research is required to improve immune response–inducing capability of centralized immunogens and to evaluate new modes of vaccine delivery, but currently, establishing protective immunity is only a distant hope.

12.5 INTRAHOST EVOLUTION AND HIV TRANSMISSION

Different evolutionary and population genetic processes shape the viral diversity within the host and between hosts.[66] While evolutionary dynamics among hosts are mainly determined by selectively neutral, epidemiological processes,[98] within-host HIV dynamics are a complex interplay of selection, recombination, demography and

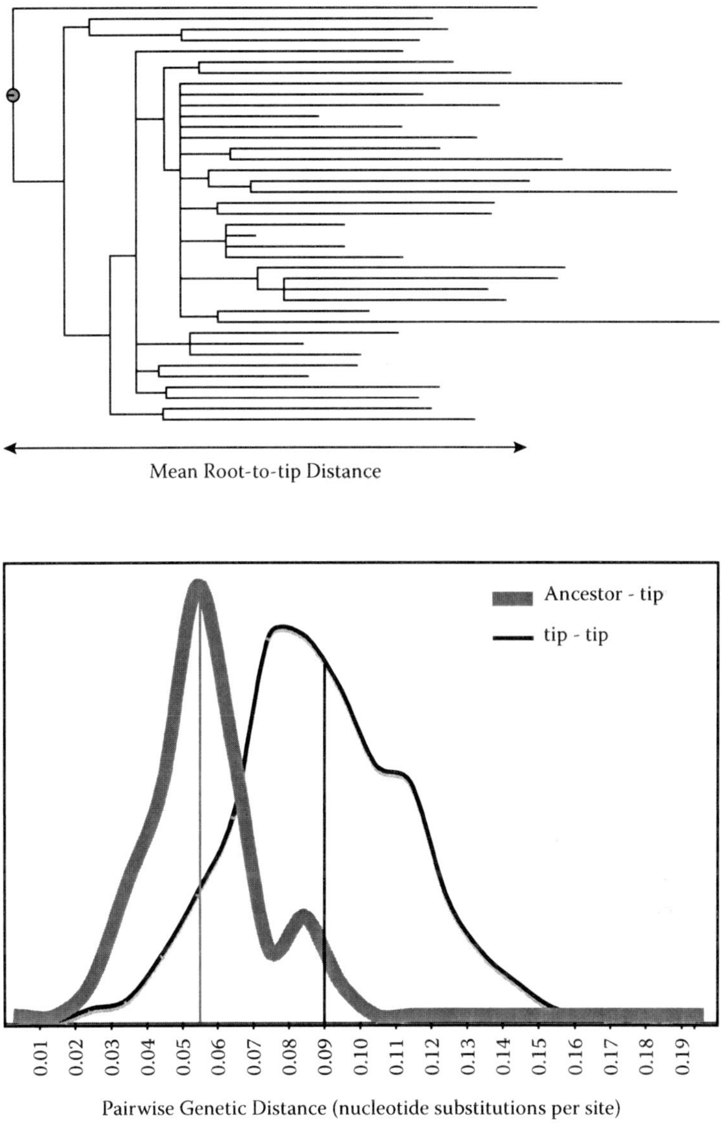

FIGURE 12.4 Schematic representation of the principle of ancestral sequences as centralized immunogens. On the left is a phylogenetic tree for HIV *env* gene sequences, sampled from the U.S. epidemic.[59] The ancestral sequence for the most recent common ancestor, indicated at the root of the tree, was inferred using maximum likelihood analyses. On the right, two pairwise genetic distance distributions are shown: the distribution of pairwise genetic distances between each contemporary sequences and all other contemporary sequences (thin black line) and the distribution of pairwise genetic distances between the ancestral sequence and all contemporary sequences (thick gray line). The mean is indicated with a vertical bar.

migration. These intrahost processes are central to our understanding of many clinically relevant issues, like the development of drug resistance (see Section 12.6) and immune escape and vaccine design (see Section 12.4.2). Although several investigations have focused on within-host HIV evolution, there is still considerable uncertainty regarding the extent to which different population genetic processes shape viral diversity.[66] To resolve this, there is a need to build complex models that allow coestimation of parameters of several processes (since their interactions are complex and cannot be ignored). Population genetic approaches are usually more suitable for this purpose than phylogenetic methods. For example, population genetic methods have now been developed to estimate the differential action of selection among protein sites, like the ones used to investigate selection in host proteins (see Section 12.3), in the presence of recombination.[99]

Transmission of HIV is the interface between evolution within hosts and among hosts, and characterization of this process is pivotal to understanding transitions in HIV evolution at different epidemiological scales. Sequence data sampled from HIV transmission pairs have shown that transmission is typically associated with a strong genetic bottleneck (e.g., see Derdeyn et al.[100]), which can at least partly be responsible for genetic drift in HIV evolution among hosts. Phylogenetic analysis has also been used to reconstruct the transmission history for known HIV transmission chains.[101–103] This research has shown that transmission histories can be fairly accurately reconstructed,[101,102] provided that homoplasies resulting from strong selective pressure are not confounding the analysis.[101] It has been shown that genealogy-based population genetic approaches can be used to quantify the genetic bottleneck associated with transmission, revealing a loss in diversity of about 99%.[104] However, it remains to be established whether transmission is selectively neutral.

12.6 DATA-MINING TECHNIQUES FOR GENETIC ANALYSIS OF DRUG RESISTANCE

12.6.1 Obtaining HIV Drug Resistance Data

Resistance testing is performed to assess the activity of all available drugs in the face of resistance mutations.[105] Resistance testing can therefore assist a clinician in combining active drugs in an effective treatment when treatment failure occurs.[106] Because of possible transmission of resistant virus to a new patient, resistance testing is also recommended increasingly before start of a first treatment.[107] Ideally, the ability of the virus to replicate in the presence of treatment, that is, the *fitness in presence of treatment*, needs to be determined or estimated for all possible treatment combinations. In addition, it is desirable to choose not only an effective treatment but also a treatment with a high genetic barrier to resistance to avoid the development of resistance. The genetic barrier not only differs from drug to drug but also may depend on the viral genome. For example, the presence of a resistance mutation V82A in PRO, selected by treatment with indinavir, does not reduce the susceptibility of the virus to lopinavir, an inhibitor with a high genetic barrier for which HIV needs to develop several other mutations in addition to V82A to become resistant. Acquisition of V82A may, however, reduce the genetic barrier to lopinavir resistance. In addition, the large natural variation of HIV may be reflected in both synonymous and nonsynonymous mutations that can change the genetic barrier toward resistance.

The fitness of the virus in the presence of treatment in vivo cannot be measured, but a number of in vitro assays have been developed that provide useful information.[108] In a *phenotypic drug resistance assay*, the viral genes that are targeted by the drugs PRO, RT, and gp41 are recombined in an HIV lab strain, and resistance against each of the drugs is quantified as fold-change in IC_{50}, the concentration of drug needed to inhibit 50% of the virus replication. For a resistant virus, a higher concentration of drug will be needed for an equal inhibition in virus replication. In a *genotypic drug resistance assay*, the gene regions coding for PRO, RT, and gp41 are sequenced, and an interpretation of the observed genotypic changes is made.[109] Commercial and academic systems are available for both types of assays.

The phenotypic assays have the advantage of objectively and directly measuring resistance as a continuous number. However, the interpretation of this quantity with respect to clinical response is not straightforward and differs for each drug. Furthermore, differences between *in vivo* and *in vitro* environments may cause biased results, in particular, the composition of deoxynucleoside triphosphate pools that compete with nucleoside analogue RT inhibitors. Finally, some of the observed discordances between measured resistance phenotype and treatment response are attributed to *replication capacity*, which is compromised by some resistance mutations. This has prompted the use of *fitness assays* that measure the ability of the virus to replicate in a drug-free environment.

The genotypic assay has the potential to predict any phenotypic changes that may affect both the short-term fitness in the presence of treatment and the long-term genetic barrier to resistance. However, interpreting the viral genotype is challenging, in part because of the large amount of HIV-1 natural variation and in part because of the lack of a single gold standard for these phenotypes (short-term and long-term treatment response). The interpretation problem is evident from discordances between a multitude of available genotypic resistance interpretation algorithms.[110,111] The design of genotypic resistance interpretation algorithms remains a challenging and an ongoing application of comparative genomics.

Because a genotypic assay is cheaper and faster than a phenotypic assay, and interpretation algorithms have been shown to perform well at predicting short-term treatment response in a clinical setting (e.g., see Van Laethem[109]), this technique is widely used. Phenotypic assays are mostly used for new drugs, for which genotypic information is scarce, and to complement the genotypic assay in the presence of a high number of resistance mutations and few remaining treatment options. As a result of genotyping efforts, a vast amount of sequence data has become available, and comparative genomics techniques are assisting interpretation of these data and improving their clinical use.

12.6.2 Sources of Data

To design genotypic drug resistance interpretation systems, several sources of information are available. Each of these sources may be used individually to infer a resistance interpretation system, but it is hard to combine these data in an objective way.

This may explain why expert systems, which combine all this information in a subjective way, are popular approaches and perform fairly well.

12.6.2.1 Genotype–Phenotype

By determining both the genotype of a viral isolate and measuring phenotypic drug resistance for available drugs, a direct comparison is available of the effect of observed substitutions in the genome and phenotypic changes.

Large data sets of these genotype–phenotype pairs have been collected to develop algorithms that predict phenotype from genotype using various statistical and machine-learning methods (Figure 12.5a). Predictive models are found to be reasonably accurate, resulting in R^2 values over 0.8; linear models perform surprisingly well.[112] With the benefit of reduced cost and faster results compared to determining the phenotype, these *virtual phenotype* prediction systems are a popular class of genotypic drug resistance systems. However, they inherit all other disadvantages from the phenotypic assays.

12.6.2.2 Genotype: Treatment Response

A measure of treatment response, such as the drop in viral load after a number of weeks, is the variable of direct interest to the clinician. Therefore, data relating genotype at baseline with observed treatment response is an appealing source of information. However, clinical data are much harder to obtain, and treatment response is confounded by many other factors that cannot be easily measured (such as metabolism kinetics and, importantly, treatment adherence) or are simply unknown. Moreover, because treatments are composed of several drugs, it is not straightforward to untangle the contribution of activity of each single drug to the observed response. Therefore, attempts to derive a treatment prediction system entirely from this kind of data have had limited success.[113]

12.6.2.3 Genotype: Observed Selection

Since mutations are fixed during treatment in an environment under strong selective pressure, observed substitutions are expected to increase the fitness of the virus

FIGURE 12.5 (Opposite) (A) Decision tree for predicting resistance against zidovudine, learned from matched genotype–phenotype data.[121] Gray leaves (circles) classify as resistant and white leaves as susceptible based on a biological cutoff for the in vitro IC_{50} fold change. (B) Using a phylogenetic tree to differentiate between (a) observed substitutions caused by a single ancient mutation event versus (b) substitutions resulting from convergent evolution. Note that, in both cases, the mutation is prevalent in an equal amount of contemporary sequences. (C) Dendogram, as obtained from average linkage hierarchical linkage clustering, showing clustering of NRTI resistance mutations.[114] (D) Mutagenic tree mixture model for the development of zidovudine resistance. The mixture has three components, of which the first component is a "noise" component. The other two components define an ordered accumulation of mutations: a mutation develops with the given probability in presence of its parent and with zero probability in absence of its parent.[116]

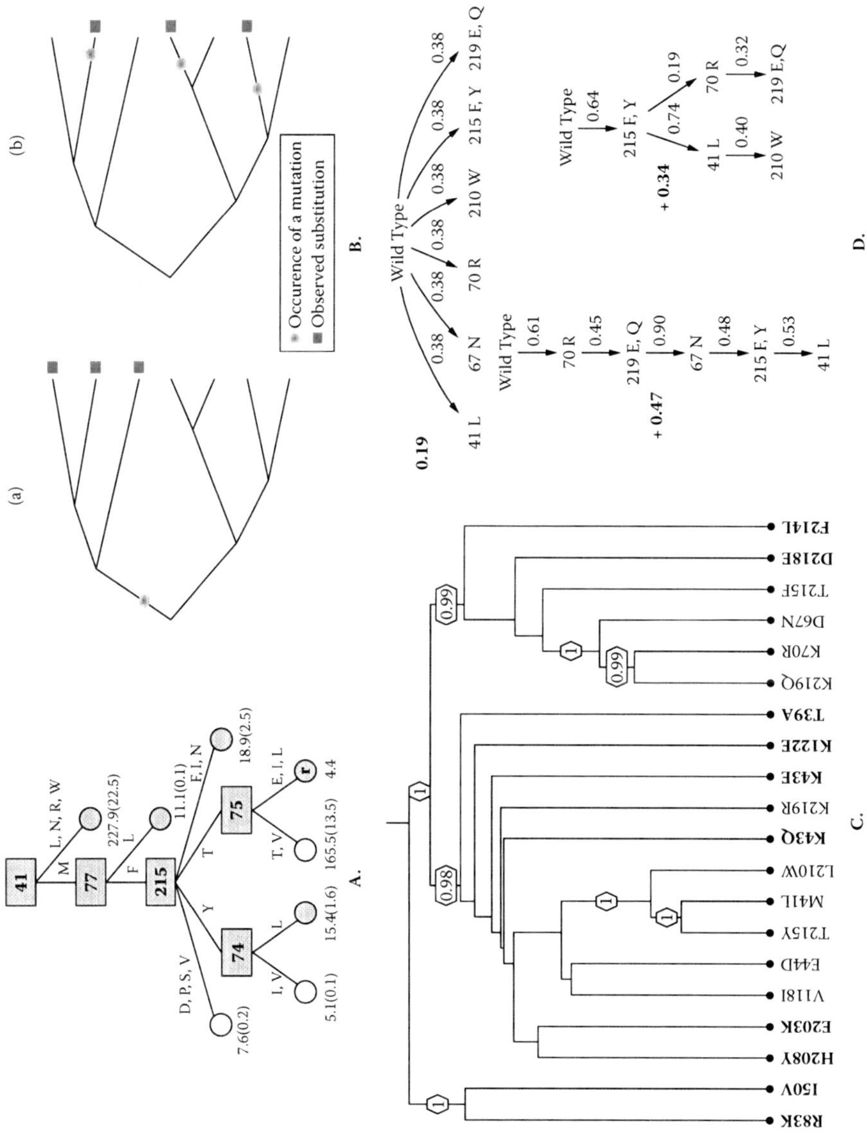
(a)
(b)
Occurence of a mutation
Observed substitution
A.
B.
C.
D.
Wild Type
0.19
+ 0.47
+ 0.34
41 L
67 N
70 R
210 W
215 F, Y
219 E, Q
R83K
I50V
H208Y
E203K
V118I
E44D
T215Y
M41L
L210W
K43Q
K219R
K43E
K122E
T39A
K219Q
K70R
D67N
T215F
D218E
F214L

during treatment. Therefore, a statistical analysis of observed mutations provides at least a qualitative measure of their role in resistance. In addition, a structured accumulation of mutations has been observed in many cases, revealing information on drug resistance pathways. The structure and length of these resistance pathways may be used to derive information on the genetic barrier to resistance. In the remainder of this chapter, we focus on techniques to extract knowledge from these observed genotypic changes.

12.6.3 Learning from Observed Selection

Several techniques have been proposed to extract information from observed substitutions during treatment. *Longitudinal data sets* consist of sequence pairs with a baseline and follow-up sequence during a particular treatment and provide direct information on substitutions observed during that treatment. *Cross-sectional data sets*, on the other hand, use populations of unrelated sequences for which each population has a specific treatment history. They provide information on mutations associated with treatment by observing differences in prevalence of substitutions. The latter data sets are popular because they are generally much larger than longitudinal data sets, which require monitoring a single patient through time.

Genotypic changes that are associated with resistance against a drug may be determined by comparing an "experienced" population of sequences from patients only treated with that drug within that drug class to a "naïve" population of sequences from patients without exposure to a drug from that drug class. A difference in prevalence of a particular amino acid mutation between these two populations may indicate a role in resistance. However, not all differences need to be the consequence of evolution of drug resistance. These populations will undoubtedly share common ancestry because of the epidemiology of HIV infection, implying that differences may also reflect *evolutionary drift* of distinct HIV-1 populations throughout the HIV pandemic, for example, through repetitive bottleneck events, or differences in evolution of immune escape because of different host immune responses.

By stratifying the data sets according to HIV-1 subtype or limiting the study to one epidemiological cluster, the confounding effect of evolutionary drift may be reduced but not completely eliminated. A more appropriate approach uses phylogenetic techniques. By reconstructing the evolutionary history of sequences, one may determine whether the observed difference in prevalence of a mutation is an indication of multiple independent cases of *convergent evolution*, occurring at the tips of the phylogenetic tree, and thus most probably is a consequence of evolution of resistance versus an indication of inherited substitutions occurring at internal branches deeper in the phylogenetic tree (Figure 12.5b).

When more background knowledge on HIV intrahost evolution is incorporated, more detailed information on drug resistance may be learned, with increasing levels of sophistication. In the simplest approach, an individual mutation may be tested for association with treatment by comparing its prevalence in the treated and naïve populations. A higher prevalence of a mutation in the treated population may be of little clinical predictive value if the mutation further increases resistance only in presence of other mutations that already compromise clinical response.

Pronounced associations among treatment-associated mutations are an indication of structured evolution toward drug resistance and ultimately resistance pathways. Pairwise *covariation* of mutations provides a first indication on possible antagonistic or synergistic interactions between mutations.[114] *Clustering* techniques provide a more detailed analysis of covariation and are more informative than pairwise associations (Figure 12.5c).[114] Svicher et al. applied these methods to show that novel treatment-associated mutations were likely to be involved in PRO resistance since they associated and clustered with known PRO resistance mutations.[115] Distinct clusters of resistance mutations may indicate different resistance pathways. However, no statement can be made about the order of mutations because no evolutionary model is assumed.

To assess the accumulation of resistance mutations, mixtures of mutagenic trees have been proposed as an elegant graphical technique with an underlying evolutionary model (Figure 12.5d). Here, a probabilistic model is constructed from the data based on restricted Bayesian networks. The model is a tree structure in which nodes are mutations, and a "child" mutation only develops in the presence of the parent mutation. Beerenwinkel et al. applied the technique to model resistance pathways against most available drugs.[116] A strict ordering of resistance mutations, however, is not always appropriate to describe the stochastic effects that apply to HIV evolution.

A more general use of Bayesian network learning allows simultaneously untangling three different kinds of biological interactions: (1) interactions between treatment and selection of major resistance mutations, (2) interactions between major and minor resistance mutations leading to resistance pathways, and (3) interactions between background polymorphisms and resistance mutations (Figure 12.6). Using Bayesian network learning, Deforche et al. demonstrated that the polymorphism L89M interacted with the major nelfinavir resistance mutation D30N, explaining its subtype-dependent prevalence.[117] Abecasis et al. used Bayesian network learning to hypothesize the role of PRO mutations M89I/V, which are seen frequently in subtype G at treatment failure but not in subtype B.[118]

12.6.4 Combining Information

To predict the response of an HIV patient to antiviral treatment, successful predictive systems based on comparative genomics have been developed. The challenge remains to combine all available information from *in vitro* measurements, treatment response, and observed in vivo evolution. Ultimately, the in vivo fitness during treatment drives evolution of HIV to escape the treatment-selective pressure. Observed evolution in clinical sequences thus provides the potential to estimate this in vivo fitness landscape. For this purpose, a biologically accurate model of HIV evolution in the presence of selection is needed, and this would enable at the same time the use of the estimated fitness landscape to predict evolutionary aspects such as the genetic barrier to resistance through simulations on the estimated fitness landscape.

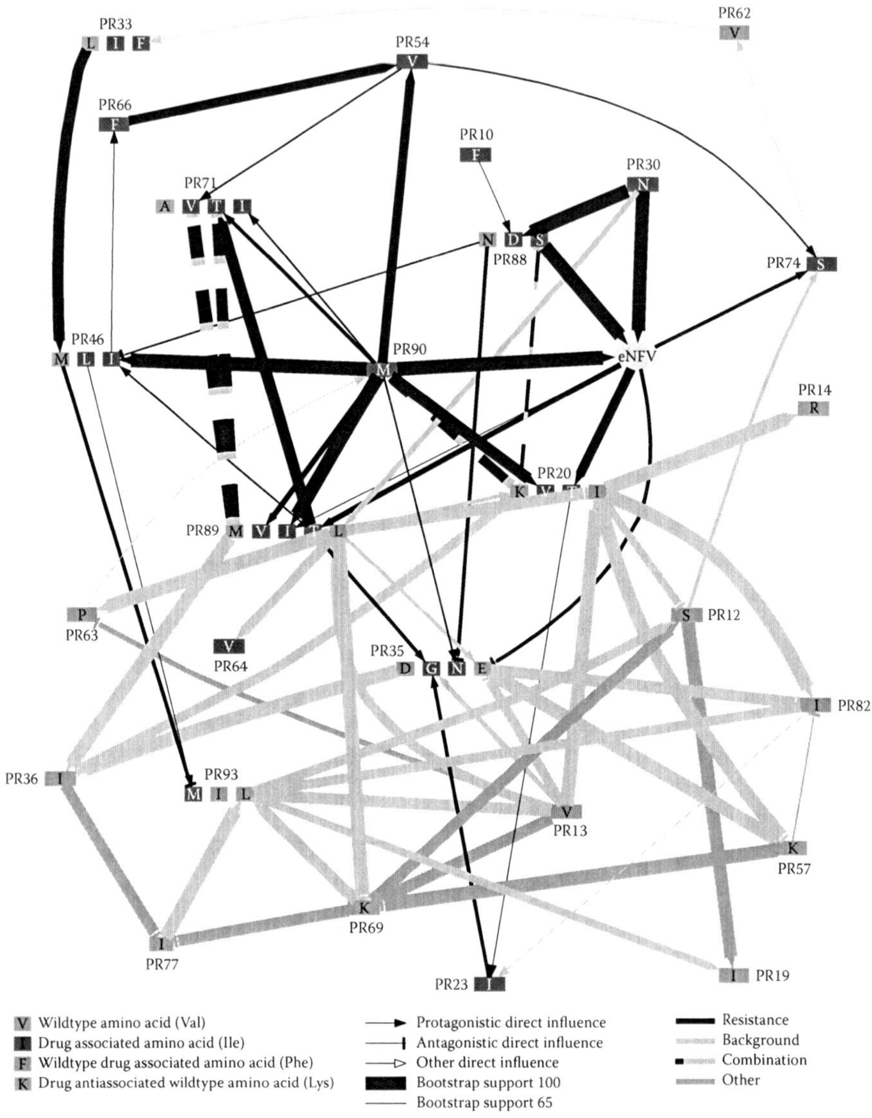

FIGURE 12.6 (See color figure in the insert following page 48.) Bayesian network model for drug resistance against nelfinavir visualizes relationships between exposure to treatment (eNFV), drug resistance mutations (red), and background polymorphisms (green).[117]

12.7 CONCLUSION

In a relatively short time frame, HIV research field has generated an unprecedented amount of genetic information. In combination with molecular biology research, comparative genomics techniques are used extensively in an attempt to understand the epidemic history of HIV, the role of the different HIV genes in viral replication, and the interaction with the host. In this chapter, we focused on only a few aspects of these applications. Increasingly, different types of bioinformatics approaches are combined to analyze HIV genetic information. One of the challenges of this research is to integrate different data-mining and molecular epidemiological techniques to support or design better therapeutic strategies. Armed with these novel approaches, we try to obtain useful insights that can assist in the struggle against HIV infection.

ACKNOWLEDGMENTS

A long-term fellowship from the European Molecular Biology Organization (EMBO) supported the work of P. L., and K. D. was funded by a doctoral grant of the Institute for the Promotion of Innovation through Sciences and Technology in Flanders (IWT).

REFERENCES

1. Centers for Disease Control and Prevention. Pneumocystis pneumonia — Los Angeles. *Morb. Mortal. Wkly.* Rep. 30, 250–252 (1981).
2. Barre-Sinoussi, F. et al. Isolation of a T-lymphotropic retrovirus from a patient at risk for acquired immune deficiency syndrome (AIDS). *Science* 220, 868–871 (1983).
3. Poiesz, B. J. et al. Detection and isolation of type C retrovirus particles from fresh and cultured lymphocytes of a patient with cutaneous T-cell lymphoma. *Proc. Natl. Acad. Sci. U. S. A.* 77, 7415–7419 (1980).
4. Kalyanaraman, V. S. et al. A new subtype of human T-cell leukemia virus (HTLV-II) associated with a T-cell variant of hairy cell leukemia. *Science* 218, 571–573 (1982).
5. Gallo, R. C. & Montagnier, L. The discovery of HIV as the cause of AIDS. *N. Engl. J. Med.* 349, 2283–2285 (2003).
6. Coffin, J. et al. Human immunodeficiency viruses. *Science* 232, 697 (1986).
7. Coffin, J. et al. What to call the AIDS virus? *Nature* 321, 10 (1986).
8. Trkola, A. HIV–host interactions: vital to the virus and key to its inhibition. *Curr. Opin. Microbiol.* 7, 555–559 (2004).
9. Leigh Brown, A. In: *The Evolutionary Biology of Viruses* (Ed. Morse, S. S.) (Raven Press, New York, 1994).
10. Salminen, M. O. et al. Recovery of virtually full-length HIV-1 provirus of diverse subtypes from primary virus cultures using the polymerase chain reaction. *Virology* 213, 80–86 (1995).
11. McCutchan, F. E. Global epidemiology of HIV. *J. Med. Virol.* 78 Suppl 1, S7–S12 (2006).
12. Coffin, J. M. In: *Fields Virology* (Eds. Fields, B. N., Knipe, D. M. & Howley, P. M.) (Lippencott-Raven, Philadelphia, 1996).
13. Luciw, P. A. In: *Fields Virology* (Eds. Fields, B. N., Knipe, D. M. & Howley, P. M.) (Lippencott-Raven, Philadelphia, 1996).
14. Frankel, A. D. & Young, J. A. HIV-1: 15 proteins and an RNA. *Annu. Rev. Biochem.* 67, 1–25 (1998).
15. Turner, B. G. & Summers, M. F. Structural biology of HIV. *J. Mol. Biol.* 285, 1–32 (1999).
16. Cann, A. J. & Chen, I. S. Y. In: *Fields Virology* (Eds. Fields, B. N., Knipe, D. M., & Howley, P. M.) (Lippencott-Raven, Philadelphia, 1990).
17. Coffin, J. M. In: *The Retroviridae* (Ed. Levy, J. A.), pp. 19–49 (Plenum Press, New York, 1992).

18. Dalgleish, A. G. et al. The CD4 (T4) antigen is an essential component of the receptor for the AIDS retrovirus. *Nature* 312, 763–767 (1984).
19. Wain-Hobson, S. The fastest genome evolution ever described: HIV variation in situ. *Curr. Opin. Genet. Dev.* 3, 878–883 (1993).
20. Mansky, L. M. & Temin, H. M. Lower in vivo mutation rate of human immunodeficiency virus type 1 than that predicted from the fidelity of purified reverse transcriptase. *J. Virol.* 69, 5087–5094 (1995).
21. Wei, X. et al. Viral dynamics in human immunodeficiency virus type 1 infection. *Nature* 373, 117–122 (1995).
22. Ho, D. D. et al. Rapid turnover of plasma virions and CD4 lymphocytes in HIV-1 infection. *Nature* 373, 123–126 (1995).
23. Perelson, A. S., Neumann, A. U., Markowitz, M., Leonard, J. M. & Ho, D. D. HIV-1 dynamics in vivo: virion clearance rate, infected cell life-span, and viral generation time. *Science* 271, 1582–1586 (1996).
24. Levy, D. N., Aldrovandi, G. M., Kutsch, O. & Shaw, G. M. Dynamics of HIV-1 recombination in its natural target cells. *Proc. Natl. Acad. Sci. U. S. A.* 101, 4204–4209 (2004).
25. Drummond, A. J., Pybus, O. G., Rambaut, A., Forsberg, R. & Rodrigo, A. G. Measurably evolving populations. *Trends Ecol. Evol.* 18, 481–488 (2003).
26. Drummond, A. J., Nicholls, G. K., Rodrigo, A. G. & Solomon, W. Estimating mutation parameters, population history and genealogy simultaneously from temporally spaced sequence data. *Genetics* 161, 1307–1320 (2002).
27. Drummond, A. J., Pybus, O. G. & Rambaut, A. Inference of viral evolutionary rates from molecular sequences. *Adv. Parasitol.* 54, 331–358 (2003).
28. Rambaut, A. Estimating the rate of molecular evolution: incorporating non-contemporaneous sequences into maximum likelihood phylogenies. *Bioinformatics* 16, 395–399 (2000).
29. Rodrigo, A. G. & Felsenstein, J. In: *The Evolution of HIV* (Ed. Crandall, K. A.) (John Hopkins University Press, Baltimore, MD, 1999).
30. Pauwels, R. Aspects of successful drug discovery and development. *Antiviral Res.* 71, 77–89 (2006).
31. Towers, G. J. et al. Cyclophilin A modulates the sensitivity of HIV-1 to host restriction factors. *Nat. Med.* 9, 1138–1143 (2003).
32. Stremlau, M. et al. The cytoplasmic body component TRIM5alpha restricts HIV-1 infection in Old World monkeys. *Nature* 427, 848–853 (2004).
33. Stremlau, M. et al. Specific recognition and accelerated uncoating of retroviral capsids by the TRIM5alpha restriction factor. *Proc. Natl. Acad. Sci. U. S. A.* 103, 5514–5519 (2006).
34. Liu, H. L. et al. Adaptive evolution of primate TRIM5alpha, a gene restricting HIV-1 infection. *Gene* 362, 109–116 (2005).
35. Sawyer, S. L., Wu, L. I., Emerman, M. & Malik, H. S. Positive selection of primate TRIM5alpha identifies a critical species-specific retroviral restriction domain. *Proc. Natl. Acad. Sci. U. S. A.* 102, 2832–2837 (2005).
36. Song, B. et al. The B30.2(SPRY) domain of the retroviral restriction factor TRIM5alpha exhibits lineage-specific length and sequence variation in primates. *J. Virol.* 79, 6111–6121 (2005).
37. Emerman, M. How TRIM5alpha defends against retroviral invasions. *Proc. Natl. Acad. Sci. U. S. A.* 103, 5249–5250 (2006).
38. Sawyer, S. L., Wu, L. I., Akey, J. M., Emerman, M. & Malik, H. S. High-frequency persistence of an impaired allele of the retroviral defense gene TRIM5alpha in humans. *Curr. Biol.* 16, 95–100 (2006).
39. Mangeat, B. et al. Broad antiretroviral defence by human APOBEC3G through lethal editing of nascent reverse transcripts. *Nature* 424, 99–103 (2003).

40. Harris, R. S. et al. DNA deamination mediates innate immunity to retroviral infection. *Cell* 113, 803–809 (2003).
41. Mehle, A. et al. Vif overcomes the innate antiviral activity of APOBEC3G by promoting its degradation in the ubiquitin-proteasome pathway. *J. Biol. Chem.* 279, 7792–7798 (2004).
42. Sawyer, S. L., Emerman, M. & Malik, H. S. Ancient adaptive evolution of the primate antiviral DNA-editing enzyme APOBEC3G. *PLoS. Biol.* 2, E275 (2004).
43. Zhang, J. & Webb, D. M. Rapid evolution of primate antiviral enzyme APOBEC3G. *Hum. Mol. Genet.* 13, 1785–1791 (2004).
44. Ortiz, M., Bleiber, G., Martinez, R., Kaessmann, H. & Telenti, A. Patterns of evolution of host proteins involved in retroviral pathogenesis. *Retrovirology* 3, 11 (2006).
45. Zheng, Y. H. et al. Human APOBEC3F is another host factor that blocks human immunodeficiency virus type 1 replication. *J. Virol.* 78, 6073–6076 (2004).
46. Yu, Q. et al. APOBEC3B and APOBEC3C are potent inhibitors of simian immunodeficiency virus replication. *J. Biol. Chem.* 279, 53379–53386 (2004).
47. Xu, H. et al. A single amino acid substitution in human APOBEC3G antiretroviral enzyme confers resistance to HIV-1 virion infectivity factor-induced depletion. *Proc. Natl. Acad. Sci. U. S. A.* 101, 5652–5657 (2004).
48. Leitner, T. In: *The Molecular Epidemiology of Human Viruses* (Ed. Leitner, T.) (Kluwer Academic, Boston, 2002).
49. Salemi, M. & Vandamme, A. M. *The Phylogenetic Handbook: A Practical Approach to DNA and Protein Phylogeny* (Cambridge University Press, Cambridge, 2003).
50. Page, R. D. M. & Holmes, E. C. Molecular Evolution. *A Phylogenetic Approach* (Blackwell Science, Oxford, UK, 1998).
51. Swofford, D., Olsen, G. J., Waddell, P. J. & Hillis, D. M. In: *Molecular Systematics* (Eds. Hillis, D. M., Moritz, C., & Mable, B. K.), pp. 407–514 (Sinauer, Sunderland, MA, 1996).
52. Hahn, B. H., Shaw, G. M., De Cock, K. M. & Sharp, P. M. AIDS as a zoonosis: scientific and public health implications. *Science* 287, 607–614 (2000).
53. Corbet, S. et al. *env* sequences of simian immunodeficiency viruses from chimpanzees in Cameroon are strongly related to those of human immunodeficiency virus group N from the same geographic area. *J. Virol.* 74, 529–534 (2000).
54. Gao, F. et al. Origin of HIV-1 in the chimpanzee *Pan troglodytes troglodytes. Nature* 436–441 (1999).
55. Korber, B. et al. Timing the ancestor of the HIV-1 pandemic strains. *Science* 288, 1789–1796 (2000).
56. Salemi, M. et al. Dating the common ancestor of SIVcpz and HIV-1 group M and the origin of HIV-1 subtypes using a new method to uncover clock-like molecular evolution. *FASEB* J. 15, 276–278 (2001).
57. Rambaut, A., Robertson, D. L., Pybus, O. G., Peeters, M. & Holmes, E. C. Human immunodeficiency virus. Phylogeny and the origin of HIV-1. *Nature* 410, 1047–1048 (2001).
58. Rambaut, A., Posada, D., Crandall, K. A. & Holmes, E. C. The causes and consequences of HIV evolution. *Nat. Rev. Genet.* 5, 52–61 (2004).
59. Robbins, K. E. et al. Human immunodeficiency virus type 1 epidemic: date of origin, population history, and characterization of early strains. *J. Virol.* 77, 6359–6366 (2003).
60. Buve, A., Bishikwabo-Nsarhaza, K. & Mutangadura, G. The spread and effect of HIV-1 infection in sub-Saharan Africa. *Lancet* 359, 2011–2017 (2002).
61. Walker, P. R., Worobey, M., Rambaut, A., Holmes, E. C. & Pybus, O. G. Epidemiology: sexual transmission of HIV in Africa. *Nature* 422, 679 (2003).
62. Kingman, J. F. C. The coalescent. *Stochastic Proc. Appl.* 13, 235–248 (1982).
63. Griffiths, R. C. & Tavare, S. Sampling theory for neutral alleles in a varying environment. *Philos. Trans. R. Soc. Lond. B. Biol. Sci.* 344, 403–410 (1994).

64. Yusim, K. et al. Using human immunodeficiency virus type 1 sequences to infer historical features of the acquired immune deficiency syndrome epidemic and human immunodeficiency virus evolution. *Philos. Trans. R. Soc. Lond. B. Biol. Sci.* 356, 855–866 (2001).
65. Lemey, P. et al. Tracing the origin and history of the HIV-2 epidemic. *Proc. Natl. Acad. Sci. U. S. A.* 100, 6588–6592 (2003).
66. Lemey, P., Rambaut, A. & Pybus, O. G. HIV evolutionary dynamics within and among hosts. *AIDS Rev.* 8, 155–170 (2006).
67. Salminen, M. O., Carr, J. K., Burke, D. S. & McCutchan, F. E. Identification of breakpoints in intergenotypic recombinants of HIV type 1 by bootscanning. *AIDS Res. Hum. Retroviruses* 11, 1423–1425 (1995).
68. Robertson, D. L., Hahn, B. H. & Sharp, P. M. Recombination in AIDS viruses. *J. Mol. Evol.* 40, 249–259 (1995).
69. Robertson, D. L., Sharp, P. M., McCutchan, F. E. & Hahn, B. H. Recombination in HIV-1. *Nature* 374, 124–126 (1995).
70. McCutchan, F. E. Understanding the genetic diversity of HIV-1. *AIDS* 14 Suppl 3, S31–S44 (2000).
71. Peeters, M., Toure-Kane, C. & Nkengasong, J. N. Genetic diversity of HIV in Africa: impact on diagnosis, treatment, vaccine development and trials. *AIDS* 17, 2547–2560 (2003).
72. Lole, K. S. et al. Full-length human immunodeficiency virus type 1 genomes from subtype C-infected seroconverters in India, with evidence of intersubtype recombination. *J. Virol.* 73, 152–160 (1999).
73. Martin, D. P., Posada, D., Crandall, K. A. & Williamson, C. A modified bootscan algorithm for automated identification of recombinant sequences and recombination breakpoints. *AIDS Res. Hum. Retroviruses* 21, 98–102 (2005).
74. Kuhner, M. K., Yamato, J. & Felsenstein, J. Maximum likelihood estimation of recombination rates from population data. *Genetics* 156, 1393–1401 (2000).
75. McVean, G., Awadalla, P. & Fearnhead, P. A coalescent-based method for detecting and estimating recombination from gene sequences. *Genetics* 160, 1231–1241 (2002).
76. Shriner, D., Rodrigo, A. G., Nickle, D. C. & Mullins, J. I. Pervasive genomic recombination of HIV-1 in vivo. *Genetics* 167, 1573–1583 (2004).
77. Sharp, P. M. In: *Conference on Retroviruses and Opportunistic Infections* (Eds. Calmy, A., Gayet-Ageron, A., B., H. & Telenti, A.) (Denver, 2006).
78. Hirsch, V. M. What can natural infection of African monkeys with simian immunodeficiency virus tell us about the pathogenesis of AIDS? *AIDS Rev.* 6, 40–53 (2004).
79. O'Donovan, D. et al. Maternal plasma viral RNA levels determine marked differences in mother-to-child transmission rates of HIV-1 and HIV-2 in the Gambia. MRC/Gambia Government/University College London Medical School working group on mother-child transmission of HIV. *AIDS* 14, 441–448 (2000).
80. Marlink, R. et al. Reduced rate of disease development after HIV-2 infection as compared to HIV-1. *Science* 265, 1587–1590 (1994).
81. Schindler, M. et al. Nef-mediated suppression of T cell activation was lost in a lentiviral lineage that gave rise to HIV-1. *Cell* 125, 1055–1067 (2006).
82. Asquith, B., Edwards, C. T., Lipsitch, M. & McLean, A. R. Inefficient cytotoxic T lymphocyte-mediated killing of HIV-1-infected cells in vivo. *PLoS. Biol.* 4, e90 (2006).
83. Finzi, D. et al. Identification of a reservoir for HIV-1 in patients on highly active antiretroviral therapy. *Science* 278, 1295–1300 (1997).
84. Girard, M. P., Osmanov, S. K. & Kieny, M. P. A review of vaccine research and development: the human immunodeficiency virus (HIV). *Vaccine* 24, 4062–4081 (2006).

85. Frost, S. D. et al. Neutralizing antibody responses drive the evolution of human immunodeficiency virus type 1 envelope during recent HIV infection. *Proc. Natl. Acad. Sci. U. S. A.* 102, 18514–18519 (2005).
86. Cohen, J. Public health. AIDS vaccine trial produces disappointment and confusion. *Science* 299, 1290–1291 (2003).
87. Mascola, J. R et al. Immunization with envelope subunit vaccine products elicits neutralizing antibodies against laboratory-adapted but not primary isolates of human immunodeficiency virus type 1. The National Institute of Allergy and Infectious Diseases AIDS Vaccine Evaluation Group. *J. Infect. Dis.* 173, 340–348 (1996).
88. Koup, R. A. et al. Temporal association of cellular immune responses with the initial control of viremia in primary human immunodeficiency virus type 1 syndrome. *J. Virol.* 68, 4650–4655 (1994).
89. Carrington, M. et al. HLA and HIV-1: heterozygote advantage and B*35-Cw*04 disadvantage. *Science* 283, 1748–1752 (1999).
90. Trachtenberg, E. et al. Advantage of rare HLA supertype in HIV disease progression. *Nat. Med.* 9, 928–935 (2003).
91. Markel, H. The search for effective HIV vaccines. *N. Engl. J. Med.* 353, 753–757 (2005).
92. Doria-Rose, N. A. et al. Human immunodeficiency virus type 1 subtype B ancestral envelope protein is functional and elicits neutralizing antibodies in rabbits similar to those elicited by a circulating subtype B envelope. *J. Virol.* 79, 11214–11224 (2005).
93. Gaschen, B. et al. Diversity considerations in HIV-1 vaccine selection. *Science* 296, 2354–2360 (2002).
94. Kothe, D. L. et al. Ancestral and consensus envelope immunogens for HIV-1 subtype C. *Virol.* 352, 438–449 (2006).
95. Andre, S. et al. Increased immune response elicited by DNA vaccination with a synthetic gp120 sequence with optimized codon usage. *J. Virol.* 72, 1497–1503 (1998).
96. Gao, F. et al. Codon usage optimization of HIV type 1 subtype C *gag, pol, env,* and *nef* genes: in vitro expression and immune responses in DNA-vaccinated mice. AIDS Res. Hum. *Retroviruses* 19, 817–823 (2003).
97. Gao, F. et al. Antigenicity and immunogenicity of a synthetic human immunodeficiency virus type 1 group m consensus envelope glycoprotein. *J. Virol.* 79, 1154–1163 (2005).
98. Grenfell, B. T. et al. Unifying the epidemiological and evolutionary dynamics of pathogens. *Science* 303, 327–332 (2004).
99. Wilson, D. J. & McVean, G. Estimating diversifying selection and functional constraint in the presence of recombination. *Genetics* 172, 1411–1425 (2006).
100. Derdeyn, C. A. et al. Envelope-constrained neutralization-sensitive HIV-1 after heterosexual transmission. *Science* 303, 2019–2022 (2004).
101. Lemey, P. et al. Molecular footprint of drug-selective pressure in a human immunodeficiency virus transmission chain. *J. Virol.* 79, 11981–11989 (2005).
102. Leitner, T., Escanilla, D., Franzen, C., Uhlen, M. & Albert, J. Accurate reconstruction of a known HIV-1 transmission history by phylogenetic tree analysis. *Proc. Natl. Acad. Sci.* U. S. A. 93, 10864–10869 (1996).
103. Leitner, T. & Fitch, W. In: *The Evolution of HIV* (Ed. Crandall, K. A.), pp. 315–345 (Johns Hopkins University Press, Baltimore, MD, 1999).
104. Edwards, C. T. et al. Population genetic estimation of the loss of genetic diversity during horizontal transmission of HIV-1. *BMC Evol. Biol.* 6, 28 (2006).
105. Vandamme, A. M., Van Laethem, K. & De Clercq, E. Managing resistance to anti-HIV drugs: an important consideration for effective disease management. *Drugs* 57, 337–361 (1999).
106. Vandamme, A. M. et al. Updated European recommendations for the clinical use of HIV drug resistance testing. *Antivir. Ther.* 9, 829–848 (2004).

107. Wensing, A. M. et al. Prevalence of drug-resistant HIV-1 variants in untreated individuals in Europe: implications for clinical management. *J. Infect. Dis.* 192, 958–966 (2005).
108. Vandamme, A. M. et al. In: *Antiviral Methods and Protocols* (Eds. Kinchington, D. & Schinazi, R. F.) (Humana Press, Totowa, NJ, 1999).
109. Van Laethem, K. et al. A genotypic drug resistance interpretation algorithm that significantly predicts therapy response in HIV-1-infected patients. *Antivir. Ther.* 7, 123–129 (2002).
110. Ravela, J. et al. HIV-1 protease and reverse transcriptase mutation patterns responsible for discordances between genotypic drug resistance interpretation algorithms. *J. Acquir. Immune Defic. Syndr.* 33, 8–14 (2003).
111. Snoeck, J. et al. Discordances between interpretation algorithms for genotypic resistance to protease and reverse transcriptase inhibitors of human immunodeficiency virus are subtype dependent. *Antimicrob. Agents Chemother.* 50, 694–701 (2006).
112. Wang, K., Jenwitheesuk, E., Samudrala, R. & Mittler, J. E. Simple linear model provides highly accurate genotypic predictions of HIV-1 drug resistance. *Antivir. Ther.* 9, 343–352 (2004).
113. DiRienzo, G. & DeGruttola, V. Collaborative HIV resistance-response database initiatives: sample size for detection of relationships between HIV-1 genotype and HIV-1 RNA response using a non-parametric approach. *Antivir. Ther.* 7, S71 (2002).
114. Sing, T. et al. In: *Knowledge Discovery in Databases: PKDD 2005* (Eds. Jorge, A., Togo, L., Brazdil, P., Camacho, R. & Gama, J.) (Springer, New York, 2005).
115. Svicher, V. et al. Novel human immunodeficiency virus type 1 protease mutations potentially involved in resistance to protease inhibitors. *Antimicrob. Agents Chemother.* 49, 2015–2025 (2005).
116. Beerenwinkel, N. et al. In: *RECOMB* 36–44 (ACM Press, San Diego, CA, 2004).
117. Deforche, K. et al. Analysis of HIV-1 pol sequences using Bayesian networks: implications for drug resistance. *Bioinformatics* 22, 2975–2979 (2006).
118. Abecasis, A. B. et al. Protease mutation M89I/V is linked to therapy failure in patients infected with the HIV-1 non-B subtypes C, F or G. *AIDS* 19, 1799–1806 (2005).
119. Voght, P. K. In: *Retroviruses* (Eds. Coffin, J. M., Hughes, S. H. & Varmus, H. E.) (Cold Spring Harbor Laboratory Press, New York, 1997).
120. Huelsenbeck, J. P. & Ronquist, F. MRBAYES: Bayesian inference of phylogenetic trees. *Bioinformatics* 17, 754–755 (2001).
121. Beerenwinkel, N. et al. Geno2pheno: estimating phenotypic drug resistance from HIV-1 genotypes. *Nucleic Acids Res.* 31, 3850–3855 (2003).

13 Detailed Comparisons of Cancer Genomes

Timon P. H. Buys, Ian M. Wilson, Bradley P. Coe, Eric H. L. Lee, Jennifer Y. Kennett, William W. Lockwood, Ivy F. L. Tsui, Ashleen Shadeo, Raj Chari, Cathie Garnis, and Wan L. Lam

CONTENTS

ABSTRACT

While heritable DNA polymorphisms and genetic mutations may be associated with cancer predisposition, the accumulation of somatic DNA alterations is thought to drive the clonal evolution of cancer cells.[1,2] The identification of such genetic events will provide molecular targets for developing biomarkers and novel therapies. Detailed comparisons of cancer genomes will facilitate gene discovery. This chapter describes the role of tumor DNA profiling in cancer research.

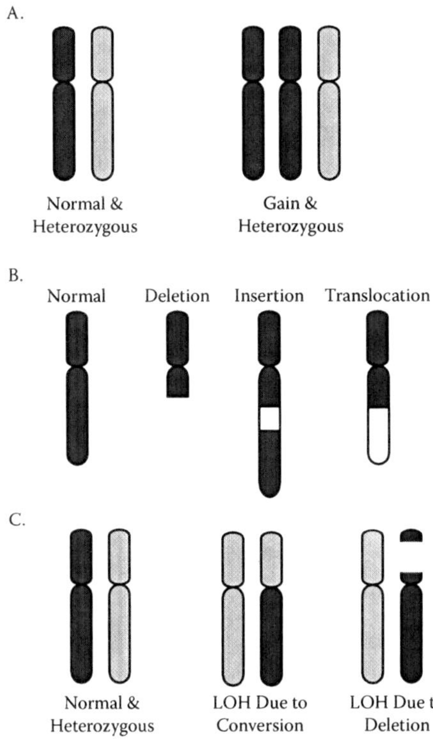

FIGURE 13.1 Genomic aberrations. (A) Alterations affecting normal allelic balance and DNA dosage. One of the common genomic alterations that may occur is the generation of an aneuploid or polyploid genome through gain or loss of chromosomes. This can be detected by copy number–sensitive technologies such as CGH, quantitative PCR, and cytogenetic evaluation. (B) Segmental copy number alterations. DNA copy number alterations and structural rearrangements are commonly observed in cancer genomes and affect only part of a chromosome. These may include the loss of DNA material, duplication of chromosomal segments, or translocation of chromosomal ends by recombination. (C) Allelic imbalance. LOH can arise from a deletion event or gene conversion during mitosis.

13.1 TECHNOLOGIES FOR CANCER GENOME COMPARISON

The disruption of tumor suppressors and oncogenes is caused by a variety of genetic mechanisms, including loss of heterozygosity (LOH), DNA copy number change, sequence mutation, and chromosomal rearrangement (see Figure 13.1). Genome-wide comparison of tumor samples typically involves the detection of regions of loss of heterozygosity or allelic imbalance, the molecular cytogenetic evaluation of chromosomal aberrations and rearrangements, or the identification of segmental DNA copy number changes to identify key genetic features contributing to disease phenotypes[3] (Figure 13.2).

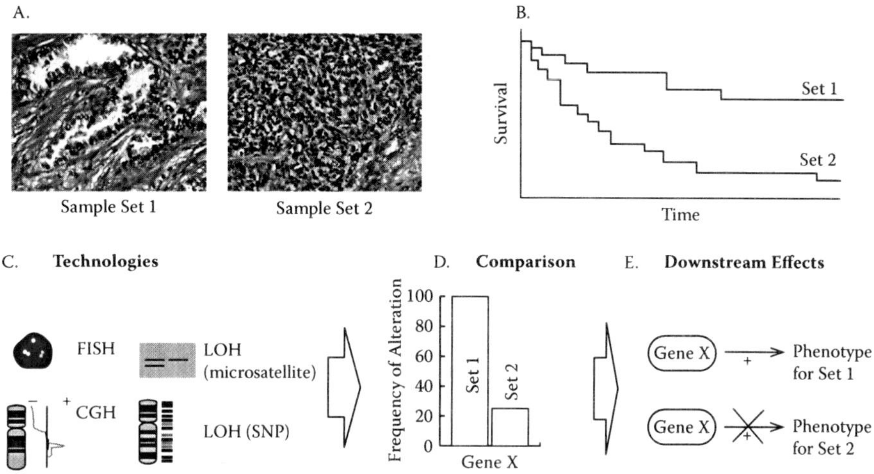

FIGURE 13.2 Comparative genomics strategy and utility. When analyzing a tumor data set, the complexity of the genomic alterations present can impose difficulty in determining which alterations are truly related to tumor biology and which are by-products of genomic instability. By comparing sets of samples with distinct histology (A) or prognosis (B), alterations associated with disease phenotypes can be identified. In this case, a genomic technology (C) is selected to screen both sample sets. Results show genomic loci that behave differently between the two sample sets, and these values from each sample set can be compared (D). Further analysis may yield mechanistic insight into how the genetic alteration may lead to phenotypic changes (E).

13.1.1 Loss of Heterozygosity

Microsatellite analysis employs simple sequence repeat (SSR) polymorphisms as markers for detecting LOH. In an individual with heterozygous alleles, Analysis based on polymerase chain reaction (PCR) with primers flanking a specific SSR should yield two signals, one for each allele. When the signal intensity ratio of the alleles for the tumor differs from that seen for a normal specimen from the same individual, LOH is inferred. Microarray-based surveys of single-nucleotide polymorphisms (SNPs) offer the advantage of simultaneous high-resolution analysis of both genotype and relative gene copy number for each sample profiled.[4,5]

13.1.2 Cytogenetics

Molecular cytogenetic techniques such as G-banding and spectral karyotyping (SKY) enable global surveys for large-scale chromosomal rearrangements and DNA ploidy status. In G-banding, metaphase chromosome spreads are stained to detect rearrangements and gain or loss of chromosome bands. In SKY, a mixture of differentially labeled chromosome-specific probes are used to generate a virtual karyogram, with each chromosome displayed in a different color to facilitate the detection of chromosomal rearrangements.[6] These techniques are often used in clinical settings for the analysis of cancer cells, especially in hematological cancers. The Mitelman

Database of Chromosome Aberrations in Cancer is one of the most comprehensive databases of cytogenetic information.[7]

Fluorescence in situ hybridization (FISH) uses locus-specific DNA probes to evaluate genetic alterations on a cell-by-cell basis as tissue heterogeneity in a tumor may mask detection of features unique to a subpopulation of cells. Gain and loss of hybridization signals reflect DNA duplication and deletion, while split signals indicate a translocation event. Multicolor FISH (M-FISH) using probes that fluoresce at different wavelengths enables the examination of several loci in the same experiment.[8]

13.1.3 Comparative Genomic Hybridization

Comparative genomic hybridization (CGH) detects segmental gains and losses. Tumor and reference DNA are differentially labeled and competitively hybridized to metaphase chromosomes, and the copy number profile is deduced from the signal ratio between the two images.[9] The adaptation of CGH to display discrete DNA targets (with known position on the human genome map) in a matrix or array format has greatly enhanced the resolution of this technology.[5,10]

For genome-wide analysis, the pioneering studies were performed on complementary DNA microarrays.[11] The need for representation of unannotated genes, regulatory regions, and intergenic sequences was achieved by the development of array platforms comprised of large insert clones (e.g., bacterial artificial chromosomes [BACs]).[12,13] These arrays have improved detection sensitivity and reduced DNA input requirements, also offering an effective means of profiling formalin-fixed paraffin embedded (FFPE) tissues from hospital archives.[5] DNA copy number analysis with oligonucleotide-based platforms, such as those used for SNP analysis and representative oligonucleotide microarray analysis (ROMA), offers marked improvements in the number of loci interrogated in a single experiment relative to earlier platforms.[14,15] Moreover, SNP arrays allow determination of LOH and DNA copy number status on the same platform,[4,16] although some SNP loci will be uninformative for allelic status due to homozygosity. In comparison to large-insert clone arrays, current oligonucleotide platforms show limited success in profiling archival FFPE specimens. The reliance of some of the platforms on genomic reduction and whole-genome amplification steps, which contribute to experimental variability, amplification bias, and loss of details,[16,17] is a key consideration in selecting a suitable platform for specific application.

Comprehensive analysis of tumor genomes has been greatly improved by the increasing resolution of array CGH platforms, including development of arrays comprised of targets that span the entire human genome with overlapping or tiling DNA segments.[18,19] Such coverage facilitates an unbiased analysis of the whole genome without the need for inferring copy number status between genetic markers.

Ultimately, the type of study undertaken will dictate platform selection. Minimum DNA quality and quantity requirements vary for different array CGH platforms, as does the ability to detect genetic alterations in heterogeneous tumor samples.[17,20] In addition, the uniformity of array element distribution throughout the genome also inevitably influences the probability of detecting genetic alterations.[17]

13.1.4 DNA Sequencing-Based Technologies

The utility of emerging DNA sequencing-based technologies has been demonstrated in copy number profiling. In digital karyotyping, relative DNA copy number is deduced by enumerating sequence tags representing loci throughout the genome.[21] This method is comparable to the serial analysis of gene expression (SAGE) technique,[22] except that genomic DNA is used to generate concatenated DNA tags for sequence analysis. To date, this technique has been used to uncover multiple activating alterations in ovarian cancer[23,24] and has been adapted to assess epigenetic alterations in tumors.[25] In end sequence profiling (ESP), the tumor genome is represented by fosmid or BAC clones, and the sampling of clones by end sequencing identifies copy number changes and chromosomal rearrangements (e.g., inversions or translocations) throughout the genome.[26–28]

Sequencing-based screens are also available for mutational analysis of tumor genomes. Mutational status for more than 13,000 protein-encoding genes was ascertained in individual colorectal and breast tumors by a Sanger-sequencing-based approach.[29] Recurring mutations (nonsense mutations, missense mutations, etc.) were identified at hundreds of novel candidate loci, underscoring the complexity of tumorigenic processes. Emerging high-throughput sequencing technologies promising reduced costs and increased speed (e.g., pyrosequencing, multiplex polony sequencing) will facilitate detailed analysis of tumor genomes on a large scale.[30–32]

13.2 COMPARISON OF TUMOR TYPES

13.2.1 Disease-Specific Genetic Alterations

Comparative analysis of tumor genomes can be used to classify malignancies (e.g., different types of cancer that arise in the same organ) (Figure 13.2). Cancer can be broadly divided into solid (epithelial and connective tissue) and hematologic (blood and lymph system) malignancies.[33] Hematological cancers often exhibit signature genetic events that drive disease. The t(9;22) Philadelphia chromosome translocation in chronic myeloid leukemia creates a *BCR-ABL* fusion gene.[34,35] The t(11;14) translocation in mantle cell lymphoma fuses *IgG Heavy Chain Locus* with *Cyclin D1*. The t(14;18) translocation, which is frequently observed in follicular lymphoma, results in *immunoglobulin H (IgH)–Bcl2* fusion.[36] Signature genetic alterations not only facilitate clinical diagnosis but also provide the opportunity for developing targeted therapy in hematological cancers.[37]

In solid tumors, there is a high degree of variability in the number and location of alterations, making it difficult to distinguish between causal genetic events and consequences of genomic instability.[38,39] Comparison of multiple tumors of the same tissue origin is a means to identify disease-specific genetic alteration, while cross-tissue comparison may reveal genetic mechanisms common in cancer.

In addition to differentiating between broad tumor classes, genomic profiling can also be used to define organ-specific tumor subtypes. One example is the identification of distinguishing genetic features of disease subtypes within lung cancer. Small cell lung cancer (SCLC) demonstrates a more aggressive phenotype than non–small cell lung cancer (NSCLC), yet the two subtypes share many common genomic alterations. Analysis of the differences between these groups identified distinct causal mechanisms for each subtype.[40] Specifically, NSCLC cell lines demonstrate many alterations to upstream components of the cell cycle pathways (e.g., the *EGFR* pathway), while SCLC amplifies and overexpresses downstream components such as the *E2F2* transcription factor (which activates transcription of various proproliferative elements). This comparison also identified the presence of an amplicon in SCLC lines that contained multidrug resistance genes that were also overexpressed, potentially accounting for the chemotherapy-resistant phenotype of SCLC. This study illustrates the utility of comparative genomics in identifying alterations responsible for tumor-specific phenotypes.

13.2.2 Genomic Changes during Cancer Progression

The association between genetic instability (accumulating DNA alterations) and the histopathological progression model in cancer was first observed in colorectal cancer.[41] This concept has since been demonstrated in many other cancer types.[3] Premalignant lesions harbor key initiating genetic alterations that may be masked by the widespread genomic instability of later-stage disease; therefore, their analysis is essential to understanding the initiating events in tumorigenesis (see Figure 13.3). Interrogation of the genomes of minute premalignant lesions has been made possible by the development of high-density genomic microarray platforms with very low input DNA requirements. For example, examining preinvasive and invasive lung cancer using an array displaying DNA elements in a tiling path manner showed that genomic instability escalates with progression, masking early causal genomic events.[42] Similarly, a study in bladder cancer showed that the fraction of the tumor genome that was altered appeared to be significantly increased with tumor stage.[43]

Defining the genomic alterations responsible for disease progression may also overcome ambiguity in determining which morphologically similar premalignant lesions carry a significant risk of progression. As an example, based on specific genomic alterations, histologically indistinguishable oral precancerous lesions can be categorized into those that progress to invasive cancer and those that do not.[44] Rapid LOH surveys have yielded similar findings in other cancers.[45]

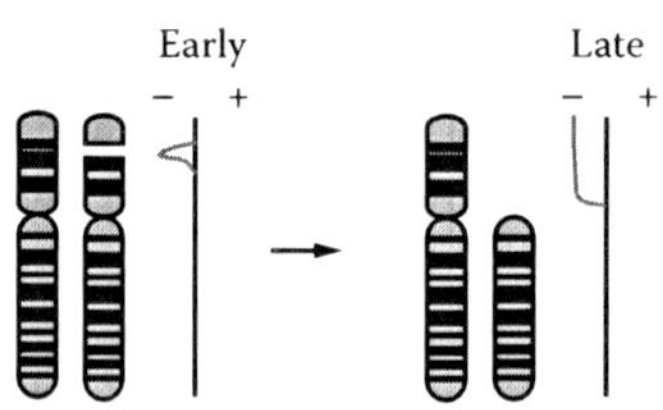

FIGURE 13.3 Masking of early genetic events. During the progression of neoplasias from early-stage disease to invasive cancer, the number and complexity of DNA copy number alterations often increase. The accumulated alterations of later disease stages may mask earlier causal alterations. For example, a focal deletion is masked by a later loss of an entire chromosome arm. Analysis of early-stage lesions represents the best means of identifying initiating genetic events in tumorigenesis.

13.3 DETERMINING CLONAL RELATIONSHIPS

13.3.1 CLONAL EVOLUTION VERSUS MULTIPLE PRIMARY TUMORS

Patients can present with multiple tumors (*synchronous* or *metachronous*). It is important to distinguish cases of multiple primary cancers from cases for which there is a shared progenitor (e.g., metastasis). The frequency of multiple primary tumors varies among cancer types: approximately 1% incidence for synchronous lung tumors, 3%–5% for breast tumors, greater than 30% in prostate cancer, and about 20% in hepatocellular cancer.[46–49] Establishing the relationship between such tumors is essential for understanding underlying tumor biology and will have an impact on disease staging and patient management.

In general, clinical diagnosis of multiple primary tumors relies on differences in location, histology, and staging. Unfortunately, these criteria may not reflect the genetic reality underlying disease behavior. Not only may histologically similar synchronous tumors exhibit genetic evidence of diverse clonal origin,[50] histologically distinct tumors may show common genetic alterations indicative of shared ancestry.[51]

Analysis of singular genetic features, such as the mutational status of the tumor suppressor gene *p53* or the loss of a chromosome arm, is often used to determine clonality.[52] Recent application of multiloci assays to this problem has offered a more detailed description of the similarities and differences among synchronous tumors.[51,53,54] For example, a case report used the detection of shared genetic alteration features identified by array CGH (e.g., amplification of 17ptel-17p13.1) to establish that leiomyosarcomas within the same patient were in fact metastatic recurrences.[54] Differences between genomic profiles for invasive ductal carcinomas for this same patient were used to infer that these tumors were in fact multiple primary lesions. Future application of high-resolution technologies (e.g., whole-genome tiling path array CGH) that allow the precise alignment of the boundaries of genetic alterations will improve the ability to determine clonal relationships. Such technology will improve studies determining the root causes of multifocal disease (e.g., examination of the field effect[55,56]).

13.3.2 METASTASIS

Metastasis occurs when a cell or cells from a primary tumor break away and settle in a new location in the body. Although metastases are understood to follow the emergence of invasive disease, there are reports that suggest a nonsequential progression model in which prometastatic genetic alterations occur prior to invasion.[57,58]

Preliminary efforts to predict the metastatic potential of tumors focused on the morphology of the primary tumors and on biological markers such as hormone levels. More rigorous and informative techniques have evolved with the advent of genomic analysis and gene expression testing. Work employing genomic screening techniques has uncovered chromosomal regions of alteration associated with the likelihood of metastasis. For example, in an array CGH study of squamous cell carcinomas of the esophagus, gain of 8q23-qter and loss of 11q22-qter were shown to predict lymph node metastasis, while other common alterations such as gain of 3q were less predictive.[59]

The application of gene expression microarrays and SAGE technology to investigate metastasis-associated changes have identified tumor suppressors, protease inhibitors, cell adhesion molecules, angiogenesis-related genes, and oncogenes with roles in metastasis.[60,61] In particular, the loss of E-cadherin is a hallmark that is strongly associated with invasive/metastatic phenotypes in many cancer types, including bladder, breast, pancreatic, and gastric cancers. Ultimately, the ability to determine the likelihood of metastasis and the clonality of multifocal disease will help predict whether a given treatment regime will effectively target both primary tumor and metastases.

13.4 PREDICTING DISEASE OUTCOME AND PATIENT SURVIVAL

Whole-genome surveys will play a growing role in prognosis and personalized medicine, with patient management based on genomic and gene expression profiles. Studies have examined the role of genomic alterations in response to specific treatments and in determining relative survival time and likelihood of recurrence. Genomic features that can predict disease outcome or drug response will have immediate clinical utility.

13.4.1 PREDICTING OUTCOME

Comparative analysis of tumor genome profiles can identify genetic signatures useful in delineating prognostic groupings (see Figure 13.2). Correlating genomic profiles with clinical features such as progression and metastasis will yield predictive markers for developing risk models, even if the role of the genetic alteration in disease mechanisms is not fully understood. Genetic features are used in the same way that histology and staging information have been used in predicting outcome. Previously, gene expression studies were used to identify signatures predictive of outcome.[62–65] The approach of using high-resolution genomic analyses to identify DNA alterations as prognostic markers has been applied to a variety of tumor types (e.g., chondrosarcoma, diffuse large B-cell lymphoma, mantle cell lymphoma, and bladder, gastric, breast, and liver cancers).[43,66–71] Specific breast cancer biomarkers (e.g., concurrent amplification of *TOP2A*, *ERBB2*, and *EMS1*) were validated in a sample set comprised of hundreds of tumors, demonstrating the immediate clinical utility for findings from such surveys.[67] Qualitative genomic differences identified by large-scale screens have also been correlated to outcome successfully. For example, genomic instability — defined by "fraction of genome altered," determined by array CGH — was found to correlate strongly with outcome in bladder cancer.[43] As high-resolution platforms become more robust and affordable, such whole-genome analyses — which do not require a priori knowledge of important regions altered in a given type of cancer — will become widely used.

13.4.2 DRUG RESPONSE

Genomic alterations can drive resistance to chemotherapy. Resistance mechanisms may either act directly against a drug (e.g., limiting intracellular drug accumulation, increasing drug detoxification, or failing to convert drug precursors into active form) or act by compensating for drug-induced effects (e.g., altering amounts or activities of drug targets, activating analogous pathways not targeted by drugs, or increasing DNA repair and antiapoptotic signaling).[72] These resistance mechanisms can be generated by alteration in gene dosage (DNA copy number) and gene sequence. For example, increased gene copy number leads to P-glycoprotein overexpression, and the resulting increase in drug efflux causes a multidrug resistance phenotype.[73,74] Genome-wide surveys have identified additional genes involved in resistance. LOH analysis identified *PTEN* loss in chemotherapy resistance in gastric cancer, while CGH analysis implicated *PDZK1* gain in the resistance to different drugs in multiple myeloma cells.[75,76] Gene discoveries are anticipated as the application of high-resolution microarray platforms has begun to yield insights into drug response.[77–84]

13.5 CANCER SUSCEPTIBILITY AND DRUG SENSITIVITY

Recent work such as the HapMap project promises to uncover heritable genome features that are predictive of susceptibility to cancer and drug response for cancer patients.[85] Numerous heritable cancer susceptibility loci have already been identified, with key examples including *BRCA1*, *BRCA2*, *VHL*, and *APC*.[86] Widespread application of high-throughput platforms will facilitate the discovery of mutations and polymorphisms that predispose for cancer.

In terms of drug response, profiling of constitutional DNA will identify polymorphisms influencing responsiveness to drug therapy. Ultimately, this knowledge will lead to the tailoring of treatment to individual patients. One example of this is the identification of *UGT1A1* polymorphisms that have an impact on the efficacy of the chemotherapeutic agent irinotecan. This drug is applied to many common types of cancer, and the *UGT1A1* genotype is used to guide drug dosing.[87] Another example is the family of cytochrome P450 enzymes, which are key components in drug metabolism. Numerous SNPs have been identified that can have an impact on drug response, and these are in use to guide treatment choices.[88] These examples illustrate the impact of comparative genomics in developing personalized medicine.

13.6 INTEGRATION OF MULTIDIMENSIONAL GENOMIC DATA

Dysregulation in cancer cells occurs at many levels, meaning that genomic analysis using multiple complementary platforms will provide a more comprehensive description of the tumor genome. For example, an integrative study identifying alterations in DNA and messenger RNA expression patterns uncovered causal genetic events and their downstream effects.[89] Similarly, matching DNA copy number status with DNA methylation profiles may identify genes disrupted in both alleles and predict silencing of gene expression. The need for multidimensional profiling of tumors has prompted the development of integrative software catering to the display and analysis of complementary data sets. Programs such as Magellan, ACE-it, and VAMP are able to integrate DNA alteration and gene expression data,[90–92] while recently developed SIGMA (System for Integrative Genomic Microarray Analysis) is a user interface for direct mining of multidimensional data.[93] The ability to merge data from various genomic profiling platforms will facilitate cancer gene discovery and

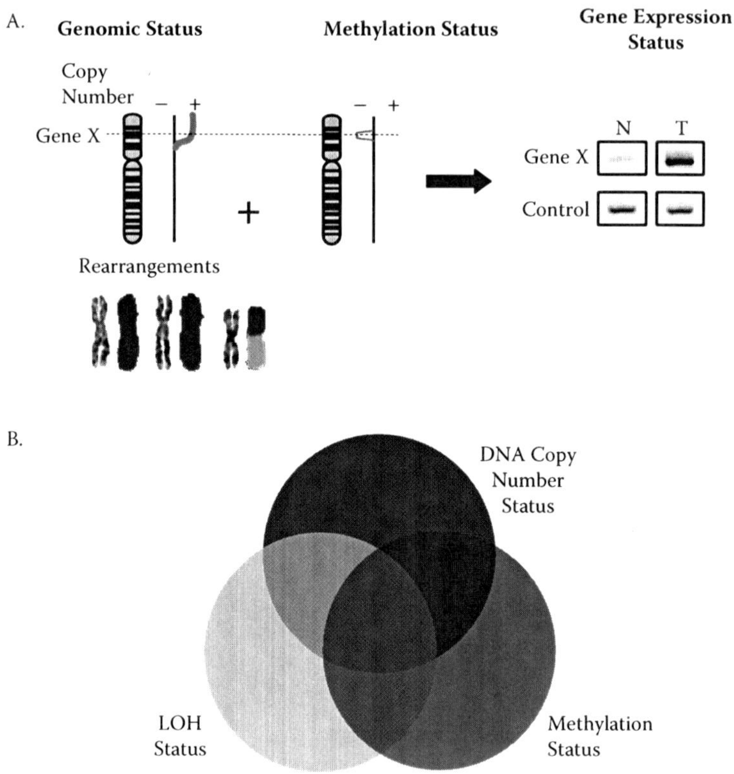

FIGURE 13.4 Integrative analysis of tumor activation. (A) Modes of activation for a specific gene. For example, activation of gene X is synergistic, driven by hypomethylation and amplification that resulted from a duplication event. Understanding the exact mechanisms governing activation of specific genes yields greater insight into the processes of cancer initiation and progression. (B) Integrating data from various global surveys for alteration in tumors identifies key oncogenes and tumor suppressors. Those loci with multiple types of alteration "hits" (i.e., loci falling within the overlapping regions of the Venn diagram) are more likely to represent causal events.

contribute to the understanding of the underlying causes for the diversity of existing cancer phenotypes (Figure 13.4).

13.7 SUMMARY

The emergence of high-resolution whole-genome profiling techniques is enabling the discovery of key genetic alterations that would have escaped detection by conventional molecular cytogenetic methods. Integration of multidimensional genomic profiles will provide comprehensive characterization of the molecular basis of disease phenotypes. This chapter conveys the need for detailed analysis of cancer genomes and emphasizes the advantages of using integrative approaches to describe tumor behavior. Recent advances in cancer genome profiling have fueled much optimism for establishing a mechanistic basis for cancer subclassification, identifying molecular targets for rational therapy design, and moving cancer management toward personalized medicine.

REFERENCES

1. Hanahan, D. & Weinberg, R. A. The hallmarks of cancer. *Cell* **100**, 57–70 (2000).
2. Hahn, W. C. & Weinberg, R. A. Rules for making human tumor cells. *N Engl J Med* **347**, 1593–1603 (2002).
3. Garnis, C., Buys, T. P. & Lam, W. L. Genetic alteration and gene expression modulation during cancer progression. *Mol Cancer* **3**, 9 (2004).
4. Zhao, X. et al. An integrated view of copy number and allelic alterations in the cancer genome using single nucleotide polymorphism arrays. *Cancer Res* **64**, 3060–3071 (2004).
5. Lockwood, W. W., Chari, R., Chi, B. & Lam, W. L. Recent advances in array comparative genomic hybridization technologies and their applications in human genetics. *Eur J Hum Genet* **14**, 139–148 (2006).
6. Bayani, J. M. & Squire, J. A. Applications of SKY in cancer cytogenetics. *Cancer Invest* **20**, 373–386 (2002).
7. Mitelman, F., Johansson, B. & Mertens, F. (Eds.). Mitelman Database of Chromosome Aberrations in Cancer. 2006. Available at: http://cgap.nci.nih.gov/Chromosomes/Mitelman.
8. Gray, J. W. et al. Applications of fluorescence in situ hybridization in biological dosimetry and detection of disease-specific chromosome aberrations. *Prog Clin Biol Res* **372**, 399–411 (1991).
9. Kallioniemi, A. et al. Comparative genomic hybridization for molecular cytogenetic analysis of solid tumors. *Science* **258**, 818–821 (1992).
10. Solinas-Toldo, S. et al. Matrix-based comparative genomic hybridization: biochips to screen for genomic imbalances. *Genes Chromosomes Cancer* **20**, 399–407 (1997).
11. Pollack, J. R. et al. Genome-wide analysis of DNA copy-number changes using cDNA microarrays. *Nat Genet* **23**, 41–46 (1999).
12. Snijders, A. M. et al. Assembly of microarrays for genome-wide measurement of DNA copy number. *Nat Genet* **29**, 263–264 (2001).
13. Greshock, J., Naylor, T. L. & Margolin, A. 1-Mb resolution array-based comparative genomic hybridization using a BAC clone set optimized for cancer gene analysis. *Genome Res* **14**, 179–187 (2004).
14. Lucito, R. et al. Representational oligonucleotide microarray analysis: a high-resolution method to detect genome copy number variation. *Genome Res* **13**, 2291–2305 (2003).
15. Matsuzaki, H. et al. Genotyping over 100,000 SNPs on a pair of oligonucleotide arrays. *Nat Methods* **1**, 109–111 (2004).
16. Bignell, G. R. et al. High-resolution analysis of DNA copy number using oligonucleotide microarrays. *Genome Res* **14**, 287–295 (2004).
17. Davies, J. J., Wilson, I. M. & Lam, W. L. Array CGH technologies and their applications to cancer genomes. *Chromosome Res* **13**, 237–248 (2005).
18. Bertone, P. et al. Global identification of human transcribed sequences with genome tiling arrays. *Science* **306**, 2242–2246 (2004).
19. Ishkanian, A. S. et al. A tiling resolution DNA microarray with complete coverage of the human genome. *Nat Genet* **36**, 299–303 (2004).
20. Garnis, C., Coe, B. P., Lam, S. L., Macaulay, C. & Lam, W. L. High-resolution array CGH increases heterogeneity tolerance in the analysis of clinical samples. *Genomics* **85**, 790–793 (2005).

21. Wang, T. L. et al. Digital karyotyping. *Proc Natl Acad Sci USA* **99**, 16156–16161 (2002).
22. Velculescu, V. E., Zhang, L., Vogelstein, B. & Kinzler, K. W. Serial analysis of gene expression. *Science* **270**, 484–487 (1995).
23. Park, J. T. et al. Notch3 gene amplification in ovarian cancer. *Cancer Res* **66**, 6312–6318 (2006).
24. Shih, I. M. et al. Amplification of a chromatin remodeling gene, Rsf-1/HBXAP, in ovarian carcinoma. *Proc Natl Acad Sci USA* **102**, 14004–14009 (2005).
25. Hu, M. et al. Distinct epigenetic changes in the stromal cells of breast cancers. *Nat Genet* **37**, 899–905 (2005).
26. Volik, S. et al. End-sequence profiling: sequence-based analysis of aberrant genomes. *Proc Natl Acad Sci USA* **100**, 7696–7701 (2003).
27. Tuzun, E. et al. Fine-scale structural variation of the human genome. *Nat Genet* **37**, 727–732 (2005).
28. Volik, S. et al. Decoding the fine-scale structure of a breast cancer genome and transcriptome. *Genome Res* **16**, 394–404 (2006).
29. Sjoblom, T. et al. The consensus coding sequences of human breast and colorectal cancers. *Science* **314**, 268–294 (2006).
30. Shendure, J. et al. Accurate multiplex polony sequencing of an evolved bacterial genome. *Science* **309**, 1728–1732 (2005).
31. Metzker, M. L. Emerging technologies in DNA sequencing. *Genome Res* **15**, 1767–1776 (2005).
32. Costabile, M., Quach, A. & Ferrante, A. Molecular approaches in the diagnosis of primary immunodeficiency diseases. *Hum Mutat* **27**, 1163–1173 (2006).
33. Parkin, D. M., Bray, F., Ferlay, J. & Pisani, P. Global cancer statistics, 2002. *CA Cancer J Clin* **55**, 74–108 (2005).
34. Nowell, P. C. & Hungerford, D. A. Chromosome studies on normal and leukemic human leukocytes. *J Natl Cancer Inst* **25**, 85–109 (1960).
35. Rowley, J. D. Letter: a new consistent chromosomal abnormality in chronic myelogenous leukaemia identified by quinacrine fluorescence and Giemsa staining. *Nature* **243**, 290–293 (1973).
36. Kuppers, R. Mechanisms of B-cell lymphoma pathogenesis. *Nat Rev Cancer* **5**, 251–262 (2005).
37. Taki, T. & Taniwaki, M. Chromosomal translocations in cancer and their relevance for therapy. *Curr Opin Oncol* **18**, 62–68 (2006).
38. Hoglund, M., Frigyesi, A., Sall, T., Gisselsson, D. & Mitelman, F. Statistical behavior of complex cancer karyotypes. *Genes Chromosomes Cancer* **42**, 327–341 (2005).
39. Frigyesi, A., Gisselsson, D., Mitelman, F. & Hoglund, M. Power law distribution of chromosome aberrations in cancer. *Cancer Res* **63**, 7094–7097 (2003).
40. Coe, B. P. et al. Differential disruption of cell cycle pathways in small cell and non-small cell lung cancer. *Br J Cancer* **94**, 1927–1935 (2006).
41. Vogelstein, B. et al. Genetic alterations during colorectal-tumor development. *N Engl J Med* **319**, 525–532 (1988).
42. Garnis, C. et al. Chromosome 5p aberrations are early events in lung cancer: implication of glial cell line-derived neurotrophic factor in disease progression. *Oncogene* **24** (2005).
43. Blaveri, E. et al. Bladder cancer stage and outcome by array-based comparative genomic hybridization. *Clin Cancer Res* **11**, 7012–7022 (2005).
44. Rosin, M. P. et al. Use of allelic loss to predict malignant risk for low-grade oral epithelial dysplasia. *Clin Cancer Res* **6**, 357–362 (2000).
45. Tuziak, T. et al. High-resolution whole-organ mapping with SNPs and its significance to early events of carcinogenesis. *Lab Invest* **85**, 689–701 (2005).

46. Martini, N. & Melamed, M. R. Multiple primary lung cancers. *J Thorac Cardiovasc Surg* **70**, 606–612 (1975).
47. Matsumoto, Y., Fujii, H., Matsuda, M. & Kono, H. Multicentric occurrence of hepatocellular carcinoma: diagnosis and clinical significance. *J Hepatobiliary Pancreat Surg* **8**, 435–440 (2001).
48. Imyanitov, E. N. et al. Concordance of allelic imbalance profiles in synchronous and metachronous bilateral breast carcinomas. *Int J Cancer* **100**, 557–564 (2002).
49. Demandante, C. G., Troyer, D. A. & Miles, T. P. Multiple primary malignant neoplasms: case report and a comprehensive review of the literature. *Am J Clin Oncol* **26**, 79–83 (2003).
50. Dacic, S., Ionescu, D. N., Finkelstein, S. & Yousem, S. A. Patterns of allelic loss of synchronous adenocarcinomas of the lung. *Am J Surg Pathol* **29**, 897–902 (2005).
51. Nyante, S. J., Devries, S. & Chen, Y. Y. Array-based comparative genomic hybridization of ductal carcinoma in situ and synchronous invasive lobular cancer. *Hum Pathol* **35**, 759–763 (2004).
52. Pateromichelakis, S., Farahani, M., Phillips, E. & Partridge, M. Molecular analysis of paired tumours: time to start treating the field. *Oral Oncol* **41**, 916–926 (2005).
53. Wang, Z. C., Buraimoh, A., Iglehart, J. D. & Richardson, A. L. Genome-wide analysis for loss of heterozygosity in primary and recurrent phyllodes tumor and fibroadenoma of breast using single nucleotide polymorphism arrays. *Breast Cancer Res Treat* **97**, 301–309 (2006).
54. Wa, C. V., DeVries, S., Chen, Y. Y., Waldman, F. M. & Hwang, E. S. Clinical application of array-based comparative genomic hybridization to define the relationship between multiple synchronous tumors. *Mod Pathol* **18**, 591–597 (2005).
55. Slaughter, D. P., Southwick, H. W. & Smejkal, W. Field cancerization in oral stratified squamous epithelium; clinical implications of multicentric origin. *Cancer* **6**, 963–968 (1953).
56. Braakhuis, B. J., Tabor, M. P., Kummer, J. A., Leemans, C. R. & Brakenhoff, R. H. A genetic explanation of Slaughter's concept of field cancerization: evidence and clinical implications. *Cancer Res* **63**, 1727–1730 (2003).
57. Ramaswamy, S., Ross, K. N., Lander, E. S. & Golub, T. R. A molecular signature of metastasis in primary solid tumors. *Nat Genet* **33**, 49–54 (2003).
58. Fidler, I. J. & Kripke, M. L. Metastasis results from preexisting variant cells within a malignant tumor. *Science* **197**, 893–895 (1977).
59. Tada, K. et al. Gains of 8q23-qter and 20q and loss of 11q22-qter in esophageal squamous cell carcinoma associated with lymph node metastasis. *Cancer* **88**, 268–273 (2000).
60. Bogenrieder, T. & Herlyn, M. Axis of evil: molecular mechanisms of cancer metastasis. *Oncogene* **22**, 6524–6536 (2003).
61. Dennis, J. L. & Oien, K. A. Hunting the primary: novel strategies for defining the origin of tumours. *J Pathol* **205**, 236–247 (2005).
62. Sorlie, T. et al. Gene expression patterns of breast carcinomas distinguish tumor subclasses with clinical implications. *Proc Natl Acad Sci USA* **98**, 10869–10874 (2001).
63. van de Vijver, M. J. et al. A gene-expression signature as a predictor of survival in breast cancer. *N Engl J Med* **347**, 1999–2009 (2002).
64. Beer, D. G. et al. Gene-expression profiles predict survival of patients with lung adenocarcinoma. *Nat Med* **8**, 816–824 (2002).
65. Golub, T. R. et al. Molecular classification of cancer: class discovery and class prediction by gene expression monitoring. *Science* **286**, 531–537 (1999).

66. Katoh, H. et al. Genetic profile of hepatocellular carcinoma revealed by array-based comparative genomic hybridization: identification of genetic indicators to predict patient outcome. *J Hepatol* **43**, 863–874 (2005).
67. Callagy, G. et al. Identification and validation of prognostic markers in breast cancer with the complementary use of array-CGH and tissue microarrays. *J Pathol* **205**, 388–396 (2005).
68. Weiss, M. M. et al. Genome wide array comparative genomic hybridisation analysis of premalignant lesions of the stomach. *Mol Pathol* **56**, 293–298 (2003).
69. Rubio-Moscardo, F. et al. Characterization of 8p21.3 chromosomal deletions in B-cell lymphoma: TRAIL-R1 and TRAIL-R2 as candidate dosage-dependent tumor suppressor genes. *Blood* **106**, 3214–3222 (2005).
70. Chen, W. et al. Array comparative genomic hybridization reveals genomic copy number changes associated with outcome in diffuse large B-cell lymphomas. *Blood* **107**, 2477–2485 (2006).
71. Morrison, C. et al. MYC amplification and polysomy 8 in chondrosarcoma: array comparative genomic hybridization, fluorescent in situ hybridization, and association with outcome. *J Clin Oncol* **23**, 9369–9376 (2005).
72. Yasui, K. et al. Alteration in copy numbers of genes as a mechanism for acquired drug resistance. *Cancer Res* **64**, 1403–1410 (2004).
73. Juliano, R. L. & Ling, V. A surface glycoprotein modulating drug permeability in Chinese hamster ovary cell mutants. *Biochim Biophys Acta* **455**, 152–162 (1976).
74. Bradley, G., Naik, M. & Ling, V. P-glycoprotein expression in multidrug-resistant human ovarian carcinoma cell lines. *Cancer Res* **49**, 2790–2796 (1989).
75. Oki, E. et al. Akt phosphorylation associates with LOH of PTEN and leads to chemoresistance for gastric cancer. *Int J Cancer* **117**, 376–380 (2005).
76. Inoue, J. et al. Overexpression of PDZK1 within the 1q12-q22 amplicon is likely to be associated with drug-resistance phenotype in multiple myeloma. *Am J Pathol* **165**, 71–81 (2004).
77. O'Toole, S. A. et al. Analysis of DNA in endometrial cancer cells treated with phytoestrogenic compounds using comparative genomic hybridisation microarrays. *Planta Med* **71**, 435–439 (2005).
78. Irving, J. A. et al. Loss of heterozygosity in childhood acute lymphoblastic leukemia detected by genome-wide microarray single nucleotide polymorphism analysis. *Cancer Res* **65**, 3053–3058 (2005).
79. Wilson, C. et al. Overexpression of genes on 16q associated with cisplatin resistance of testicular germ cell tumor cell lines. *Genes Chromosomes Cancer* **43**, 211–216 (2005).
80. Bernardini, M. et al. High-resolution mapping of genomic imbalance and identification of gene expression profiles associated with differential chemotherapy response in serous epithelial ovarian cancer. *Neoplasia* **7**, 603–613 (2005).
81. van de Wiel, M. A. et al. Expression microarray analysis and oligo array comparative genomic hybridization of acquired gemcitabine resistance in mouse colon reveals selection for chromosomal aberrations. *Cancer Res* **65**, 10208–10213 (2005).
82. Goldstein, M. et al. Combined cytogenetic and array-based comparative genomic hybridization analyses of Wilms tumors: amplification and overexpression of the multidrug resistance associated protein 1 gene (MRP1) in a metachronous tumor. *Cancer Genet Cytogenet* **141**, 120–127 (2003).
83. Snijders, A. M. et al. Shaping of tumor and drug-resistant genomes by instability and selection. *Oncogene* **22**, 4370–4379 (2003).
84. Simon, R. & Wang, S. J. Use of genomic signatures in therapeutics development in oncology and other diseases. *Pharmacogenomics J* **6**, 166–173 (2006).

85. The International HapMap Consortium. A haplotype map of the human genome. *Nature* **437**, 1299–1320 (2005).
86. Futreal, P. A. et al. A census of human cancer genes. *Nat Rev Cancer* **4**, 177–183 (2004).
87. Marsh, S. & McLeod, H. L. Pharmacogenomics: from bedside to clinical practice. *Hum Mol Genet* **15 Spec No 1**, R89–R93 (2006).
88. Rodriguez-Antona, C. & Ingelman-Sundberg, M. Cytochrome P450 pharmacogenetics and cancer. *Oncogene* **25**, 1679–1691 (2006).
89. Pollack, J. R. et al. Microarray analysis reveals a major direct role of DNA copy number alteration in the transcriptional program of human breast tumors. *Proc Natl Acad Sci USA* **99**, 12963–12968 (2002).
90. van Wieringen, W. N., Belien, J. A., Vosse, S. J., Achame, E. M. & Ylstra, B. ACE-it: a tool for genome-wide integration of gene dosage and RNA expression data. *Bioinformatics* **22**, 1919–1920 (2006).
91. Kingsley, C. B., Kuo, W. L., Polikoff, D., Berchuck, A., Gray, J. W. & Jain, A. N. Magellan: A Web based system for the integrated analysis of heterogeneous biological data and annotations; application to DNA copy number and expression data in ovarian cancer. *Cancer Informatics* **2**, 10–21 (2006).
92. Rosa, P. L. et al. VAMP: visualization and analysis of array-CGH, transcriptome and other molecular profiles. *Bioinformatics* **22**, 2066–2073 (2006).
93. Chari, R. L., Lockwood, W. W., Coe, B. P., Chu, A., Macey, D., Thomson, A., Davies, J. J., MacAulay, C. & Lam, W. L. SIGMA: a system for integrative genomic microarray analysis of cancer genomes. *BMC Genomics* **7**, 324 (2006).

14 Comparative Cancer Epigenomics

Alice N. C. Kuo, Ian M. Wilson, Emily Vucic, Eric H. L. Lee, Jonathan J. Davies, Calum MacAulay, Carolyn J. Brown, and Wan L. Lam

CONTENTS

ABSTRACT

The study of epigenomics includes the analysis of changes in DNA methylation and histone protein modification states. Recent technical advances allow analysis of epigenetic features in a high-throughput manner. This has resulted in accelerated discovery of candidate disease-causing epigenetic changes and fueled development of novel epigenetic therapeutics. We describe the current understanding of the role epigenomics plays in normal developmental processes and tumorigenesis; we address the current technologies for analyzing these changes.

14.1 BACKGROUND

Epigenomics refers to the genome-wide study of heritable changes other than those alterations found in the DNA sequence.[1] Imprinting and X-chromosome inactivation are examples of epigenetic changes that occur due to DNA methylation of cytosines and posttranslational modification of histones affecting chromatin condensation and DNA packaging.[2]

14.1.1 DNA METHYLATION

In mammalian cells, DNA methylation involves the cytosine in CpG dinucleotide sequences. The C5 position of the base is modified to become 5′-methylcytosine (5mC). The spontaneous deamination of 5mC to uracil results in an underrepresention of CpG dinucleotides in the genome. In normal tissues, 3% to 4% of all cytosines are methylated.[3] CpG islands are regions rich in CpG dinucleotides that are often conserved through evolution and associated with gene promoter regions.[4] Cancer cells display abnormal DNA methylation by which DNA is globally hypomethylated with focal hypermethylation at CpG islands.[5] Global hypomethylation may lead to genomic instability; hypermethylation of CpG islands is linked with the transcriptional silencing of associated genes.[5]

14.1.2 HISTONE MODIFICATION

Another significant epigenetic event is posttranslational histone modification. Histones are proteins that enable the condensation of double-stranded supercoiled eukaryotic DNA into nucleosomes, thus allowing for further folding of the DNA into chromatin structures. The histone core of nucleosomes consists of two copies each of H2A, H2B, H3, and H4.[6] Posttranslational modifications to the histone tails, including acetylation, methylation, and phosphorylation, determine whether the chromatin exists as euchromatin or heterochromatin.[6] Euchromatin is loosely compacted and represents active transcription, while heterochromatin is tightly compacted and is associated with transcriptional silencing, as illustrated in Figure 14.1. The level of chromatin compaction is ultimately regulated by modifications to both the protein and DNA components. The term *histone code* was proposed to describe distinct combinations of histone modifications that regulate specific downstream events.[7,8]

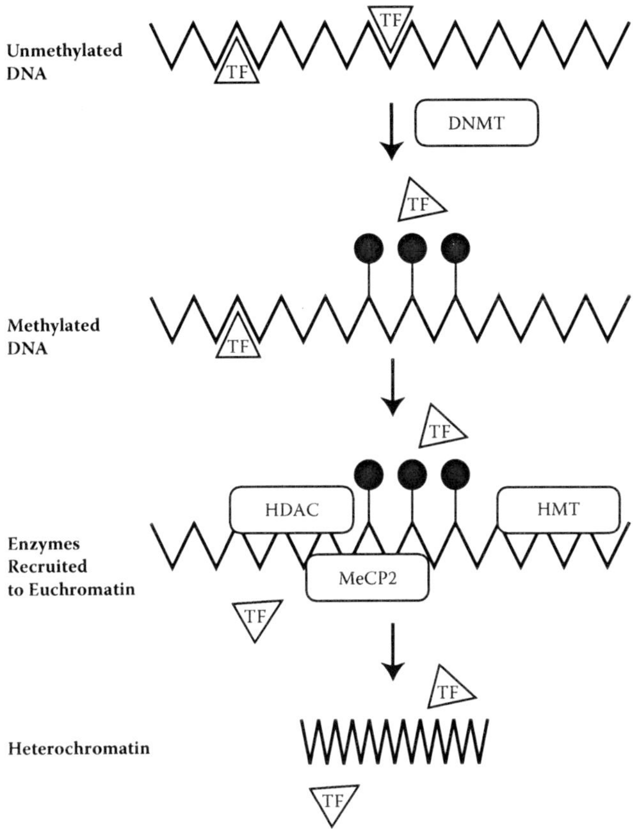

FIGURE 14.1 DNA is methylated (represented as filled lollipops) via DNA methyltransferases (DNMTs). Methylated DNA blocks the access of some transcription factors (TFs) to DNA. Methyl CpG binding protein 2 (MeCP2) and enzymes, including histone deacetylases (HDACs) and histone methyltransferases (HMTs), are recruited to the loosely compacted DNA (euchromatin), forming a more tightly compacted DNA (heterochromatin). The condensed chromatin blocks TFs, resulting in gene silencing.

14.1.3 Chromatin Condensation Regulates Gene Expression

In a synergistic manner, DNA methylation and histone modifications determine the level of chromatin condensation, which in turn regulates gene transcription. Figure 14.1 is an illustration of this process. DNA is methylated by DNA methyltransferases (DNMTs), and methylated DNA is recognized by methyl-binding proteins such as methyl CpG binding domain protein 2 (MeCP2) and methyl-binding domain protein (MBD2).[5] Heterochromatin is then formed by the removal of acetyl groups from the histone tails by histone deacetylases (HDACs), and the addition of methyl groups by histone methyltransferases (HMTs) with the transcriptional corepressor Sin3a. In contrast, histone acetyltransferases (HATs) are responsible for maintaining the open structure of chromatin for active transcription.

A family of DNMTs is involved in de novo methylation (DNMT3a and DNMT3b) and maintenance of methylation *patterns* (DNMT1).[5] In particular, DNMT1 also mediates transcriptional repression together with HDAC2 when acetylated histones are deacetylated just prior to DNA methylation.[9] At least 18 HDAC enzymes of three classes have been identified based on homology to yeast HDACs.[10] HDACs target not only histones but also nonhistone proteins that regulate gene expression and proteins involved in regulation of cell cycle progression and cell death.[10] The classical HDAC family involves class I and class II HDACs.[11] Class I HDACs reside in the nucleus, while class II HDACs are transported in and out of the nucleus in response to certain cellular signals, such as muscle cell differentiation.[10–12] Individual HDACs perform different functions. For example, disruption of HDAC1 leads to embryonic lethality as well as reduced proliferation, whereas disruption of HDACs 4, 5, and 7 may affect muscle cell differentiation.[10,12] Class III HDACs are distinct from the classical HDACs and are discussed together with drug potentiality in Section 14.6.3.

Similar to the classification of HDACs, HATs can be classified as HAT-B and HAT-A. HAT-Bs are involved in acetylation events that are linked to transporting newly synthesized histones from the cytoplasm to the nucleus onto newly replicated DNA.[13] On the other hand, HAT-As are more involved in acetylation events related to transcription, ensuring open structures of chromatin.[13] HATs may be specific for certain residue. For example, Gcn5 (general control nonderepressible 5) a HAT involved in transcription, is specifically targeted[14] to H3K14, H4K8, and K16. Likewise, there are several classes of HMTs, with lysine-specific HMTs and arginine-specific HMTs the major classes.[6] For example, SUV39H1 is a HMT that specifically methylates the lysine 9 residue of histone H3 (so-called H3K9).[15]

To date, histone H3 and H4 modifications have been most widely studied. For example, methylation of H3K9 is associated with methylated DNA and transcriptional repression, whereas acetylation of this residue corresponds to unmethylated DNA and transcriptional activation.[16] Histone modifications can also result in de novo methylation of DNA.[5] H3K9 may be methylated by HMTs, creating a binding site that allows a heterochromatin protein (HP1) to recruit DNMTs, resulting in methylation of DNA.[5]

14.1.4 Imprinting

Genomic imprinting is the differential epigenetic marking of parental chromosomes to achieve monoallelic expression.[17] Imprinted genes play an important role in embryonic development and are largely regulated by DNA methylation.[18]

An example of imprinting is the epigenetic regulation of insulin-like growth factor II (*IGF2*) and *H19*. *IGF2* promotes growth and may play a role in fetal development. *H19* is an untranslated messenger RNA (mRNA). *IGF2* and *H19* are only expressed from the paternal and maternal chromosome, respectively.[19] The expression of these genes is regulated by allele-specific DNA methylation.

At the maternal *IGF2* allele, binding of the protein factor CTCF to the unmethylated imprinting control region (ICR) activates an insulator.[19] The insulator prevents the promoter of *IGF2* from interacting with enhancers downstream of *H19*.[19] Figure 14.2 illustrates how methylation of the ICR prevents CTCF from binding on

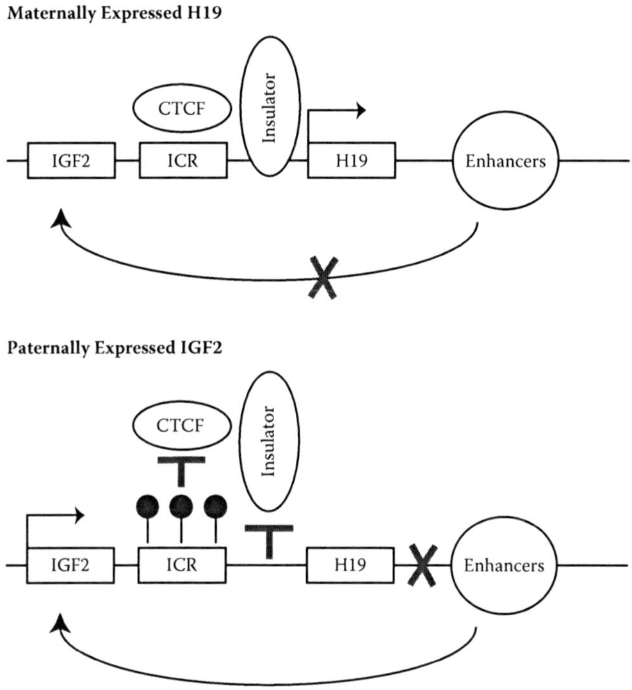

FIGURE 14.2 The imprinted IGF2/H19 locus. Methylation ensures that IGF2 and H19 are each normally expressed in paternal and maternal genome, respectively. Genomic instability may lead to loss or duplication of either allele, which will in turn result in gene dosage disequilibrium. For example, duplication of the paternal *IGF2* allele is linked with overexpression of IGF2 and tumorigenesis.

the paternal *IGF2* allele, thus preventing insulator activation. As a result, IGF2 is paternally expressed.

At the repressed paternal *H19* allele, MeCP2 recognizes methylation at the ICR, resulting in HDAC and Sin3a recruitment. HDACs deacetylate the tails of histones near *H19*, leading to chromatin condensation and silencing of the *H19* gene. This does not occur in the maternal allele, and *H19* is expressed.[20] Errors in this system lead to loss of imprinting (LOI), which is discussed in Section 14.3.2.

14.1.5 X-Chromosome Inactivation

Silencing of one of the X chromosomes in females is a well-established epigenetic event. One of the two X chromosomes (Xi) is randomly silenced early in female development to achieve gene dosage compensation with males.[21] Inactivation of Xi is linked with DNA hypermethylation, recruitment of the histone variant Macro2A, as well as hypoacetylation and methylation at histone residues H3K9 and H2K27.[21] The process of X-chromosome inactivation involves the random silencing of one X chromosome. Once silenced, the same X chromosome is inactivated throughout all subsequent mitotic divisions, making females mosaic for two epigenetically different cell populations.

The X-chromosome inactivation is a complex process dependent on both *cis*- and *trans*-regulatory factors. *XIST* encodes a functional RNA necessary in *cis* for inactivation; that is, *XIST* is only transcribed from and localized to the inactivated chromosome. The mechanisms allowing regulation of preferential *XIST* expression are still not clear, although the promoter on the active X (Xa) is methylated.[21] How methylation spreads along Xi is not clear. However, it is thought that the relative overabundance of the L1 class of long interspersed nuclear elements (LINEs) on the X chromosome may influence the spread of silencing, including DNA methylation, by functioning as "boosting stations."[22]

14.1.6 Small Interfering RNAs

Small interfering RNA, or sometimes referred to as short interfering RNA, (siRNA) is another epigenetic mechanism of gene regulation. RNA interference (RNAi) was first discovered in plants and lower eukaryotes and has been a tool for studying gene function.[23] RNAi is a naturally occurring, posttranscriptional process in which short double-stranded RNAs (average length 22 base pairs) induce the degradation of homologous mRNA transcripts. Normal roles of siRNA-induced transcriptional gene silencing (TGS) include transposon silencing, mutated gene silencing, and protection against RNA viruses.[24]

The silencing effects of small RNA molecules led scientists to correlate this event to methylation status in humans and plants. In humans, it was previously thought that RNAi-induced TGS only occurred via mRNA degradation.[25] Interestingly, gene-specific DNA methylation has been linked to siRNA-induced TGS of three genes: *EF1A*, *ERBB2*, and *RASSF1A*.[24,26,27] In *Arabidopsis thaliana*, extensive methylation has been observed 1 kb downstream of the microRNA (miRNA)–binding sites of *phabulosa* and *phavoluta*, genes that regulate adaxial–abaxial polarity in *Arabidopsis*.[28] As mutation in these regions leads to decreased methylation, miRNA-mediated DNA methylation models were proposed.[28,29] One of these models speculates that when mRNA is transcribed, an miRNA binds to the complementary sequence on the mRNA. During this time, a "chromatin-remodeling" machinery is recruited to the DNA to accomplish methylation.[28,29] However, the role of RNA-mediated gene-specific methylation in TGS remains controversial and is complicated by reports demonstrating that TGS is independent of DNA methylation.[30]

14.2 EPIGENETICS IN NORMAL DEVELOPMENT

14.2.1 Developmental Biology

The role of DNA methylation and other epigenetic marks in normal development is complex and important. Serious defects, ranging from sterility to early embryonic death, have been demonstrated in mice using double knockout models of genes involved in the establishment and maintenance of DNA methylation.[31] Knockouts of histone-remodeling proteins also show a wide range of defects, ranging from failure to implant to behavioral disturbances. Studies of *Dnmt3a/b/l*-deficient mice have shown that establishment of maternal/paternal imprinting is of obvious importance in development, and lack of *Dnmt3l* inhibits proper oocyte and sperm formation.[31] The involvement of DNMTs in cellular differentiation is demonstrated by spatial and temporal differences in *DNMT3a* and *DNMT3b* expression in olfactory receptors.[32] For example, Dnmt3b is present in a narrow window of time during embryonic development, while Dnmt3a is present uniformly, implying that distinct roles exist for different members of the gene family in development.[32,33]

14.2.2 Tissue Specificity

Methylation levels may vary between tissue types. This variation may contribute to tissue-specific gene expression. The Human Epigenome Project has been launched to identify, catalog, and interpret the DNA methylation patterns of all human genes in all major tissues through out the genome (http://www.epigenome.org).[34] This project has so far studied gene specificity in seven tissues (adipose, brain, breast, liver, lung, muscle, and prostate) from different individuals.[35] One of the tissue-specific methylation patterns observed was the CpG island within the *tenascin-XB* (TNXB) gene.[35] This gene is only hypomethylated in muscle samples, correlating to its role in limb, muscle, and heart development.[36] Studies in mouse models have also identified tissue-specific methylation patterns. An example of their findings is the promoter region–CpG island of DEAD-box protein 4 (Ddx4), which is densely methylated in most tissues except for the testes.[37]

14.2.3 Epigenetic Contributions to Phenotypic Diversity

Although monozygotic twins share a common genotype, as they age phenotypic differences become progressively more apparent. It has been proposed that epigenetics may be one possible contributor to the observed phenotypic diversity.[38]

Global and locus-specific differences in DNA methylation and histone acetylation of peripheral lymphocytes in twins were studied. It was concluded that both external factors and internal cellular factors such as the transmission of epigenetic information, management of methylation patterns, and aging processes can influence and be responsible for the differences in epigenetic patterns in monozygotic twins.[38] The observed epigenetic differences are distributed throughout the genome and can influence gene expression as repeat DNA sequences and single-copy genes might be affected as a result of methylation and histone modification events.[38] It was also reported that, in older twins, epigenetic discretion is more distinct. This finding shows the impact of environmental factors and their contribution to similar genotypes in the expression of different phenotypes. Nutrition also plays an important role in the maintenance of methylation pattern in normal cells. For example, the intake of folates can restore normal methylation levels in patients.[39] *Paramutation*, a term that describes *trans*-interactions that lead to heritable changes in a phenotype, has been associated with many genome models, including mouse and humans.[40]

14.3 CANCER EPIGENOMICS

Epigenetic events such as gene silencing, LOI, skewed X-chromosome inactivation, and hypomethylation of parasitic DNA sequences can contribute to tumorigenesis.

14.3.1 Gene Silencing

Hypermethylation in cancer is associated with the silencing of tumor suppressor genes (TSGs). Normally, most CpG islands are unmethylated. In cancer cells, CpG islands can become hypermethylated, resulting in the silencing of certain TSGs. Aberrant promoter hypermethylation is an early event that may drive tumorigenesis.[3,41,42] For example, silencing of *CDKN2A* contributes to the bypass of early mortality checkpoints in the cell cycle. This event has been shown in several experimental systems of carcinogenesis and early stages of naturally occurring tumors.[43] The timing of promoter hypermethylation makes CpG islands a potential target for early tumor detection, while tissue-specific methylation patterns may be useful in subclassifying specific tumor types and determining tissue of origin in metastases.[44–47] Genes commonly hypermethylated in human cancer are listed in Table 14.1. In addition, it has been shown that, in colorectal cancer cells, some CpG islands over a large chromosomal region may have similar methylation levels.[48] This suggests that epigenetic events may affect a whole genome "neighborhood" and may not be just a focal event.

14.3.2 Loss of Imprinting

Given the importance of imprinting in normal cells, it is not surprising that LOI is associated with developmental diseases and cancers. Imprinted genes are expressed monoallelically; however, due to the genomic instability in cancer, the active or inactive allele may be duplicated. Thus, LOI can include activation of a normally silent gene or silencing of a normally active gene.[5] This imbalance in gene dosage may contribute to tumorigenesis. An example of LOI in cancer is at 11p15.5, affecting the *H19/IGF2* locus. Increased dosage of IGF2 is thought to promote tumor formation.[5] LOI at this region has been shown in neuroblastoma, acute myeloblastic leukemia, childhood Wilms tumor, prostate cancer, lung adenocarcinomas, osteosarcoma, colorectal carcinomas, head-and-neck squamous cell carcinoma, adenocarcinomas, and epithelial ovarian cancer.[49,50]

14.3.3 Skewed X-Chromosome Inactivation

Selection of a specific X chromosome for inactivation is normally a random process. Nonrandom or skewed X inactivation denotes a consistent abnormal inactivation of one X preferentially over another. Skewed X inactivation has been noted in many tumor types.[51,52] Nonrandom X inactivation in cancer may be a somatic phenomenon or may be an artifact of clonal expansion in the tumor. Causes of skewed X inactivation include parental imprinting effects, mutations in *XIST*, reduced progenitor populations, and selective processes.[21]

14.3.4 Hypomethylation of Parasitic DNA Sequences

With the exception of CpG islands, the CpG dinucleotides throughout the genome are normally methylated. The bulk of 5mC is found in repetitive/parasitic DNA sequences, such as LINEs and short interspersed nuclear elements (SINEs), and in

TABLE 14.1
Hypermethylated Genes in Cancer

Function	Genes
DNA repair	
DNA repair	hMLH1[a,b,c,d], Hmsh2[a], MGMT[b,c,d], GSTP1[c,d]
Cell cycle/evasion apoptosis	
Inhibits transcription	HIC-1[a,b,c,d], HLTF[b]
Maintains telomere ends	hTERT[a]
Regulates proliferation	ER-α/β[b]
Proliferation and apoptosis	FHIT[a]
Inhibits cell growth	HIN1[a]
Growth regulation	PR[c], PR A/B[d]
Growth suppression	LOT1[a]
Cell cycle	TGFbRII[a], 14-3-3sigma[a], BRCA1[a], CCND2[a,d], CDKN2A[a,b,c,d], CDKN1A[d], PAX5a[c], PAX5b[c], RB1[d], CHFR[c]
Cell cycle and apoptosis	APC[a,b,c,d], ZAC[a]
Apoptosis	DAPK[a,c,d], GPC3[a], HOXA5[a], TP53[a], RARB[a,b,c,d], RASSF1A[a,b,c,d], SOCS1[a], TMS1[a], TWIST[a], CACNA1G[b], ARF[b,c,d], CDKN1B[d], TP73[c,d], TRAILR[c] TSLC1[c,d], FAS[c], Caspase-8[c], TNFRSF6[d]
Cell cycle, differentiation, apoptosis	RUNX3[d]
Cell cycle, multiple functions	PTEN[d]
Contact inhibition/metastasis	
Invasiveness	BCSG1[a]
Inhibits metastasis	HNm23-H1[a]
Inhibits invasion	PRSS8[a], SYK[a], THBS1[a], TIMP3[a]
Inhibits angiogenesis	SERPINB5[a]
Cell adhesion	CDH1 (E-Cad)[a,c], CDH13 (H-Cad)[a,c,d], LAMA3[c,d], LAMB3[c,d], LAMC2[c,d], CAV1[d], CD44[d]
Cell motility	GSN[a], CSPG2[b]
Inhibits invasion	THBS1d/2[b,d], TIMP3[c,d]
Against Ca accumulation	S100A2[c]
Others	
Cellular uptake of methotrexate	RFC[a]
Detoxification	GSTP1[a,c], ESR1[c,d], ESR2[c,d], GDF10[c], ZNF185[d]
Inhibit tumor formation	NES1[a]
Interact with BRCA1	SRBC[a]
Ras signaling	NORE1[a]
Tumor suppressor	DUTT1[a], NOEY2[a], RIZ1[a,b,c], LKBI/STK11[b], HOXB[c]
Differentiation	EGFR[b,c]
Tumor growth regulation	PTGS2[b]/COX2[b]
Differentiation and apoptosis	TIG1[d]
Fibroblast differentiation	MYOD1[c]
Regulation differentiation	PTHRP[c]
Unknown	HPP1/TPEF[b], IGF2[b], MYOD1[b,c], PAX6[b]

[a] Breast cancer.[96,97]
[b] Colorectal cancer.[98,99]
[c] Lung cancer.[68,100]
[d] Prostate cancer.[101,102]

centromeric satellite DNA.[53] Methylation of these sequences is thought to be important for suppressing retrotransposition events, illegitimate recombination events, and inappropriate gene transcription from retroelement promoters/enhancers.

The genomes of many cancer types become globally hypomethylated. This has a large effect on repeat DNA sequences. For example, there are approximately 400,000 L1 retrotransposons, composing approximately 18% of the genome. Of those, 60 to 100 are still functionally able to retrotranspose.[54] In cancer cell lines, 70% to 80% of the CpG sites in L1 elements have been shown to be demethylated. This lack of methylation may lead to increased genomic instability via double-stranded DNA breaks from retrotransposons and increased rates of homologous recombination. In addition, gene regulation may be directly affected by either the antisense promoter in L1 elements, which may drive the aberrant transcription of neighboring genes, or direct insertional mutagenesis.[55,56] Although CpG island hypermethylation has largely been the focus of cancer research in the past, global hypomethylation may prove to play a significant role.

14.4 GENOME-WIDE TECHNOLOGIES FOR EPIGENETIC ANALYSIS

Many techniques have been developed for studying methylation at both locus-specific and genome-wide levels. Current methods used to study the epigenome are as follows:

1. **Methods based on polymerase chain reaction (PCR).** Methylated DNA can be differentiated based on susceptibility to digestion by restriction enzymes and their 5mC-sensitive isoschizomers. A commonly used enzyme pair is *Hpa* II and *Msp* I. *Msp* I is not sensitive to DNA methylation; however, *Hpa* II is. Using primers flanking restriction cut sites, PCR will only generate product in methylated samples that are digested with *Hpa* II.[57,58]
2. **Restriction landmark genomic scanning (RLGS).** RLGS combines the use of labeled genomic DNA digested with restriction enzymes and high-resolution two-dimensional gel electrophoresis. It can measure the DNA methylation level quantitatively in thousands of CpG islands separated based on restriction sites.[59]
3. **Methylation-specific digital karyotyping (MSDK).** MSDK uses the methylation-sensitive enzyme Asc I, which yields large DNA fragments. Linker-ligation-mediated enrichment for these long fragments is followed by Nla III digestion. Sequence tags adjacent to the Nla III sites are concatenated and sequenced to quantify methylated sites in the genome.[60]
4. **Bisulfite conversion.** Sodium bisulfite treatment converts unmethylated cytosine to uracil, while methylated cytosine is not affected. Sequencing of untreated and treated DNA identifies the 5mC positions. Alternatively, this technique can be used in a PCR application to distinguish between unmethylated and methylated loci.
5. **Methylation-specific oligonucleotide (MSO) microarrays.** Microarrays allow for the simultaneous examination of multiple loci. Arrays include those that cover whole chromosomes or the entire genome in interval or

tiling fashions, as well as specifically designed arrays such as promoter and CpG island arrays. To analyze DNA samples, experimental and control DNA are each labeled with different fluorescent dyes. They are cohybridized to the microarray and scanned, after which image analysis software is used to determine the ratio of the experimental and control dyes relative to the background. The MSO microarrays combine PCR-amplified bisulfite-treated DNA fragments with an oligonucleotide array that is designed to differentiate methylated and unmethylated CpG islands.[61]

6. **Methylation-dependent immunoprecipitation (MeDIP).** MeDIP is a recently developed method that uses anti-5mC antibodies to enrich for methylated genomic DNA fragments. The immunoprecipitated DNA is compared with untreated DNA by competitive cohybridization to a whole-genome resolution tiling path array[62,63] (see description in Figure 14.3).
7. **Chromatin immunoprecipitation (ChIP).** ChIP is a method that identifies the DNA sequence associated with a specific protein. This is achieved using an antibody against the protein–DNA complex of interest. Chromosomal CGH and CpG island microarrays have been used to localize ChIP-captured MBD protein–DNA complexes to their genomic locations.[64] Another application of ChIP is for the study of the global distribution of histone modifications using specific antibodies coupled with CpG island microarrays, complementary DNA arrays, and tiling arrays.[65]

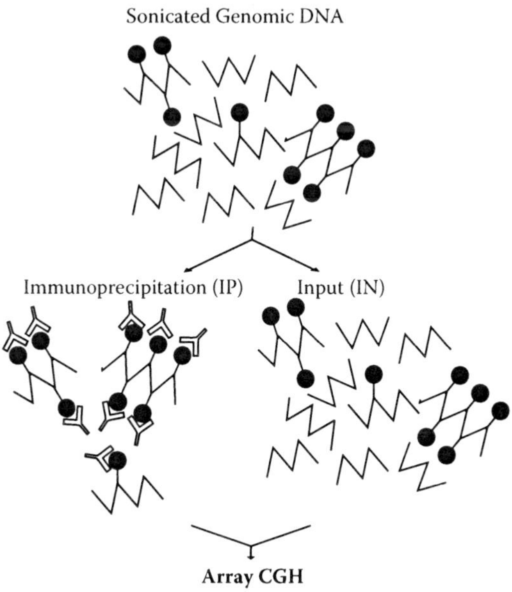

FIGURE 14.3 Methylation-dependent immunoprecipitation (MeDIP) uses anti-5mC antibodies to immunocapture methylated fragments of DNA. The immunoprecipitated DNA (IP DNA) and input reference DNA (IN DNA) are differentially labeled with different cyanine dyes, cohybridized onto genomic targets on microarrays.

14.5 COMPARATIVE EPIGENOMICS IN CANCER

14.5.1 Early Detection and Cancer Progression Using Epigenetic Markers

Promoter methylation status may serve as a marker for cancer detection. In a study that analyzed patient sputum, it was found that methylation of *CDKN2A*, *MGMT*, *PAX5*-β, *DAPK*, *GATA5*, and *RASSF1A* is associated with increased lung cancer risk.[66]

Promoter methylation is also associated with cancer progression. For example, in lung cancer, *CDKN2A* promoter methylation was present in 17% of hyperplasias and 60% to 70% of adenocarcinomas and squamous cell carcinomas.[67] Similarly, *MGMT* methylation levels increase with tumor stage in lung adenocarcinoma.[68] Overexpression of HDAC proteins is also related to progression in non–small cell lung cancer.[11] Furthermore, in esophageal cancers, deacetylation of histone 4 (H4) has been linked with metastasis and poor prognosis.[11]

14.5.2 CpG Island Methylator Phenotype and Colon Cancer

Epigenetic changes in colon cancer have been well documented.[41,48] Nonrandom methylation of multiple CpG islands has been observed in individual colon cancers, leading to the discovery of a phenomenon known as CpG island methylator phenotype (CIMP).[69,70] Although not all methylated genes are reliable identifiers of the CIMP phenomenon, five marker genes (*CACNA1G, IGF2, NEUROG1, RUNX3,* and *SOCS1*) have improved the classification of the methylator phenotype in colorectal cancer.[71]

14.5.3 Epigenetic Changes in Stromal Cells of Breast Cancers

Methylation changes are not restricted to cancer cells. Comparison of methylation patterns using the MSDK technique (described in Section 14.4) in specific breast cell types (epithelial, myoepithelial, and stromal cells) of normal and tumor specimens revealed distinct methylation levels of *PRDM14, HOXD4, SLC9A3R1, CDC42EP5, LOC389333,* and *CXorf12*. For example, methylation of *PRDM14* and *LOC389333* is only observed in epithelial cells and not in myoepithelial and stromal cells. Conversely, in stromal cells, *HOXD4, SLC9A3R1, CDC42EP5,* and *CXorf12* are more methylated than in epithelial and myoepithelial cells.[60] Among these genes, *CXorf12* is differentially methylated in tumor specimens, while very little methylation was observed in normal specimens. Further studies of cell type–specific methylated genes will greatly aid the identification of methylated genes during tumorigenesis and the effects of tumors on the epigenomes of normal cells in the microenvironment.

14.6 EPIGENOMIC-BASED THERAPEUTICS

14.6.1 DNA Demethylating Drugs

The reversibility of DNA methylation has raised the potential for "epigenetic drug" development. Nucleoside analog drugs aim to reactivate genes aberrantly silenced in cancer through the demethylation of hypermethylated DNA. 5-Azacytidine (5-aza/Vidaza) covalently interacts with DNMTs. This drug was approved by the U.S. Food and Drug Administration for treatment of patients with myelodysplastic syndromes.[72] Genes critical for differentiation and proliferation are reactivated after treatment.[73] 5-Aza-2-deoxycytidine (5-aza-CdR/Decitabine) is an S-phase-specific agent that induces terminal differentiation of human leukemic cells.[74] In aqueous solution, 5-aza and 5-aza-CdR are known to be highly unstable and sensitive to pH and may be prone to rapid inactivation by liver cytidine deaminase.[73–75]

5-Fluoro-deoxycytidine (Zebularine) functions as both a cytidine deaminase and a DNMT inhibitor and is currently in clinical trials.[76] Gene reexpression patterns generated by this drug are similar to those produced by 5-aza and 5-aza-CdR. Zebularine restores expression of *CDKN2A* in various cancer cell models as well as tumor cells grown in mice. Unlike 5-aza and 5-aza-CdR, Zebularine may modify DNA such that it cannot be remethylated.[77] Zebularine is exceedingly more stable in aqueous, acidic, and neutral environments and is less toxic than 5-aza and 5-aza-CdR.[76] Due to its stability, zebularine is showing promise as an orally administered mechanism-based DNMT inhibitor and is currently in clinical trials.[76]

DNA methylation inhibitors are not restricted to nucleoside analogs.[78,79] For example, hydralazine is a vasodilator found to decrease DNMT1 and DNMT3a expression, and procainamide is an antiarrhythmic drug shown to inhibit DNMT activity, resulting in DNA hypomethylation.[80,81] EGCG [(−)-epigallocatechin-3-gallate], a major polyphenol in green tea, has been reported to inhibit DNMT enzymes and reactivate genes such as *RAR*-β and *CDKN2A*, which are commonly silenced via methylation.[82]

14.6.2 Histone Deacetylase Inhibitors

Histone deacetylase inhibitors (HDACIs) aim to relax chromatin, allowing access by HATs and transcription factors, to restore normal cell proliferation. A variety of HDACIs are under consideration for cancer treatment. For example, valproic acid (VPA) induces apoptosis in the presences of kinase inhibitors or in conjunction with NF-kβ inhibitors.[83] Hydroxamic acid derivative HDACIs, such as suberoylanilide (SAHA) and NVP-LAQ824, affect the expression of p21, presumably through promoter reactivation.[84–86] Several other HDACIs, including trichostatin A (TSA), phenylbutyrate, depsipeptide (FK-22), and the cyclic tetrapeptide depsipeptides MS-275 and CI-994, are also in clinical trials.[87,88] HDACIs are found to be most effective when used in conjunction with DNMT inhibitors.[89] For example, combined treatment targeting DNMTs using vidaza or decitabine preceding the administration of an HDACI shows significant reexpression of *CDKN2A, CDKN2B, MLH-1*, and *TIMP3*.[3] Tamoxifen sensitivity in estrogen receptor-negative breast cancer patients was regained after treatment with decitabine and TSA.[90]

14.6.3 Class III HDACs as a Potential Anticancer Drug Agent

As discussed in the initial sections, the classical HDACs involve classes I and II. There is a class III HDAC family, the Sir2 family, that is distinct from the classical HDACs in that histones are not their main substrates.[10,91] SIRT1 is the mammalian homolog of yeast Sir2. This enzyme normally binds to several transcription factors and is known[91] to deacetylate a lysine residue of the tumor suppressor protein p53. In a recent report,[91] a small molecule called EX-527 was shown to increase lysine 382 residue acetylation of p53 through inhibition of SIRT1 enzymatic activity without affecting the normal function of p53.

14.6.4 Small RNAs as Epigenetic Therapies

As RNA-mediated gene silencing can be considered an epigenetic phenomenon, the use of siRNA qualifies as epigenetic therapy.[92] RNA-directed DNA methylation can regulate transcription and nuclear domain organization and therefore may be involved in the inheritance of chromatin states.[24,93] siRNA induction of apoptosis targeting the M-BCR/ABL fusion gene has been demonstrated in chronic myeloid leukemia cells.[94] Although specific dosage, vector design, and methods of delivery are still in development, siRNA-directed gene silencing is a promising concept in cancer therapy.

14.7 CONCLUSION

Similar to how the human genome project led to rapid improvements in technology for mapping and sequencing the genome, our growing understanding of the importance of epigenetic change has led to the development of a human epigenome project as well as many new approaches trying to unravel the complexity of epigenetic modifications.[95] Unlike the genome, which is relatively static between cell types, the major challenge in studying cancer epigenomics is defining the "normal" epigenetic marks in the precursor cell. Most important, unlike genetic mutations, epigenetic changes may be reversible, and thus the therapeutic potential of epigenetic drugs has raised great expectations.

REFERENCES

1. Callinan, P. A. & Feinberg, A. P. The emerging science of epigenomics. *Hum Mol Genet* 15 Spec No 1, R95–R101 (2006).
2. Keshet, I., Lieman-Hurwitz, J. & Cedar, H. DNA methylation affects the formation of active chromatin. *Cell* 44, 535–543 (1986).
3. Baylin, S. B. DNA methylation and gene silencing in cancer. *Nat Clin Pract Oncol* 2 Suppl 1, S4–S11 (2005).
4. Ushijima, T. et al. Establishment of methylation-sensitive-representational difference analysis and isolation of hypo- and hypermethylated genomic fragments in mouse liver tumors. *Proc Natl Acad Sci USA* 94, 2284–2289 (1997).
5. Feinberg, A. P. & Tycko, B. The history of cancer epigenetics. *Nat Rev Cancer* 4, 143–153 (2004).
6. Shilatifard, A. Chromatin modifications by methylation and ubiquitination: implications in the regulation of gene expression. *Annu Rev Biochem* 75, 243–269 (2006).
7. Valley, C. M., Pertz, L. M., Balakumaran, B. S. & Willard, H. F. Chromosome-wide, allele-specific analysis of the histone code on the human X chromosome. *Hum Mol Genet* 15, 2335–2347 (2006).
8. Strahl, B. D. & Allis, C. D. The language of covalent histone modifications. *Nature* 403, 41–45 (2000).
9. Jones, P. A. & Baylin, S. B. The fundamental role of epigenetic events in cancer. *Nat Rev Genet* 3, 415–428 (2002).

10. Dokmanovic, M. & Marks, P. A. Prospects: histone deacetylase inhibitors. *J Cell Biochem* 96, 293–304 (2005).
11. Bowman, R. V., Yang, I. A., Semmler, A. B. & Fong, K. M. Epigenetics of lung cancer. *Respirology* 11, 355–365 (2006).
12. de Ruijter, A. J., van Gennip, A. H., Caron, H. N., Kemp, S. & van Kuilenburg, A. B. Histone deacetylases (HDACs): characterization of the classical HDAC family. *Biochem J* 370, 737–749 (2003).
13. Roth, S. Y., Denu, J. M. & Allis, C. D. Histone acetyltransferases. *Annu Rev Biochem* 70, 81–120 (2001).
14. Struhl, K. Histone acetylation and transcriptional regulatory mechanisms. *Genes Dev* 12, 599–606 (1998).
15. Martin, C. & Zhang, Y. The diverse functions of histone lysine methylation. *Nat Rev Mol Cell Biol* 6, 838–849 (2005).
16. Esteller, M. Aberrant DNA methylation as a cancer-inducing mechanism. *Annu Rev Pharmacol Toxicol* 45, 629–656 (2005).
17. Esteller, M. & Herman, J. G. Cancer as an epigenetic disease: DNA methylation and chromatin alterations in human tumours. *J Pathol* 196, 1–7 (2002).
18. Li, E., Beard, C. & Jaenisch, R. Role for DNA methylation in genomic imprinting. *Nature* 366, 362–365 (1993).
19. Weber, M. et al. Genomic imprinting controls matrix attachment regions in the Igf2 gene. *Mol Cell Biol* 23, 8953–8959 (2003).
20. Drewell, R. A., Goddard, C. J., Thomas, J. O. & Surani, M. A. Methylation-dependent silencing at the H19 imprinting control region by MeCP2. *Nucleic Acids Res* 30, 1139–1144 (2002).
21. Chang, S. C., Tucker, T., Thorogood, N. P. & Brown, C. J. Mechanisms of X-chromosome inactivation. *Front Biosci* 11, 852–866 (2006).
22. Lyon, M. F. X-chromosome inactivation: a repeat hypothesis. *Cytogenet Cell Genet* 80, 133–137 (1998).
23. Fire, A. et al. Potent and specific genetic interference by double-stranded RNA in *Caenorhabditis elegans*. *Nature* 391, 806–811 (1998).
24. Morris, K. V., Chan, S. W., Jacobsen, S. E. & Looney, D. J. Small interfering RNA-induced transcriptional gene silencing in human cells. *Science* 305, 1289–1292 (2004).
25. Zeng, Y. & Cullen, B. R. RNA interference in human cells is restricted to the cytoplasm. *RNA* 8, 855–60 (2002).
26. Xia, W. et al. Regulation of survivin by ErbB2 signaling: therapeutic implications for ErbB2-overexpressing breast cancers. *Cancer Res* 66, 1640–1647 (2006).
27. Castanotto, D. et al. Short hairpin RNA-directed cytosine (CpG) methylation of the RASSF1A gene promoter in HeLa cells. *Mol Ther* 12, 179–183 (2005).
28. Ronemus, M. & Martienssen, R. RNA interference: methylation mystery. *Nature* 433, 472–473 (2005).
29. Bao, N., Lye, K. W. & Barton, M. K. MicroRNA binding sites in Arabidopsis class III HD-ZIP mRNAs are required for methylation of the template chromosome. *Dev Cell* 7, 653–662 (2004).
30. Ting, A. H., Schuebel, K. E., Herman, J. G. & Baylin, S. B. Short double-stranded RNA induces transcriptional gene silencing in human cancer cells in the absence of DNA methylation. *Nat Genet* 37, 906–910 (2005).
31. Li, E. Chromatin modification and epigenetic reprogramming in mammalian development. *Nat Rev Genet* 3, 662–673 (2002).
32. MacDonald, J. L., Gin, C. S. & Roskams, A. J. Stage-specific induction of DNA methyltransferases in olfactory receptor neuron development. *Dev Biol* 288, 461–473 (2005).

33. Feng, J., Chang, H., Li, E. & Fan, G. Dynamic expression of de novo DNA methyltransferases Dnmt3a and Dnmt3b in the central nervous system. *J Neurosci Res* 79, 734–746 (2005).
34. Bird, A. DNA methylation patterns and epigenetic memory. *Genes Dev* 16, 6–21 (2002).
35. Rakyan, V. K. et al. DNA methylation profiling of the human major histocompatibility complex: a pilot study for the human epigenome project. *PLoS Biol* 2, e405 (2004).
36. Burch, G. H., Bedolli, M. A., McDonough, S., Rosenthal, S. M. & Bristow, J. Embryonic expression of tenascin-X suggests a role in limb, muscle, and heart development. *Dev Dyn* 203, 491–504 (1995).
37. Song, F. et al. Association of tissue-specific differentially methylated regions (TDMs) with differential gene expression. *Proc Natl Acad Sci USA* 102, 3336–3341 (2005).
38. Fraga, M. F. et al. Epigenetic differences arise during the lifetime of monozygotic twins. *Proc Natl Acad Sci USA* 102, 10604–10609 (2005).
39. Feil, R. Environmental and nutritional effects on the epigenetic regulation of genes. *Mutat Res* 600, 46–57 (2006).
40. Rassoulzadegan, M. et al. RNA-mediated non-mendelian inheritance of an epigenetic change in the mouse. *Nature* 441, 469–474 (2006).
41. Suzuki, H. et al. Epigenetic inactivation of SFRP genes allows constitutive WNT signaling in colorectal cancer. *Nat Genet* 36, 417–422 (2004).
42. Fong, K. M., Sekido, Y., Gazdar, A. F. & Minna, J. D. Lung cancer. 9: Molecular biology of lung cancer: clinical implications. *Thorax* 58, 892–900 (2003).
43. Brenner, A. J., Stampfer, M. R. & Aldaz, C. M. Increased p16 expression with first senescence arrest in human mammary epithelial cells and extended growth capacity with p16 inactivation. *Oncogene* 17, 199–205 (1998).
44. Baylin, S. B. et al. Aberrant patterns of DNA methylation, chromatin formation and gene expression in cancer. *Hum Mol Genet* 10, 687–692 (2001).
45. Costello, J. F. et al. Aberrant CpG-island methylation has non-random and tumour-type-specific patterns. *Nat Genet* 24, 132–138 (2000).
46. Esteller, M. CpG island hypermethylation and tumor suppressor genes: a booming present, a brighter future. *Oncogene* 21, 5427–5440 (2002).
47. Esteller, M., Corn, P. G., Baylin, S. B. & Herman, J. G. A gene hypermethylation profile of human cancer. *Cancer Res* 61, 3225–3229 (2001).
48. Frigola, J. et al. Epigenetic remodeling in colorectal cancer results in coordinate gene suppression across an entire chromosome band. *Nat Genet* 38, 540–549 (2006).
49. Falls, J. G., Pulford, D. J., Wylie, A. A. & Jirtle, R. L. Genomic imprinting: implications for human disease. *Am J Pathol* 154, 635–647 (1999).
50. Ohlsson, R. Loss of IGF2 imprinting: mechanisms and consequences. *Novartis Found Symp* 262, 108–121; discussion 121–124, 265–268 (2004).
51. McDonald, H. L., Gascoyne, R. D., Horsman, D. & Brown, C. J. Involvement of the X chromosome in non-Hodgkin lymphoma. *Genes Chromosomes Cancer* 28, 246–257 (2000).
52. Guo, Z., Li, Q., Wilander, E. & Ponten, J. Clonality analysis of multifocal carcinoid tumours of the small intestine by X-chromosome inactivation analysis. *J Pathol* 190, 76–79 (2000).
53. Ehrlich, M. DNA methylation in cancer: too much, but also too little. *Oncogene* 21, 5400–5413 (2002).
54. Brouha, B. et al. Hot L1s account for the bulk of retrotransposition in the human population. *Proc Natl Acad Sci USA* 100, 5280–5285 (2003).
55. Speek, M. Antisense promoter of human L1 retrotransposon drives transcription of adjacent cellular genes. *Mol Cell Biol* 21, 1973–1985 (2001).

56. Morse, B., Rotherg, P. G., South, V. J., Spandorfer, J. M. & Astrin, S. M. Insertional mutagenesis of the myc locus by a LINE-1 sequence in a human breast carcinoma. *Nature* 333, 87–90 (1988).
57. Gonzalgo, M. L. et al. Identification and characterization of differentially methylated regions of genomic DNA by methylation-sensitive arbitrarily primed PCR. *Cancer Res* 57, 594–599 (1997).
58. Huang, T. H. et al. Identification of DNA methylation markers for human breast carcinomas using the methylation-sensitive restriction fingerprinting technique. *Cancer Res* 57, 1030–1034 (1997).
59. Hatada, I., Hayashizaki, Y., Hirotsune, S., Komatsubara, H. & Mukai, T. A genomic scanning method for higher organisms using restriction sites as landmarks. *Proc Natl Acad Sci USA* 88, 9523–9527 (1991).
60. Hu, M. et al. Distinct epigenetic changes in the stromal cells of breast cancers. *Nat Genet* 37, 899–905 (2005).
61. Gitan, R. S., Shi, H., Chen, C. M., Yan, P. S. & Huang, T. H. Methylation-specific oligonucleotide microarray: a new potential for high-throughput methylation analysis. *Genome Res* 12, 158–164 (2002).
62. Weber, M. et al. Chromosome-wide and promoter-specific analyses identify sites of differential DNA methylation in normal and transformed human cells. *Nat Genet* 37, 853–862 (2005).
63. Wilson, I. M. et al. Epigenomics: mapping the methylome. *Cell Cycle* 5, 155–158 (2006).
64. Ballestar, E. et al. Methyl-CpG binding proteins identify novel sites of epigenetic inactivation in human cancer. *EMBO J* 22, 6335–6345 (2003).
65. Wu, J., Smith, L. T., Plass, C. & Huang, T. H. ChIP-chip comes of age for genome-wide functional analysis. *Cancer Res* 66, 6899–6902 (2006).
66. Belinsky, S. A. et al. Promoter hypermethylation of multiple genes in sputum precedes lung cancer incidence in a high-risk cohort. *Cancer Res* 66, 3338–3344 (2006).
67. Belinsky, S. A. Silencing of genes by promoter hypermethylation: key event in rodent and human lung cancer. *Carcinogenesis* 26, 1481–1487 (2005).
68. Belinsky, S. A. Gene-promoter hypermethylation as a biomarker in lung cancer. *Nat Rev Cancer* 4, 707–717 (2004).
69. Toyota, M. et al. CpG island methylator phenotype in colorectal cancer. *Proc Natl Acad Sci USA* 96, 8681–8686 (1999).
70. Issa, J. P. CpG island methylator phenotype in cancer. *Nat Rev Cancer* 4, 988–993 (2004).
71. Weisenberger, D. J. et al. CpG island methylator phenotype underlies sporadic microsatellite instability and is tightly associated with BRAF mutation in colorectal cancer. *Nat Genet* 38, 787–793 (2006).
72. Kaminskas, E. et al. Approval summary: azacitidine for treatment of myelodysplastic syndrome subtypes. *Clin Cancer Res* 11, 3604–3608 (2005).
73. Fenaux, P. Inhibitors of DNA methylation: beyond myelodysplastic syndromes. *Nat Clin Pract Oncol* 2 Suppl 1, S36–S44 (2005).
74. Momparler, R. L. Epigenetic therapy of cancer with 5-aza-2′-deoxycytidine (decitabine). *Semin Oncol* 32, 443–451 (2005).
75. Kuykendall, J. R. 5-azacytidine and decitabine monotherapies of myelodysplastic disorders. *Ann Pharmacother* 39, 1700–1709 (2005).
76. Marquez, V. E. et al. Zebularine: a unique molecule for an epigenetically based strategy in cancer chemotherapy. *Ann NY Acad Sci* 1058, 246–254 (2005).
77. Cheng, J. C. et al. Inhibition of DNA methylation and reactivation of silenced genes by zebularine. *J Natl Cancer Inst* 95, 399–409 (2003).

78. Cornacchia, E. et al. Hydralazine and procainamide inhibit T cell DNA methylation and induce autoreactivity. *J Immunol* 140, 2197–2200 (1988).
79. Chuang, J. C. et al. Comparison of biological effects of non-nucleoside DNA methylation inhibitors versus 5-aza-2′-deoxycytidine. *Mol Cancer Ther* 4, 1515–1520 (2005).
80. Deng, C. et al. Hydralazine may induce autoimmunity by inhibiting extracellular signal-regulated kinase pathway signaling. *Arthritis Rheum* 48, 746–756 (2003).
81. Lin, X. et al. Reversal of GSTP1 CpG island hypermethylation and reactivation of pi-class glutathione S-transferase (GSTP1) expression in human prostate cancer cells by treatment with procainamide. *Cancer Res* 61, 8611–8616 (2001).
82. Fang, M. Z. et al. Tea polyphenol (–)-epigallocatechin-3-gallate inhibits DNA methyltransferase and reactivates methylation-silenced genes in cancer cell lines. *Cancer Res* 63, 7563–7570 (2003).
83. Yeow, W. S. et al. Potentiation of the anticancer effect of valproic acid, an antiepileptic agent with histone deacetylase inhibitory activity, by the kinase inhibitor staurosporine or its clinically relevant analogue UCN-01. *Br J Cancer* 94, 1436–1445 (2006).
84. Catley, L. et al. NVP-LAQ824 is a potent novel histone deacetylase inhibitor with significant activity against multiple myeloma. *Blood* 102, 2615–2622 (2003).
85. Gui, C. Y., Ngo, L., Xu, W. S., Richon, V. M., & Marks, P. A. Histone deacetylase (HDAC) inhibitor activation of p21WAF1 involves changes in promoter-associated proteins, including HDAC1. *Proc Natl Acad Sci USA* 101, 1241–1246 (2004).
86. Marks, P. A., Miller, T. & Richon, V. M. Histone deacetylases. *Curr Opin Pharmacol* 3, 344–351 (2003).
87. Yoshida, M., Kijima, M., Akita, M. & Beppu, T. Potent and specific inhibition of mammalian histone deacetylase both in vivo and in vitro by trichostatin A. *J Biol Chem* 265, 17174–17179 (1990).
88. Kelly, W. K. & Marks, P. A. Drug insight: histone deacetylase inhibitors — development of the new targeted anticancer agent suberoylanilide hydroxamic acid. *Nat Clin Pract Oncol* 2, 150–157 (2005).
89. Garcia-Manero, G. & Gore, S. D. Future directions for the use of hypomethylating agents. *Semin Hematol* 42, S50–S59 (2005).
90. Sharma, D., Saxena, N. K., Davidson, N. E. & Vertino, P. M. Restoration of tamoxifen sensitivity in estrogen receptor-negative breast cancer cells: tamoxifen-bound reactivated ER recruits distinctive corepressor complexes. *Cancer Res* 66, 6370–6378 (2006).
91. Solomon, J. M. et al. Inhibition of SIRT1 catalytic activity increases p53 acetylation but does not alter cell survival following DNA damage. *Mol Cell Biol* 26, 28–38 (2006).
92. Dykxhoorn, D. M., Palliser, D. & Lieberman, J. The silent treatment: siRNAs as small molecule drugs. *Gene Ther* 13, 541–552 (2006).
93. Santoro, R. & De Lucia, F. Many players, one goal: how chromatin states are inherited during cell division. *Biochem Cell Biol* 83, 332–343 (2005).
94. Wilda, M., Fuchs, U., Wossmann, W. & Borkhardt, A. Killing of leukemic cells with a BCR/ABL fusion gene by RNA interference (RNAi). *Oncogene* 21, 5716–5724 (2002).
95. Esteller, M. The necessity of a human epigenome project. *Carcinogenesis* 27, 1121–1125 (2006).
96. Szyf, M., Pakneshan, P. & Rabbani, S. A. DNA methylation and breast cancer. *Biochem Pharmacol* 68, 1187–1197 (2004).
97. Widschwendter, M. & Jones, P. A. DNA methylation and breast carcinogenesis. *Oncogene* 21, 5462–5482 (2002).

98. Jubb, A. M., Bell, S. M. & Quirke, P. Methylation and colorectal cancer. *J Pathol* 195, 111–134 (2001).
99. Kondo, Y. & R Issa, J. P. Epigenetic changes in colorectal cancer. *Cancer Metastasis Rev* 23, 29–39 (2004).
100. Tsou, J. A., Hagen, J. A., Carpenter, C. L. & Laird-Offringa, I. A. DNA methylation analysis: a powerful new tool for lung cancer diagnosis. *Oncogene* 21, 5450–5461 (2002).
101. Li, L. C., Okino, S. T. & Dahiya, R. DNA methylation in prostate cancer. *Biochim Biophys Acta* 1704, 87–102 (2004).
102. Bastian, P. J. et al. Molecular biomarker in prostate cancer: the role of CpG island hypermethylation. *Eur Urol* 46, 698–708 (2004).

15 G Protein-Coupled Receptors and Comparative Genomics

Steven M. Foord

CONTENTS

ABSTRACT

In the next few years, the genomes of many more mammalian species will be sequenced. We will be able to compare gene complements, protein sequences, and selection pressures. What might these comparisons suggest? This chapter discusses what we might learn from G protein-coupled receptors and their ligands in particular and from these approaches in general.

15.1 INTRODUCTION

Traditionally, a discussion of comparative genomics would refer to those cellular systems that are universal or at least appear to be so. These discussions are usually driven from data generated using model organisms and genetic approaches. Studies using fruit flies or yeast as model organisms have contributed to our understanding of the cell cycle, the secretory pathway, and G protein-coupled receptor (GPCR) signaling, to name just three. The study of differences between species gets less attention. We now have more genomes available, particularly among the mammals.

Although these genomes will not be sequenced to completion (the return on investment beyond twofold sequence coverage is relatively poor), the volume of data at our disposal is going some way toward completing the inevitable gaps in sequence.

Some genes matter more than others to the pharmaceutical industry. The GPCRs matter more than most. About 40% of marketed drugs act via GPCRs, and so they get considerable attention. The GPCRs are activated by ligands as diverse as light, odorants, lipids, monoamines, peptides, or proteins. Discovery of the ligand for a GPCR is significant because it provides biological context and sometimes an effective tool that can reveal biology. The complement of GPCRs (and their ligands) within the genomes of rodents and humans is of particular interest to the pharmaceutical industry because of their mandated role in toxicological testing. The mouse genome is essentially complete, and the rat genome is approaching completion. If the target gene is not present in the rodent genomes, then the most amenable and best-characterized experimental models are generally not available. Just as serious is the unsuitability of rodents as models of toxicology; if the target is absent, then it is harder to judge potential toxicology in the absence of efficacy. A second major genome comparison of interest is that between humans and primates. There are many disorders (and obviously other traits) that appear to manifest only in humans. The sequencing of the genomes of other primates has suggested mechanisms of evolution that were not obvious from examining the genomes of more distant relatives.

This chapter assesses how close we are to recognition of the differences between genomes and understanding how they might have an impact on our approach to drug discovery. A focus on GPCRs has the advantage of pharmaceutical relevance, size (at over 700 members, they are the largest gene family), and knowledge (as more is known about this particular family than most others, it is more likely that the examples will remain current for longer).

15.2 THE GPCR COMPLEMENT OF DIFFERENT SPECIES

The broad categories of GPCRs that we find in the genomes of mammals (families A, B, C, and Frizzled receptors) arose about 530 million years ago with the evolution of multicellular organisms such as nematodes and insects, typified by the genomes of *Caenorhabditis elegans* and *Drosophila melanogaster*, respectively.[1] Although there are different classes of GPCRs, in "lower" organisms that share similar signal transduction machinery their receptors have little homology (e.g., the GPCRs in yeast and slime molds). The conservation of GPCRs throughout so many millennia and across so many species has provided us with a vast collection of sequences to refer to and draw from. Families B, C, and Frizzled also have conserved sequence motifs in their amino termini. In addition, all of the receptors have seven transmembrane domains, which enables candidate sequences to be evaluated on the basis of the properties of their amino acids even if there is little sequence homology.

Putative orthologs can be identified between different species using reciprocal BLAST (Basic Local Alignment Search Tool).[2] This is to say a human gene sequence (sequence X) is searched using a sequence-matching program such as BLAST against the mouse genome. The best match from the mouse genome should find sequence X as its top hit in the human genome when the reverse process is performed. If the rat genome is included, then three-way reciprocal BLAST searches generate "trios" of genes shared among the three species. Reciprocal BLAST does sometimes mislead; a more accurate method for determining orthologs is to use phylogenetic methods (there are a number, such as neighbor-joining, Bayesian, or maximum parsimony approaches) that compare multiple sequences.[3] But, these are computationally more expensive and require the initial "collection" of candidate sequences. Gene duplications are one of the most common reasons why genomes differ. If the rodent genes have duplicated, then reciprocal BLAST searches may only identify one of the two duplicated genes, but phylogenetic analysis would identify both sequences.

A directory of all the GPCRs in the human genome (except for olfactory receptors) is maintained on the International Union of Basic and Clinical Pharmacology (IUPHAR) Web site (http://www.iuphar-db.org/GPCR/index.html) along with the mouse and rat orthologs associated with those receptors.[4,5] The list has been assembled from the literature and is updated by IUPHAR correspondents. It currently lists 82 human GPCRs that do not have complete mouse or rat orthologs. The database is updated every 6 months, and now there are another 37 murine sequences to add to the receptor list (leaving 26 mouse and 56 rat sequences missing). This suggests that the mouse genome is still more complete than that of the rat.

Examination of this list reveals many features common to cross-genome comparisons between any species.

1. **Incomplete genomes.** It is clear that any conclusions drawn can be invalidated by the discovery of a gene that had been missed previously because of gaps in genome sequence. Genomes from two similar species help to ensure against this (mouse/rat, chimpanzee/rhesus). However, there are differences even between species as close as these examples.
2. **Pseudogenes.** GPR42 is unusual in having a complete open reading frame but no detectable expression or function.[5] Pseudogenes usually show clearer evidence of their disruption. There is evidence that two receptors, GnRH2 and EMR4, are disrupted in humans but not necessarily in other primates.[6,7] The most subtle pseudogenes are those that exist only in some individuals. There are three reported GPCRs for which this appears to hold: Trace amine 3, GPR33, and CCR5.[8–10] The resistance of certain individuals to human immunodeficiency virus (HIV) has been attributed to these individuals having a deletion in the chemokine receptor CCR5. This renders the receptor relatively ineffective as both a receptor and an HIV cellular entry point. To detect this type of event then, the genotypes from many individuals have to be determined, which is clearly more likely to happen for humans than any other species.
3. **Gene fusions.** The reported "fusion" of P2RY11 appears to be human specific.[11] Species-specific gene fusion is one mechanism for species-specific genome changes.[12,13]
4. **Gene duplication.** All of the genes in Table 15.1 represent relatively recent primate gene duplications, but most represent even greater expansions. For the FPR family, rodents show significant expansion, the MRG family appears to be expanding in rodents and primates, whereas the EMR

family represents a primate-only expansion.[14–17] It is important to note that most in the list have been defined as human specific after detailed phylogenetic analysis. These receptor families are under strong positive selection pressure, but it is not clear what that selection pressure is.

The 18 receptors listed in Table 15.1 are unlikely to have rodent orthologs as the genes are missing in both mouse and rat. It is worth pointing out that the association of a gene with a duplication event does not prevent it from existence as an effective drug target. For example, rodents have duplicated angiotensin AT1 receptors, and yet AT1 receptor antagonists are effective antihypertensives in the clinic. The absence of a rodent ortholog is a much greater impediment to drug discovery than the presence of gene duplications.

TABLE 15.1
GPCRs Present in Primate but Not in Rodent Genomes

	Macaca Mulatta	Pan Troglodites
5HT1E	XP_001090804	
GPR148	XP_001094021	ENSPTRG00000029194
GPR78	XP_001090919	XP_526521
MLNR	XP_001101857	XP_001149683*
OXER1	XP_001110986	XP_001139923
MAS1L	ENSMMUG00000020688	XP_518317
MRGPRX2	NP_001035512	XP_521864
MRGPRX3	NP_001035511	XP_521853
MRGPRX4	NP_001035708	XP_521855
MCHR2	NP_001028120	XP_527461
NPBWR2	XP_001113051	XP_514795
FPRL2	XP_001116463	XP_524363
GPR32		XP_001173894
GPR42		
P2RY8	XP_001115826	XP_001175457
P2RY11	ENSMMUG00000017216	ENSPTRG00000029133
EMR2	NP_001033751	XP_512446
EMR3		ENSPTRG00000010598

Note: The table lists those "nonsensory" GPCRs that are present in the human genome but not in those of the mouse or rat. The first column lists the HUGO gene names. The majority of the receptors are in family A; EMR2 and EMR3 are in Family B. There are no differences in the family C complement. The NCBI protein sequences for the Macaque monkey and the chimp are given if available, and the Ensembl IDs are given if they are not. GPR42 is probably a pseudogene, leading to the conclusion that all human nonsensory GPCRs have a primate ortholog.

15.3 PHYLOGENETIC ANALYSIS OF GPCRs

Do the human receptors without rodent orthologs fall into any specific family group? The phylogenetic analysis that enables the definition of orthologs also provide a way of visualizing their relationships with other receptors. However, phylogenetic analyses on a large scale are computationally expensive. Figures 15.1 and 15.2 show phylogenetic analyses of family A GPCRs constructed using different computational shortcuts. In Figure 15.1, the alignments between GPCR sequences within family A were forced according to certain well-established conserved residues. These are typified by amino acid motifs such as the DRY sequence at the bottom of transmembrane 3 or the NPXXY motif within transmembrane 7. By this means, sequence alignments for the majority of human GPCRs can be obtained and a rough assessment of their phylogenetic relationships made. With about 275 input sequences, the result is complex and difficult to represent, but in broad terms the human complement of family A GPCRs (excluding olfactory receptors) falls into five main groups according to this particular analysis (and in agreement with those of others[1]). The receptors with the longest branch lengths have been labeled. It is interesting that many of these receptors either have no recognized ligands or have only just had their ligands discovered. This suggests a degree of mechanistic novelty. The main groups

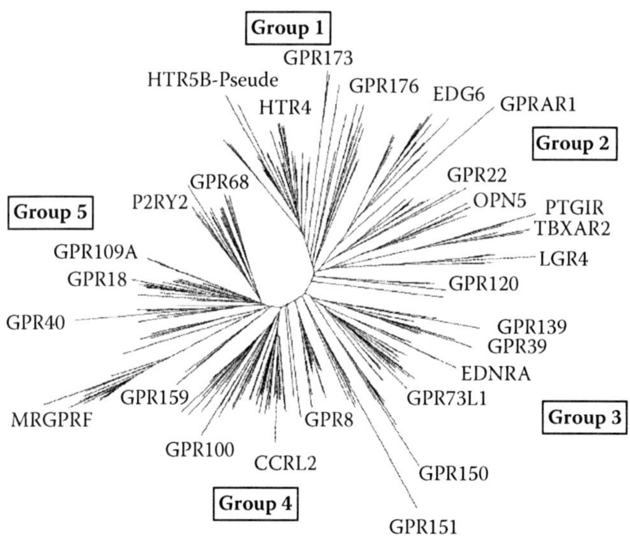

FIGURE 15.1 Phylogenetic analysis of all nonsensory human family A GPCRs after alignment forced according to INTERPRO signatures such as DRY in transmembrane 3 and NPxxY in transmembrane 7. The neighbor-joining method was used. The receptors have been identified that represent those with the longest branch length for each major cluster. The clusters have been assigned to groups (see text). Group 1, the monoamine-like receptors; Group 2, a diverse group that contains the opsins, cannabinoid, lipid, prostaglandin, and glycoprotein/LRG-type receptors; Group 3, brain/gut peptide receptors; Group 4, chemokine receptors; and Group 5, metabolic receptors (including purinergic, thrombin, and free fatty acid receptors).

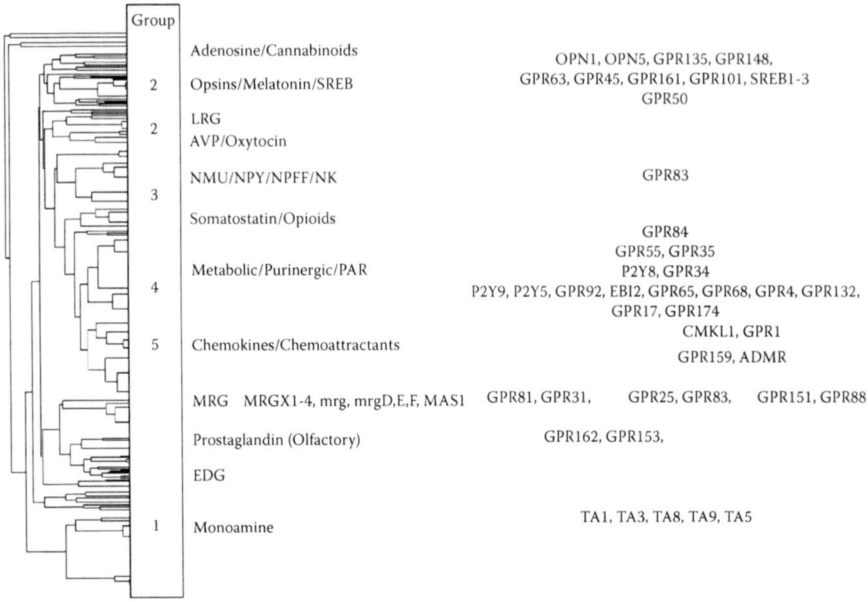

FIGURE 15.2 Phylogenetic analysis of the alignments from predicted "inward-facing" residues from the nonsensory GPCRs for family A in the genomes of humans, the mouse, and the fugu fish (*Tetroadon nigroviridis*). The detail of the figure is too complex to be clear, annotated or not, but the major branch points are shown for comparison with the analysis in Figure 15.1. The groups of nonliganded (orphan) receptors are shown for each group.

are (1) the monoamine-like receptors; (2) a diverse group that contains the opsins, cannabinoid, lipid, prostaglandin, and glycoprotein/LRG-type receptors; (3) brain/gut peptide receptors; (4) chemokine receptors; and (5) metabolic receptors (including purinergic, thrombin, and free fatty acid receptors).

When the GPCR complement of other species is viewed against this grouping, it is clear that each set shows differences, but some are more different than others.[1] Monoamine, lipid, and peptide receptors (groups 1, 2, and 3) are found in insects, nematodes, and fish. Insects show the first melatonin and significant opsinlike receptors. However, insects and nematodes do not appear to share mammalian prostaglandin receptors (from group 2) or purinergic (group 5) and chemokine receptors. Prostaglandin receptors are represented for the first time in Chordates (550 million years ago), but purinergic, olfactory, and leucine-rich repeat-bearing (LRG) receptors (group 2) appear first in fish (420 million years).

Chemokine and metabolic receptors (groups 4 and 5) are not found in insects and nematodes. This may be attributed to the evolution of acquired immunity in the former. The evolution of mechanisms that enable species to survive beyond a single breeding season may be the case for the latter. The discovery that fish (specifically, the pufferfish *Takifugu rubripes*) contained orthologs of most mammalian GPCRs prompted a further phylogenetic analysis but using a slightly different method. Instead of "forcing" the alignment using key motifs, the sequence of each GPCR was reduced to a smaller and more computationally tractable level (important when performing a phylogenetic analysis of about 275 sequences when they are diverse and at least 300 amino acids long) from human, mouse, rat, and pufferfish genomes). The GPCR sequences were progressively reduced to transmembrane domains, then to "inward-facing residues" using the rhodopsin model as a standard predictive template. This method actually removes "consensus" residues from the alignment as they do not contribute directly to the inward face of the receptor. It also removes elements from the receptors that might be conserved because of ligand recognition (for peptide ligands) and G protein coupling (through the removal of external and internal loops). It was therefore a surprise that the results had major similarities to those analyses performed using the entire sequences of the receptors.

The analysis is shown in Figure 15.2 as a rooted rather than radial tree for clarity as approximately three times more sequences are involved. Groups 1, 3, 4, and 5 remained substantially the same in the new analysis, but the second group is markedly changed. There appears to be a clear distinction in the inward-facing analysis among adenosine/cannabinoid receptors, opsin/melatonin/SREB receptors, lipid (EDG) receptors, and prostaglandin receptors. These receptors have (or are anticipated to have) small-molecule ligands. It will take significant structural insights into these receptors to determine how significant these phylogenetic analyses are, but superficially the result suggests different mechanisms of activation via disparate ligands. In contrast, the LRG and MRG receptors might be expected to be activated primarily through their amino termini and external loops (both excluded from this analysis). They fall into distinct but related groups and so may share some common ancestor or mechanism of activation.

Taken together, the analyses revealed that the receptors unique to primates over rodents do not fall into any particular group and are distributed throughout the phylogenetic groups. This is in contrast to the distribution of novel receptors in groups 4 and 5 in fish over lower organisms.

15.4 PHYLOGENETIC ANALYSIS AND THE PREDICTION OF LIGAND TYPE

The granularity of these types of analysis suggests that it is possible to predict which types of ligand would activate each receptor type. Orphan GPCRs are receptors for which the native ligand has still to be discovered. There remain about 100 of them in the human genome for family A type GPCRs, excluding olfactory receptors. Pairing GPCRs with their ligands suggests the function of the receptor. It usually provides a means of activating the receptor and so establishing an assay and can provide a pharmacological tool. The phylogenetic analysis in Figures 15.1 and 15.2 suggests that one reason for our poor performance in "deorphanizing" receptors is the small number of GPCRs for which it is possible to either make a prediction of the ligand or act on that prediction. Figure 15.2 lists the family A receptors that remain orphans in each of the groups defined by phylogenetic analysis of the inward-facing residues. Most are in the same groups as they were in the whole-sequence analysis, and most remain in the "metabolic group." Many of the newly discovered ligands for GPCRs have turned out to be in this group. Examples are purinergic ligands, carboxylic acids, and intermediates in the Krebs cycle. These ligands were known through diligent biochemistry, but it is difficult to identify such candidate molecules from scratch, even though this is the type of ligand that is most likely to activate the remaining orphan GPCRs.

Hardly any orphan GPCRs could be confidently predicted to have peptides as ligands. It is particularly difficult to predict GPCR peptide ligands because it is difficult to establish rules for defining both small bioactive peptides and small genes. We have made an attempt at predicting peptide/protein candidates on the basis of the properties of the ligands that are known to activate GPCRs.

In general, GPCR peptide ligands (1) have a signal peptide; (2) lack a transmembrane domain; (3) are no longer than 300 amino acids; (4) do not have a domain that is shared by a protein that is not a GPCR ligand; (5) have no close paralogs; and (6) show low gene expression.

The number of gene products that break these rules is shown in Figure 15.3. The input sequences were derived from the combined and nonredundant human, mouse, and rat proteomes at GlaxoSmithKline (GSK) (about 26,000 protein sequences overall). About 7 of 70 GPCR peptide ligands break these rules, whereas 25,912/26,147 of the nonredundant proteins do (leaving 235 candidate peptide GPCR ligands). Since 2004 when this work was done, only 1 of the 235 has been shown to be a GPCR ligand;

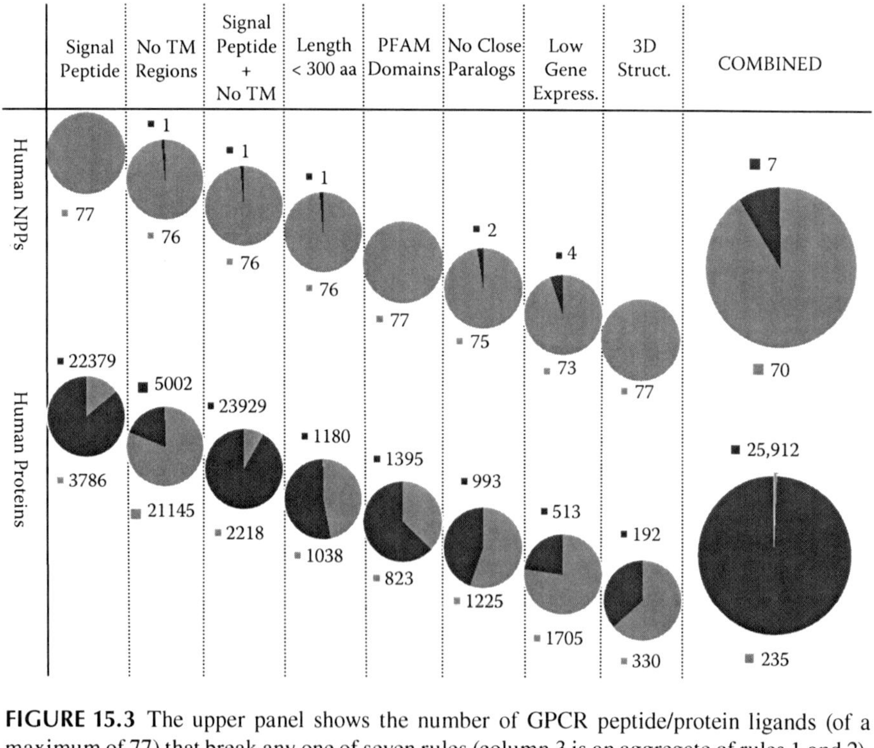

FIGURE 15.3 The upper panel shows the number of GPCR peptide/protein ligands (of a maximum of 77) that break any one of seven rules (column 3 is an aggregate of rules 1 and 2). Only seven fail any of the rules. In contrast, the lower panel shows the same rules applied to the nonredundant proteomes of the human, rat, and mouse genomes (less the GPCR ligands). Only 235 pass all seven rules, but only one of the list (Norrie disease protein) has been shown to be a GPCR ligand (for FZD4) since this analysis was performed in 2004.

Norrie disease protein has been reported to activate the Frizzled 4 receptor (and this is not a family A GPCR).[18]

Despite the apparent failure of our reductionist approach in the prediction of novel GPCR ligands, it remains possible that it might yet be effective if combined with a similar analysis of the genomes of other species. The comparative genomics approach has proved effective in the identification of GPCR ligands and particularly through the comparison with fish. One notable case concerned the discovery of a new member of the CGRP (calcitonin gene-related peptide) peptide family, intermedin.[19–21] This peptide was discovered in fish before it was identified in the human genome. The genomes of fish contain multiple copies of genes that resemble CGRP and its receptors (which consist of calcitonin-like receptors and accessory proteins called RAMPs).[19] This appears to be a biological system that is under strong selection pressure in the fish, and it is expanded to a greater extent than is the case in mammals, but it pointed the way to a hitherto unrecognized human hormone.

The methodology might also be more effective if there were consensus on the number of exons within the human (at least) genome let alone the number of genes (discussed in more detail next). Neuropeptides are particularly small genes, and their prediction is difficult.

15.5 ISSUES WITH GENE IDENTIFICATION

During the course of our attempts to predict putative GPCR peptide ligands from the human or rodent genomes, it became apparent just how variable the source material for these analyses actually was. The annotation of the human genome continues to change all the time and quite significantly. Although the completion of the Human Genome Project was celebrated in April 2003 and sequencing of the human chromosomes is essentially "finished," the exact number of genes encoded by the genome is still unknown. Most estimates are in the range 20,000–25,000, a surprisingly low number for our species. The reason for so much uncertainty is that predictions are derived from different computational methods and gene-finding programs. Some programs detect genes by looking for distinct patterns that define where a gene begins and ends (ab initio gene finding). Other programs look for genes by comparing segments of sequence with those of known genes and proteins (comparative gene finding). While ab initio gene finding tends to overestimate gene numbers by counting any segment that looks like a gene, comparative gene finding tends to underestimate since it is limited to recognizing only genes similar to those seen before. Defining a gene is problematic because small genes can be difficult to detect, one gene can code for several protein products, some genes code only for RNA, two genes can overlap, and there are many other complications.

To exemplify these approaches, Ensembl[22] and AceView[23] each reported about 1 million exons within the human genome. Ensembl tends to use predictive methods that rely on what we know genes look like. AceView tends to rely on physical evidence such as expressed sequence tags (ESTs) that reflect gene expression. Only about 50% of the exons these two systems describe are common to both lists. About 37% are completely identical, and 12% are identical but with some degree of imprecision regarding the exon boundary. Of the remaining calls, 13% are unique to Ensembl and 28% unique to AceView. Given this degree of discrepancy, it is not surprising that comparative genomics is difficult when the object is to spot the differences rather than the similarities.

Gene predictions will have to be verified by labor-intensive work in the laboratory before the scientific community can reach any real consensus. However, there are some computational approaches that can be used to evaluate these genes. Specifically, similarity searches should establish whether the sequences have matches to others in the databases (and in which reading frame) even though they may not have a strict ortholog; even in the absence of a homologous sequence that can be detected by BLAST, it may be possible to identify discrete motifs. The Vertebrate Genome Annotation (VEGA) database is a central repository for high-quality, frequently updated, manual annotation of vertebrate finished genome sequence.[24] The manual curation and high sequence fidelity found in these regions facilitates gene calling. It is worthwhile listing some of the approaches that contribute to this type of curation. Duplicated genes (or those that are not duplicated) can be investigated using the BLAST-like alignment tool (BLAT). This matches sequences against the reference human genome and so provides clarity regarding what is a distinct gene and what represents a sequencing error.[25] EST analyses provide some indication of the frequency of the source sequences (and if they are only evidence by prediction).[23]

Affymetrix chip[26] hybridization (if probe sets represent these sequences) can also provide evidence of expression. The use of the new "array chips" also provides a potential data source that will lend support (or otherwise) to genes that are predominantly supported by EST data. At the genomic level, syntenic relationships (the order in which genes appear on chromosomes) may reveal ancestral pseudogenes in some species.[27] Single-nucleotide polymorphism (SNP) databases may provide evidence of significant variation within a given region of DNA, and dN/dS ratios (the ratio of nucleotide changes to resulting amino acid changes) provide some indication of the selection pressure detected at a given locus. Genes are generally conserved above nongenes.[28,29]

All of these approaches can be accessed from the public domain using any standard Web browser. The relevant sites usually support batch queries, so even quite large data sets can be processed.

15.6 GENE COMPARISONS

A gene list is important because so many technologies now operate at the whole-genome scale. Genetic, expression array, and inhibitory RNA technologies all generate data that require an index of all the genes in the human genome with some representation of the confidence that supports their existence. Here at GSK we draw human genomic information from many disparate sources, principally the National Center for Biotechnology Information (NCBI)[28]; and University of California, Santa Cruz (UCSC)[25]; and Ensembl.[22] At present, there are only 19,928 genes in the GSK human genome. The list represents many genes that are unique to each of the main public domain sources. This is a conservative list, but it is still adequate for representing whole-genome studies (such as expression data derived from Affymetrix chips).[30] In the spring of 2005, using reciprocal BLAST, these 19,928 genes were checked for orthologs in any of 10 published mammalian genomes; 14,598/19,928 human genes had trios of mouse and rat orthologs. A further 4,289 had at least one rodent ortholog by reciprocal blast. There were approximately 1,041 human genes that did not appear to have orthologs in any of the mammalian genomes examined. This was a larger number than we expected, and so we looked at the list in more detail.

15.6.1 Analysis of "Human-Only" GPCRs

In the GPCR analysis outlined above, there was no human GPCR gene (of 375 examined) that was not assigned either a chimpanzee (*Pan troglodytes*) or rhesus monkey (*Macaca mulatta*) ortholog by either Entrez Genome or Ensembl. At the human/rodent level, there were only 18 human (nonsensory) GPCR genes without rodent orthologs. For a family of 375 genes, 5% is a relatively small number. However, to find the same percentage of human genes without an ortholog in any mammalian species (1,041/19,928 = 5%) is surprising (given we only found one GPCR unique to humans over primates), and it was examined in more detail.

Of the 1,041 human genes that did not appear to have orthologs in any of the mammalian genomes examined, 74 appeared to be olfactory GPCRs. How accurate is this number? It is surprising that human-only olfactory receptors exist as the general consensus is that a deterioration of the olfactory repertoire occurred during primate evolution, with a particularly sharp decline in the human lineage.[31] Olfactory receptors are not well represented by either Ensembl or NCBI. We combined these sources with that published by Niimura and Nei[32] and our own analysis of the genome. Build 41 of HORDE (the public olfactory receptor compendium) lists 391 human olfactory receptors and 464 related pseudogenes.[33] The GSK analysis was almost identical (384 genes and 462 pseudogenes). However accurate the number of human-only olfactory genes is, the number of human-only genes is definitely less than 74. A total of 40 receptors had Ensembl IDs, while 34 did not.

Ensembl uses a process for ortholog calling that involves a phylogenetic analysis step. It shows broad agreement with simple reciprocal BLAST, but it is also able to find more complex one-to-many and many-to-many relations.[34] Of the 40 olfactory genes with Ensembl ID numbers, reexamination of their current status revealed only one receptor annotated as human only (ENSG00000181017, HsOR11.3.79). ENSG00000180494 was annotated as having a many-to-many ortholog relationship to the chimp. ENSG00000180477, ENSG00000186483, ENSG00000184055, ENSG00000171481, and ENSG00000181927 were described as having one-to-many relationships with their chimp orthologs. Eight genes (ENSG00000181950, ENSG00000183444, ENSG00000185074, ENSG00000184321, ENSG00000185074, ENSG00000181950, ENSG00000177381, and ENSG00000173285) had no annotation regarding their orthologs. This makes only 13/40 receptors that are likely to have no ortholog or complex ortholog relationships. Another view on these data is that they suggest that fewer than 25% of the genes we thought to be human specific are likely to be so one year later. Detailed phylogenetic analysis of human and chimp olfactory sequences has revealed (within just four families) about 30 human olfactory receptors that do not have simple chimp orthologs.[31] When sequences are as homologous as the olfactory receptors, phylogenetic analysis is required to enable definitive conclusions.

It is possible that the differences between human individuals are substantial enough to make a definitive list of human olfactory receptors (and the number of olfactory receptors specific to humans) impossible. The number of olfactory receptors in the human genome may well vary from individual to individual. Sensory receptors are among those genes that exhibit significant copy number variation.[35] This means that the individuality of those people who contributed DNA to genomic sequence studies is more marked for olfactory receptors than most genes. Olfactory receptors exist in different parts of the genome that have exactly the same sequence. OR2J3 (HUGO name) is an example for which three identical or almost-identical sequences lie in tandem at chromosome location 6p22.1. It is possible that these may not be present in all individuals. Many olfactory receptors are so similar to each other that each has to be mapped to the genome to differentiate (or not) between sequence duplication and nonsynonymous SNPs, but it is possible that the genome is not reliable in this respect because of copy number variation.

There is evidence to suggest that olfactory repertoire is one of the greatest distinguishing features between genomes (and not just between individuals). For olfactory receptors, it is not because of loss of function in the human repertoire (although there is a relaxed constraint on the human rather than chimp set) but more because each species has selection pressure that leads to the expansion of different sets.

Bitter taste receptors show relatively little difference in the proportion of genes/pseudogenes between species but lineage-specific expansions in each and evidence for significant selection pressure exist.[36] In time, it may be possible to identify which olfactory "talents" associate with each set of receptors in each species in a manner similar to that published for bitter taste receptors. A possible candidate might be the ability of humans to smell the metabolites of asparagus in urine, a trait that is now thought to relate more to olfaction than metabolism.[37,38]

15.6.2 Human-Specific Genes?

As described, the list of about 1,041 human-specific genes contained 74 olfactory receptors, of which fewer than 25% were estimated to be really human specific when the genes were looked at in detail.

The immune system contributes a significant number of human-only genes just as it contributes species-only and individual-only genes. What is the extent of the immune system? Many of the human-only genes encode surface antigens and the enzymes that control glycosylation (cell surface complement through another route). Before the list of 1,041 human-specific genes was reached, 77 sequences were removed because they represented genes that are inherently variable even within a given species, such as immunoglobulins, T-cell receptors, and major histocompatibility antigens.

A further 20 sequences were removed that represented genes that appeared to have been formed as the result of recombination with human retroviral elements. LOC113386 is one of a number of genes that appear to be real yet share significant (>80%) identity with mobile repetitive elements. A clearly significant difference between species is their susceptibility to different agents that can alter their genomes.[12,13]

Detailed examination of the remaining genes produced a similar reduction. First, 455 genes were excluded because they had been removed by either Ensembl or Entrez Genome as they were no longer considered genes (in the sense that they no longer considered them encoding proteins, functional proteins, or reliable predictions). On this basis, it is likely that a significant number of genes will have been added to both of these databases since our analysis in 2005, and that these have not been included in what is described here.

Table 15.2 shows the breakdown of the remaining 583 genes that appeared to be human only. Of these, 49 genes turned out to be "variable" genes or those associated with retroviral sequences that had not been screened out earlier. The majority of the genes (333) were represented in the Ensembl database and over the past year most (249) have become annotated with a clear ortholog in the Ensembl database. Some 52 genes were annotated as human only and 32 genes possibly unique to humans but with a complex relationship. For those genes that are not in Ensembl then, the best estimate of the current status of their orthologs comes from the Homologene database within the NCBI suite of tools.[28] This database uses algorithms that define "best match" rather than true ortholog status. Having said this, the results are similar. There are 83 sequences without the benefits of the analysis in either the Ensembl or Homologene databases. All of these human-only or complex relationship genes were checked against the InParanoid database (InParanoid) run by the Swedish Bioinformatics Centre,[39] and those that remained in that category are listed in Table 15.3. This is a short list considering the phenotypic differences between humans and chimpanzees. It is notable that many genes on the list contribute to intercellular recognition even though they might not be formally considered elements of

TABLE 15.2
Analysis of Human-Specific Genes

Source	Total	Now with Ortholog	Complex Ortholog Relationship	Human Only
Ensembl	333	249	32	52
Entrez Genome	118	41 (65 not in Homologene)	4	8
Other source	83			
	583		36	60

Notes: The breakdown of 1,041 supposedly human-specific genes produced from an analysis performed in 2005. There were 458 genes withdrawn from their original source databases (Entrez Genome or Ensembl), leaving 583. There were 49 other genes withdrawn from the analysis as they turned out to be "variable" genes or those associated with retroviral sequences that had not been screened out earlier. Fewer than 100 human-specific genes remain, and almost half of those have a complex relationship to one or many primate genes. Over 100 genes remain to be analyzed, and it is also probable that many other genes will have been entered into the human databases since this study.

TABLE 15.3
Homology of Human-Only Genes Based on InParanoid Database[39] Analysis

Genes with Little Homology to Others	Genes with Homologs	Genes that Appear Duplicated	Gene Families with Many Human-Only Genes (>75 total)
AF130093	FLJ42953	LOC390688	DAZ family
BC026043	LOC389857	LOC440872	GAGE family
C11orf72	LOC392242	LOC642669	MAGE family
C17orf55	LOC441294	LOC646177	SSX family
C18orf56	LOC641922	LOC653114	SPANX family
FLJ25102	HCA25a	LOC727848	KERATIN family
FLJ31659	BPY2B	RBMY1F	ZNF family
FLJ45121	LOC440839	OPN1MW2	TRIM family
FLJ46300	CFHR4	PLGLB2	Ribosomal proteins
FLJ26056	OBP2A	LOC390033	Golgi antigen family
ATXN3L	DEFB107B	LOC644054	Olfactory receptors
BC006438	ICEBERG	OR2J3	Other sensory receptors
PBOV1	MT1G		
LOC653363	VCY		
LOC653483	LOC196120		
	POTE15		
	LOC440776		
	LOC644038		
	LOC644739		
	LOC653441		
	LOC727773		
	LOC727851		
	LOC727858		

Note: This table details the derivation of 60 human-only genes, 36 genes that could be human only and that share complex ortholog relationships with those of other species, and 83 genes that appear human only from our analysis but are not represented in the Entrez Genome or Ensembl databases. Table 15.3 shows the output of looking these genes up in the InParanoid database and subsequent sequence analysis; 15 genes appeared to have no homologs, 23 had at least one homolog, 12 genes appear to be exactly duplicated (one version listed), but the majority of human–only genes fell into just 12 large gene families.

the immune system. For example, the GAGE, MAGE, SSX, and SPANX families are all cancer-associated antigens.[40] Other proteins on the list may have a role in maintaining host defense. The TRIM family has been linked to immune function and resistance to retroviral infection. The Golgi antigen family contributes to cell surface glycosylation and so to the recognition of self.[41]

Probably the most important thing to take from the short list of human-only genes is its shortness. Although the list contains many transcription factors that probably control many other genes (the ZNF zinc finger containing transcription factors, for example), it is clear that even subtle changes can produce major effects, and that these can happen with genes that are almost entirely conserved. The best-known example of this is the mutation in the transcription factor FOXP2 that associates with disorders in speech[42] but is not human specific — although the region of the protein that harbors the crucial mutation is. In some instances, phenotypes (such as hair type) can be associated with genotype (type of keratin),[43] but in the majority of cases the effect of the change is subtle. For example, FOXP2 is not a "gene that controls speech" but a widely expressed transcription factor that has a complex role in development, and one of these roles appears to be in the development of the centers that process speech. Varki and Altheide[40] listed about 30 genes that have human-specific alternations and their associated phenotypic changes, but few of these genes are actually human specific (even though their mutations and associated diseases unfortunately are). Some of those genes have already been discussed as they are GPCRs (olfactory receptors, bitter taste receptors, and EMR4), but most of the list is anonymous. This is mostly because about half of the list is comprised of genes that have names that indicate very little, if anything, is known about their function.

There is an increasing body of literature that focuses on selection pressure between orthologs,[29,31,40,42–44] which illustrates evolutionary pressures. Lists like that represented in Table 15.3 do not contribute genes to these analyses because they do not have orthologs and so represent the final product of selection pressure (even though there is never a "final product" of evolution). However, such lists do prompt questions to be asked when selection pressure is evaluated between orthologs at the whole-genome level.

15.6.3 Limitations of This Analysis

The first limitation of this study is the likelihood that significant parts of it can be disproved by the simple discovery of another primate gene, the realization that a human gene is no longer likely to have a functional product and a reasonably simple phylogenetic analysis (which is likely to reveal a closely related gene is human specific, not the one listed). With regard to GPCRs, there is enough background information to feel confident in the numbers presented, but a whole-genome analysis is a different prospect (nonetheless facilitated by this review).

A second limitation of this chapter is its narrow scope. Concentrating on coding genes alone is a gross oversimplification. The discovery of micro RNAs (miRNAs) highlights the importance of noncoding genes. The differential expression of genes leading to altered developmental patterns or different physiology will be more important than the absolute number and exact nature of the genes we have. For example, comparisons of human and chimpanzee brains on the basis of which genes showed coordinated changes in expression revealed that the patterns recapitulated evolutionary hierarchies, with white matter > cerebellum > caudate nucleus > caudate nucleus + anterior cingulate cortex > cortex.[45] This was not evident if simple gene expression

30. Affymetrix. Available at: http://www.affymetrix.com/index.affx.
31. Giladl, Y., Man, O. & Glusman, G. A comparison of the human and chimpanzee olfactory receptor gene repertoires. *Genome Res.* **15**, 224–230 (2005)
32. Niimura, Y. & Nei, M. Evolution of olfactory receptor genes in the human genome. *Proc. Natl. Acad. Sci. U. S. A.* **100**, 12235–12240 (2003).
33. The Human Olfactory Receptor Data Exploratorium (HORDE). Available at: http://bioportal.weizmann.ac.il/HORDE/.
34. Gene Orthology/Paralogy predection method at Ensembl. Available at: http://www.ensembl.org/info/data/compara/homology_method.html.
35. Wong, K.K. et al. A comprehensive analysis of common copy-number variations in the human genome. *Am. J. Hum. Genet.* **80**, 91–104 (2007).
36. Fischer, A., Gilad, Y., Man, O. & Paabo, S. Evolution of bitter taste receptors in humans and apes. *Mol. Biol. Evol.* **22**, 432–436 (2005).
37. Mitchell, S.C. Asparagus and malodorous urine. *Br. J. Clin. Pharmacol.* **27**, 641–642 (1989).
38. Richer, C., Decker, N., Belin, J., Imbs, J.L., Montastruc, J.L. & Giudicelli, J.F. Odorous urine in man after asparagus. *Br. J. Clin. Pharmacol.* **27**, 640–641 (1989).
39. InParanoid: Eukaryotic Ortholog Groups. Available at: http://inparanoid.sbc.su.se/.
40. Varki, A. & Altheide, T.K. Comparing the human and chimpanzee genomes: searching for needles in a haystack. *Genome Res.* **15**, 1746–1758 (2005).
41. Varki, A. Nothing in glycobiology makes sense, except in the light of evolution. *Cell.* **126**, 841–845 (2006).
42. Dorus, S. et al. Accelerated evolution of nervous system genes in the origin of *Homo sapiens. Cell.* **119**, 1027–1040 (2004).
43. Clark, A.G. et al. Inferring nonneutral evolution from human-chimp-mouse orthologous gene trios. *Science.* **302**, 1960–1963 (2003)
44. Fisher, S.E. Tangled webs: tracing the connections between genes and cognition. *Cognition.* **101**, 270–297 (2006)
45. Oldham, M.C., Horvath, S. & Geschwind, D.H. Conservation and evolution of gene coexpression networks in human and chimpanzee brains. *Proc. Natl. Acad. Sci. U. S. A.* **103**, 17973–17978 (2006)

16 Comparative Toxicogenomics in Mechanistic and Predictive Toxicology

Joshua C. Kwekel, Lyle D. Burgoon, and Tim. R. Zacharewski

CONTENTS

ABSTRACT

The availability of complete genomic sequences for multiple model species provides unprecedented opportunities for comprehensive comparative analysis in support of mechanistic and predictive toxicology and quantitative safety assessments. More specifically, comparison studies can be used to inform and define the limits of use of surrogate models used for human risk assessment, drug discovery, and basic research. Moreover, comparative approaches support functional annotation efforts of orthologous genes. However, several factors affect how comparative data will be used, including study design issues such as the array format and experimental design, analysis methods dealing with normalization and the definitions of orthologs and orthologous expression profiles, and the computational identification and empirical verification of *cis*-regulatory elements responsible for species-specific or conserved expression. This chapter reviews the available genomic resources and bioinformatic tools and discusses several of the limitations that hinder the full realization of comparative genomics in mechanistic and predictive toxicology and quantitative safety assessments.

16.1 INTRODUCTION

Whole-genome sequencing has advanced biomedical research by providing the nucleotide sequence of entire genomes for a number of model organisms. These advances were preceded by decades of research investigating the roles of individual genes, proteins, and metabolites in a variety of processes. The functional significance of each gene, protein, and metabolite can now be investigated in the context of their global interactions and relationships and associated with outcomes such as disease and toxicity. The common basis for biology (DNA → messenger RNA [mRNA] → protein) allows research tools and methodology to be shared between models, producing a wealth of information across organisms. This includes comprehensive comparative analyses to identify conserved aspects important in development, homeostasis, disease, and toxicity as well as divergent responses that impart species–species advantages or sensitivities. This chapter focuses on comparative gene expression and its emerging role in mechanistic and predictive toxicology as well as quantitative risk assessment.

16.1.1 SEQUENCING IS NOT ENOUGH: THE ROLE OF TRANSCRIPTOMICS

Comparative analysis assumes that important biological properties and responses are conserved across species and share common mechanisms.[1] This includes the structure and function of coding regions as well as associated regulatory elements (Figure 16.1). *Transcriptomics* (Table 16.1) characterizes the spatiotemporal changes in gene expression, providing information on when and where genes are expressed. Global expression can be monitored using open platforms, such as differential display and serial analysis of gene expression (SAGE), which require little to no a priori knowledge about the genomic sequence of an organism. Alternatively, closed platforms, such as microarray technology, require discrete sequence information prior to experimentation.

16.1.2 WHAT IS FUNCTIONAL ORTHOLOGY?

Functional annotation establishes relationships between the nucleotide sequence and the biological role of the putative gene (Table 16.1). Although focused biochemical assays are the gold standard for determining function, many fail to consider the possibility of a gene product having multiple functions dependent on location or interactions with other proteins. Consequently, a gene product involved in more than one biological process may require different approaches to characterize all of its potential functions. Nucleotide sequence similarity provides preliminary data for the

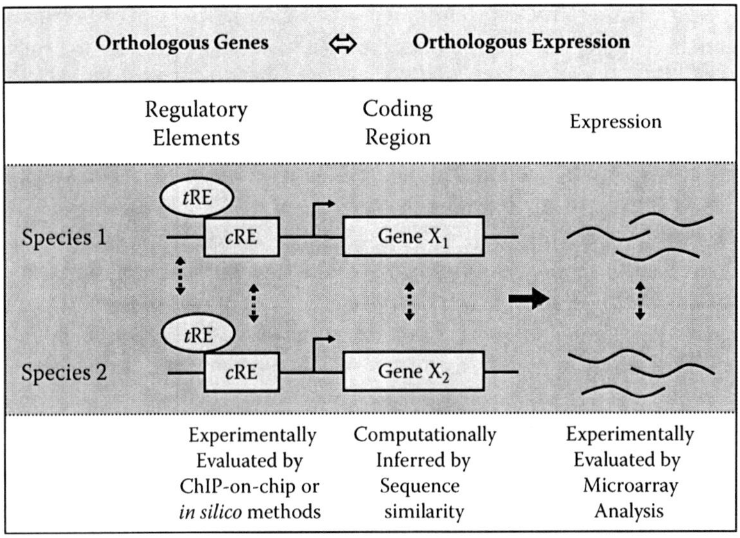

FIGURE 16.1 Functional orthology. Orthology designations based on coding region sequence homology in addition to other criteria are evaluated by expression analysis. Orthologous expression would suggest similar regulatory mechanisms, whereas differential expression of orthologous genes suggests either incorrect orthology designations or divergent regulation.

extrapolation of experimentally based functional annotations across species for orthologous genes. Implicit in this extrapolation is the concept of *functional orthology.*

Although debated,[2] *homology* is commonly defined as the relationship between structurally related genes descendant from a common ancestor (Table 16.1). However,

TABLE 16.1
Key Terminology

Term	Definition
Transcriptomics	Assessing global gene expression at the mRNA level (e.g., microarray analysis, SAGE, differential display, etc.)
Functional annotation	Attributing molecular function, biological process, or tissue location to a specific gene
Functional orthology	Property of orthologs that exhibit similar molecular function, biological process, and tissue location
Homolog	Structurally related gene descendant from a common ancestor
Paralog	Homolog within the same species
Ortholog	Homolog between species
Orthologous co-expression	Property of two orthologs exhibiting similar gene expression patterns across experiment parameters
Experimental parameter	Independent variable that is tested (e.g., treatment, time, dose, disease state, developmental stage, tissue location, etc.)

it is not clear whether structural similarity or common ancestry of orthologous genes also extends to functional equivalence, in terms of both function and regulation. In general, there is insufficient information on a gene-by-gene basis to accurately determine the timing of speciation and gene duplication events that gave rise to the contemporary slate of genomes. In particular, the analysis of structure–function relationships among highly divergent proteins usually proceeds in the absence of this information. Consequently, it cannot be determined with certainty whether two contemporary proteins are orthologs or paralogs (Gerlt and Babbit in Jensen[2]). In many cases, this uncertainty can be mitigated by comparing the structural similarity of the genes to define orthologous relationships (Homologene, http://www.ncbi.nlm.nih.gov/entrez/query.fcgi?db=homologene; Ensembl, http://www.ensembl.org/index.html).

These measures of similarity can also be supplemented with spatiotemporal expression data to assess *orthologous expression,* defined as putative orthologs exhibiting comparable patterns of regulation and expression. *Differentially expressed genes* are those that exhibit a significant change in response to different experimental parameters, such as treatment (e.g., vehicle, chemical, drug, other manipulations); dose (level of the experimental manipulation); time; developmental stage; or disease state. If orthologous genes are regulated in a comparable manner under the same conditions, then we refer to this as *orthologous expression,* providing further compelling evidence, in addition to sequence similarity, of the functional and regulatory equivalence of the putative orthologous genes.

This chapter examines comparative gene expression analysis and its utility in mechanistic and predictive toxicology and quantitative risk assessment. Different experimental and comparative methods as well as the available annotation and interpretative tools and resources are also presented. Furthermore, an assessment of current limitations and needs is discussed to frame the current challenges associated with cross-species comparisons.

16.2 OBJECTIVES

It is generally assumed that biological information collected in one species is transferable to others, including humans, which has far-reaching implications when evaluating the safety and risk of drugs, chemicals, and pollutants to human health and environmental quality (Figure 16.2). Comparative toxicogenomics can be used to assess and refine the relevance of surrogate species in elucidating mechanisms involved in development, homeostasis, disease, and toxicity to improve risk prediction and product development. Fundamental to these efforts is the ability to transfer gene annotation from one species to another with confidence based not only on sequence similarity but also on comparable function and regulation.

Policies regarding product safety, including those for drugs, chemicals, and food derivatives or additives, are largely based on established regulatory testing using model organisms. When extrapolating data between species, uncertainty factors are applied to account for incomplete information regarding the similarity of response between species. These data gaps can be attributed to differences between species in absorption, distribution, metabolism, excretion (ADME), regulation (i.e., DNA regulatory elements, protein–protein interactions, methylation), or protein function

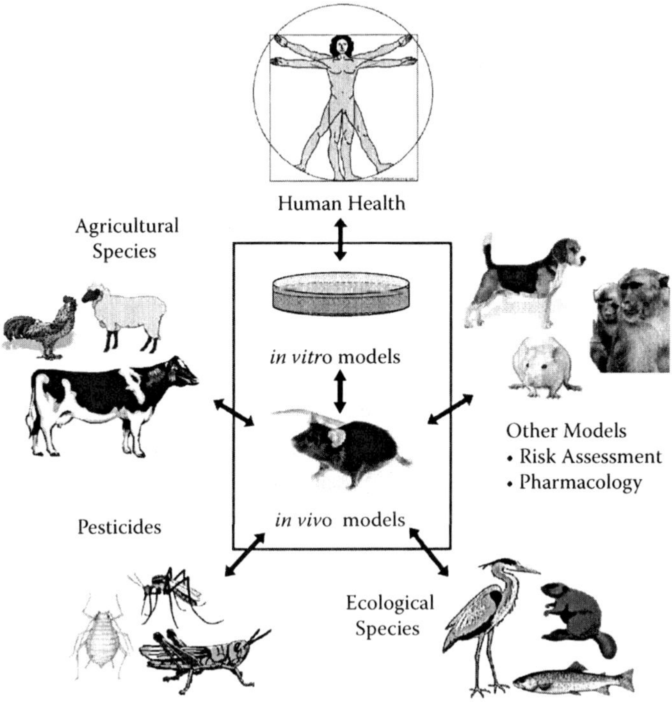

FIGURE 16.2 Importance and applications of species comparisons. Cross-species comparisons hold the potential to extend knowledge to human medicine, agriculture, pesticides, ecology, toxicology, and risk assessment.

(i.e., binding affinities, enzyme kinetics) (Figure 16.1). For example, hamsters are exquisitely sensitive to 2,3,7,8-tetrachlorodibenzo-*p*-dioxin (TCDD), whereas the guinea pig exhibits limited to no toxicity. The effects of TCDD are mediated by the aryl hydrocarbon receptor (AhR), and sequence comparisons between hamster and guinea pig AhRs identified an expanded glutamine-rich domain in the C-terminus that correlates with sensitivity.[3] Differences in avian TCDD toxicity can also be partially attributed to differences in TCDD–AhR binding affinity[4] but do not completely explain the broad range of species sensitivities.

In addition, viral transmission across species barriers has been an important research area, especially regarding the spread of human immunodeficiency virus/acquired immunodeficiency syndrome (HIV/AIDS) and more recently with avian influenza. Cross-species examination of the apoptosis genes involved in species-specific cytomegalovirus infection revealed that intrinsic pathway caspase-9 activation and counteraction with Bcl-2 guards the boundary between human and murine forms.[5] Such investigations into the molecular functions and expression patterns of pathologically relevant mechanisms will have direct impact on public health in the future.

Pharmacokinetic (PK) and pharmacodynamic (PD) studies facilitate the interpretation of toxicity findings and support refinements in mechanistically based risk assessments. PK data minimize uncertainties inherent in route-to-route, high-to-low-dose, and species-to-species extrapolations.[6,7] Genes involved in regulating ADME are important in elucidating such toxicological and pharmacological effects. To this end, a large compendium of hepatic gene expression profiles was compiled[8] to assess changes in ADME-related genes for AhR, constitutive androstane receptor (CAR), and pregnane X-receptor (PXR) ligands between the mouse and rat. Species-specific profiles for each family of ligands were characterized across the transcriptome, providing a comprehensive comparison of ADME differences. These cross-species comparisons support further assessments of functional orthology and not only identify important conserved responses but also reduce uncertainties associated with extrapolating model data to humans.

16.3 CONSIDERATIONS

To conduct informative comparisons, orthologous gene relationships must be established based on sequence similarity, synteny, phylogenetic tree matching, and functional complementation (Table 16.2). Several resources are available (Table 16.3) that utilize different algorithms and stringency levels to provide ortholog predictions.

A confounding factor in comparative genomics is the one-to-many or many-to-many relationships between orthologs and paralogs, which is further complicated when complete genome sequence information is not available. Although a reciprocal best-hit "ortholog" can always be identified, without a complete genome sequence, the true ortholog may not yet be sequenced. To optimize comparisons, a tiered approach can be implemented that uses loosely set criteria to identify all possible relationships. False positives can be subsequently ruled out by further filtering and identifying divergent responses using more stringent criteria, assuming that orthologs will exhibit comparable expression patterns. However, discretion is needed in balancing the tradeoff between the number of genomes to be compared and the size and veracity of identified orthologs, using a consistent mapping strategy to minimize error. In general, the more species included in the comparisons, the fewer orthologs identified. Alternatively, more focused gene-specific, hypothesis-driven investigations that use more stringent ortholog determinations may further validate cross-species extrapolations. Nevertheless, ortholog assignments will continue to improve as

TABLE 16.2
Orthology Criteria

Criteria	Description or Method	Information Source
Sequence similarity	Reciprocal BLAST best hit	Nucleotide sequence Amino acid sequence
Synteny	Conserved order of genes in the genome	Whole-genome sequence
Phylogenetic tree matching	Organism-level relatedness based on nonmolecular data	Taxonomy
Functional complementarity	Conservation of molecular function	Biochemical evidence

TABLE 16.3
Orthology Resources

Resource	Sequence Similarity	Synteny	Phylogeny	Functional Complementation	Cluster Algorithm	Number of Species
HomoloGene	RBH[a]	Yes	Sequence	—	Yes	≥3
Ensembl	RBH	Yes	Species	—	Yes	≥3
EGO (Eukaryotic Gene Orthology)	RBH	No	No	—	Yes	≥3
InParanoid	RBH	Yes	No	—	Yes	Pairwise
PhIGs (Phylogenetically Inferred Groups)	—	No	Species	—	Yes	≥3
OrthoMCL	RBH	No	No	—	Markov clustering	≥3
HCOP (HGNC[b] Comparison of Orthology Predictions)	RBH	Yes	Species	—	Yes	≥3
KOBAS (KO[c]-Based Annotation System)	—	No	No	GO[d] terms	Yes	≥3

[a] Reciprocal Best BLAST Hit.

[b] HUGO Gene Nomenclature Committee.

[c] Kegg Orthology.

[d] Gene Ontology.

genome sequences are completed and refined, gene annotation improves, and additional sequence information form other species becomes available.

In addition to sequence similarity, the degree to which putative orthologs exhibit similar behavior across different experimental conditions provides further evidence of orthology based on conserved regulation. Defining orthologous expression may include comparisons of tissue- or cell-type-specific gene expression profiles, in which (1) direction, (2) magnitude, (3) time and duration, and (4) the shape of response curves are considered. Correlation analyses can be used to quantitatively assess similarities in direction, magnitude, and time, depending on the distance metric utilized. Currently, there is no consensus regarding which quantitative measures or how many must be satisfied to be defined as orthologous expression. Nevertheless, if conserved regulation of gene expression defines orthologous expression, then gene expression regulation under several conditions and in response to different stimuli (Figure 16.3) provides more robust determinations. This requires some knowledge regarding which types of stimuli effect changes in specific gene families or molecular processes. A distinction must also be made regarding the basal level of expression across tissues in response to a stimulus or environmental change. Significant differences in the constitutive gene expression across models may alter a response and subsequent comparison. Therefore, basal levels are required to properly assess orthologous expression between models.

Comparative microarray-based gene expression studies across species include

1. **Same-species hybridization, cross-platform comparison:** Comparing (one to one) data between two or more species-specific array experiments (e.g., mouse liver on mouse arrays compared to rat liver on rat arrays)
2. **Cross-species hybridization, same-platform comparisons:** Hybridizing (many to one) biological samples from multiple species to array targets of a single species (e.g., human liver, rhesus monkey liver, and canine liver individually hybridized on human arrays)

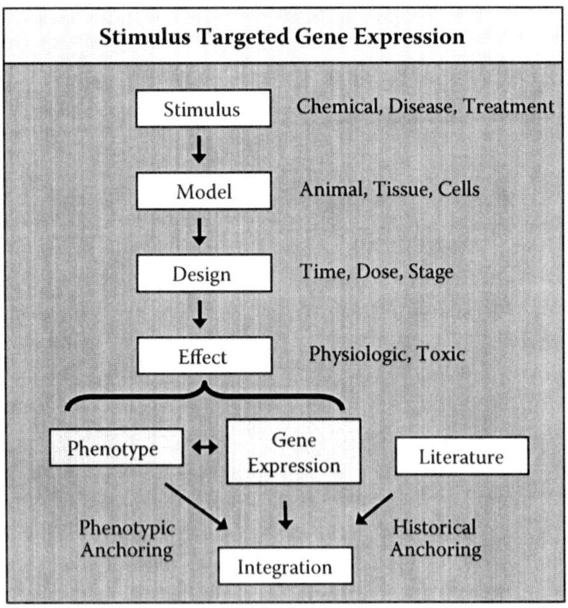

FIGURE 16.3 Stimulus-targeted workflow. Microarray data derived from responses to stimuli as opposed to correlation across tissues will result in more physiologically based determinations of orthologous expression. Important and integral steps involve merging phenotypic and histomorphological endpoints with specific gene expressions to phenotypically link the profiles.

3. **Mixed-species hybridization, same-platform comparison:** Hybridizing (one to many) biological samples from one species to array targets of multiple species (e.g., human, mouse, and rat probes arrayed on a single platform)

Most comparisons are made between data sets derived from same-species hybridizations, for example, mouse samples hybridized to mouse-based arrays compared to a human data set obtained using human arrays. An important consideration is to determine whether normalization occurs independently or following data set merging. For example, a novel strategy compared the expression of human breast tumor and chemically induced rat mammary tumor samples to validate the rat mammary tumor model.[9] In this study, 2,305 rat orthologs were used to classify human tumors derived from array data suggesting that rat primary tumors share comparable signatures with low- to intermediate-grade estrogen-receptor-positive human breast cancer, thus validating chemically induced rat mammary tumors as a model of human disease.

Many factors, including differing study designs, experimental timing, platforms, and coverage, confound the merging and normalization of raw microarray data. Independent normalization was used to examine orthologous uterine gene expression during uterotrophy in rats and mice treated with ethynyl estradiol, an orally active estrogen common in contraceptives.[10] Parallel but species-specific statistical analyses identified 153 orthologous pairs that exhibited conserved temporal responses. Compelling evidence supported the transfer of functional annotation from characterized mouse genes to 44 previously unannotated rat expressed sequence tags (ESTs) based not only on sequence homology but also on orthologous time-dependent expression, demonstrating a novel utility of cross-species analysis.

To circumvent the problem of limited microarray resources for nontraditional models studies using direct cross-hybridization experiments of labeled complementary DNAs (cDNAs) from one species (ape, pig, cow, mouse, salmon)[11–18] hybridized to arrays developed for a related organism with more developed annotation have been conducted. This approach assumes that cDNA probes are of sufficient length, and that homology will overcome interspecies differences in gene sequence but still exhibit specificity. For example, rabbit RNA samples have been cross-hybridized to mouse.[19] Other studies[11,14,20] have cross-hybridized various species using multiple biological and technical replicates and validated the responses with independent, quantitative, real-time PCR with moderate success.

Oligonucleotide arrays (Affymetrix, Agilent, CodeLink) raise concerns regarding the species specificity of smaller probes. Cross-hybridizations between mouse and human samples on human oligonucleotide arrays were conducted to examine a dual-species chimeric tissue model of transplanted human hepatocytes in mouse liver. This study investigated the degree to which incidental and undesired mouse tissue would contribute to the human sample hybridizations to human arrays.[21] Specific cross-reactive probes were identified, and a method to monitor species-specific contributions to the expression data was developed.

Cross-species hybridization can also involve printing orthologous cDNAs from multiple species onto a single array. Samples from represented species are then hybridized to identify same-species and cross-species interactions on the same array. Analysis of oocyte expression in the cow, mouse, and frog found that cross-species hybridizations are highly reproducible, and that the expression of a number of orthologs is conserved.[22] These results were verified by gene- and species-specific quantitative real-time PCR and further species-specific array experiments. Although cross-hybridization experiments make interspecies comparisons easier, there still remains a lack of consensus regarding their reliability. Furthermore, their long-term utility is likely to decrease as more genomes are sequenced, allowing for the development of species-specific arrays.

Conservation of gene sequence and its regulation in a number of pathways is expected for comparable responses. However, given the increasing number of gene expression studies and screening algorithms that select for conserved responses, there will inevitably be examples of divergent orthologous expression (i.e., one ortholog is induced while the other is not responsive or is repressed) that requires further investigation to exclude orthology misclassifications, artifacts, and false negatives. Overall, it is easier to identify conserved orthologous expression as opposed to divergent regulation. Divergent orthologous expression may be due to species differences in *trans*-acting factors or ribonucleases (RNases) that modulate transcription rates or mRNA stability. The degeneracy of *cis*-acting regulatory elements (*c*REs) such as transcription factor binding sites may also result in divergent regulation. In addition, differences in methylation status, chromatin structure, and other epigenetic modifications may be a factor. It is therefore important to further investigate divergent expression patterns to elucidate the regulatory mechanisms involved to assess the designation of functional orthology.

Due to the relationship between gene expression and regulatory motifs, the role of *c*REs in orthologous expression can also be examined. Computational genomic sequence search algorithms and experimental approaches have been developed to identify and associate *c*REs with gene regulation. Supervised methods involve the identification of known response elements by computationally scanning proximal, regulatory genomic sequences for consensus response elements based on a position weight matrix (PWM).[23] For example, a PWM approach was used to search human, mouse, and rat genomes for dioxin response elements (DREs), the *c*RE that binds activated AhR complexes.[24] This identified 48 genes with conserved putative DREs in their respective proximal promoters; 19 were positionally conserved between all three species. Furthermore, fewer than 40% of the mouse–rat orthologs possessing conserved putative DREs also had a human ortholog, suggesting moderate-to-low conservation of *c*REs between rodents and humans. Transcription factor–binding site databases and Web resources (i.e., TRANSFAC, http://www.gene-regulation.com/pub/databases.html) provide consensus motifs for PWM development. The ENCODE (Encyclopedia of DNA Elements) project[25] seeks to identify and characterize all functional elements in the human genome. Alternatively, unsupervised approaches that generate unique 5- to 15-nucleotide "words" from proximal/regulatory genomic sequences (i.e., total genome or upstream promoter regions) can be used to determine the frequency of overrepresented putative regulatory motifs/words within the regulatory sequence of genes exhibiting comparable expression patterns when compared to random sequences.[26]

Protein–DNA interactions can be examined experimentally using chromatin immunoprecipitation (ChIP) to identify genomic regions bound by transcription factors. Following computational screening for consensus or near-consensus estrogen response elements (EREs) in the human and mouse genomes, a list of orthologs with conserved EREs was generated.[27] ChIP analysis demonstrated estrogen receptor (ER) binding at distal promoter sites, suggesting the long-range activity of ER for these orthologs is species conserved in vivo. Genome-wide ChIP analysis, commonly referred to as ChIP-on-chip or ChIP-chip, uses a microarray strategy to identify protein–DNA interactions.[28] Alternatively, a SAGE-like approach that uses high-throughput sequencing of imunoprecipitated chromatin provides an unsupervised strategy to identify protein–DNA interactions.[29–31] However, there is poor correlation between protein–DNA interactions and transcriptional activity. Consequently, the integration of complementary gene expression profiling, computational regulatory motif searches, and protein–DNA interactions facilitate a more comprehensive interpretation of these data to elucidate the affected regulatory networks.

Further examination of divergent ortholog expression will depend largely on the resources available. Bioinformatic approaches require genomic sequence data, computing power, and programming capabilities, while protein–DNA interaction approaches require antibodies to transcription factors of interest as well as specialized array platforms or access to high-throughput sequencing facilities. These complementary methods are crucial for identifying comparable patterns of gene expression involving conserved mechanisms of regulation that will support conclusions regarding the orthologous expression.

16.4 RESOURCES

The computational identification of orthologous genes begins with a list of putative relationships that requires verification. This section describes available database and computational resources for (1) obtaining gene annotation and expression data, (2) identifying orthologous relationships, and (3) mapping gene regulatory sequences, and it provides examples of ortholog comparison and verification.

16.4.1 Genome-Level Databases

The Ensembl database,[32,33] Entrez Genome database,[34] and the University of California, Santa Cruz (UCSC) Genome Browser[35] provide sequence data in the genomic context but differ in their integration of other types of data and often in their assignment of computationally predicting genes and gene structures (e.g., untranslated regions [UTRs], regulatory regions, introns, and exons) (Figure 16.4).

Ensembl utilizes several different, complex methods for the prediction of genes and gene structures; these methods are biased toward the alignment of species-specific proteins and cDNAs as well as orthologous protein and cDNA alignments.[36] The use of the protein and cDNA alignments against the genome sequence facilitates the identification of exonic and intronic sequences, UTRs, and a putative transcription start site (TSS) (Figure 16.5). In contrast, the National Center for Biotechnology Information (NCBI) Entrez Genome database annotates genes based on outside reference information; however, NCBI provides annotation for the human and mouse genome projects. NCBI also provides RefSeq records that represent the genome

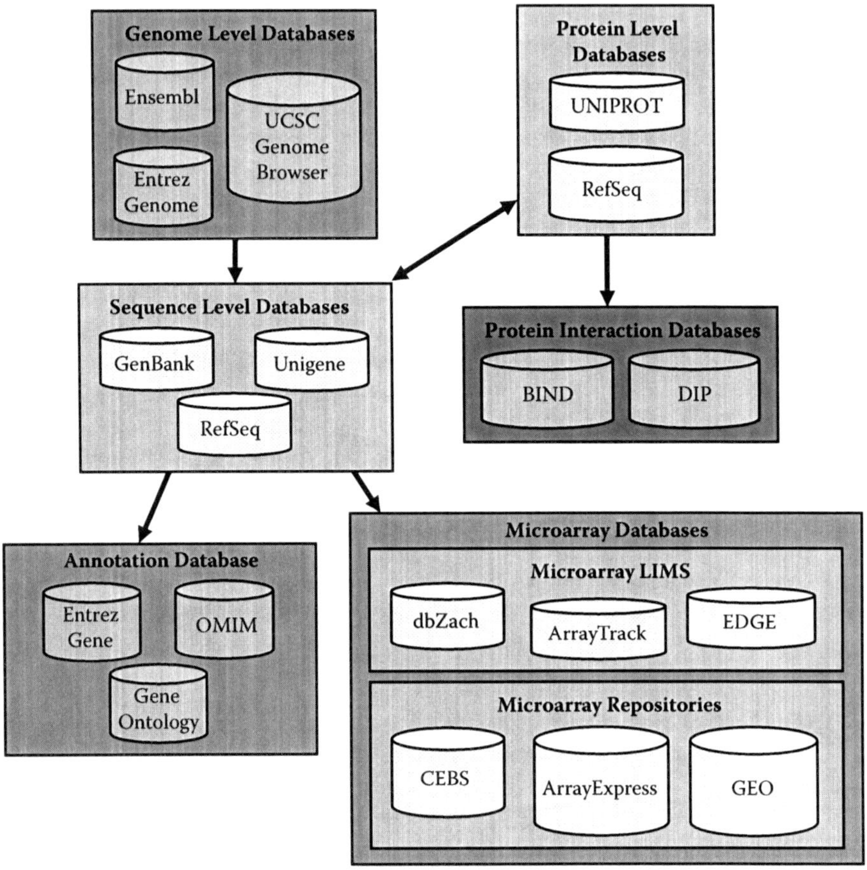

FIGURE 16.4 The biological database universe. Six biological database levels are depicted as they pertain to genomic data analysis and interpretation. Genome-level databases catalog data with respect to the sequence of the full genome. Sequence-level databases catalog sequence reads from cells, including genomic sequence and expressed sequence tags (ESTs). Annotation databases provide functional information about genes and their products. Protein-level databases provide information on protein sequences, families, and domain structures. Protein interaction databases provide interaction data concerning proteins, genes, chemicals, and small molecules. Microarray databases include local laboratory information management systems (LIMS) and data repositories. The arrows depict possible interactions between different database domains; information from one level may exist in another to allow for cross-domain integration.

assemblies, as well as the proteins and transcripts associated with them, via the RefSeq database (http://www.ncbi.nlm.nih.gov/genome/guide/build.html#contig; accessed April 5, 2005).

The UCSC browser uses the NCBI human genome build, which is part of the human genome sequencing project; therefore, there are no differences between the human genome builds. However, prior to the December 2001 human genome build, UCSC created its own genome annotation builds, separate from the NCBI.

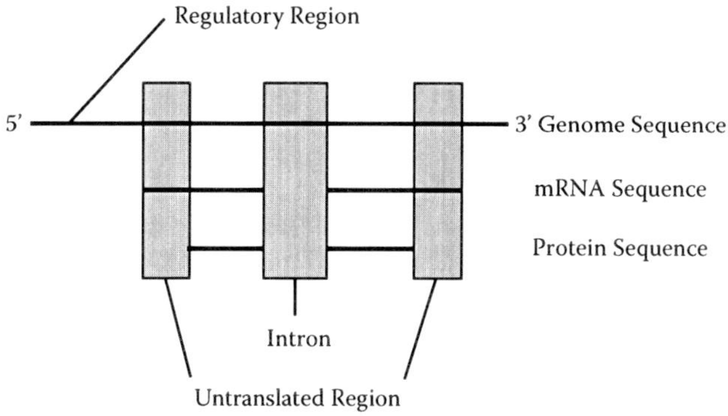

FIGURE 16.5 Ensembl genome annotation. This simplified view illustrates the method used by the Ensembl genome annotation system for identifying gene structures, such as the untranslated region (UTR), exons, and introns by combining genome, mRNA, and protein alignments.

The UCSC mouse genome is for the C57BL/6 strain only and does not report other available mouse genomes[37] (see http://genome.ucsc.edu/FAQ/FAQreleases for further details).

Despite using the same genome build, annotation of the genome (i.e., assignment of genes and functions to the genomic sequence) may still differ. Whereas NCBI uses a variant of the BLAST algorithm for alignment of mRNA, EST, and RefSeq sequences to the genome, UCSC uses BLAT (BLAST-like alignment tool) for alignments to the same genome potentially resulting in different annotations (i.e., assignment of genes and functions to the genomic sequence). Furthermore, the UCSC Genome Browser also incorporates gene predictions from other sources, such as Ensembl and Acembly,[35] and offers the flexibility of uploading investigator annotations for display in the browser.

16.4.2 Sequence-Level Databases

Sequence-level databases manage data with respect to a particular sequence read of an EST or cDNA. They may deal with sequences directly, as is the case for GenBank and RefSeq, or may manage larger data sets, for which multiple sequences are clustered, as in UniGene. Generally, these databases provide the first level of annotation for microarray studies as the sequences are directly represented on the microarrays.

Sequence reads are generally submitted to GenBank and assigned an accession number, a unique identifier that can be used to represent that sequence. GenBank Accessions are the most reliable and commonly used identifiers for microarray probes. The GenBank Accession matches the probe to one sequence within the GenBank database,[34] a database of submitted sequences (ESTs, cDNAs, etc.). UniGene creates nonredundant clusters by aligning GenBank sequences, which may then be annotated based on overall sequence alignment to genes in the Entrez Genome database. UniGene clusters are collections of GenBank sequences that most likely describe the same gene.

The RefSeq database provides exemplary transcript and protein sequences based on either manual curation or information from a genome authority (e.g., Jackson Labs).[34,38] RefSeq accession numbers follow a PREFIX_NUMBER format (e.g., NM_123456 or NM_123456789). All curated RefSeq transcript accessions are prefixed by an NM, while XM prefixes represent accessions that have been generated using automated methods. Although some NM transcript accessions have been generated by automated methods, they are more mature and stable and have undergone some level of review. Illustrating the state of maturity of the annotation, RefSeq records also contain one of seven status codes: (1) genome annotation, (2) inferred, (3) model, (4) predicted, (5) provisional, (6) validated, and (7) reviewed. See http://www.ncbi.nlm.nih.gov/RefSeq/key.html#status for further information regarding the status codes currently in use by RefSeq (Table 16.4).

16.4.3 Protein-Level Databases

Recently, the Swiss-Prot, TrEBML, and PIR-PSD databases were merged into the Universal Protein Resource (UniProt), consisting of the UniProt Archive (UniParc), the UniProt Knowledgebase (UniProt), and the UniRef reference database.

UniParc is a database of nonredundant protein sequences obtained from (1) translated sequences within the gene sequence-level databases (e.g., GenBank); (2) RefSeq; (3) FlyBase; (4) WormBase; (5) Ensembl; (6) the International Protein Index; (7) patent applications; and (8) the Protein Data Bank.[39] UniProt provides functional annotation of the sequences within UniParc, including the protein name, listing of domains and families from the InterPro database (http://www.ebi.ac.uk/interpro/),[40] Enzyme Commission identifier, and Gene Ontology identifiers. Proteins represented within the UniParc and UniProt are computationally gathered to create UniRef, a

TABLE 16.4
RefSeq Status Codes

Code	Level of Annotation
Genome annotation	Records that are aligned to the annotated genome
Inferred	Predicted to exist based on genome analysis, but no known mRNA/EST exists within GenBank
Model	Predicted based on computational gene prediction methods; a transcript sequence may or may not exist within GenBank
Predicted	Sequences from genes of unknown function
Provisional	Sequences represent genes with known functions; however, they have not been verified by NCBI personnel
Validated	Provisional sequences that have undergone a preliminary review by NCBI personnel
Reviewed	Validated sequences that represent genes of known function that have been verified by NCBI personnel

Source: http://www.ncbi.nlm.nih.gov/RefSeq/Key.html#status; accessed April 7, 2005.

database of exemplary sequences based on sequence identity. Three different UniRef versions exist (i.e., UniRef100, UniRef90, and UniRef50); the number denotes the percent identity required for sequences to be merged across all species represented in the parent databases into a single reference protein sequence. Thus, UniRef50 requires only 50% identity for proteins to be merged together. The UniRef50 and 90 databases provide faster sequence searches for identifying probable protein domains and functions by decreasing the size of the search space. RefSeq also contains reference protein sequences, similar in concept to the reference mRNA sequences that are available through the Entrez Genome system.

16.4.4 Annotation Databases

Annotation databases provide functional gene information, including the structure of a gene, thus serving as a launching point for mechanistic interpretation and hypothesis generation based on microarray data. Several specific annotation databases, such as the Mouse Genome Database, exist that focus on particular species.[41]

Entrez Genome is a part of NCBI's Entrez suite of bioinformatic tools that provide information on annotated genes for different genomes, including human, mouse, rat, and dog.[42] Annotated genes within the Entrez Genome either have a RefSeq identifier or have been annotated by a genome annotation authority (e.g., Jackson Labs for mice). Thus, Entrez Genome entries may or may not have a RefSeq associated with them and are classified as either the NM (mature) or the XM (nonreviewed) series. Consequently, it is possible for an Entrez Genome record not to have an exemplary RefSeq sequence associated with it.

Entrez Genome integrates data from diverse sources on the gene detail page or provides hyperlinks to outside databases (Table 16.5). It provides gene names, aliases, and abbreviations required for further annotation through the literature and integrates data from the RefSeq, Gene Ontology (GO), Gene Expression Omnibus (GEO), Gene References into Function (GeneRIF), and GenBank databases. RefSeq sequences, both mRNA and protein, facilitate sequence-based searches for identifying homologous genes or gene functions based on protein domains. GO catalogs the molecular function, cellular location, and biological process of genes. Tissue expression information can be obtained from GenBank, in which the tissue source for an EST is recorded, as well

TABLE 16.5
Entrez Genome Annotation

Annotation Categories	Source
Gene names and abbreviations/symbols	Publications and genome authorities
RefSeq sequence	RefSeq database
Genome position and gene structures	Genome databases
Gene function	Gene Ontology (GO) database, Gene References into Function (GeneRIF)
Expression data	Gene Expression Omnibus (GEO), EST tissue expression from GenBank

as GEO, NCBI's gene expression repository.[34] GeneRIFs provide curated functional data and primary references regarding the functional information about a particular gene but may not deliver the most up-to-date functional annotation from the literature. Investigators can facilitate GeneRIF updates by submitting suggestions directly to the NCBI through their update form: http://www.ncbi.nlm.nih.gov/RefSeq/update.cgi.

The Online Mendelian Inheritance in Man (OMIM) database[43] provides information regarding links between human genes and diseases.[44] OMIM is searchable through the NCBI Entrez system with links in Entrez Genome query output pages. OMIM includes a synopsis of the clinical presentation in addition to links to genes associated with the disease. References are also made available and have hyperlinks to the PubMed database entries. In addition, OMIM contains information on known allelic variants and some polymorphisms.[44]

Another source of gene functional annotative information[45] is GO (http://www.geneontology.org). It consists of an ontology (i.e., a catalog of existents/ideas/concepts and their interrelationships) in which terms exist within a graphical structure leading from a high to a low level, referred to as a *directed acyclic graph* (DAG) (Figure 16.6). In a DAG, a child node (i.e., an object or concept) may not serve as its own predecessor (i.e., parent node, etc.). Any child node within a DAG may have multiple parents, and a number of paths lead to the child. For example, GO:0045814: negative regulation of gene expression, epigenetic, has two paths leading to the same child (Figure 16.6). This epigenetic negative regulation of gene expression is both a regulation process and developmentally critical. GO entries that exist at the same level relative to the root, or starting node, do not necessarily reflect the same level of specificity. The level of specificity afforded must be taken on a per DAG basis, and not relative to the other DAGs. Thus, a fourth-order node (a node that is four levels below the root node) in one DAG has no specificity relationship regarding a fourth-order node in a different DAG. At each node within the GO, there may exist a list of genes. As gene annotation improves, a gene may change node associations. For example, if gene X were previously GO:0040029 (regulation of gene expression, epigenetic), and new data suggested gene X was a negative regulator of gene expression through an epigenetic mechanism, then it would be reassigned to GO:0045814 (negative regulation of gene expression, epigenetic).

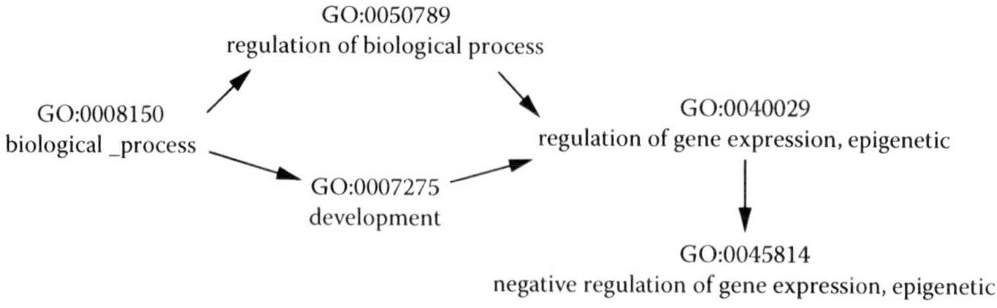

FIGURE 16.6 Example of a Gene Ontology (GO) directed acyclic graph (DAG). This DAG shows two paths to reach the same GO entry, GO:0045814. It is important to note that the DAG travels from the most general case and becomes more specific with entries that are farther down the DAG.

The GO Consortium maintains the mappings between genes and GO terms (http://www.geneontology.org). Note that genes may have multiple associated GO terms, and that the assignment of a GO number has no other significance than as a unique identifier.

16.4.5 Protein Interaction Databases

Protein interaction databases capture data on the interaction of proteins with other proteins, genes, and small molecules. For example, the Biomolecular Interaction Network Database (BIND)[46] and the Database of Interacting Proteins (DIP)[47] manage data from protein interaction experiments, including yeast-two-hybrid and coimmunoprecipitation experiments typically available in the Protein Standards Initiative (PSI) Molecular Interaction (PSI-MI) XML format. Visualization of these data sets into putative interaction pathways is possible using Osprey[48] and Cytoscape, which also facilitates overlaying gene expression data onto protein interaction maps.[49]

16.4.6 Global Orthology Mapping

Several resources are available for globally mapping orthologs between species to facilitate comparative analyses (Table 16.3). These resources differ in the criteria used to identify orthologs but have comparable numbers based on comparisons to available genomes. HGCN Comparison of Orthology Predictions (HCOP) provides comparisons across several of the resources to derive consensus orthology mappings.

16.4.7 Microarray Resources

Microarray databases typically include laboratory information management systems (LIMS) and data repositories. LIMS manage data within a laboratory or a consortium by ensuring proper data management, facilitating analysis, and archiving data in accordance with the Minimum Information About a Microarray Experiment (MIAME).[50]

Data repositories are warehouses that store data from multiple sites and investigators and facilitate data dissemination to the public. Repositories also facilitate the comparison of data sets across laboratories, and the independent reanalysis data can complement the interpretation of nontranscriptomic studies. Several journals require microarray submissions to repositories such as NCBI's GEO[51] and ArrayExpress[52] at the European Bioinformatics Institute (EBI), using the MIAME standard as a condition of publication, similar to requirements that novel sequences be submitted to GenBank prior to publication.[53] Other specialized repository efforts have also been undertaken, such as the Chemical Effects in Biological Systems (CEBS) Knowledgebase,[54] which catalogs gene expression data from drug and chemical exposures with associated pathology data.

16.4.8 Regulatory Element Searching

Regulatory elements are sequences bound by transcription factors to regulate gene expression. PWMs can be developed from known functional regulatory elements to computationally search genomic sequences and provide a functionality score for putative transcription factor binding sites relative to a consensus sequence. However, many regulatory elements are unknown or degenerative, thus requiring an unsupervised search. PWM strategies assume that the transcription factor will bind most favorably to its consensus sequence as determined in functional assays and less favorably to divergent sequences. The PWM itself is an $n \times 4$ matrix, where n is the number of bases within the site, and 4 represents each nucleotide. Each cell within the matrix represents the occurrence of each base at that location or the relative percentage (percentages are generally represented as whole numbers, so if a base were present at that location 5 of 10 times, the percentage would be represented as 50 and not 0.50). Note that the consensus is based on known functional sequences and may change as additional binding sites are characterized. TRANSFAC (http://www.gene-regulation.com/pub/databases.html; free for noncommercial users) is the most widely used database for characterized response elements and PWMs for a number of species.

Several approaches for response element prediction exist that do not require a priori information about the binding site.[55] However, these approaches may (1) only be validated on a limited number of data sets (e.g., algorithm may be organ, cell type, or species biased due to data sets available for development); (2) not consider more complex protein–protein interactions and their effect on transcription factor binding; (3) not consider more complex DNA structures, such as methylation and histone acetylation; and (4) not take into account changes in the DNA-binding domains induced by ligand structure, protein–protein interactions, or other posttranslational modifications that influence DNA-binding specificity and affinity. Therefore, computational response element search and prediction algorithms tend to exhibit high false-positive and false-negative rates that require empirical verification.[55]

Computational predictions can be verified using ChIP assays that identify interactions between proteins and DNA. For example, a transcription factor can be immunopreciptated as a complex bound to DNA and then PCR amplified, labeled, and hybridized to a microarray to identify the region of interaction. Thus, the integration of gene expression data with complementary computational response element search data and ChIP results provides comprehensive information regarding the cascade of events involved in the elicited effects. Response element, protein–DNA interaction, and gene expression conservation across species that can be phenotypically anchored to physiological outcomes provides compelling mechanistic information that not only supports more refined testable hypotheses to further elucidate the mechanism of action but also provides compelling evidence that the model is relevant to humans.

16.5 LIMITATIONS

Several factors limit cross-species analyses, including (1) incomplete genome sequence data, (2) incomplete and unstable gene annotation with complementary functional annotation, (3) the complexities of orthology mapping, (4) inconsistent reporting standards and the lack of compliance, (5) limited relevant human gene expression data, and (6) inadequate tools and resources that integrate disparate data from different sources. For instance, incomplete sequence information compromises the ability to identify orthologous genes with certainty, thus limiting comprehensive comparisons. For many species, such as those with low genomic sequencing coverage (e.g., cow, pig, sheep, chicken, dog, and horse), annotation is limited to a few hundred genes, consisting mainly of ESTs and computationally predicted mRNA.[33–35] Thus, orthology mapping against genomes with mature annotation (e.g., human, mouse, or rat) is frequently performed to interpret expression data.

In general, there is no consensus on the most appropriate way to determine orthology. The presence of large paralogous gene families resulting in one-to-many or many-to-many relationships between species also confounds orthology assignments. Ambiguities in orthology mapping resulting from poor resolution of homologous gene families and isotypes further compromise the ability to assess cross-species responses.

Comparative gene expression studies also require appropriate study designs that include sufficient replication to support statistically rigorous comparisons. Although microarray costs continue to decrease, this is mitigated by the increasing cost of newer technologies, higher genome coverage, and required QA/QC robustness. Furthermore, several journals have adopted problematic reporting standards as a condition of publication to facilitate the accessibility of expression data. Ambiguities in the definition and description of proposed standards (i.e., MIAME) have resulted in different interpretations and a lack of consensus regarding implementation, resulting in MIAME-compliant public repositories with different reporting requirements, which confounds comparisons.[56,57]

While direct comparisons of specific tissue or organ responses between species are desirable, genetic heterogeneity and the availability of appropriate human samples limit comparisons that include humans. Studies with model species also allow more precise control with greater latitude regarding treatment regimens and dose ranges to obtain a more comprehensive assessment. While there is a preference for rodent models (mouse and rat) due to platform availability, genome sequence coverage, and annotation maturity, other mammalian models are more valued for toxicological and pharmacological screening, testing, and regulatory review. In particular, fully sequenced and annotated chimpanzee, dog, pig, and rabbit genomes will be of particular use for these purposes. Although human cell culture systems are available, the ability of in vitro systems to accurately model in vivo responses has not been adequately demonstrated.

Despite increasing access to software packages as well as free Web-based tools for data mining, analysis, annotation, and visualization, few of these solutions explicitly address cross-species comparisons, facilitate orthology designations, or address orthologous expression. The lack of robust, publicly available cross-species data sets may contribute to the paucity of comparative analysis tools. However, some tools with inter- or cross-species functionalities (Comparative Toxicogenomics Database, http://ctd.mdibl.org/; Integrative Array Analyzer, http://zhoulab.usc.edu/iArrayAnalyzer.htm; Resourcerer, http://www.tigr.org/tigr-scripts/magic/r1.pl; yMGV, http://transcriptome.ens.fr/ymgv/) have been made available or are in development.

16.6 CONCLUSIONS

Cross-species comparisons can provide compelling information that significantly advances our understanding of the mechanisms of action of disease, drug efficacy, and toxicity. More comprehensive knowledge of species-specific responses and conserved mechanisms not only will increase the efficiency of drug development but also will significantly improve our ability to assess potential risks to human health based on data from model species. These efforts will be facilitated with the development of the required infrastructure and resources needed to support comparative studies. This includes increasing array platform options, coverage, and reliability; facilitating public access to toxicogenomic data; compliance with consensus reporting standards; the maturation of annotation; and improvements in integrative and comparative bioinformatics tools and resources. These advances will facilitate future comparative that will improve human health.

REFERENCES

1. Zhou, X.J. & Gibson, G. Cross-species comparison of genome-wide expression patterns. *Genome Biol* **5**, 232 (2004).
2. Jensen, R.A. Orthologs and paralogs — we need to get it right. *Genome Biol* **2**, INTERACTIONS1002 (2001).
3. Hahn, M.E. Aryl hydrocarbon receptors: diversity and evolution. *Chem Biol Interact* **141**, 131–160 (2002).
4. Karchner, S.I., Franks, D.G., Kennedy, S.W. & Hahn, M.E. The molecular basis for differential dioxin sensitivity in birds: role of the aryl hydrocarbon receptor. *Proc Natl Acad Sci U S A* **103**, 6252–6257 (2006).
5. Jurak, I. & Brune, W. Induction of apoptosis limits cytomegalovirus cross-species infection. *EMBO J* **25**, 2634–2642 (2006).
6. Andersen, M.E. Physiologically based pharmacokinetic (PB-PK) models in the study of the disposition and biological effects of xenobiotics and drugs. *Toxicol Lett* **82–83**, 341–348 (1995).
7. Leung, H.W. & Paustenbach, D.J. Physiologically based pharmacokinetic and pharmacodynamic modeling in health risk assessment and characterization of hazardous substances. *Toxicol Lett* **79**, 55–65 (1995).
8. Slatter, J.G. et al. Microarray-based compendium of hepatic gene expression profiles for prototypical ADME gene-inducing compounds in rats and mice in vivo. *Xenobiotica* **36**, 902–937 (2006).
9. Chan, M.M., Lu, X., Merchant, F.M., Iglehart, J.D. & Miron, P.L. Gene expression profiling of NMU-induced rat mammary tumors: cross species comparison with human breast cancer. *Carcinogenesis* **26**, 1343–1353 (2005).
10. Kwekel, J.C., Burgoon, L.D., Burt, J.W., Harkema, J.R. & Zacharewski, T.R. A cross-species analysis of the rodent uterotrophic program: elucidation of conserved responses and targets of estrogen signaling. *Physiol Genomics* **23**, 327–342 (2005).
11. Chismar, J.D. et al. Analysis of result variability from high-density oligonucleotide arrays comparing same-species and cross-species hybridizations. *Biotechniques* **33**, 516–518, 520, 522 passim (2002).
12. Medhora, M., Bousamra, M., 2nd, Zhu, D., Somberg, L. & Jacobs, E.R. Upregulation of collagens detected by gene array in a model of flow-induced pulmonary vascular remodeling. *Am J Physiol Heart Circ Physiol* **282**, H414–H422 (2002).
13. Shah, G., Azizian, M., Bruch, D., Mehta, R. & Kittur, D. Cross-species comparison of gene expression between human and porcine tissue, using single microarray platform — preliminary results. *Clin Transplant* **18 Suppl 12**, 76–80 (2004).
14. Adjaye, J. et al. Cross-species hybridisation of human and bovine orthologous genes on high density cDNA microarrays. *BMC Genomics* **5**, 83 (2004).

15. Robert, C., Hue, I., McGraw, S., Gagne, D. & Sirard, M.A. Quantification of cyclin B1 and p34(cdc2) in bovine cumulus-oocyte complexes and expression mapping of genes involved in the cell cycle by complementary DNA macroarrays. *Biol Reprod* **67**, 1456–1464 (2002).
16. Huang, G.S., Yang, S.M., Hong, M.Y., Yang, P.C. & Liu, Y.C. Differential gene expression of livers from ApoE deficient mice. *Life Sci* **68**, 19–28 (2000).
17. Grigoryev, D.N. et al. In vitro identification and in silico utilization of interspecies sequence similarities using GeneChip technology. *BMC Genomics* **6**, 62 (2005).
18. Tsoi, S.C. et al. Use of human cDNA microarrays for identification of differentially expressed genes in Atlantic salmon liver during *Aeromonas salmonicida* infection. *Mar Biotechnol (NY)* **5**, 545–554 (2003).
19. Cavallaro, S., Schreurs, B.G., Zhao, W., D'Agata, V. & Alkon, D.L. Gene expression profiles during long-term memory consolidation. *Eur J Neurosci* **13**, 1809–1815 (2001).
20. Walker, S.J., Wang, Y., Grant, K.A., Chan, F. & Hellmann, G.M. Long versus short oligonucleotide microarrays for the study of gene expression in nonhuman primates. *J Neurosci Methods* **152**, 179–189 (2006).
21. Walters, K.A. et al. Application of functional genomics to the chimeric mouse model of HCV infection: optimization of microarray protocols and genomics analysis. *Virol J* **3**, 37 (2006).
22. Vallee, M., Robert, C., Methot, S., Palin, M.F. & Sirard, M.A. Cross-species hybridizations on a multi-species cDNA microarray to identify evolutionarily conserved genes expressed in oocytes. *BMC Genomics* **7**, 113 (2006).
23. Henikoff, S. & Henikoff, J.G. Position-based sequence weights. *J Mol Biol* **243**, 574–578 (1994).
24. Sun, Y.V., Boverhof, D.R., Burgoon, L.D., Fielden, M.R. & Zacharewski, T.R. Comparative analysis of dioxin response elements in human, mouse and rat genomic sequences. *Nucleic Acids Res* **32**, 4512–4523 (2004).
25. Thomas, D.J. et al. The ENCODE Project at UC Santa Cruz. *Nucleic Acids Res* **35**, D663–D667 (2007).
26. Lee, K. et al. Identification and characterization of genes susceptible to transcriptional cross-talk between the hypoxia and dioxin signaling cascades. *Chem Res Toxicol* **19**, 1284–1293 (2006).
27. Bourdeau, V. et al. Genome-wide identification of high-affinity estrogen response elements in human and mouse. *Mol Endocrinol* **18**, 1411–1427 (2004).
28. Kim, T.H. & Ren, B. Genome-wide analysis of protein–DNA interactions. *Annu Rev Genomics Hum Genet* **7**, 81–102 (2006).
29. Wei, C.L. et al. A global map of p53 transcription-factor binding sites in the human genome. *Cell* **124**, 207–219 (2006).
30. Loh, Y.H. et al. The Oct4 and Nanog transcription network regulates pluripotency in mouse embryonic stem cells. *Nat Genet* **38**, 431–440 (2006).
31. Kobayashi, M., Takahashi, E., Miyagawa, S., Watanabe, H. & Iguchi, T. Chromatin immunoprecipitation-mediated target identification proved aquaporin 5 is regulated directly by estrogen in the uterus. *Genes Cells* **11**, 1133–1143 (2006).
32. Clamp, M. et al. Ensembl 2002: accommodating comparative genomics. *Nucleic Acids Res* **31**, 38–42 (2003).
33. Hubbard, T. et al. Ensembl 2005. *Nucleic Acids Res* **33**, D447–D453 (2005).
34. Wheeler, D.L. et al. Database resources of the National Center for Biotechnology Information: update. *Nucleic Acids Res* **32**, D35–D40 (2004).
35. Karolchik, D. et al. The UCSC Genome Browser Database. *Nucleic Acids Res* **31**, 51–54 (2003).

36. Curwen, V. et al. The Ensembl automatic gene annotation system. *Genome Res* **14**, 942–950 (2004).
37. Rouchka, E.C., Gish, W. & States, D.J. Comparison of whole genome assemblies of the human genome. *Nucleic Acids Res* **30**, 5004–5014 (2002).
38. Pruitt, K.D. & Maglott, D.R. RefSeq and LocusLink: NCBI gene-centered resources. *Nucleic Acids Res* **29**, 137–140 (2001).
39. Bairoch, A. et al. The Universal Protein Resource (UniProt). *Nucleic Acids Res* **33**, D154–D159 (2005).
40. Mulder, N.J. et al. The InterPro Database, 2003 brings increased coverage and new features. *Nucleic Acids Res* **31**, 315–318 (2003).
41. Eppig, J.T. et al. The Mouse Genome Database (MGD): from genes to mice — a community resource for mouse biology. *Nucleic Acids Res* **33**, D471–D475 (2005).
42. Maglott, D., Ostell, J., Pruitt, K.D. & Tatusova, T. Entrez Gene: gene-centered information at NCBI. *Nucleic Acids Res* **33**, D54–D58 (2005).
43. McKusick, V.A. & Amberger, J.S. The morbid anatomy of the human genome: chromosomal location of mutations causing disease. *J Med Genet* **30**, 1–26 (1993).
44. Hamosh, A., Scott, A.F., Amberger, J.S., Bocchini, C.A. & McKusick, V.A. Online Mendelian Inheritance in Man (OMIM), a knowledgebase of human genes and genetic disorders. *Nucleic Acids Res* **33**, D514–D517 (2005).
45. Harris, M.A. et al. The Gene Ontology (GO) database and informatics resource. *Nucleic Acids Res* **32**, D258–D261 (2004).
46. Alfarano, C. et al. The Biomolecular Interaction Network Database and related tools 2005 update. *Nucleic Acids Res* **33**, D418–D424 (2005).
47. Xenarios, I. et al. DIP, the Database of Interacting Proteins: a research tool for studying cellular networks of protein interactions. *Nucleic Acids Res* **30**, 303–305 (2002).
48. Breitkreutz, B.J., Stark, C. & Tyers, M. Osprey: a network visualization system. *Genome Biol* **4**, R22 (2003).
49. Shannon, P. et al. Cytoscape: a software environment for integrated models of biomolecular interaction networks. *Genome Res* **13**, 2498–2504 (2003).
50. Brazma, A. et al. Minimum information about a microarray experiment (MIAME) — toward standards for microarray data. *Nat Genet* **29**, 365–371 (2001).
51. Edgar, R., Domrachev, M. & Lash, A.E. Gene Expression Omnibus: NCBI gene expression and hybridization array data repository. *Nucleic Acids Res* **30**, 207–210 (2002).
52. Brazma, A. et al. ArrayExpress — a public repository for microarray gene expression data at the EBI. *Nucleic Acids Res* **31**, 68–71 (2003).
53. Ball, C. et al. Standards for microarray data: an open letter. *Environ Health Perspect* **112**, A666–A667 (2004).
54. Waters, M. et al. Systems toxicology and the chemical effects in biological systems (CEBS) knowledge base. *EHP Toxicogenomics* **111**, 15–28 (2003).
55. Tompa, M. et al. Assessing computational tools for the discovery of transcription factor binding sites. *Nature Biotechnol* **23**, 137–144 (2005).
56. Edgar, R. Challenge of choosing right level of microarray detail. *Nature* **443**, 394 (2006).
57. Microarrays: Share and share alike. *Nature* **442**, 1069–1069 (2006).

17 Comparative Genomics and Crop Improvement

Michael Francki and Rudi Appels

CONTENTS

ABSTRACT

Gene and genome sequence similarity is a popular strategy for predicting gene function across plant species. However, the release of genome sequences from two model species, *Arabidopsis thaliana* (thale cress) and *Oryza sativa* (rice), and subsequent comparison of genome-wide sequence similarity have revealed that gene content is different. It is now evident that over evolutionary time there has been an increase or decrease in gene copy number by duplications and rearrangement of different multigene families during independent speciation of the lineages. Furthermore, chromosomal rearrangements cause a convoluted organization of gene content and order even across closely related species. This chapter summarizes our knowledge of gene content and order within and across plant species and provides examples highlighting successful applications and limitations of comparative genomics for predicting gene function in crop species.

17.1 INTRODUCTION

Plant improvement programs are constantly challenged to develop crop plants adaptable to abiotic stress (drought, salinity, microelement, and heavy metal toxicities), with the ability to resist infection from a suite of biotic influences (viruses, fungi, insect pests, and bacteria), and carrying quality attributes suitable for end-product requirements. The development of high-yielding crops producing quality end products for human consumption with added health benefits is necessary to maintain the world's increasing food supply. Plant breeding programs have access to a wide gene pool through cultivated germplasm, land races, or wild relatives, providing a source of genetic variability to develop high-yielding crops that meet the demands of the world's food supply. However, plant breeding programs alone are not positioned to meet these ever-increasing demands and require the integration of new technologies to improve their efficiency in developing food crops that adapt to changing environmental conditions. Plant genomics is one area that will enable identification of genes and allelic variants that control the agronomic performance of crops and their adaptation to a range of environmental conditions. Identifying all genes and their function for one species is a key focus where comparative genomics can deploy knowledge in model species to identify genes controlling similar traits across a range of crops.

Comparative genomics can be broadly defined as gene and genome similarity between two or more species that may or may not share a taxonomic lineage. Much of our fundamental understanding of comparative relationships in the past 15 years has been at the level of similarity of gene content and order on chromosomes (synteny) within taxonomically related species. Nucleotide and protein sequence similarity is the basic tool for comparative genomics. DNA sequences are compared within a species to identify duplicated but diverged genes (paralogs) or between species that are derived from a common ancestor (orthologs). DNA probes from a model species representing coding regions can identify paralogs and orthologs in a particular species and can form the basis for alignment of whole chromosomes (macrosynteny or collinearity) across species.

In plants, macrosynteny has been well studied in the grasses, and a simplified summary of genome relatedness across species in the subfamilies Ehrhartoideae, Panicoideae, and Pooideae has been formulated.[1,2] The concept that gene content and order remained conserved across species during evolution provided a means by which genes controlling trait variation in one species could be directly related to corresponding genes in a related species.[3–5] The concept that gene content and order remained conserved across species is now extensively modified as more large-scale genome sequencing has become available. It is evident that expansion or contraction of gene families occurs frequently, and that the presence of intervening nonsyntenic genes (microrearrangements) can disrupt macrosynteny into smaller chromosomal blocks (microsynteny or microcollinearity). Therefore, the translation of gene function from one species to another based on macrosynteny is difficult due to the evolution of new genes during independent speciation of the plant lineages. The sequencing of genomes from model plant species has provided the templates for these investigations.

Advances made in obtaining more fundamental knowledge on comparative gene content and order within the model species *Arabidopsis thaliana* (thale crest) and *Oryza sativa* (rice) are summarized in this chapter and limitations for comparative genomic studies in more complex crop genomes assessed. *Arabidopsis* is of particular importance in plant biology because of the large volumes of knowledge in plant development, physiology, biochemistry, and disease resistance generated over several decades and the availability of an entire sequenced genome. Rice has been the preferred model system for comparative genomics in monocots, and its sequenced genome is the first completed for one of the world's major grain crops. Comparative genomics is discussed with the specific aim of capturing the exciting developments in gene and genome organization in model species with respect to the breeding of crop plants.

17.2 GENE AND GENOME EVOLUTION

A major factor influencing the evolution of genomes is gene duplication. The duplicated genes and genome regions provide new genetic material for mutation, drift, and selection to act on and meet the demands of changing environments in which plants survive.[6] The duplicated gene copies can either be lost as a result of functional redundancy or provide functional diversity by which new genes are retained as a part of the natural selection process — the concept of "use it or lose it."[7] The recent advances in generating near-complete genome sequences for *Arabidopsis*[8] and rice[9,10] provide opportunities for a genome-wide analysis and examination of the occurrence of families of repetitive DNA and gene paralogs and orthologs that shaped these genomes.

17.2.1 *Arabidopsis:* Gene and Whole-Genome Duplications

The international genome sequencing consortium (the Arabidopsis Information Resource, TAIR) has reported that *Arabidopsis* contains 25,500 genes[8,11]; more than 60% of these are represented by duplicated loci.[12–16] Although this is a significant increase from earlier studies predicting less than 15% of the *Arabidopsis* genome as represented by duplicated loci,[17,18] there remain conflicting reports whether the proportion of duplicated loci are under- or overestimated. Blanc et al.[15] proposed that evolution of paralogous sequences was at a massive scale prior to the evolution of the modern-day *Arabidopsis* genome, leading to diversification by which many duplication events are no longer discernible as related sequences. The details of the number of unrelated genes that have evolved from the ancestral genome and the rates and timings of these events since divergence from the progenitor genome need to be clarified. A particular challenge is the accurate annotation of genomes,[19] and current analyses may still represent an overestimation of gene content as a result of ambiguity in defining active and relevant DNA sequences. Evolutionary events have led to a complex array of closely related and distinct genes in *Arabidopsis,* and identification of these features in the genome sequence provides a starting point for understanding functional attributes.

Since the release of the genome sequence of *Arabidopsis*, several studies have analyzed the extent of duplication events at the whole-genome level. The genome evolved from its ancestor as a result of at least two whole-genome duplication events. It is estimated that this occurred about 100–200 million years ago (MYA),[15,20–22] with 58% of the genome representing duplicated segments larger than 100 kbp.[8] The *Arabidopsis* genome is therefore a result of a tetraploid ancestral genome in which interchromosomal recombination, reciprocal transposition, translocations, and inversions played a significant role in giving rise to the present-day genome.[15,23,24]

The different levels of expansion of repetitive sequence arrays and transposable elements (TEs) have served to further differentiate duplicated regions of the genome as well as confound the annotation of genes. An extreme example is the differentiation of classically defined heterochromatin from euchromatin (analysis of chromosome 4 by Lippman et al.[25]). The high concentration of repetitive sequence arrays and TEs provides targets for DNA methylation[26] and increases the number of DNA sequences coding for short interfering RNAs (siRNAs) and microRNAs.[27] The availability of genomic tiling microarrays for *Arabidopsis*[28] provides the basis for the genome-wide mapping of epigenetic features such as DNA methylation by mapping the distribution of methylated cytosine in genomic DNA digested with McrBC enzyme, an endonuclease which cleaves DNA-containing methyl cytosine in one or both strands, or treated with bisulfite (converts unmethylated cytosine to uracil). Since the impacts of DNA methylation[28] and siRNAs[29] on transcription are well established, the restructuring of the genome as discussed in this section is therefore also a major factor in modifying the transcriptome of the plant.

17.2.2 Rice Genome Sequence Variation

It is generally accepted that diploid species have similar gene contents but vary in genome size due to the abundance of noncoding repetitive DNA in the intergenic regions. Based on this assumption, we would expect that other diploid species would have similar gene content to *Arabidopsis*. The release of the rice genome sequence in 2002 identified between 32,000 and 55,000 genes for rice,[9,10] an estimate that was larger than the gene content predicted in *Arabidopsis*. A key issue here is the annotation methodologies used for assigning sequences as genes.[19,30] For example, reannotation of the rice genome taking into consideration retrotransposon content provides a more conservative estimate of fewer than 40,000 genes,[19] but nevertheless substantially more than *Arabidopsis*. It seems reasonable, therefore, to assume that paralogous genes and multigene families that evolved as a result of duplication-divergence events are confounding estimates of gene numbers.

Genomic tiling microarrays analogous to those established for *Arabidopsis* have also been analyzed in rice using[30] the genome sequence of rice chromosome 10 in addition to standard microarrays.[31] The study of the rice transcriptome using genomic tiling microarrays provided a new technology to map the location of complementary DNA (cDNA) sequences derived from polyA-plus RNA and assisted in confirming gene content. In addition to highlighting significant errors in the current annotation of the rice 10 genome, the Li et al. study[30] also identified some potentially interesting features of the transcriptome originating from the classically defined heterochromatin regions. These regions appeared to become more transcriptionally active in tissues under stress.

It was initially predicted that 15%–20% of the rice genome is represented by duplicated segments.[32] However, it appears that this proportion is an underestimation. Paterson et al.[33] reported that up to 62% of the rice transcriptome is represented by duplicated loci; a similar figure (65.7%) was corroborated by Yu et al.,[34] but a more conservative estimate of 45% of the total predicted genes has been reported by Wang et al.[35] Guyot and Keller[36] estimated 53% of the rice genome was present as segmental duplications generally greater than 1 Mb. Regardless of the frequency of duplicated rice segments, all reports indicated that whole-genome duplication arose as a result of the evolution of an ancient polyploid of the rice ancestor, similar to that seen for the events that shaped the *Arabidopsis* genome. The evolutionary events in rice are predicted to have occurred as recently as 66–70 MYA[33,35] and around the time of grass speciation.[37] Similar to *Arabidopsis*, it appears that the rice lineage has experienced more than one round of whole-genome duplication, and that the events are part of an ongoing process,[38,39] with segmental duplications possibly occurring as recently as 5 MYA.[39]

The evidence available clearly shows that gene and whole-genome duplications account for a substantial proportion of the rice and *Arabidopsis* genomes. Based on similar duplication and rearrangement events and evolutionary trends in crop plants, we can depict a general model for how plant genomes have evolved (Figure 17.1).

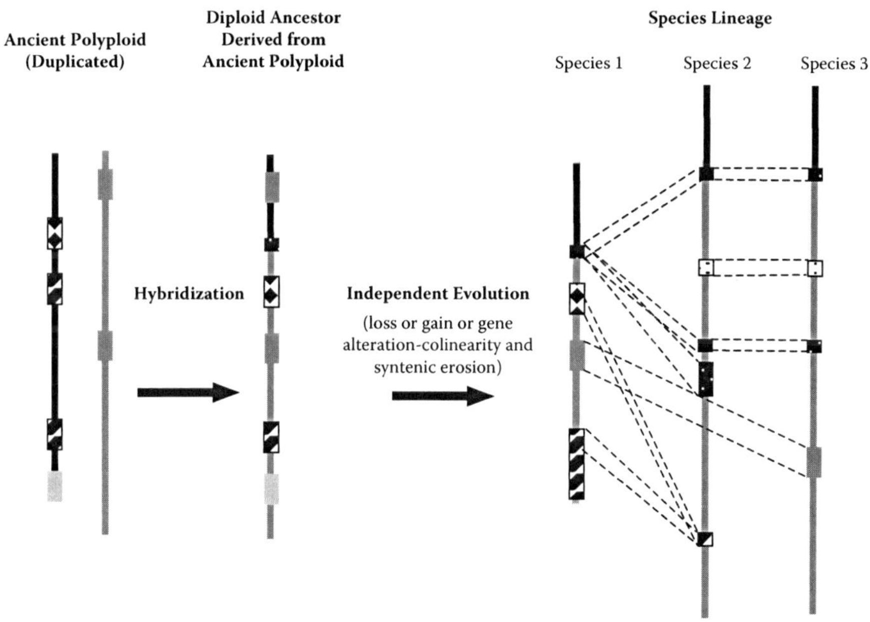

FIGURE 17.1 A simplified model highlighting events common during plant genome evolution based on independent analysis of *Arabidopsis* and rice genome sequences. The model has taken into consideration the hybridization of ancient polyploid species (converged diploidization) and gene and genome rearrangements during independent evolution of plant lineages. The patterned boxes represent genes that are unchanged or have undergone gain, loss, or alteration during evolution from a common ancestor. Dashed lines highlight similar gene origins and the syntenic and nonsyntenic relationships between genomes of modern-day plant species.

17.2.3 Cereal Genome Variation

Conservation of gene order on a broad level is generally recognized among cereals, but extensive variation has been documented at a detailed level when specific chromosomes or chromosome segments were studied.[40–44] The repetitive elements, combined with deletions, insertions, duplications, and rearrangements, in cereal genomes account for extensive variation in genome structure. Retrotransposable elements in cereals have been reviewed[45,46] and represent the major proportion (>70%) of the genome. Expressed genes are at a relatively low density among the retrotransposable element/repetitive DNA sequences, and the latter provide a distinctive DNA sequence environment in which the genes need to function. The repetitive elements, such as retrotransposons in cereal genomes, account for most of the variation in genome structure.

Singh et al.[47] argued that, because the genome duplications identified in rice occurred well before the evolutionary divergence of rice and wheat (*Triticum* spp.), then these duplications should be observable in wheat. Although the resolution in wheat is not as great as in rice to confirm this proposition, the authors did find examples that were consistent with this concept. Using only low- or single-copy rice gene sequences to probe the mapped wheat expressed sequence tag (EST) sequences, Singh et al.[47] demonstrated, for example, that a large segment of rice chromosome 1 that is duplicated in rice chromosome 5 is identifiable on wheat group 3 and 1 chromosomes, respectively. A number of similar examples were detailed by Singh et al., although they did note that in some cases duplications in rice (one describing a second duplication between rice 1 and 5 and one between rice 4 and 10) did not have syntenic equivalents in wheat.

In addition to identifying ancient genome duplications, wheat is a recent polyploid and provides an interesting model for studying events that must have occurred early in the whole-genome duplication events described in plants such as rice and *Arabidopsis*.[48] Deletions of regions containing homoeologous loci have been common events, and a well-characterized sample is that of the *Ha* locus (moderates the grain texture or hardness) at the distal end of the short arm of group 5 chromosomes. In hexaploid wheat, only the 5D genome has the *Ha* locus, and the homoeologous loci on chromosomes 5A and 5B are absent. Consistent with this situation in hexaploid wheat, the *Ha* locus is also missing from the tetraploid progenitor (with the genome designations AABB), although present in the diploid progenitors[49] — a major deletion event is therefore assumed to have occurred after the polyploidization event that generated the AABB tetraploid wheat. The *Ha* locus is defined by three genes: grain softness protein (*Gsp*), puroindoline a (*Pina*), and puroindoline b (*Pinb*). Extensive sequence analyses on the region were carried out by Chantret et al.[50,51] Based on genomic DNA sequences identifiable in tetraploid wheat, the 5′ boundary of the *Ha* locus was defined by the *Gsp* gene since this is present in the A, B genomes of tetraploid wheat. The 3′ boundary was defined by a block of repeated genes (called *Gene7* and *Gene8*) that were also present in A, B genomes of tetraploid wheat. The *Ha* locus was therefore defined by an approximately 55 kb segment of genomic DNA and contained *Pina, Pinb*, two degenerate copies of *Pinb*, *Gene 3* (present only in the D genomes), and *Gene 5*. *Gene 3* and *Gene 5* are of unknown function. The study by

Chantret et al.[51] indicated major differences between the D genome progenitor locus and the D genome locus in hexaploid wheat, and these included the deletion of about 38 kb of DNA sequence in the hexaploid locus relative to the diploid locus.

Furthermore, rearrangements were identified and correlated with the location of TEs. Duplications, expansion of repetitive sequences, and deletions also characterize the difference between the low molecular weight glutenin genes on 1A of the A genome progenitor and 1A of hexaploid wheat[52]; the *Wx* (granule-bound starch synthase) genes on chromosomes 7A and 7D of hexaploid wheat[53]; the *wPBF* transcription factor genes on chromosomes 5A, 5B, and 5D of hexaploid wheat[54]; and the *Wknox* genes on 4A, 4B, and 4D of hexaploid wheat.[55] Some of these structural changes can lead to differential changes in the expression of homoeologous genes present on all three chromosome groups of hexaploid wheat.[56] The differential expression of homolgous genes demonstrated that some homoeologous loci on the A, B, and D genomes (identified by single-nucleotide polymorphisms [SNPs]) were expressed differentially depending on the tissue that was assayed.

The changes in genome structure discussed above also occur within diploid crop plants. The rapid divergence of equivalent *Rph1* loci in cultivars of barley (*Hordeum vulgare* L.), for example, has been shown to be due to changes in the number and type of repetitive elements,[57] resulting in so-called haplotype variability. A gene sequence (*Hvhel1*) located near one of the conserved gene sequences (*HvHGA2*) was also present in cultivar Morex but missing in cultivar Cebada Capa. This variation occurred against a background of conserved collinearity of five gene sequences (*Hvgad1, Hvpg1, Hvpg4, HvHGA1, HvHGA2*).

Studies on the helitron elements in maize (*Zea Mays* L.)[58–60] have provided an interesting blurring of the gene "space" and the mobile element space within the genome. Helitrons coding for proteins related to those required for potentially undergoing transposition (a helicase and replication protein A) are defined as autonomous, in contrast to nonautonomous helitrons that are missing these elements. The helitron elements have 5′ TCT and 3′ CTAG ends that are preceded by an 18- to 25-hairpin region and an AT target site[60] with variable lengths of DNA sequence between these characteristic features (6–20 kb).[58] Some of the elements characterized to date also house multiple portions of pseudogenes.[58] The helicase and replication protein A-like genes present in autonomous helitrons also occur in bacterial transposons that transpose via a rolling circle mechanism,[61] but evidence for this mechanism operating in plants or other eukaryotes has not been reported to date.

The feature of helitrons that is particularly interesting in the context of the cereal genome and understanding of its structure relevant to breeding is that it has been speculated that, during the course of transposition, helitrons can acquire exon segments from genes. The evidence for the duplication of segments of different genes rather than simply large genome regions comes from the analysis of duplicated loci in maize.[59,60] Comparison of the B73 and Mo17 maize inbred lines identified the so-called NOPQ9002 cluster in different chromosomal locations (1L in B73 and 9L in M17). Additional loci were located on chromosome 6S in B73 and 1L in M17 (not in the equivalent location relative to the locus on 1L in B73). Structural analysis indicated that the exon clusters were flanked by the sequence elements identified for helitrons and led to their identification as nonautonomous helitrons.[59] The interpretation of the structural relationships between the nonallelic loci suggested that a process of acquisition of additional exon segments from low copy expressed genes can occur during the transposition events originating from an ancestral copy of the cluster on 9L. Furthermore, Brunner et al.[59] demonstrated that a full-length, polyadenylated transcript originating from the proposed helitron genic DNA could be identified in RNA prepared from a mixed-tissue sample.

It is evident from the individual analysis of plant genome that mechanisms causing gene and genome rearrangements evolved during independent speciation of the plant lineages. Figure 17.2 summarizes the evolutionary timescale for plant lineages arising from a progenitor plant species. Based on the individual analysis of plant genomes, we can estimate gene and genome rearrangements that occurred during evolutionary time and those mechanisms that acted on genomes to evolve the modern-day species. However, it is unclear whether mechanisms identified in one species may or may not have occurred during genome evolution in other species. For example, new exon combinations through helitron activity have been identified in maize (Figure 17.2) but have not yet been studied for rice, wheat, or *Arabidopsis* in as much detail. Nevertheless, common mechanisms (such as gene duplications) have been identified that occur during independent evolution of plant species.

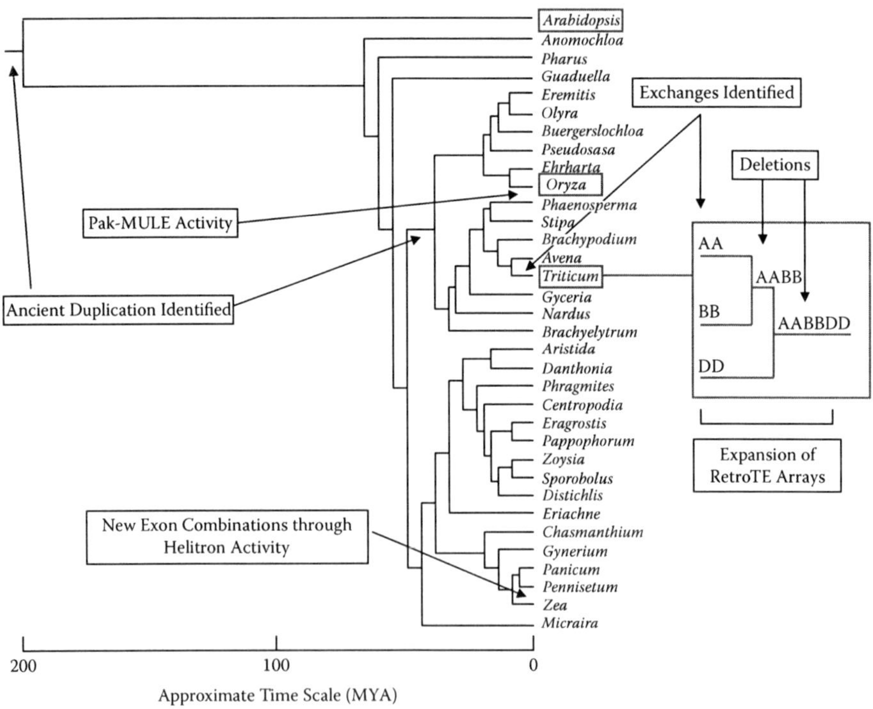

FIGURE 17.2 Taxonomic relationships of plant species and approximate time of divergence during evolution. (Adapted from Kellog, E.A., *Plant Physiol* 125, 1198–1205, 2001.). Gene and genome rearrangements and their timing with species divergence are indicated. Pack-MULE activity relates to a class of retrotransposable elements described in particular detail by Jiang et al.[133]

17.3 *ARABIDOPSIS* AND RICE: BRIDGING THE DICOT–MONOCOT DIVIDE USING COMPARATIVE GENOMICS

Rice and *Arabidopsis* represent sequenced genomes from monocot and dicot species, respectively, and separated from a common lineage 150–200 MYA.[62] Direct comparisons between these genomic resources provides the basis for comparing and contrasting genes and genomes between taxonomically diverged species, interpreting the data, and developing new approaches for the functional characterization of complex crop genomes.

17.3.1 DICOT–MONOCOT COMPARATIVE GENE ANALYSIS

The reason for comparing gene orthologs is to assess the level of conservation and hence increase the probability of accurately predicting gene function across species. Since the release of the genome sequence for *Arabidopsis* and rice, this can now be achieved at the whole-genome level, and the analysis of specific gene families has been possible. For example, the GRAS,[63] receptor-like kinases,[64] transcription factors,[65] Dof,[66] and gene families related to cell wall accumulation[67] have been analyzed to some detail in *Arabidopsis* and rice. Some of the gene families have similar copy numbers between species, whereas others have fewer, consistent with gene duplication occurring as a result of the independent expansion of gene families since divergence of the dicot and monocot lineages. Detailed interpretation of these observations needs to consider that the sequenced rice genome is from a species that has been subjected to hundreds of years of intensive breeding and for which artificial selection for domestication may have resulted in variation of copy number for a particular gene family over and above what may have occurred in a natural population. For example, the analysis of the receptor-like kinase gene family shows estimated 600 copies in *Arabidopsis* and in excess of 1,100 in rice.[64] It is thought that higher copy numbers in rice may reflect the increasing role of these enzymes in a variety of pathogen responses that have been intensively selected in rice breeding and domestication.[68] An analysis of the same gene family in the complete genome sequence of a wild species of rice and comparison with *Arabidopsis* could provide some clues regarding the effects of artificial selection on retaining or eliminating gene duplications and paralogous sequences in domesticated species.

Although differences in copy number within multigene families are evident in *Arabidopsis* and rice, other related gene families have similar copy numbers and thereby are presumed to have an evolutionary role in maintaining plant survival through conserved gene function. For example, the 32 gene families encoding enzymes and proteins for cell wall synthesis show no significant difference in copy number between *Arabidopsis* and rice.[67] It is evident that these gene families have been maintained throughout the *Arabidopsis* and rice lineages from the ancestral angiosperm genome, possibly in relation to their roles in maintaining plant cell function. Given the conserved evolutionary nature of some sequences, genes that are vital in determining plant growth and development, such as those encoding enzymes involved in cell wall synthesis, would be excellent candidates for predicting biological function based on comparative genomic approaches.

17.3.2 Similarities and Differences between *Arabidopsis* and Rice Genomes

Although the direct comparison of genes and gene families is fundamental for comparative genomics, the conservation of chromosomal segments between genomes provides an alternative approach to predicting gene orthologs based on synteny. Initial sequence analysis of portions from *Arabidopsis* and rice genomes revealed low levels of microsynteny between species,[69,70] and this was confirmed by the comparative analysis of both sequenced genomes. The few collinear regions between genomes are often represented by regions of less than 3 cM and are frequently interrupted with noncollinear genes.[32,71–73] Interestingly, the duplicated chromosomal regions identified within species were not collinear between genomes.[73] Therefore, the collective analysis within and between species clearly indicated that diversification of genes and genomes was not a static event but rather a dynamic process during the independent evolution of monocots and dicots, revealing a mosaic of similar and unique genes with orders more extensively rearranged than originally predicted.[74–76] Knowledge of genes controlling basic biological processes (such as plant cell growth and development) can benefit from studying taxonomically diverse species through gene family comparisons. However, more specialized niche functions (such as adaptability to extreme environments) will be best addressed in species that share a closer evolutionary lineage in which both comparative gene analysis and chromosomal synteny may provide additional strategies to discover genes that control trait variation.

17.3.3 Future Direction for Comparative Genomics between *Arabidopsis* and Rice

The sequenced genomes of *Arabidopsis* and rice have provided significant contributions to our understanding of gene and genome organization within and between species, but we have only reached the periphery of how plant genomes function. Multidisciplinary research in gene expression and the relationship with the proteome and phenotypic variation are now combined with high-throughput gene expression through microarray technologies to analyze the expressed portion of the *Arabidopsis* and rice genomes.[77] Although some attempts have been made to compare transcript profiles between *Arabidopsis* and rice,[31] it is likely that *Arabidopsis* and rice transcriptomes will continue to be analyzed individually and data integrated with genome sequence data to compare gene relatedness and expression profiles across species. Similarly, the integration of the *Arabidopsis* and rice proteome with phenotypic effects through TILLING (target-induced local lesion in genomes), quantitative trait loci (QTL) mapping, and transgenics will add to the tools that will collectively determine the function of unique genes and complex multigene families.[78]

17.4 COMPARATIVE GENOMICS FOR CROP IMPROVEMENT

The primary objective of using comparative genomics is to identify genes that control trait variation in one species and translate this information so that it will benefit crops, particularly to adapt in different environmental conditions. As noted, a proportion of crop genomes is polyploids that have either originated through intergeneric hybridization and contain different genomes (allopolyploids) or arose from a single species (autopolyploids). Given that the converged hybridization of an ancestral polyploid genome resulted in the evolution of *Arabidopsis* and rice, it is reasonable to assume that similar events had evolved in the diploid progenitors before hybridization to form the polyploid crop genomes. Also, genome restructuring is apparently more rapid and extensive in polyploids,[1,79,80] leading to further genome rearrangements compared to their diploid progenitors or more distantly related species. *Brassica napus* L. (allotetraploid) and *Triticum aestivum* L. (allohexaploid) are typical crop species with complex allopolyploid (similar but different) genomes for which the translation of information from model species can be confounded by further gene and genome expansion, inevitably leading to more complicated analysis and interpretations.

17.4.1 *Arabidopsis* and Other Model Species for Crop Improvement

Brassica species provide a significant proportion of the world's edible foods, targeting oilseed, vegetable, and condiment markets and, since they are members of the same Crucifereae family as *Arabidopsis*, are the immediate beneficiary of the sequenced *Arabidopsis* genome.[81,82] *Arabidopsis* and *Brassica* species are taxonomically classified into different tribes for which divergence from their ancestral species was a recent event, estimated at between 14.5 and 20.4 MYA.[83] The close evolutionary relationship and the importance of *Brassica* crops in the world's diet provide the opportunity to exploit the genome analyses outcomes from the *Arabidopsis* sequencing project for *Brassica* crop improvement. Based on initial comparative genomics studies, it was estimated that a significant portion of *Arabidopsis* and *Brassica* genomes are syntenic.[84–87] However, gene and chromosomal disruption by multiple rearrangements is evident[87] even though *Brassica* species have evolved from the same lineage as *Arabidopsis* as a relatively recent event.

Although synteny between chromosomes is disrupted by nonrelated genes, genome shotgun sequencing represented 0.44X the *Brassica oleraceae* genome, and its comparison with *Arabidopsis* identified a high proportion of gene sequence similarity, with an average 71% sequence conservation between coding regions.[88] Interestingly, it was also noted in the study that the sequencing of a portion of the *B. oleraceae* and its comparison improved the annotation and identification of new genes in the *Arabidopsis* genome,[88] highlighting annotation improvements as a side benefit of comparative genomic studies. The loss of protein-coding genes in *B. oleraceae* compared to *Arabidopsis* is widespread throughout the genome.[89] A successful example of applying comparative genomics from model to crop species has been the cloning of duplicated *Brassica rapa* homologs of the MADS-box flowering time regulator gene, having a similar function as its *Arabidopsis* counterpart *FLC*[90,91] in moderating flowering time. The impacts of this study on *Brassica* improvement are yet to be fully realized but hold promising aspects for developing early- or late-maturing *Brassica* varieties by the strategic application of gene variants through either transgenic or conventional breeding approaches.

In some instances, the application of comparative genomics was extended beyond close relatives of *Arabidopsis*. For example, the *Arabidopsis GA1* gene provided the basis for the isolation of the wheat *Rht1* gene and, in turn, maize-dwarfing genes *D8* and *D9*.[92,93] The height-reducing gene *Rht1* was the basis for the so-called green revolution in the 1960s[93] through its introduction into the CIMMYT breeding program. The assay for this gene has been implemented to optimize parental selection for crossing and as a selection tool in modern breeding programs.

Studies have also extended comparative sequence analysis to include horticultural and other crops important in agriculture, particularly species in the Solanaceae[94,95] and Fabaceae.[96,97] However, the Crucifereae, Solonaceae, and Fabaceae are widely separated,[98] limiting opportunities for outcomes from comparative genomics to translate information from model to commercially important horticultural crops.[97,99,100] In addition, certain plant species have evolved unique biological processes for which the sequenced genome of *Arabidopsis* may not be relevant. For example, legumes have developed the ability to establish symbiotic relationships with Rhizobia by which novel biochemical pathways provide the innate ability to fix nitrogen, providing necessary nutrients required for increased yields during cereal production. Therefore, model species other than *Arabidopsis* are favored for comparative genomics in legumes.

In particular, *Medicago truncatula* and *Lotus japonicus* have been the model species of choice for commercial legumes such as soybean, beans, field peas, and alfalfa and the genome sequencing projects for *M. truncatula* and *L. japonicus* are in progress.[101–103] In some instances, model legume species are in use as a "bridging" species to close the evolutionary gap between *Arabidopsis* and legumes (estimated divergence about 90 MYA[104,105]) even though comparisons are often limited to small networks of microsynteny[96,97,106] and are subjected to high proportions of selective gene loss.[97] As an example of the small, specialized networks, a study by Allen[11] identified 545 genes from *M. truncatula* that did not have a detectable ortholog in *Arabidopsis*. Genome rearrangements have also been assayed between *M. truncatula* and *Glycine max*, for which microsynteny was interrupted with lineage expansion/contraction of gene families.[107,108]

17.4.2 Rice Genome Sequence for Crop Improvement in Cereals and Other Grasses

The small size relative to grass species was one of the incentives for sequencing the rice genome, for which comparative genomics would play a pivotal role in deciphering gene and genome function of wheat (*Triticum aestivum* L.), barley (*Hordeum vulgare* L.), maize (*Zea mays* L.), and sorghum (*Sorghum bicolour* L.). Draft sequences of these large cereal genomes are still some years from completion.[109–112] Therefore, comparative genomics between the sequenced rice genome and the increasing resources (ESTs and full-length cDNAs) from grass species of commercial significance are currently important in deciphering genome organization and function.

Comparative gene and genome organization within grass genomes has relied predominantly on heterologous DNA probes and recombination mapping, setting the benchmark for macrosyntenic relationships within crop species and between rice.[40] The high-throughput sequencing of large EST collections has refined comparative gene and genome analysis across members of the Poaceae family, which represent the majority of cereal crops. For example, there are more than 875,000 *Triticum* ESTs represented in public domain databases,[113] of which more than 7,600 ESTs, representing greater than 16,000 loci, have been assigned to chromosomal regions by deletion bin mapping.[114–117] The allocation of a large set of ESTs to specific chromosomal regions in wheat and the comparative analysis with the sequenced genome of rice provides a first detailed comparison of genes and genomes between species. Interestingly, based on nucleotide and protein sequence similarity, only 43%–60% of the wheat ESTs mapped in wheat had significant sequence similarity with rice genes.[41–43,118–120] This indicated that gene gain or loss occurred since the separation of wheat and rice about 30–60 MYA.[121] It is yet uncertain whether the genes that share high sequence similarity represent gene families affecting the same plant phenotypes. The assignment of genes to specific regions of the wheat genome has enabled the detailed alignment of genomic segments with rice chromosomes and confirms macrosyntenic relationships between species, but identified microrearrangements, including insertions/deletions, inversions, duplications, and translocations causing erosion of collinearity between species.[43,47]

The rearrangements and disruptions in gene content and order between rice and wheat can have significant implications when attempting to identify candidate genes controlling specific traits in wheat. For example, the identification of a candidate gene controlling resistance for a major pathogen of wheat, *Fusarium* head blight (FHB), on wheat chromosome 3BS could not be readily achieved[122] by analyzing macrosyntenic regions and sequence annotations on rice chromosome 1. However, a resistance-like gene with scant similarity to a region on rice chromosome 11 shared common origins with the barley *Rpg1* gene for rust resistance on chromosome 7H and mapped to a major QTL controlling FHB resistance on wheat 3BS.[123–125] In some instances, conservation in gene content between rice and cereals can be used effectively to identify candidate genes that may be related to trait variation. In a study by Li et al.,[42] a gibberellic acid (GA) 20 oxidase gene annotated on rice chromosome 3 and syntenic with barley chromosome 5H aligned with a major QTL controlling variation to preharvest tolerance. Adkins et al.[126] have shown that GA may be involved in seed dormancy, giving the opportunity to further investigate the possible role of GA 20 oxidase in controlling seed dormancy and preharvest sprouting tolerance in barley.

Since taxonomic relationships are an important consideration for the effective use of comparative genomics, crop species more closely related to each other than their relationship to rice can also serve to compare gene content and order for shared traits and metabolic processes. It is estimated that perennial ryegrass (*Lolium perenne* L.) has been shown to have significant macrosynteny with other Poaceae species in comparative genetic mapping[127] and can be effectively used for candidate gene discovery for similar traits of interest. QTL have been identified as controlling variation for herbage quality on ryegrass chromosome 3, and wheat genes with similarity to lignin biosynthetic genes from ryegrass, *LpCAD2* and *LpCCR1*, have been mapped on wheat chromosome 3BL.[128] Interestingly, variation controlling stem solidness in wheat has also been mapped in the same region on 3BL,[129] where cell wall lignification is presumed to contribute to trait expression. The lignin-related sequences provide an indication of wheat orthologs for *LpCAD2* and *LpCCR1* as potential candidates influencing variability in solid stem trait through the lignin biosynthetic pathway.

The study of fructan accumulation in cereals is of particular interest for crop improvement as it is associated with drought and cold stress tolerance.[130,131] The study of fructan synthesis and accumulation during plant development is consequently of interest to researchers studying the physiological, biochemical, and molecular basis of abiotic stress tolerance in commercial grass species. Numerous reports have shown that the fructosyltransferase genes of the fructan biochemical pathway from perennial ryegrass (*LpFT*) have a close evolutionary relationship with rice invertase genes[131] even though rice does not accumulate fructans as carbohydrate reserves. A study by Francki et al.[132] showed that invertase and fructosyltransferase genes in rice and perennial ryegrass, respectively, constitute multigene families as a result of gene duplication and divergence from a single progenitor gene. Furthermore, in wheat, it appears that each member of multigene families has further duplicated and diverged from their rice and ryegrass counterparts either as haplotypes or insertion/deletion gene variants prior to or after polyploidization of the hexaploid wheat genome.[132]

17.5 CONCLUSIONS

The concept of comparative genomics to identify genes that control trait variation and the translation of genomic information from one organism to another is an exciting concept to accelerate gene discovery for crop improvement. As the sequence information from more plant genomes becomes available, our knowledge of the convoluted arrangement of gene and genomes will have a significant bearing on how we apply comparative genomics. The genome of the model plant organisms *Arabidopsis* and rice have allowed an in-depth analysis of how plant genomes evolved, and there are examples of gene function discovery. The analysis and integration of the large databases derived from proteomics, transcriptomics, and phenomics (high-throughput technologies to determine phenotypes) will ensure that comparative genomics based on model species can provide accurate predictions of gene functions that control specific traits in major crop species.

REFERENCES

1. Gale, M.D. & Devos, K.M. Plant comparative genetics after 10 years. *Science* 282, 656–659 (1998).
2. Devos K.M. Updating the "crop circle." *Curr Opin Plant Biol* 8, 155–162 (2005).
3. King, G.J. Through a genome, darkly: comparative analysis of plant chromosomal DNA. *Plant Mol Biol* 48, 5–20 (2002).
4. Feuillet, C. & Keller, B. Comparative genomics in the grass family: molecular characterization of grass genome structure and evolution. *Ann Bot* 89, 3–10 (2002).
5. Paterson, A.H., Freeling, M. & Sasaki, T. Grains of knowledge: genomics of model cereals. *Genome Res* 15, 1643–1650 (2005).
6. Crow, K.D. & Wagner, G.P. What is the role of genome duplication in the evolution of complexity and diversity? *Mol Biol Evol 23*, 887–892 (2006).
7. Blanc, G. & Wolfe, K.H. Functional divergence of duplicated genes formed by polyploidy during *Arabidopsis* evolution. *Plant Cell* 16, 1679–1691 (2004).

8. The *Arabidopsis* Genome Initiative Analysis of the genome sequence of the flowering plant *Arabidopsis thaliana*. *Nature* 498, 796–815 (2000).
9. Goff, S.A. et al. A draft sequence of the rice genome (*Oryza sativa* L. ssp. japonica). *Science* 296, 92–100 (2002).
10. Yu, J. et al. A draft sequence of the rice genome (*Oryza sativa* L. ssp. indica). *Science* 296, 79–91 (2002).
11. Allen, K.D. Assaying gene content in *Arabidopsis*. *Proc Natl Acad Sci USA* 99, 9568–9572 (2002).
12. Lin, X. et al. Sequence and analysis of chromosome 2 of the plant *Arabidopsis thaliana*. *Nature* 402, 761–772 (1999).
13. Mayer, K. et al. Sequence and analysis of chromosome 4 of the plant *Arabidopsis thaliana*. *Nature* 402, 761–772 (1999).
14. Terryn, N., Rouze, P., & Van Montagu, M. Plant genomics. *FEBS Lett* 452, 3–6 (1999).
15. Blanc, G., Barakat, A., Guyot, R., Cooke, R. & Delseny, M. Extensive duplication and reshuffling in the *Arabidopsis* genome. *Plant Cell* 12, 1093–1101 (2000).
16. Seoighe, C. & Gehring, C. Genome duplication led to highly selective expansion of the *Arabidopsis thaliana* proteome. *Trends Genet* 20, 461–464 (2004).
17. McGrath, J.M., Jansco, M.M. & Pichersky, E. Duplicate sequences with similarity to expressed genes in the genome of *Arabidopsis thaliana*. *Theor Appl Genet* 86, 880–888 (1993).
18. Kowalski, S.P., Lan, T.H., Feldmann, K.A. & Paterson, A.H. Comparative mapping of *Arabidopsis thaliana* and *Brassica oleraceae* chromosomes reveals islands of conserved organization. *Genetics* 138, 499–510 (1994).
19. Bennetzen, J.L., Coleman, C., Liu, R., Ma, J., & Ramakrishna, W. Consistent over-estimation of gene number in complex plant genomes. *Curr Opin Plant Biol* 7, 732–736 (2004).
20. Vision, T.J., Brown D.G. & Tanksley, S.D. The origins of genomic duplications in *Arabidopsis*. *Science* 290, 2114–2117 (2000).
21. Simillion, C., Vandepoele, K., Van Montagu, M.C.E., Zabeau, M. & Van de Peer, Y. The hidden duplication past of *Arabidopsis thaliana*. *Proc Natl Acad Sci USA* 99, 13627–13632 (2002).
22. De Bodt, S., Maere, S. & Van de Peer Y. Genome duplication and the origin of angiosperms. *Trends Ecol Evol* 20, 591–597 (2005).
23. Ziolkowski, P.A., Blanc, G., & Sadowski, J. Structural divergence of chromosomal segments that arose from successive duplication events in the *Arabidopsis* genome. *Nucl Acids Res.* 31, 1339–1350 (2003).
24. Henry, Y., Bedhomme, M. & Blanc, G. History, protohistory and prehistory of the *Arabidopsis thaliana* chromosome complement. *Trends Plant Sci* 11, 267–273 (2006).
25. Lippman, Z.L. et al. Role of transposable elements in heterochromatin and epigenetic control. *Nature* 430, 471–476 (2004).
26. Gendrel, A.V. et al. Dependence of heterochromatic histone H3 methylation patterns on the *Arabidopsis* gene DDM1. *Science* 297, 1871–1873 (2002).
27. Llave, C. et al. Endogenous and silencing-associated small RNAs in plants. *Plant Cell* 14, 1605–1619 (2002).
28. Martienssen, R.A., Doerge, R.W. & Colot, V. Epigenomic mapping in *Arabidopsis* using tiling microarrays. *Chrom Res* 13, 299–308 (2005).
29. Millar, A.A. & Waterhouse, P.M. Plant and animal microRNAs: similarities and differences. *Funct Integr Genomics* 5, 129–135 (2005).
30. Li, L. et al. Tiling microarray analysis of rice chromosome 10 to identify the transcriptome and relate its expression to chromosomal architecture. *Genome Biol* 6, R52.1–R52.17 (2005).

31. Ma, L. et al. A microarray analysis of the rice transcriptome and its comparison to *Arabidopsis Genome Res* 15, 1274–1283 (2006).
32. Vandepoele, K., Saeys, Y., Simillion, C., Raes, J. & Van de Peer, Y. The automatic detection of homologous regions (*ADHoRe*) and its application to microcolinearity between *Arabidopsis* and rice. *Genome Res* 12, 1792–1801 (2002).
33. Paterson, A.H., Bowers, J.E., Chapman, B.A. Ancient polyploidization predating divergence of the cereals, and its consequence for comparative genomics. *Proc Natl Acad Sci USA* 101, 9903–9908 (2004).
34. Yu, J. et al. The genomes of *Orya sativa*: a history of duplications. *PLoS Biol.* 3, e38 (2005).
35. Wang, H., Yu, L., Lai, F., Liu, L. & Wang, J. Molecular evidence for asymmetric evolution of sister duplicated blocks after cereal polyploidy. *Plant Mol Biol* 162, 63–74 (2005).
36. Guyot, R. & Keller, B. Ancestral genome duplication in rice. *Genome* 47, 610–614 (2004).
37. Kellog, E.A. Evolutionary history of the grasses. *Plant Physiol* 125, 1198–1205 (2001).
38. Vandepoele, K., Simillion, C. & Van de Peer, Y. Evidence that rice and other cereals are ancient aneuploids. *Plant Cell* 15, 2192–2202 (2003).
39. Wang, X., Shi, X., Hao, B., Ge, S. & Luo, J. Duplication and DNA segmental loss in the rice genome: implications and diploidization. *New Phytol* 165, 937–946 (2005).
40. Appels, R., Francki, M. & Chibbar, R. Advances in cereal functional genomics. *Funct Integr Genomics* 3, 1–24 (2003).
41. Francki, M. et al. Comparative organization of wheat homoeologous group 3S and 7L using wheat–rice synteny and identification of potential markers for genes controlling xanthophyll content in wheat. *Funct Integr Genomics* 4, 118–130 (2004).
42. Li C. et al. Genes controlling seed dormancy and pre-harvest sprouting in a rice-wheat barley comparison. *Funct Integr Genomics* 4, 84–93 (2004).
43. La Rota, M. & Sorrells, M.E. Comparative DNA sequence analysis of mapped wheat ESTs reveals the complexity of genome relationships between rice and wheat. *Funct Integr Genomics* 4, 34–46 (2004).
44. Lu, H. & Faris, J.D. Macro- and microcolinearity between the genomic region of wheat chromosome 5B containing the Tsn1 gene and the rice genome. *Funct Integr Genomics* 6, 90–103 (2006).
45. Feschotte, C., Jiang, N. & Wessler, S.R. Plant transposable elements: where genetics meets genomics. *Nat Rev* 3, 329–341 (2002).
46. Schulman, A.H. & Kalendar, R. A movable feast: diverse retrotransposons and their contribution to barley genome dynamics. *Cytogenet Genome Res* 110, 598–606 (2005).
47. Singh, N.K. et al. Single-copy genes define a conserved order between rice and wheat for understanding differences caused by duplication, deletion, and transposition of genes. *Funct Integr Genomics* in press (2006).
48. Chen, Z.J. & Ni, Z. Mechanisms of genomic rearrangements and gene expression changes in plant polyploids. *Bioessays* 28, 240–252 (2006).
49. Gautier, M.F., Cosson, P., Guirao, A., Alary, R. & Joudrier, P. Puroindoline genes are highly conserved in diploid ancestor wheats and related species but absent in tetraploid *Triticum* species. *Plant Sci* 153, 81–91 (2000).
50. Chantret, N., Cenci, A., Sabot, F., Anderson, O. & Dubcovsky, J. Sequencing of the *Triticum monococcum* hardness locus reveals good microcolinearity with rice. *Mol Genet Genomics* 271, 377–386 (2004).
51. Chantret, N. et al. Molecular basis of evolutionary events that shaped the hardness locus in diploid and polyploidy wheat species (*Triticum* and *Aegilops*). *Plant Cell* 17, 1033–1045 (2005).

52. Wicker T. et al. Rapid genome divergence at orthologous low molecular weight glutenin loci of the A and A^m genomes of wheat. *Plant Cell* 15, 1186–1197 (2003).
53. Shariflou, M.R. & Sharp, P.J. A polymorphic microsatellite in the 3′ end of "waxy" genes of wheat *Triticum aestivum*. *Plant Breeding* 118, 275–277 (1999).
54. Ravel, C. et al. Single nucleotide polymorphisms, genetic mapping and expression of genes coding for the DOF wheat prolamin-box binding factor. *Funct Integr Genomics* 6, 310–321 (2006).
55. Morimoto, R., Kosugi, T., Nakamura, C. & Takumi, S. Intragenic diversity and functional conservation of the three homoeologous loci of the KN1-type homeobox gene *Wknox1* in common wheat. *Plant Mol Biol* 57, 907–924 (2005).
56. Mochida, K., Yamazaki, Y. & Ogihara, Y. Discrimination of homoeologous gene expression in hexaploid wheat by SNP analysis of contigs groups from a large number of expressed sequence tags. *Mol Genet Genomics* 270, 371–377 (2003).
57. Scherrer, B. et al. Large intraspecific haplotype variability at the *Rph7* locus results from rapid and recent divergence in the barley genome. *Plant Cell* 17, 361–374 (2005).
58. Gupta, S., Gallvotti, A., Stryker, G.A., Schmidt, R.J. & Lal, S.K. A novel class of Helitron-related elements in maize contain portions of multiple pseudogenes. *Plant Mol Biol* 57, 115–127 (2005).
59. Brunner, S., Pea, G. & Rafalski, A. Origins, genetic organization and transcription of a family of non-autonomous helitron elements in maize. *Plant J* 43, 799–810 (2005).
60. Morgante, M. et al. Gene duplication and exon shuffling by helitron-like transposons generate intraspecies diversity in maize. *Nat Genet* 37, 997–1002 (2005).
61. Kapitonov, V.V. & Jurka, J. Rolling-circle transposons in eukaryotes. *Proc Natl Acad Sci USA* 98, 8714–8719 (2001).
62. Wolfe, K.H., Gouy, M., Yang, Y.W., Sharp, P.M. & Li, W.H. Date of the monocot–dicot divergence estimated from chloroplast DNA sequence data. *Proc Natl Acad Sci USA* 86, 6201–6205 (1989).
63. Tiang, C., Wan, P., Sun, S., Li, J. & Chen, M. Genome-wide analysis of the GRAS family in rice and *Arabidopsis*. *Plant Mol Biol* 54, 519–532 (2004).
64. Shiu, S.-H. et al. Comparative analysis of the receptor-like kinase family in *Arabidopsis* and rice. *Plant Cell* 16, 1220–1234 (2004).
65. Xiong, Y. et al. Transcription factors in rice: a genome wide comparative analysis between monocots and eudicots. *Plant Mol Biol* 59, 191–203 (2005).
66. Lijavetzky, D., Carbonero, P. & Vicente-Carbajosa, J. Genome wide comparative phylogenetic analysis of the rice and *Arabidopsis Dof* gene families. *BMC Evol Biol* 3, 17 (2003).
67. Yokoyama, R. & Nishitani, K. Genomic basis for cell-wall diversity in plants. A comparative approach to gene families in rice and *Arabidopsis*. *Plant Cell Physiol* 45, 1111–1121 (2004).
68. Morillo, S.A., & Tax, F.E. Functional analysis of receptor-like kinases in monocots and dicots. *Curr Opin Plant Biol* 9, 460–469 (2006).
69. Devos, K.M., Beales, J., Nagamura, Y. & Sasaki, T. Arabidopsis-rice: will colinearity allow gene prediction across the eudicot–monocot divide? *Genome Res* 148, 435–443 (1999).
70. Van Dodeweerd, A.-M. et al. Identification and analysis of homoeologous segments of the genomes of rice and *Arabidopsis thaliana*. *Genome* 42, 887–892 (1999).
71. Liu, H., Sachidanandam, R. & Stein, L. Comparative genomics between rice and *Arabidopsis* shows scant collinearity in gene order. *Genome Res* 11, 2020–2026 (2001).
72. Mayer, K. et al. Conservation of microstructure between a sequenced region of the genome of rice and multiple segments of the genome of *Arabidopsis thaliana*. *Genome Res* 11, 1167–1174 (2001).

73. Salse, J., Piegu, B., Cooke, R. & Delseny, M. Synteny between *Arabidopsis thaliana* and rice at the genome level: a tool to identify conservation in the ongoing rice genome sequencing project. *Nucleic Acids Res* 11, 2316–2328 (2002).
74. Kumar, A. & Bennetzen, J.L. Plant retrotransposons. *Annu Rev Genet* 33, 355–365 (1999).
75. Federoff, N. Transposons and genome evolution in plants. *Proc Natl Acad Sci USA* 97, 7002–7007 (2000).
76. Wendel, J.F. Genome evolution in polyploids. *Plant Mol Biol* 42, 225–249 (2000).
77. Galbraith, D.W. & Birnbaum, K. Global studies of cell type-specific gene expression in plants. *Annu Rev Plant Biol* 57, 451–475 (2006).
78. Sappl, P.G., Heazlewood, J.L. & Millar, A.H. Untangling multi-gene families in plants by integrating proteomics and functional genomics. *Phytochemistry* 65, 1517–1530 (2004).
79. Soltis, D.E. & Soltis, P.S. Polyploidy: recurrent formation and genome evolution. *Trends Ecol Evol* 14, 348–352 (1999).
80. Soltis, P.S. Ancient and recent polyploidy in angiosperms. *New Phytol* 166, 5–8 (2005).
81. Paterson, A.H., Lan, T.-H, Amasino, R., Osborn, T.C. & Quiros, C. *Brassica* genomics: a complement to, and early beneficiary of, the *Arabidopsis* sequence. *Genome Biol* 2, 10111–10114 (2001).
82. Quiros, C.F. et al. *Arabidopsis* and *Brassica* comparative genomics: sequence, structure and gene content in the ABI-Rps2-Ck chromosomal segment and related regions. *Genetics* 157, 1321–1330 (2001).
83. Yang, Y.-W., Lai, K.N., Tai, Y. & Li, W.-H. Rates of nucleotide substitution in Angiosperm mitochondrial DNA sequences and dates of divergence between *Brassica* and other angiosperm lineages. *J Mol Evol* 48, 597–604 (1999).
84. Lagercrantz, U. & Lydiate, D. Comparative genome mapping in *Brassica*. *Genetics* 144, 1903–1910 (1996).
85. Lan, T.H. et al. An EST-enriched comparative map of *Brassica oleraceae* and *Arabidopsis thaliana*. *Genome Res* 10, 776–788 (2000).
86. Babula, D. et al. Chromosomal mapping of *Brassica oleraceae* based on ESTs from *Arabidopsis thaliana*: complexity of the comparative map. *Mol Genet Genomics* 268, 656–665 (2003).
87. Suwabe, K. et al. Simple sequence repeat-based comparative genomics between *Brassica rapa* and *Arabidopsis thaliana*: the genetic origin of clubroot resistance. *Genetics* 173, 309–319 (2006).
88. Ayele, M. et al. Whole genome shotgun sequence of *Brassica oleracea* and its application to gene discovery and annotation in *Arabidopsis*. *Genome Res* 15, 487–495 (2005).
89. Town, C.D. et al. Comparative genomics of *Brassica oleracea* and *Arabidopsis thaliana* reveal gene loss, fragmentation and dispersal after polyploidy. *Plant Cell* 18, 1348–1359 (2006).
90. Michaels, S.D. & Amasino, R.M. *FLOWERING LOCUS C* encodes a novel MADS domain protein that acts as a repressor of flowering. *Plant Cell* 11, 949–956 (1999).
91. Schranz, M.E. et al. Characterization and effects of the replicated flowering time gene *FLC* in *Brassica rapa*. *Genetics* 162, 1457–1468 (2002).
92. Peng, J.R. et al. "Green revolution" genes encode mutant gibberellin response modulators. *Nature* 400, 256–261 (1999).
93. Hedden, P. The genes of the green revolution. *Trends Genet* 19, 5–9 (2003).
94. Ku, H.M., Doganlar, S. & Tanksley, S.D. Exploitation of *Arabidopsis*–tomato synteny to construct a high resolution map of the ovate containing region in tomato chromosome 2. *Genome* 44, 470–475 (2001).

95. Rossberg, M. et al. Comparative sequence analysis reveals extensive microcolinearity in the lateral suppressor regions of the tomato, *Arabidopsis* and *Capsella* genomes. *Plant Cell* 13, 979–988 (2001).
96. Yan, H.H. et al. Estimates of conserved microsynteny among the genomes of *Glycine max, Medicago truncatula* and *Arabidopsis thaliana. Theor Appl Genet* 106, 1256–1265 (2003).
97. Zhu, H. et al. Syntenic relationships between *Medicago truncatula* and *Arabidopsis* reveal extensive divergence of genome organization. *Plant Physiol* 131, 1018–1026 (2003).
98. Palmer, J.D., Soltis, D.E. & Chase, M.W. The plant tree of life: an overview and some points of view. *Am J Bot* 91, 1437–1445 (2004).
99. Mudge, J. et al. Highly syntenic regions in the genomes of soybean, *Medicago truncatula* and *Arabidopsis thaliana. BMC Plant Biol* 5, 15 (2005).
100. Kevei, Z. et al. Significant microsynteny with new evolutionary highlights is detected between *Arabidopsis* and legume model plant despite the lack of macrosynteny. *Mol Genet Genomics* 274, 644–657 (2005).
101. Bell, C.J. et al. The *Medicago* genome initiative: a model legume database. *Nucl Acids Res* 29, 114–117 (2001).
102. Young, N.D. Sequencing the genespace of *Medicago truncatula* and *Lotus japonicus. Plant Physiol* 137, 1174–1181 (2005).
103. Udvardi, M.K., Tabata, S., Parniske, M. & Stougaard, J. *Lotus japonicus*: legume research in the fast lane. *Trends Plant Sci* 10, 222–228 (2005).
104. Gandolfo, M., Nixon, K. & Crepet, W. A new fossil flower from the Turonian of New Jersey: *Dressiantha bicarpellata* gen. Et sp. Nov. (Capparales). *Am J Bot* 85, 964–974 (1998).
105. Lee, J.M., Grant, D., Vallejos, C.E. & Shoemaker, R.C. Genome organization in dicots II. *Arabidopsis* as a "bridging species" to resolve genome evolution events among legumes. *Theor Appl Genet* 103, 765–773 (2001).
106. Grant, D., Cregan, P. & Shoemaker, R.C. Genome organization in dicots: genome duplication in *Arabidopsis* and synteny between soybean and *Arabidopsis. Proc Natl Acad Sci USA* 97, 4168–4173 (2000).
107. Choi, H.-K. et al. Estimating genome conservation between crop and model legume species. *Proc Natl Acad Sci USA* 101, 15289–15294 (2004).
108. Zhu, H., Choi, H.-K., Cook, D.R. & Shoemaker, R.C. Bridging model and crop legumes through comparative genomics. *Plant Physiol* 137, 1189–1196 (2005).
109. Gill, B.S. et al. A workshop report on wheat genome sequencing: international genome research on wheat consortium. *Genetics* 168, 1087–1096 (2004).
110. Sorghum Genomics Planning Workshop Participants. Toward sequencing the sorghum genome. A U.S. National Science Foundation-sponsored workshop report. *Plant Physiol* 138, 1898–1902 (2005).
111. Rabinowicz, P.D. & Bennetzen, J.L. The maize genome as a model for efficient sequence analysis of large plant genomes. *Curr Opin Plant Biol* 9, 149–156 (2006).
112. Maize Genome Sequencing Projects. Available at: http://maizegenome.org.
113. National Center for Biotechnology Information. Available at: http://www.ncbi.nlm.nih.gov/.
114. Zhang, D. et al. Construction and evaluation of cDNA libraries for large-scale expressed sequence tag sequencing in wheat (*Triticum aestivum* L). *Genetics* 168, 595–608 (2004).
115. Lazo, G.R. et al. Development of an expressed sequence tag (EST) resource for wheat (*Triticum aestivum* L): EST generation, unigene analysis, probe selection and bioinformatics for a 16,000-locus bin-delineated map. *Genetics* 168, 585–593 (2004).

116. Qi, L.L. et al. A chromosome bin map of 16,000 expressed sequence tag loci and distribution of genes among the three genomes of polyploid wheat. *Genetics* 168, 701–712 (2004).
117. Qi, L.L., Echalier, Friebe, B. & Gill, B.S. Molecular characterization of a set of wheat deletion stocks for use in chromosome bin mapping of ESTs. *Funct Integr Genomics* 3, 39–55 (2003).
118. Munkvold, J.D. et al. Group 3 chromosome bin maps of wheat and their relationship to rice chromosome 1. *Genetics* 168, 639–650 (2004).
119. Miftahudin, K. et al. Analysis of expressed sequence tag loci on wheat chromosome group 4. *Genetics* 168, 651–663 (2004).
120. Randhawa, H.S. et al. Deletion mapping of homoeologous group 6-specific wheat expressed sequence tags. *Genetics* 168, 677–686 (2004).
121. Soreng, R.J. & Davis, J.I. Phylogenetics and character evolution in the grass family. *Bot Rev* 64, 1–47 (1998).
122. Liu, S. & Anderson, J.A. Targeted molecular mapping of a major wheat QTL for *Fusarium* head blight resistance using wheat ESTs and synteny with rice. *Genome* 46, 817–823 (2003).
123. Killian, A. et al. Rice-barley synteny and its application to saturation mapping of the barley *Rpg1* region. *Nucl Acids Res* 23, 2729–2733 (1995).
124. Brueggeman, R. et al. The barley stem rust-resistance gene *Rpg1* is a novel disease-resistance gene with homology to receptor kinases. *Proc Natl Acad Sci USA* 99, 9328–9333 (2002).
125. Shen, X., Francki, M.G. & Ohm, H.W. A resistance-like gene identified by EST mapping and its association with a QTL controlling *Fusarium* head blight infection on wheat chromosome 3BS. *Genome* 49, 631–635 (2006).
126. Adkins, S.W., Bellairs, S.M. & Loch, D.S. Seed dormancy mechanisms in warm season grass species. *Euphytica* 126, 13–20 (2002).
127. Jones, E.S. et al. (2002). An enhanced molecular marker based genetic map of perennial ryegrass (*Lolium perenne*) reveals comparative relationships with other Poaceae genomes. *Genome* 45, 282–295.
128. Cogan, N.O.I. et al. QTL analysis and comparative genomics of herbage quality traits in perennial ryegrass (*Lolium perenne* L.). *Theor Appl Genet* 110, 364–380 (2005).
129. Cook, J.P., Wichman, D.M., Martin, J.M., Bruckner, P.L. & Talbert, L.E. Identification of microsatellite markers associated with a stem solidness locus in wheat. *Crop Sci* 44, 1397–1402 (2004).
130 Vijn, I. & Smeekens, S. Fructan: more than a reserve carbohydrate? *Plant Physiol* 120, 351–359 (1999).
131. Chalmers, J. et al. Molecular genetics of fructan metabolism in perennial ryegrass. *Plant Biotech J* 3, 459–474 (2005).
132. Francki, M.G., Walker, E., Forster, J.W., Spangenberg, G. & Appels, R. Fructosyltransferase and invertase genes evolved by gene duplication and rearrangements: rice, perennial ryegrass and wheat gene families. *Genome* 49, 1081–1091 (2006).
133. Jiang, N., Bao, Z., Zhang, X., Eddy, S.R. & Wessler, S.R. Pack-MULE transposable elements mediate gene evolution in plants. *Nature* 431, 569–573 (2004).

18 Domestic Animals
A Treasure Trove for Comparative Genomics

Leif Andersson

CONTENTS

ABSTRACT

Domestic animals provide unique opportunities for exploring genotype–phenotype relationships due to their long history of selective breeding and since their population structures often facilitate powerful genetic analysis. The emerging genome sequences and dense marker maps now provide the means to fully utilize the potential of domestic animals for comparative genomics. Strategies for genetic analysis of both monogenic and multifactorial traits are reviewed and exemplified in this chapter.

18.1 INTRODUCTION

Genome research in domestic animals is justified due to the potential practical applications in animal breeding programs. However, domestic animals have also an important role to play in comparative genomics. They will contribute to our understanding of genotype–phenotype relationships and the evolution of phenotypic traits. The long history of phenotypic selection in domestic animals has led to a rich phenotypic diversity that can now be exploited for comparative genomics. No model organisms have been genetically modified to the same extent as domestic animals. Furthermore, detailed pedigree records are maintained for many domestic animal populations, and phenotypic data are collected as part of the breeding activities. These circumstances provide excellent opportunities for powerful genetic analysis.

18.2 RICH PHENOTYPIC DIVERSITY

The development of domestic animals has a long history (~10,000 years) compared with the short time (~100 years) we have studied experimental organisms. Since domestication, humans have been monitoring the phenotype of domestic animals and genetically adapted them to new environments and different production systems. As an example, the red junglefowl (the wild ancestor of chickens) lives in the jungles in Southeast Asia, but the domestic chicken has been spread across the world and selected for the production of eggs or meat in a variety of environments and production systems. This has led to dramatic changes in growth patterns, behavior, fertility, metabolism, and resistance to various pathogens. This has been accomplished by altering the frequencies of mutations with phenotypic effects. Some of these allelic variants pre-date domestication, whereas others arose subsequent to domestication. For a long period of time, breeding practices were based on individual selection; that is, the animals that were best adapted and most fertile in the new environment were used for breeding. However, the development of the quantitative genetics theory, pioneered by Sir Ronald Fisher and Sewall Wright, during the last century revolutionized animal breeding, and increasingly sophisticated statistical tools for selecting the very best breeding animals have been developed.[1] This is possible by collecting phenotypic data from a large number of progenies from each potential breeding animal and using information on genetic relationships to accurately predict the ability to transmit favorable allelic variants to their progeny.

The genetic variants that have been enriched in domestic animals provide a valuable complement to the repertoire of genetic variants that is usually detected in humans or model organisms. Human genetics provides excellent opportunities to identify deleterious mutations that cause monogenic disorders. For instance, more than 1,000 different mutations in *CFTR* (cystic fibrosis transmembrane conductance regulator) causing cystic fibrosis have been described to date (Human Gene Mutation Database, http://www.hgmd.cf.ac.uk). Similarly, mutagenesis screening in rodents is an excellent tool to generate collections of deleterious mutations for a first characterization of gene functions.[2] In fact, domestic animals are rather poor models for studying deleterious mutations since there is strong purifying selection against deleterious mutations in most populations of domestic animals. However, the domestication of animals can be considered a huge screen for mutations with phenotypic effects in which millions of humans have monitored millions of animals for thousands of years. This screen is enriched for mutations with favorable phenotypic effects on traits under selection (e.g., milk production) but with no or only mild deleterious effects on other traits. Thus, we expect that the mutations underlying phenotypic diversity in domestic animals will to some extent differ from the ones detected in a mutagenesis screening in mice because it is a much deeper screen. This may include novel gain-of-function mutations, and because of the rather long history some alleles may reflect the combined effect of two or more subsequent mutations that have occurred in the same gene. The development of domestic animals by artificial selection provides an excellent model for the evolution of species by means of natural selection as recognized already by Darwin.[3]

18.3 POWERFUL GENETICS

It is possible to collect very large full-sib or half-sib families in domestic animals since breeding males may have hundreds or thousands of progeny. This creates opportunities to map quantitative trait loci (QTLs) with tiny effects. It is also possible to take advantage of the detailed phenotypic records that are collected in breeding programs. Detailed pedigree records, including information from many generations, are available in many populations of domestic animals. For instance, thoroughbred horses have complete pedigree records that trace back to the 18th century. This makes it possible to use identity-by-descent (IBD) mapping and take advantage of historical recombination events that have taken place as a haplotype has been transmitted from a common ancestor to subsequent generations.[4]

Breeds of domestic animals share common ancestors in the near or distant past, and there is often some gene flow between populations. Therefore, it is the rule rather than the exception that the same allele affecting a phenotypic trait is shared between breeds. This can be utilized in the search for causative mutation by defining the minimum shared haplotype associated with a certain allelic variant. A recent example of this concerns the Silver plumage color in the chicken. Gunnarsson et al.[5] showed that Silver is caused by mutations in *SLC45A2,* and that five different breeds fixed for the *347M* allele shared a minimum haplotype less than 35 kb in size. Similarly, Van Laere et al.[6] found that the same porcine *IGF2* haplotype associated with high muscularity was present in four different populations selected for lean growth. In this case, the minimum shared haplotype was as small as about 20 kb. This was a very important step toward the identification of the causal mutation for this major QTL.

18.4 SELECTIVE SWEEPS: GENOMIC FOOTPRINTS OF SELECTION

Selection in domestic animals as well as in natural populations leads to the fixation of favorable alleles. Selective sweeps are a consequence of this process and imply that closely linked polymorphisms also become fixed in the population due to hitchhiking.[7] This happens because there is not sufficient time to disrupt linkage between the causal mutation and closely linked polymorphisms. The genomic footprint of this process is a high degree of homozygosity in the region flanking a causal mutation favored by selection. The size of a region affected by a selective sweep depends on the local recombination rate and the number of generations that have passed from the appearance of the mutation until its fixation. This process can be fast in domestic animals due to the strong selection. As a consequence, the ancestral haplotype (on which the causal mutation occurred) may still be segregating in some populations.

The *IGF2* locus in pigs provides a classical example of a selective sweep in domestic animals.[6] The favorable allele at this QTL increases muscle content by 3%–4%, and the locus was first detected using cross-breeding experiments between wild boar and Large White pigs[8] and between Large White and Pietrain pigs.[9] An increase in muscularity by 3%–4% may appear tiny compared with the type of phenotypic effects normally detected in a mutagenesis screen in mice, but it is a huge effect from an agricultural perspective, and this QTL allele has experienced a dramatic selective sweep in many breeds used for commercial pork production in the Western world: Duroc, Hampshire, Pietrain, Large White, and Landrace.[6] Thus, in many populations of these breeds, there is basically no sequence variation around *IGF2* since the haplotype carrying the favorable substitution has gone to fixation or is close to fixation. Interestingly, genetic evidence for the causative nature of a single nucleotide substitution was obtained because an ancestral haplotype was identified that only differed by a single nucleotide substitution from the causative haplotype, and it did not show the QTL effect. Similarly, Milan et al.[10] found that the *PRKAG3* haplotype associated with a dominant mutation increasing the glycogen content in skeletal muscle only differed from one of the haplotypes associated with the wild-type allele by a single missense mutation (R225Q), which turned out to be the causal mutation. So, the possible coexistence of a mutant haplotype and its ancestral haplotype should not be ignored. This opportunity will be particularly important for the challenging task of detecting and proving the causative nature of regulatory mutations.

18.5 FACTS AND MISCONCEPTIONS

A common misconception is that domestic animals in general are highly inbred. The fact is that most populations of domestic animals show low levels of inbreeding, and the different species of domestic animals globally represent an amazing genetic diversity. It is correct that some populations of domestic animals, in particular those kept as pets, are inbred due to founder effects or small effective population sizes, but in most populations inbreeding is avoided. Let us first consider the process of domestication.

It is now clear that domestication did not involve severe population bottlenecks. The emerging picture is that domestication often involved multiple events in different geographic regions, and it is likely that there has been considerable gene flow between the early populations of domestic animals and their wild ancestors.[11–14] Thus, domestication may have captured a considerable amount of the diversity present in the wild ancestors. Furthermore, until the last few hundred years, there were no well-defined breeds of domestic animals. It was rather a diffuse population structure with gene flow between regions due to the trading of livestock. This is exemplified by the introduction of humped cattle into Africa[15] and the introduction of Asian pigs into Europe during the 18th and 19th centuries.[16] Thus, during most of the evolutionary history of domestic animals, the effective population sizes have been large due to this gene flow between populations.

Therefore, it is not surprising that estimates of genetic diversity are as high or even higher in domestic animals compared with that observed in humans.[11] It is only during the last few hundred years that well-defined and more specialized breeds have been established, including breeds developed for egg or meat production in chicken, milk or meat production in cattle, or wool or meat production in sheep. This has led to reduced genetic diversity within breeds, particularly in closed populations in which no gene flow into the population is allowed, but the ambition in all serious breeding programs is to maintain a relatively high effective population size to ensure a future selection response.

Domestic animals show dramatic phenotypic differences compared with their wild ancestors, but these changes have occurred within a short period of time (~10,000 years) from an evolutionary perspective. This is clearly shorter than the time since divergence of major population groups of humans. The genome sequences of domestic animals are therefore essentially indistinguishable from their wild ancestors. This is well illustrated by a study in chicken in which partial genome sequences (0.25X coverage) from three different breeds of domestic chicken (White Leghorn, a Broiler, and Silkie) were compared with the near-complete genome sequence (6.5X coverage) of the red junglefowl, the wild ancestor.[11] The nucleotide diversity between breeds of domestic chickens was as high as between any domestic breed and the red junglefowl, and on average there was a 0.5% sequence difference in any pairwise comparison among these four populations. This single-nucleotide polymorphism (SNP) frequency is five times higher than that observed in humans when comparing across populations.[17] Furthermore, if one compares this nucleotide difference of 0.5% between populations that have been separated for fewer than 10,000 years with the 1.2% average sequence difference between humans and chimpanzee that has evolved separately for about 5 million years, it becomes clear that most of the sequence diversity in domestic chicken (and in other domestic animals) pre-dates domestication. There has not been sufficient time to evolve distinct sequence differences.

Random DNA sequences from a domestic animal and its wild ancestor (if they are still present) will appear as allelic variants drawn from the same population. Thus, it is a paradox that any laypeople can distinguish a wild boar from a domestic pig, but it is difficult to distinguish them at the DNA level unless one studies genes that have been under strong selection during domestication. To the best of my knowledge, no specific mutation has yet been detected in any domestic animal that unequivocally distinguishes a domestic animal from its wild ancestor.

18.6 GENOME SEQUENCES AND DENSE SNP MAPS

The progress in domestic animal genomics has previously been hampered by the lack of genomic resources. The research funding in this area has been small compared with the resources allocated for human genomics, reflecting that human medicine has a higher priority than agriculture in the Western world. Furthermore, the limited resources for domestic animal genomics have been split on a number of species: cattle, pig, sheep, goat, horse, dog, cat, chicken, turkey, and so on. However, this situation is now rapidly improving due to the release of high-quality draft genome sequences accompanied by large collections of SNPs. The chicken was first out as the genome sequence was released[18] in 2004 together with a catalog of 2.8 million SNPs.[11] The dog genome sequence was released in December 2005 together with a list of 25 million SNPs.[19] The cattle genome sequence together with SNP information will soon be released (http://www.hgsc.bcm.tmc.edu/projects/bovine/), and a high-quality draft sequence of the horse genome has been released by the Broad Institute (http://www.broad.mit.edu/mammals/). At present, the pig genome is lagging behind, but the genome sequencing has been initiated at the Sanger Institute and a 3X coverage is expected to be available in early 2008 (http://piggenome.org/).

The access to a draft genome sequence and high-density SNP maps is a major leap forward for domestic animal genomics. The access to large panels of genetic markers facilitates linkage mapping and paves the way for whole-genome association analysis (see Section 18.7). The dense SNP maps circumvent the tedious work of developing new markers during positional cloning. Positional identification of causative genes and mutations is also greatly facilitated by the access to a draft genome sequence, which immediately provides a list of positional candidate genes in the target region and circumvents the need for *de novo* sequencing of the target region.

18.7 GENOME-WIDE ASSOCIATION ANALYSIS

Family-based linkage analysis is the classical way to map trait loci. This approach has been extremely successful for identifying genes controlling monogenic traits and disorders in experimental organisms, domestic animals, and humans. The genetic signal in a linkage experiment comes from tracing the inheritance of gametes transmitted from heterozygous parents to their progeny. This works beautifully for monogenic traits since there is a direct relationship between genotype and phenotype, making it easy to deduce which parents are heterozygous at the target locus. A panel of a few hundred highly informative markers (~1 marker/20 cM [centiMorgan]) is sufficient for an initial genome-wide scan, which is then followed up with fine mapping of the target region.

Linkage analysis of multifactorial traits controlled by QTLs is much more challenging than linkage mapping of monogenic trait loci (MTLs) (Table 18.1). This is because the phenotypic effect of each locus is small or moderate, and there is no simple one-to-one relationship between genotype and phenotype. In an outbred population, it is difficult or impossible to determine which parents are heterozygous at the QTL, and thus informative in a linkage analysis, and this must be deduced from segregation data using genetic markers. This problem can be illustrated as follows: Assume that you want to identify a locus causing type I diabetes in dogs, and you come across a half-sib family with a very high incidence of disease; you decide to make a genome scan using that family. However, the high incidence may occur

TABLE 18.1
Comparison of the Power of Family-Based Linkage Analysis and Genome-wide Association Analysis for Mapping Monogenic Trait Loci (MTLs) and Quantitative Trait Loci (QTLs)

	Linkage Analysis	Association Analysis
Material	Requires family material	Only case/control material required
Markers required for genome scan	~1 marker/20 cM	~10,000–500,000 depending on the pattern of linkage disequilibrium
Power for mapping MTLs	Very high if sufficiently large pedigree material is available	Very high if sufficient numbers of cases with the same mutation are available
Power for mapping QTLs	Requires very large pedigree materials to detect loci with small effects or unfavorable levels of polymorphism	May be difficult to distinguish true associations from spurious associations
	Poor initial mapping resolution	Excellent mapping resolution

because the sire is homozygous for a susceptibility factor, and there is no signal at all in the linkage analysis. A second problem is the poor resolution in QTL mapping since it is not possible to directly score recombinants as the QTL genotype cannot be deduced directly from their phenotype. The positional identification of mutations underlying QTLs is therefore challenging also in experimental organisms like mouse and *Drosophila*.[20]

In humans, linkage analyses of multifactorial disorders have been a frustrating experience since it is difficult to collect sufficiently large family materials that will give a reasonable power to detect susceptibility loci, and once they are detected, it is hard to identify the causal gene due to the poor map resolution. The current trend is therefore to replace the linkage approach by genome-wide association analysis (GWAA). Association analysis circumvents some of the problems associated with the linkage analysis (Table 18.1). First, there is no need to collect pedigrees; an association analysis is based on case/control materials. Ideally, the cases should be as unrelated as possible, and the controls should be well matched regarding sex, age, and population origin. Second, the map resolution is often high, which should facilitate the identification of the causal gene. Association mapping is based on the presence of linkage disequilibrium (LD) between markers and the causal polymorphism. The number of markers required for a GWAA is thus dependent on the length of haplotype blocks (regions of the genome with complete LD). In humans, the length of haplotype blocks was estimated to be about 10 kb by the HapMap project.[17] Therefore, genome scans using more than 100,000 SNPs tested on thousands of cases and controls are required for GWAA of multifactorial traits in humans. This is now feasible (although costly) due to the rapid development of efficient and cost-effective SNP screening methods.[21]

There are now many ongoing GWAA projects in humans. However, it is still uncertain how successful this huge investment will be. A successful outcome requires that a sufficient number of cases share the same causal mutation creating a significant difference in haplotype frequencies between cases and controls. Thus, genetic heterogeneity (multiple mutations in the same gene or many loci contributing to disease) will reduce the statistical power. Another major concern with association analysis is the risk of spurious associations due to population stratification or if cases and controls are not perfectly matched. For instance, if the cases have inherited a mutation from a shared ancestor, then it is difficult to avoid that they tend to be more closely related to each other than to the controls. This will create a significant correlation throughout the genome. This is not a major problem if there is a strong signal from a locus affecting the trait or disorder, but for QTLs with minor effects, it will be hard to distinguish true associations from spurious associations. Epistatic interaction between QTLs may also reduce the power in a standard association analysis.

There are good reasons to assume that GWAA will be more powerful for detecting QTLs in domestic animals than in humans. The reason for this optimism is the favorable population structure in which domestic animals are subdivided into breeds and subpopulations. The reduced effective population size within populations creates a considerable LD, and it is now established that haplotype blocks in general are considerably larger in domestic animals than in humans.[22–25] This has been studied in detail in the dog, for which the haplotype blocks within breeds can be on the order of 1 Mb.[19] Thus, the number of markers required for GWAA within a breed of domestic animals may be on the order of tens of thousands rather than hundreds of thousands. This reduces not only the cost but also the multiple testing problem by an order of magnitude. Another advantage with the reduced effective population size is that it reduces the problem with genetic heterogeneity; each segregating locus explains a larger proportion of the phenotypic variation, which further increases the statistical power.

The larger haplotype blocks are a double-edged sword. On the one hand, they facilitate the detection of association, but on the other hand the genomic region showing association will be larger, and it will be more difficult to identify the causal mutation. However, we expect that mutations at trait loci will often be shared between breeds due to the gene flow between breeds and the common ancestry of different breeds; this is particularly likely for those mutations that have been selected for in different breeds. Furthermore, haplotype blocks shared between breeds are expected to be much shorter than those within breeds, and in dogs they have been estimated to be on the order of 10 kb, that is, similar to the size of haplotype blocks in humans.[19] This suggests that a two-stage strategy in which trait loci are initially mapped by within-breed analysis and then fine mapped by between-breed analysis should be powerful for those loci for which the same causal mutation is present in at least two breeds. The identification of the mutation for the *IGF2* QTL in pigs, a single-nucleotide substitution in intron 3, is a beautiful illustration of the power of this strategy.[6] The QTL was first mapped to a broad region at the distal end of pig chromosome 2p by linkage analysis in intercross pedigrees.[8,9] The region harboring the QTL was reduced to a 250-kb region, including *IGF2* by haplotype sharing analysis within one breed.[26] Finally, a minimum shared haplotype block of only 15 kb was defined by resequencing the *IGF2* region from haplotypes representing four different pig breeds.[6]

18.8 MONOGENIC TRAITS: AN UNDERUTILIZED RESOURCE

A large number of mutations underlying monogenic traits have been selected during the course of animal domestication. The molecular identification of such mutations has to a large extent been a neglected area in farm animal genomics, which has primarily focused on multifactorial traits of agricultural significance. It is an anomaly that this resource has not been better utilized compared with the huge investments made to generate new mouse mutants using mutagenesis screening programs.[2] In the chicken, which is both an important production animal and an experimental organism,[27] a rich collection of spontaneous mutants has been maintained, but many of these have already been lost or are at risk of becoming lost due to lack of funding.[28]

Another reason for the low utilization of monogenic traits in domestic animals is that the positional identification even of these loci is a major undertaking in an organism with no genome sequence and a sparse linkage map. But, this situation has now dramatically changed with the development of draft genome sequences and high-density SNP maps, which pave the way for an efficient exploitation of this resource. GWAA will be an extremely powerful approach for mapping monogenic traits that are segregating within breeds. For a simple recessive trait, a sample size of 10 affected animals and 10 controls (20 chromosomes of each type) screened using a sufficiently dense set of SNPs (designed in accordance with the LD pattern) will be sufficient for an initial mapping, as demonstrated for two Mendelian traits in the dog.[86,87]

A complicating factor, though, is that some monogenic trait loci show no variation within breeds but fixed differences between breeds. It is not possible to just compare two breeds (one of each homozygous class) since they will show many fixed differences throughout the genome, but it may be possible to compare a set of breeds with multiple replicates of each homozygous class and deduce the location of the monogenic trait locus. An alternative approach, of course, is to make a small linkage study in an intercross pedigree for an initial mapping of the locus and then study haplotype sharing across breeds for defining a minimum shared haplotype associated with the trait. This approach was successfully used for the molecular characterization of the *Silver* plumage color locus in chicken.[5]

The database Online Mendelian Inheritance in Animals (OMIA) (http://omia.angis.org.au/) compiled by Dr. Frank Nicholas provides a comprehensive list of monogenic traits in domestic animals. Here, I give a few examples of interesting monogenic traits for which the causal mutation has been identified. I focus on some mutations affecting plumage/coat color, a classical developmental mutation in chicken, and some mutations that have reached high frequencies because they affect a production trait under selection.

18.8.1 Plumage and Coat Color Loci

Plumage and coat color have been under strong selection since the early times of animal domestication, possibly because this allowed the early farmers to distinguish their improved domesticated animals from their wild ancestors and perhaps because of our interest for novelties. At present, coat color variants are often used as breed characteristics and trademarks. As a consequence, a rich coat color diversity exists in domestic animals, and this area deserves a review by itself. Here, I discuss one gene, *PMEL17*, for which mutations have been reported in the chicken,[29] dog,[30] and horse[31]; this gene is denoted *Silver* (*SILV*) in the mouse and in humans, but I use *PMEL17* here across species because *Silver* is used as the locus designation for another gene in chicken.

The PMEL17 protein is present in melanosomes and has a crucial role for expression of black eumelanin. The precise function of PMEL17 is still poorly understood.[32] Dominant white color is widespread in commercial chicken populations and inhibits the expression of black pigment in feathers and skin.[33] Kerje et al.[29] mapped this dominant mutation using an intercross between red junglefowl (the wild ancestor) and White Leghorn chicken and then identified the causal mutation for this allele and two other alleles at the same locus, *Dun* and *Smoky*. *Dominant White* and *Dun* were associated with an in-frame insertion and deletion, respectively, in the part of *PMEL17* encoding the transmembrane region. *Smoky* is an interesting allele that arose in a line of White Leghorn (expected to be homozygous for *Dominant White*), and it partially restores a pigmented phenotype. Sequence analysis showed that it carries the insertion of nine nucleotides associated with *Dominant White* and a 12-bp deletion in a well-conserved part of the gene. This second mutation apparently compensates for the defect caused by the 9-bp insertion in *Dominant White*. This is an excellent illustration of the novel allelic diversity that may accumulate in a species like the chicken, for which the global population size is counted in billions of animals.

The *Merle* mutation in dogs shows an autosomal dominant inheritance, and it causes eumelanic areas to become pale but with scattered fully pigmented spots.[34] *Merle* homozygotes are pale with defective hearing and visually defective microphthalmic eyes. Based on the observation of fully pigmented spots in heterozygotes and reported germ-line reversions, it had been predicted that *Merle* is caused by a transposable insertion.[34] This was confirmed by the finding[30] that *Merle* is associated with an insertion of a short interspersed nuclear element (SINE) in the boundary of intron 10 and exon 11 of *PMEL17*. It is not yet clear how this mutation influences the expression of the protein.

The dominant *Silver* allele in horses causes a dilution of black eumelanin, but it has no effect on red pheomelanin, consistent with the known function of *PMEL17*. The mutation can give horses a spectacular appearance, with white mane and tail but with a dark body since the mutation has a more pronounced effect on the long hairs than on the short hairs.[31] *Silver* shows[31] a complete association with a putative causal missense mutation (R618C) in *PMEL17*. Interestingly, the same missense mutation is also found in the chicken *Dun* allele, which also possesses a deletion mentioned above,[29] and it is not clear which of these two mutations is most important for explaining the Dun phenotype.

Besides these five *PMEL17* mutations in the chicken, dog, and horse, only two other mutations have been described so far, one in the mouse (*Silver*)[35] and one in zebrafish (*fading vision*),[36] which are both due to premature stop codons. The phenotype of the *Silver* mouse is primarily an inhibition of black eumelanin, whereas *fading vision* also gives severe defects in the development of the visual system consistent with the eye phenotype observed in Merle dogs. This shows that *PMEL17* has an important function both in melanosome biogenesis and in the development of the eye. No human *PMEL17* mutation has yet been detected, but it can be predicted that such mutations may explain some forms of red hair. It is surprising that only a single *PMEL17* mutation has been detected in the mouse since there has been such an extensive screening for coat color mutations in the mouse. In contrast, more than 50 different mutations have been isolated at some other coat color loci in the mouse. As discussed, a mouse screen is an effective screen for loss-of-function mutations. This suggests that complete loss-of-function mutations of *PMEL17* in the mouse either have no phenotypic effect or are lethal.

18.8.2 TALPID3: A REGULATOR OF HEDGEHOG SIGNALING

Talpid3 is a classical chicken mutant that causes limb defects and malformations of face, skeleton, and the vascular system. Davey et al.[37] combined a genomic approach with detailed developmental characterization to reveal the causal mutation for *talpid3* and to determine its functional significance. Linkage mapping using only 110 birds assigned *talpid3* to an interval comprising five genes, and a frameshift mutation was detected in a novel vertebrate gene, *KIAA0586*, with unknown function. The causal nature of this mutation was confirmed by showing that the developmental defects in embryos could be reversed by electroporating wild-type *KIAA0586* into mutant embryos. Further, functional studies revealed that this novel protein is essential for normal Hedgehog signaling in the developing embryo. This is a beautiful demonstration of the scientific value in exploiting classical developmental mutations that have been collected in the chicken during decades of research.

18.8.3 MYOSTATIN AND MUSCLE DEVELOPMENT

Specialized cattle breeds for milk production (dairy cattle) and meat production (beef cattle) have been developed. In several breeds of beef cattle, an exceptional type of muscular hypertrophy denoted double muscling occurs, and genetic analysis of phenotypic data indicated that the condition is inherited as a simple recessive trait.[38] Linkage mapping confirmed this interpretation[39] and assigned the locus to chromosome 2. *Myostatin* (MSTN) became an obvious positional candidate gene for this condition when it was shown that *Mstn* knockout mice exhibited extreme muscular hypertrophy.[40] Shortly thereafter, several groups were able to show that double muscling in cattle is caused by homozygosity for *MSTN* loss-of-function mutations.[41–43] It turned out that at least five different disruptive mutations have been enriched by strong selection for muscular hypertrophy in different breeds of beef cattle.[44] The MSTN protein belongs to the transforming growth factor-β (TGF-β) family, and it is a negative regulator of muscle mass.[45]

Given the fact that at least five different disruptive *MSTN* mutations have been selected in cattle, it is surprising that no such mutations have yet been reported in other meat-producing animals like the pig, although the selection for muscularity has been strong also in these species. It is possible that the fetal muscle hypertrophy observed in *MSTN* knockouts is a major disadvantage in species that give birth to large litters. However, Georges and his colleagues have been able to show that a QTL allele for increased muscle mass in Texel sheep, selected for meat production, is caused by a single nucleotide substitution in the 3′ untranslated region (UTR) of *MSTN*.[46] Interestingly, the mutation occurs at a nonconserved site, but it creates a new target site for two microRNAs (miR-1 and miR-206) expressed in muscle. This leads to an inhibition of translation of mutant MSTN messenger RNA and thus a reduced production of MSTN protein. This is a much milder mutation than the disruptive mutations observed in beef cattle.

18.8.4 Selection for Lean Pigs

The main selection goal in pig breeding for the last 50 years has been to produce lean pigs because of consumer demand for a healthier diet. This has caused a dramatic change in the phenotype of the pigs used for commercial production in the Western world. This has increased the frequency of allelic variants promoting muscle growth and reducing fat deposition, like the missense mutation in *RYR1* causing malignant hyperthermia in the homozygous condition[47] and the *IGF2* QTL.[6] Another interesting example is the *RN⁻* mutation, which reached a high allele frequency (~70%) in Hampshire pigs. The existence of this major gene was first postulated on the basis of segregation analysis of meat quality data that indicated there was a dominant allele that reduced the yield of cured cooked ham.[48] Subsequent studies showed that pigs carrying this mutation had 70% more glycogen in skeletal muscle and produced "acid meat," meat with a lower pH due to the degradation of glycogen after slaughter.

The causal mutation was identified by a heroic positional cloning effort, given the limited genomic resources in the pig at that time, and was found to be a missense mutation (R225Q) in *PRKAG3* encoding a previously unknown, muscle-specific isoform of the adenosine monophosphate (AMP)–activated protein kinase (AMPK) γ-chain. AMPK exists in all eukaryotes and is a sensor of the energy status of the cell, which allows cells to adjust energy production and consumption to maintain energy homeostasis.[49] Subsequent studies showed that *PRKAG3* has a specific tissue distribution and is predominantly expressed in white skeletal muscle,[50] consistent with the muscle-specific phenotype in mutant pigs.

Furthermore, the causal nature of R225Q was confirmed when the glycogen excess in skeletal muscle was replicated in the *PRKAG3* transgenic mouse expressing the same missense mutation.[51] These transgenic mice showed a higher fat oxidation in white skeletal muscle than wild-type littermates, consistent with the lean phenotype in mutant pigs. They were also protected from developing insulin resistance when exposed to a high-fat diet, implicating PRKAG3 is a potential drug target for the treatment of type II diabetes in humans. Interestingly, *PRKAG3* knockout mice were fully viable, and resting mice had normal glycogen levels.[51] This is an illustrative example for which the disruption of a well-conserved gene does not give any obvious phenotype that would be detected in a standard phenotype screen. However, a closer examination of these knockout mice showed that they had a clear defect in glycogen resynthesis after exercise, and they also had a severe defect in AMPK-regulated glucose uptake in muscle cells, further emphasizing PRKAG3 as a potential as a drug target. The phenotypes observed in these transgenic and knockout mice led to the conclusion that the biological role of the PRKAG3 isoform is to ensure that the glycogen content in glycolytic skeletal muscles is restored after muscle work to make the individual ready for a new burst of muscle activity. It accomplishes this task by increasing fat oxidation and glucose uptake when the glycogen level is below its intrinsic set point. The R225Q mutation leads to a constitutively active enzyme that alters the set point for glycogen storage.[51]

Ciobanu et al.[52] identified a second missense mutation (V224I) as underlying a QTL for several meat quality traits, including glycogen content; further studies in several commercial pig populations confirmed the significant effect of this mutation.[52–54] V224I, located at the neighboring residue, has an opposite effect to R225Q as it reduces glycogen content and increases postmortem pH values. The functional significance of these two missense mutations is explained by the fact that they are located in the allosteric site that binds AMP and adenosine triphosphate (ATP) and thereby regulates the activity of the AMPK holoenzyme composed of three subunits.[55] Transfection experiments into COS cells with constructs expressing these two mutations showed that 225Q has a significantly higher basal activity in the absence of AMP stimulation than wild type but cannot be further activated by AMP, whereas 224I show normal basal activity and cannot be activated by AMP stimulation.[51] Thus, the ranking of AMPK activity obtained with the three constructs (225Q, wild type, and 224I) is fully consistent with the amount of skeletal muscle glycogen in pigs carrying these three alleles.

18.9 COMPARATIVE GENOMICS USING THE DOG

There is a bewildering diversity in size, form, color, and behavior among dog breeds in the world. An important explanation why the dog exhibits more phenotypic diversity than other domestic animals is that it is often bred as a pet, whereas farm animals (cattle, pig, chickens, etc.) are bred for fitness and high production efficiency. Thus, we have allowed the accumulation of deleterious mutations in some breeds of dogs since their only task has been to amuse their owners. The dog provides some unique advantages as a model for human medicine:

A favorable population structure for genetic studies. This is even more pronounced than in other domestic animals since dogs are divided into a large number of breeds with replicates in different countries. There is a considerable amount of genetic drift due to founder effects and small effective population sizes, which leads to large haplotype blocks and homozygosity for recessive disorders.

Dogs and humans share the same environment. Dogs and humans often share risk factors for metabolic disorders (diet) and inflammatory disorders (allergens), and the dog is therefore a particularly relevant animal model for genetic studies of those disorders and for testing new therapeutic treatments of such disorders.

A sick dog often ends up at the veterinary clinic. Similar to a sick human, a sick dog is often taken to the doctor for a clinical examination. This provides an opportunity to build large collections of clinical samples with diagnoses relevant for human medicine.

There are already a number of interesting cases for which a monogenic disorder has been characterized at the molecular level in the dog, and a comprehensive list is provided in the OMIA database (http://omia.angis.org.au/). For instance, narcolepsy is inherited as an autosomal recessive disorder with full penetrance in Doberman pinschers, which allowed Lin et al.[56] to identify the causal mutation by positional cloning. They found that this disorder is caused by an insertion of a SINE element in intron 4 of the hypocretin (orexin) receptor 2 gene (*HCRTR2*), leading to a splicing defect. The study was a breakthrough in the understanding of the molecular basis for sleep disorders and identified hypocretins as major sleep-modulating neurotransmitters. Epilepsy occurs in 5% of all dogs and is expected to be caused by mutations at several loci; one form of canine epilepsy is caused by a dodecamer expansion in the *EPM2B* gene.[57] The result established this canine disease as a model for Lafora disease, the most severe teenage-onset human epilepsy.

Another example of a dog disease that has developed into a useful model for a human disorder is canine leukocyte adhesion deficiency (CLAD), which previously occurred at a fairly high frequency in Irish setters.[58] Since this disease shared a similar clinical picture and other features (severe recurrent bacterial infections, defective expression of leukocyte integrins, autosomal recessive inheritance) with human leukocyte adhesion deficiency (LAD), which is caused by loss-of-function mutations in the gene for integrin β2 (*ITGB2*), this gene became the obvious candidate gene for CLAD. This was confirmed by Kijas et al.,[59] who showed that the causal mutation is a missense mutation, C36S. Based on this finding, Hickstein and colleagues at National Canine Institute (NCI), Maryland, decided to establish a colony of dogs segregating for this mutation as a model for evaluating novel hematopoietic therapies for treatment of this severe immunodeficiency in humans.[60] They have now reported that they can cure CLAD either by nonmyeloablative hematopoietic stem cell transplantation from a healthy MHC-matched dog[61] or by ex vivo retroviral-mediated hematopoietic stem cell gene therapy.[62] Thus, progress in human genetics facilitated the identification of the causative mutation for CLAD, which has effectively eliminated the disease from the Irish setter population, and the dog has now acknowledged this gift by facilitating the development of an effective therapy for a life-threatening immunodeficiency in humans.

Thanks to the development of the draft genome sequence and a dense SNP map for the dog, we will see a flow of positional identifications of genes underlying monogenic disorders, and GWAA will be a great tool to accomplish this. However, GWAA may also facilitate the identification of genes underlying multifactorial traits in the dog, and it has been estimated that a few hundred cases and controls should be sufficient for the initial mapping of a locus increasing the relative risk of developing disease two- to fivefold.[19]

18.10 GENETIC DISSECTION OF COMPLEX TRAITS

Domestic animals are particularly valuable for genetic dissection of complex multifactorial traits due to the extensive phenotypic diversity and the opportunities for powerful genetic studies.[20] Two basic approaches have been used, QTL mapping based on intercrosses or within commercial populations; they both have their merits and limitations.

18.10.1 QTL Analysis Using Experimental Crosses

A major advantage by using intercrosses is that it makes it possible to map trait loci that are fixed within breeds but show differences between breeds. QTL mapping in intercrosses is particularly powerful because the F_1 animals are all heterozygous at those trait loci that are fixed for different alleles in the founder populations. QTL experiments involving intercrosses between domestic pigs and their wild ancestor (the wild boar) and between domestic chicken and its wild ancestor (the red junglefowl) allow the mapping of those loci, which have played a crucial role in genetically adapting these species to a farm environment.[6,63–66] A similar approach is to cross breeds of domestic animals that have been selected for different purposes, such as chickens selected for egg (layers) or meat (broilers) production.[67]

There also exist a large number of experimental lines of domestic animals, in particular in chicken, that have been selected for different traits, such as growth, feed efficiency, fatness, leanness, antibody response, and so on. Many of these lines have an uncertain future due to the lack of funding,[28] which is unfortunate since many of them are excellent resources for comparative genomics. An example of such a resource is the high growth and low growth lines that have been established by divergent selection for body weight at 8 weeks for more than 40 generations by Paul Siegel at Virgina Polytechnic Institute, Blacksburg[68] (Figure 18.1). The two lines have been kept as closed populations and originate from the same founder population established by crossing seven partially inbred lines of White Plymouth Rock broilers.

An amazing selection response has been obtained given the rather narrow genetic base; the body weight at age of selection (eight weeks) showed an almost ninefold difference after 40 generations of selection. Although the sole selection criterion has been body weight, a number of interesting correlated responses have been obtained. The high line chickens are hyperphagic, and they develop obesity and metabolic disorders unless they are feed restricted, and they show low antibody response, whereas low line chickens are hypophagic and very lean and show a normal immune response.[68] An important explanation for the difference in growth patterns between the two lines is a huge difference in appetite, and the high line chickens have apparently lost appetite control genetically. This conclusion is based on the results of the following experiments.

Electrolytic lesion of the ventromedial hypothalamus leads to increased food intake in the low line but has no effect on feed intake in the high line, showing that

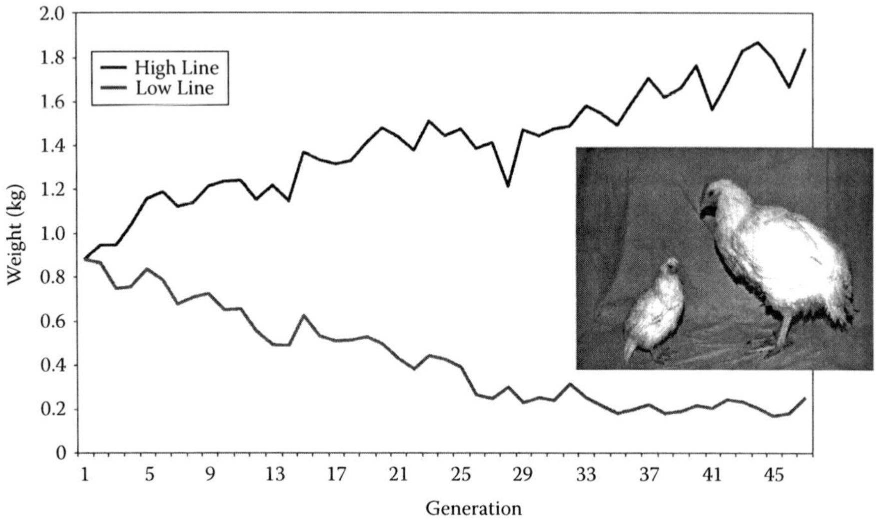

FIGURE 18.1 Body weight at 56 days of age from generations 1 to 47 of males from the high weight and low weight selection lines developed by Dr. Paul B. Siegel, Virginia Polytechnic Institute and State University. The birds illustrated are from generation 37. (The figure is from Jacobsson, L., et al., *Genet. Res.* **86**, 115–125, 2005, Genetical Research, Cambridge University Press.)

the latter has a defect in the hypothalamic satiety mechanism.[69] Food intake after intrahepatical infusion of plasma from fasted fowl was significantly increased in low line chickens, but this treatment had no effect on the already high food intake of the high line birds.[70] Finally, intracerebroventricular administration of human recombinant leptin, a satiety hormone produced by adipocytes, caused a linear decrease of food intake in low line chickens but had no effect on food intake in high line chickens, showing that the latter are leptin resistant.[71] Interestingly, no chicken leptin homolog has yet been identified in the chicken genome, although a well-conserved leptin receptor gene is present. These results demonstrate that the appetite control in the High weight line chickens is as poor as in *leptin, leptin receptor,* or *melanocortin-4 receptor* knockout mice. The low weight line chickens are as extreme in the opposite direction, and 5%–20% of the birds show an anorexic condition and do not survive to reproductive age.[68]

We decided to utilize this unique resource for genetic dissection of appetite regulation and metabolic traits by making a large intercross comprising altogether about 850 F_2 birds; an advanced intercross line (AIL)[72] is also maintained for fine mapping purposes. The 50th generation of the high and low weight selection lines and the F_{10} generation of the AIL were hatched in March 2007. A standard QTL analysis of body weight and growth traits in the F_2 generation revealed 13 loci that were considered significant, but the most striking observation was that each locus only explained a small proportion of the genetic variance (1.3%–3.1%)[73]; at each QTL, the allele from the high weight line was associated with increased growth. Thus, the extreme phenotypic difference between the two lines does not appear to involve any genetic variant with a large individual effect on growth. Combining the effect of all 13 loci could at most explain 50% of the difference in body weight between lines, implying that the remaining difference is explained by QTLs that were not detected because of a lack of marker coverage or because they have too small effect to be detected even using about 850 F_2 animals or because epistatic interaction contributes significantly to explaining the variance.

In fact, a subsequent genome-wide screen showed that epistatic interaction played an important role in this selection experiment.[74] The analysis revealed strong statistical support of a radial network comprising four interacting QTLs. Interestingly, all four loci had been detected in the standard QTL analysis, but their effect on growth had been grossly underestimated when not taking into account their interaction. An epistatic model including four interacting loci explained as much of the line difference as the combined effect of the 13 QTLs detected in a standard analysis. This result shed light on the enigma of how a steady selection response can be obtained over many generations in a rather small population such as the high and low weight lines without exhausting the genetic variance (Figure 18.1). The study by Carlborg et al.[74] provided experimental evidence that genetic variance is released during the course of a selection experiment due to changes in allele frequency at epistatic QTLs. The results obtained using this cross have important implications for the genetic analysis of multifactorial traits in humans.

18.10.2 QTL Analysis within Populations

Most QTL studies in domestic animals have been carried out using commercial populations, and this has led to the detection of numerous QTLs for myriad phenotypic traits (see Georges[75] for a recent comprehensive review). The merit of this approach is that it can take advantage of existing large multigeneration pedigrees with phenotypic data that have been collected for breeding purposes. Within-population analysis is less powerful than intercross mapping since some QTLs with major effects show all or most genetic variance in between breed comparisons, and the parental heterozygosity at QTLs must be deduced from progeny data, which reduces statistical power. However, once a QTL has been detected, further fine mapping is facilitated by the fact that it is possible to collect data from existing multigeneration pedigrees and from closely related populations or breeds.

The major challenge in QTL analysis in all organisms is the poor mapping resolution, which prohibits the molecular characterization of the underlying genes and causal mutations. Statistical methods for combining linkage and LD mapping have been developed[76,77] and given encouraging results in which QTLs in dairy cattle have been mapped to intervals of a few cM.[78–80] This approach appears attractive for QTL mapping in commercial populations of domestic animals since the LD mapping should provide high mapping resolution, while the linkage analysis should be able to rule out spurious associations due to population stratifications that often plague GWAA.

Positional identification of mutations underlying QTLs is exceedingly difficult in any organism, and there are few success stories. In domestic animals, there are three prominent examples for which the identification of causative mutations are supported by both strong genetic and functional evidence. These include a missense mutation K232A in *DGAT1* (acyl-coenzyme A:diacylglycerol acyltransferase) that has a major effect on milk fat content in cattle,[81–83] the single-nucleotide substitution in *IGF2* intron 3 affecting postnatal muscle growth in the pig,[6] and a single-nucleotide substitution in *MSTN* with a major effect on muscularity in Texel sheep.[46]

So, why is the identification of QTL mutations so difficult? One obvious reason is the difficulty in getting sufficient map resolution (<1 Mb) that makes it possible to resequence and evaluate all transcripts in the QTL region and their regulatory elements. Another complicating factor is that a good proportion of QTLs has a complex architecture in which mutations in two or more closely linked genes or two or more mutations in the same gene cause the QTL. This is well documented from QTL analysis in mice.[84] The three success stories mentioned all represent QTLs with a simple architecture; that is, the QTL is caused by a single-nucleotide substitution.

18.11 FUTURE VISIONS

The next technological revolution in genomics will be the possibility to do whole-genome resequencing as new, more cost-effective methods for DNA sequencing are developed.[85] This will be an extremely powerful tool for domestic animal genomics. The strong selection that has taken place during domestication and breed formation is expected to have led to selective sweeps when favorable mutations are approaching fixation.[20] A beautiful illustration of this phenomenon is the *IGF2* QTL in pigs, in which a good proportion of pigs used for meat production in the Western world is homozygous for the same *IGF2* haplotype.[6] Whole-genome resequencing of domestic animals will reveal such selective sweeps, but it will not be trivial to distinguish homozygosity caused by selective sweeps from those caused by genetic drift or to prove which trait a certain selective sweep has affected.

ACKNOWLEDGMENT

I am grateful to Lina Strömstedt for help with preparing Figure 18.1.

REFERENCES

1. Falconer, D. S. & Mackay, T. F. C. *Introduction to Quantitative Genetics* (Longman Science and Technology, Harlow, UK, 1996).
2. Kile, B. & Hilton, D. The art and design of genetic screens: mouse. *Nat. Rev. Genet.* **6**, 557–567 (2005).
3. Darwin, C. *On the Origins of species by Means of Natural Selection or the Preservation of Favoured Races in the Struggle for Life* (Murray, London, 1859).
4. Riquet, J. et al. Fine-mapping of quantitative trait loci by identity by descent in outbred populations: application to milk production in dairy cattle. *Proc. Natl. Acad. Sci. U. S. A.* **96**, 9252–9257 (1999).
5. Gunnarsson, U. et al. Mutations in SLC45A2 cause plumage color variation in chicken and Japanese quail. *Genetics* **175**, 867–877 (2007).
6. Van Laere, A. S. et al. Positional identification of a regulatory mutation in *IGF2* causing a major QTL effect on muscle growth in the pig. *Nature* **425**, 832–836 (2003).
7. Maynard-Smith, J. & Haigh, J. The hitch-hiking effect of a favourable gene. *Genet. Res.* **23**, 23–35 (1974).

8. Jeon, J.-T. et al. A paternally expressed QTL affecting skeletal and cardiac muscle mass in pigs maps to the *IGF2* locus. *Nat. Genet.* **21**, 157–158 (1999).
9. Nezer, C. et al. An imprinted QTL with major effect on muscle mass and fat deposition maps to the IGF2 locus in pigs. *Nat. Genet.* **21**, 155–156 (1999).
10. Milan, D. et al. A mutation in *PRKAG3* associated with excess glycogen content in pig skeletal muscle. *Science* **288**, 1248–1251 (2000).
11. International Chicken Polymorphism Map Consortium. A genetic variation map for chicken with 2.8 million single-nucleotide polymorphisms. *Nature* **432**, 717–722 (2004).
12. Larson, G. et al. Worldwide phylogeography of wild boar reveals multiple centres of pig domestication. *Science* **307**, 1618–1621 (2005).
13. Fang, M. & Andersson, L. Mitochondrial diversity in European and Chinese pigs is consistent with population expansions that occurred prior to domestication. *Proc. Biol. Sci.* **273**, 1803–1810 (2006).
14. Vila, C., Seddon, J. & Ellegren, H. Genes of domestic mammals augmented by backcrossing with wild ancestors. *Trends Genet.* **21**, 214–218. (2005).
15. Bruford, M. W., Bradley, D. G. & Luikart, G. DNA markers reveal the complexity of livestock domestication. *Nat. Rev. Genet.* **4**, 900–910 (2003).
16. Giuffra, E. et al. The origin of the domestic pig: independent domestication and subsequent introgression. *Genetics* **154**, 1785–1791 (2000).
17. The International HapMap Consortium. The international HapMap project. *Nature* **426**, 789–796 (2003).
18. International Chicken Genome Sequencing Consortium. Sequence and comparative analysis of the chicken genome provide unique perspectives on vertebrate evolution. *Nature* **432**, 695–716 (2004).
19. Lindblad-Toh, K. et al. Genome sequence, comparative analysis and haplotype structure of the domestic dog. *Nature* **438**, 803–819 (2005).
20. Andersson, L. & Georges, M. Domestic animal genomics: deciphering the genetics of complex traits. *Nat. Rev. Genet.* **5**, 202–212 (2004).
21. Syvänen, A. C. Toward genome-wide SNP genotyping. *Nat. Genet.* **37 Suppl.**, S5–S10 (2005).
22. Farnir, F. et al. Extensive genome-wide linkage disequilibrium in cattle. *Genome Res.* **10**, 220–227 (2000).
23. McRae, A. et al. Linkage disequilibrium in domestic sheep. *Genetics* **160**, 1113–1122 (2002).
24. Nsengimana, J., Baret, P., Haley, C. & Visscher, P. Linkage disequilibrium in the domesticated pig. *Genetics* **166**, 1395–1404 (2004).
25. Sutter, N. et al. Extensive and breed-specific linkage disequilibrium in *Canis familiaris*. *Genome Res.* **14**, 2388–2396 (2004).
26. Nezer, C. et al. Haplotype sharing refines the location of an imprinted QTL with major effect on muscle mass to a 250 Kb chromosome segment containing the porcine *IGF2* gene. *Genetics* **165**, 277–285 (2003).
27. Brown, W., Hubbard, S., Tickle, C. & Wilson, S. The chicken as a model for large-scale analysis of vertebrate gene function. *Nat. Rev. Genet.* **4**, 87–98 (2003).
28. Delany, M. Avian genetic stocks: the high and low points from an academia researcher. *Poult. Sci.* **85**, 223–226 (2006).
29. Kerje, S. et al. The *Dominant white*, *Dun* and *Smoky* color variants in chicken are associated with insertion/deletion polymorphisms in the *PMEL17* gene. *Genetics* **168**, 1507–1518 (2004).
30. Clark, L., Wahl, J., Rees, C. & Murphy, K. Retrotransposon insertion in SILV is responsible for merle patterning of the domestic dog. *Proc. Natl. Acad. Sci. U. S. A.* **103**, 1376–1381 (2006).

31. Brunberg, E. et al. A missense mutation in *PMEL17* is associated with the Silver coat color in the horse. *BMC Genet.* **7**, 46 (2006).
32. Theos, A., Truschel, S., Raposo, G. & Marks, M. The Silver locus product Pmel17/gp100/Silv/ME20: controversial in name and in function. *Pigment Cell Res.* **18**, 322–336 (2005).
33. Smyth, J. R. In: *Poultry Breeding and Genetics* (Ed. Crawford, R. D.), pp. 109–167 (Elsevier Science, New York, 1996).
34. Sponenberg, D. & Rotschild, M. In: *The Genetics of the Dog* (Eds. Ruvinsky, A., & Sampson, J.) (CABI, Oxon, UK, 2001).
35. Martinez-Esparza, M. et al. The mouse silver locus encodes a single transcript truncated by the silver mutation. *Mamm. Genome* **10**, 1168–1171 (1999).
36. Schonthaler, H. et al. A mutation in the *silver* gene leads to defects in melanosome biogenesis and alterations in the visual system in the zebrafish mutant fading vision. *Dev. Biol.* **284**, 421–436 (2005).
37. Davey, M. et al. The chicken *talpid*^{3} gene encodes a novel protein essential for Hedgehog signaling. *Genes Dev.* **20**, 1365–1377 (2006).
38. Hanset, R. & Michaux, C. On the genetic determinism of muscular hypertrophy in the Belgian White and Blue cattle breed. I. Experimental data. *Genet. Sel. Evol.* **17**, 359–368 (1985).
39. Charlier, C. et al. The *mh* gene causing double-muscling in cattle maps to bovine chromosome 2. *Mamm. Genome* **6**, 788–792 (1995).
40. McPherron, A. C., Lawler, A. M. & Lee, S. J. Regulation of skeletal muscle mass in mice by a new TGF-beta superfamily member. *Nature* **387**, 83–90 (1997).
41. Grobet, L. et al. A deletion in the bovine myostatin gene causes the double-muscled phenotype in cattle. *Nat. Genet.* **17**, 71–74 (1997).
42. McPherron, A. C. & Lee, S. J. Double muscling in cattle due to mutations in the myostatin gene. *Proc. Natl. Acad. Sci. U. S. A.* **94**, 12457–12461 (1997).
43. Kambadur, R., Sharma, M., Smith, T. P. & Bass, J. J. Mutations in myostatin (GDF8) in double-muscled Belgian Blue and Piedmontese cattle. *Genome Res.* **7**, 910–916 (1997).
44. Grobet, L. et al. Molecular definition of an allelic series of mutations disrupting the myostatin function and causing double-muscling in cattle. *Mamm. Genome* **9**, 210–213 (1998).
45. Tobin, J. & Celeste, A. Myostatin, a negative regulator of muscle mass: implications for muscle degenerative diseases. *Curr. Opin. Pharmacol.* **5**, 328–332 (2005).
46. Clop, A. et al. A mutation creating a potential illegitimate microRNA target site in the myostatin gene affects muscularity in sheep. *Nat. Genet.* **38**, 813–818 (2006).
47. Fujii, J. et al. Identification of a mutation in the porcine ryanodine receptor that is associated with malignant hyperthermia. *Science* **253**, 448–451 (1991).
48. Le Roy, P., Naveau, J., Elsen, J. M. & Sellier, P. Evidence for a new major gene influencing meat quality in pigs. *Genet. Res.* **55**, 33–44 (1990).
49. Kahn, B., Alquier, T., Carling, D. & Hardie, D. AMP-activated protein kinase: ancient energy gauge provides clues to modern understanding of metabolism. *Cell Metab.* **1**, 15–25 (2005).
50. Mahlapuu, M. et al. Expression profiles representing the γ-subunit isoforms of AMP-activated protein kinase suggest a major role for γ3 in white skeletal muscle fibers of mammals. *Am. J. Physiol. Endocrinol. Metab.* **286**, E194–E200 (2004).
51. Barnes, B. R. et al. The AMPK-gamma3 isoform has a key role for carbohydrate and lipid metabolism in glycolytic skeletal muscle. *J. Biol. Chem.* **279**, 38441–38447 (2004).
52. Ciobanu, D. et al. Evidence for new alleles in the protein kinase adenosine monophosphate-activated γ3-subunit gene associated with low glycogen content in pig skeletal muscle and improved meat quality. *Genetics* **159**, 1151–1162 (2001).

53. Lindahl, G. et al. A second mutant allele (V199I) at the PRKAG3 (RN) locus — I. Effect on technological meat quality of pork loin. *Meat Sci.* **66**, 609–619 (2003).
54. Lindahl, G. et al. A second mutant allele (V199I) at the PRKAG3 (RN) locus — II. Effect on colour characteristics of pork loin. *Meat Sci.* **66**, 621–627 (2003).
55. Scott, J. W. et al. CBS domains form energy-sensing modules whose binding of adenosine ligands is disrupted by disease mutations. *J. Clin. Invest.* **113**, 274–284 (2004).
56. Lin, L. et al. The sleep disorder canine narcolepsy is caused by a mutation in the hypocretin (orexin) receptor 2 gene. *Cell* **98**, 365–376 (1999).
57. Lohi, H. et al. Expanded repeat in canine epilepsy. *Science* **307**, 81 (2005).
58. Trowald-Wigh, G., Ekman, S., Hansson, K., Hedhammar, Å. & Hård af Segerstad, C. Clinical, radiological and pathological features of 12 Irish setters with canine leucocyte adhesion deficiency. *J. Small Anim. Pract.* **41**, 211–217 (2000).
59. Kijas, J. et al. A missense mutation in the b-2 integrin gene *(ITGB*2) causes canine leukocyte adhesion deficiency. *Genomics* **61**, 101–107 (1999).
60. Creevy, K. et al. Canine leukocyte adhesion deficiency colony for investigation of novel hematopoietic therapies. *Vet. Immunol. Immunopathol.* **95**, 113–121 (2003).
61. Bauer, T. J. et al. Nonmyeloablative hematopoietic stem cell transplantation corrects the disease phenotype in the canine model of leukocyte adhesion deficiency. *Exp. Hematol.* **33**, 706–712 (2005).
62. Bauer, T. J. et al. Correction of the disease phenotype in canine leukocyte adhesion deficiency using ex vivo hematopoietic stem cell gene therapy. *Blood* **108**, 3313–3320 (2006).
63. Andersson, L. et al. Genetic mapping of quantitative trait loci for growth and fatness in pigs. *Science* **263**, 1771–1774 (1994).
64. Kerje, S. et al. The two-fold difference in adult size between the red junglefowl and White Leghorn chickens is largely explained by a limited number of QTLs. *Anim. Genet.* **34**, 264–274 (2003).
65. Carlborg, Ö. et al. A global search reveals epistatic interaction between QTLs for early growth in the chicken. *Genome Res.* **13**, 413–421 (2003).
66. Keeling, L. et al. Feather-pecking and victim pigmentation. *Nature* **431**, 645–646 (2004).
67. Sewalem, A. et al. Mapping of quantitative trait loci for body weight at three, six, and nine weeks of age in a broiler layer cross. *Poult. Sci.* **81**, 1775–1781 (2002).
68. Dunnington, E. A. & Siegel, P. B. Long-term divergent selection for eight-week body weight in White Plymouth Rock chickens. *Poult. Sci.* **75**, 1168–1179 (1996).
69. Burkhart, C. A., Cherry, J. A., Van Krey, H. P. & Siegel, P. B. Genetic selection for growth rate alters hypothalamic satiety mechanisms in chickens. *Behav. Genet.* **13**, 295–300 (1983).
70. Lacy, M., Van Krey, H. P., Skewes, P., Denbow, D. & Siegel, P. B. Food intake in response of genetically selected high and low-weight line cockerels to plasma infusion from fasted fowl. *Poult. Sci.* **66**, 1224–1228 (1987).
71. Kuo, A., Cline, M., Werner, E., Siegel, P. & Denbow, D. Leptin effects on food and water intake in lines of chickens selected for high or low body weight. *Physiol. Behav.* **84**, 459–464 (2005).
72. Darvasi, A. & Soller, M. Advanced intercross lines, an experimental population for fine genetic-mapping. *Genetics* **141**, 1199–1207 (1995).
73. Jacobsson, L. et al. Many QTLs with minor additive effects are associated with a large difference in growth between two selection lines in chickens. *Genet. Res.* **86**, 115–125 (2005).
74. Carlborg, Ö., Jacobsson, L., Åhgren, P., Siegel, P. B. & Andersson, L. Epistasis and the release of genetic variation during long-term selection. *Nat. Genet.* **38**, 418–420 (2006).

75. Georges, M. Mapping, fine-mapping and cloning QTL in domestic animals. *Annu. Rev. Genomics Hum. Genet.* in press (2007).
76. Meuwissen, T. H. & Goddard, M. E. Fine mapping of quantitative trait loci using linkage disequilibria with closely linked marker loci. *Genetics* **155**, 421–430 (2000).
77. Meuwissen, T. H. & Goddard, M. E. Prediction of identity by descent probabilities from marker-haplotypes. *Genet. Sel. Evol.* **33**, 605–634 (2001).
78. Meuwissen, T. H., Karlsen, A., Lien, S., Olsaker, I. & Goddard, M. E. Fine mapping of a quantitative trait locus for twinning rate using combined linkage and linkage disequilibrium mapping. *Genetics* **161**, 373–379 (2002).
79. Olsen, H. G. et al. Mapping of a milk production quantitative trait locus to a 420-kb region on bovine chromosome 6. *Genetics* **169**, 275–283 (2005).
80. Blott, S. et al. Molecular dissection of a QTL: a phenylalanine to tyrosine substitution in the transmembrane domain of the bovine growth hormone receptor is associated with a major effect on milk yield and composition. *Genetics* **163**, 253–266 (2003).
81. Grisart, B. et al. Positional candidate cloning of a QTL in dairy cattle: identification of a missense mutation in the bovine *DGAT1* gene with major effect on milk yield and composition. *Genome Res.* **12**, 222–231 (2002).
82. Grisart, B. et al. Genetic and functional demonstration of the causality of the DGAT1 K232A mutation in the determinism of the BTA14 QTL affecting milk yield and composition. *Proc. Natl. Acad. Sci. U. S. A.* **101**, 2398–2403 (2004).
83. Winter, A. et al. Association of a lysine-232/alanine polymorphism in a bovine gene encoding acyl-CoA:diacylglycerol acyltransferase (DGAT1) with variation at a quantitative trait locus for milk fat content. *Proc. Natl. Acad. Sci. U. S. A.* **99**, 9300–9305 (2002).
84. Flint, J., Valdar, W., Shifman, S. & Mott, R. Strategies for mapping and cloning quantitative trait genes in rodents. *Nat. Rev. Genet.* **6**, 271–286 (2005).
85. Bentley, D. Whole-genome re-sequencing. *Curr. Opin. Genet. Dev.* **16**, 545–552 (2006).
86. Karlsson, E. K. et al. Efficient mapping of Mendelian traits in dogs through genome-wide association analysis. *Nat. Genet.* (in press).
87. Hillbetz, N. H. C. et al. A duplication of FGF3, FGF4, FGF9 and ORAOV1 causes the hair ridge and predisposes to dermoid sinus in Ridgeback dogs. *Nat. Genet.* (in press).

Index

Page references given in *italics* refer to figures.
Page references given in **bold** refer to tables.

B

E

H

N

O

Q

R

T

U

V

W

X

Y

Z